Gas Chromatography and Mass Spectrometry

GAS CHROMATOGRAPHY AND MASS SPECTROMETRY

A PRACTICAL GUIDE

Second Edition

O. DAVID SPARKMAN

ZELDA E. PENTON

FULTON G. KITSON

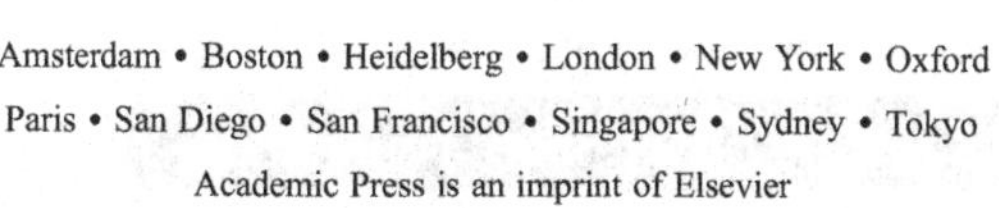
Amsterdam • Boston • Heidelberg • London • New York • Oxford
Paris • San Diego • San Francisco • Singapore • Sydney • Tokyo
Academic Press is an imprint of Elsevier

Academic Press is an imprint of Elsevier
30 Corporate Drive, Suite 400, Burlington, MA 01803, USA
525 B Street, Suite 1900, San Diego, California 92101-4495, USA
The Boulevard, Langford Lane, Kidlington, Oxford, OX51GB, UK

Second edition 2011

Library of Congress Cataloging-in-Publication Data
Sparkman, O. David (Orrin David), 1942–
Gas chromatography and mass spectrometry : a practical guide / O David Sparkman, Zelda E. Penton, Fulton G. Kitson. – 2nd ed.
p. cm.
Includes index.
ISBN 978-0-12-373628-4
1. Gas chromatography. 2. Mass spectrometry. I. Penton, Zelda E. II. Kitson, Fulton G. III. Title.
QD79.C45S66 2011
543'.85–dc22

2010027725

British Library Cataloguing in Publication Data
A catalogue record for this book is available from the British Library
ISBN–13: 978-0-12-373628-4

For information on all Academic Press publications visit our web site at elsevierdirect.com

Printed and bound in USA
11 12 9 8 7 6 5 4 3 2 1

Contents

Preface

The purpose of this book is to provide the practitioner of gas chromatography/mass spectrometry (GC/MS) with tools that will facilitate performing analyses and extracting information from the data of those analyses. To those ends, information regarding the tools available and a treatise on the evolution of the technique are also included. This book is not intended to be a detailed text on the theory of the technique of GC/MS; it includes information only on separation of components by gas chromatography (GC) followed by identification using mass spectrometry. No information is included on liquid chromatography/mass spectrometry (LC/MS).

In *Section I. The Fundamentals on GC/MS*, the available instrumentation and techniques (Chapters 1–6) have been greatly expanded over the first edition. More attention is given to available databases of electron ionization (EI) spectra, their use, and the use of programs for deconvolution of coeluting chromatographic components using the mass spectral data as well as software that can be used to develop elemental compositions from mass spectral peaks.

In *Section II. GC Conditions, Derivatization, and Mass Spectral Interpretation of Specific Compound Types* (Chapters 7–36), information pertaining to the use of packed columns has been eliminated due to the disuse of these columns and their replacement with PLOT (porous-layered open tubular) and WCOT (wall-coated open tubular) columns in the modern GC/MS laboratory. There is an overview of GC detectors included in Chapter 2 that should be useful for those considering the use of the gas chromatograph-mass spectrometer (GC-MS) for occasional GC applications. There are also some applications from which valuable information can be obtained by combining a selective GC detector with a mass spectrometer in a single analysis.

Analytes are introduced to the ion source of the GC-MS in ways other than through a GC. These techniques are carefully detailed in Chapters 2 and 4, and their advantages and disadvantages are articulated.

A good portion of the material in this book is tabular and should be used to gain information on how to perform an analysis of a specific sample category and then in the determination of the identity of the individual analytes. This is a book that should be both on the analyst's desk and on the bench next to the instrument. The original edition did an excellent job of presenting the needed tools. This edition expands mainly on the narrative section of the original book and retains and updates those tabular data that were so helpful in the previous edition. At the same time, this edition expands on the techniques of mass spectral data interpretation. Chapter 3 is new and is relatively short because it covers the interface between the gas chromatograph and the mass spectrometer. Chapter 5 goes into more depth on the interpretation of mass spectra. Chapter 6 is a completely revised treatise on the uses of GC/MS in the area of quantitation. The information in *Section III. Appendices* has been expanded and has more helpful tools for mass spectral interpretation. Additional general mass spectral interpretation information can also be found in the chapters on Specific Compound Types (*Section II*, Chapters 7–36).

Much of the material contained in the *GC Conditions, Derivatization, and Mass Spectral Interpretation of Specific Compound Types* section of the first edition was from Fulton G. Kitson's personal experiences in his 30-year career in GC and mass spectrometry. Much of this material has been retained, expanded, and its presentation somewhat modified; however, the existence of this material and the presentation style would not exist if it had not been from his experiences and efforts to organize it (Chapters 7–36 and several of the appendices).

The number of appendices has been increased from 12 to 17. The appendix on "Atomic Masses and Isotope Abundances" has been expanded to provide tools to aid in the determination of an elemental composition from isotope peak intensity ratios. An appendix with examples on "Steps to Follow in the Determination of an Elemental Composition Based on Isotope Peak Intensity Ratios" has been added. Appendices on whether to use GC/MS or LC/MS; third-party software for use in data analysis; a list of information required in reporting GC/MS data; X+1 and X+2 peak relative intensities based on the number of atoms of carbon in an ion; and a list of available EI mass spectral databases have been added. Others such as the ones on derivatization; isotope peak patterns for ions with Cl and/or Br; terms used in GC and in mass spectrometry; and tips on setting up, maintaining, and troubleshooting a gas chromatograph have all been expanded and updated as has the appendix on maintenance and troubleshooting problems in the mass spectrometer.

Since the first commercial GC/MS system was introduced in the mid-1960s by the Swedish company LKB, a number of GC and GC/MS companies have emerged only to be merged into larger companies, acquire and lose their brand identity, or just fade into the pages of scientific instrument history. In late 2009, Agilent Technologies (the noncomputer/nonprinter part of Hewlett–Packard), which was spun off in the last part of the 20th century, announced that it was acquiring Varian, Inc., which itself was a scientific multiproduct-line company that was part of an earlier breakup of Varian, Corp. In addition to vacuum technology, laboratory spectroscopy instruments (near-IR, UV-vis, atomic absorption, and inductively coupled emission spectroscopy), nuclear magnetic resonance spectroscopy, and superconducting magnets, Varian was a major provider of GC and GC/MS instruments (with the internal ionization quadrupole ion trap

(QIT) technology acquired from the then Finnigan Corp. in 1989). Varian also was a manufacturer of tandem quadrupole MS/MS instruments used in GC/MS and LC/MS. The GC business (instruments and supplies) and the ICPMS and tandem quadrupole products of the two companies were considered to be too similar, and the European Economic Community (EEC) requested that Agilent divest itself of these areas before the acquisition could go forward.

Bruker Corporation (headquartered in Billerica, Massachusetts, and the publicly traded parent company of Bruker Scientific Instruments Division (Bruker AXS, Bruker BioSpin, Bruker Daltonics, and Bruker Optics) and Bruker Energy & Supercon Technologies Division) acquired ICPMS instruments, laboratory GC instruments, and GC/MS tandem quadrupole instruments businesses of Varian from Agilent Technologies. The three acquired product lines will form the core offerings in a newly established Bruker Chemical Analysis Division. Unfortunately, this divestiture will not retain the Varian company or product name. Agilent was not required to divest itself of the Varian internal ionization QIT GC/MS technology, and Varian is the only company currently offering such an instrument commercially. Even if Agilent continues the QIT GC/MS product line (which is very likely), the Varian name will no longer be used. Throughout this book, reference is made to Varian GC, QIT, and tandem quadrupole GC/MS technology. Because of these Agilent and Bruker acquisitions, the name Varian, with respect to analytical instrumentation, will be a total unknown to practitioners entering the field in 5 years. This is why we felt this explanation was necessary. The industry will change, but the information in the book will still be valid and usable.

A very good source for peer support regarding GC and GC/MS is the Chromatography Forum at http://www.chromforum.org.

ACKNOWLEDGMENTS

The greatest thanks for the aid in the preparation of this book go to the three authors of the first edition: Fulton G. Kitson, Barbara S. Larsen, and Charles N. McEwen. Fulton Kitson's desire to share his many years of experience performing GC/MS analyses in a way that he felt would be most useful to new and experienced practitioners was carefully retained in this edition with expansion where necessary. Many people kindly provided information, encouragement, and assistance in the preparation of this book. Both of the current authors are grateful to Joan A. Sparkman, ODS's wife, for her efforts in editing and organizing the manuscript for consistency of style. Just as ODS's other three books have been all the better for her efforts, so will this one.

ZEP would like to acknowledge and give special thanks to the following:

Hal Bellows, formerly of Varian, Inc., and Valco Instruments for providing information on valves for GC and GC/MS and the thermal conductivity detector.

Professor Aviv Amirav of Tel Aviv University for being very helpful in responding to questions on the pulsed-flame photometric detector and the DSI (direct sample interface), both of which he developed.

Thomas Wampler of CDS Analytical, Inc., for providing helpful information on pyrolysis GC and the PyroProbe.

ODS wishes to offer thanks to his graduate student, Matthew E. Curtis, who assisted in the acquisition of some of the spectra and spent time proofing Appendix Q. He also wishes to express thanks to his colleague, Patrick R. Jones, at the Pacific Mass Spectrometry Facility in the University of the Pacific for his encouragement and reading of some parts of the manuscript. As with his other books, ODS must also express thanks to his two canine mass spectrometrists, Maggie and Chili, who were constantly looking over his shoulder to make sure things were being properly done.

Both authors acknowledge the support of the National Institute of Standards and Technology for their permission to use spectra from the NIST08 Mass Spectral Database (NIST08).

ACKNOWLEDGMENTS

SECTION 1

The Fundamentals of GC/MS

CHAPTER 1

Introduction and History

Gas chromatography/mass spectrometry (GC/MS) is the most ubiquitous analytical technique for the identification and quantitation of organic substances in complex matrices. The gas chromatograph-mass spectrometer (GC-MS) is indispensable in the fields of environmental science, forensics, health care, medical and biological research, health and safety, the flavor and fragrances industry, food safety, packaging, and many others. The instrumentation ranges in price from nearly 1 million dollars to just a few thousand. The size is large enough to require a 4-m × 4-m room to that of an average briefcase (Figure 1.1).

GC/MS is the synergistic combination of two powerful microanalytical techniques. The gas chromatograph separates the components of a mixture in time, and the mass spectrometer provides information that aids in the structural identification of each component. This combination has several advantages [1]. First, it separates components of a complex mixture so that mass spectra of individual compounds can be obtained for qualitative purposes; second, it can provide quantitative information on these same compounds. Mass spectrometry ionization techniques that require gas-phase analytes are ideally suited to GC/MS because sample volatility is a requirement of gas chromatography (GC). The gas chromatograph, the mass spectrometer, and the interface linking these two instruments are described in the following chapters.

GC/MS can provide a complete mass spectrum from a few femtomoles of an analyte; ideally, this spectrum gives direct evidence for the nominal mass and provides a characteristic fragmentation pattern or "chemical" fingerprint that can be used as the basis for identification along with the gas chromatograph retention time.

Mass spectrometry had its origin ca. late 1800s with the work of John Joseph Thomson [2] and Wilhelm (Willy) Carl Werner Otto Fritz Franz Wien [3]. Mass spectrometry was dominated by the measurement of the various nuclides* that made up the known elements of the time until the

* A nuclide is an atomic species characterized by the total number of neutrons and protons in the nucleus. Although not synonymous with isotope, for the purposes of this book, a nuclide is one of the many atomic species that are characterized by both their atomic number and their mass number.

Gas Chromatography and Mass Spectrometry
DOI: 10.1016/B978-0-12-373628-4.00001-0

Figure 1.1 The Thermo Scientific DFS GC/MS system (left), a high-resolution double-focusing magnetic-sector GC-MS that plays a significant role in the high-resolution (HR) selected ion monitoring (SIM) analysis of trace levels of dioxins; and the Torion Technologies, Inc., GUARDION-7™ (right) used by environmentalists and first-responders to hazardous waste spills.

mid-part of the 20th century when the mass spectrometer's use for the analysis of petroleum products and other organic compounds began to gain momentum.

Chromatography began about the same time as mass spectrometry (ca. 1900) with the seminal publication by Mikhail Semenovich Tsvet (two other papers appeared in German that are often mistakenly referenced as the beginning of chromatography: Tswett MS) [4–7]. The early practice of chromatography consisted of the application of liquid samples to short homemade columns of various absorbents or to absorbent paper. The report of partition chromatography by Archer John Porter Martin and Richard Laurence Millington Synge [8, 9] in 1941 led to the development of GC by Martin and Anthony Tarfford James [10, 11] in 1950.

Very soon after the development of GC, attempts to interface the gas chromatograph with the mass spectrometer began. This was a natural development as the gas chromatograph separates organic compounds, and they eluted from the column in a purified state in the gas phase; and the mass spectrometers of that time required pure gas-phase analytes for ionization. However, the original gas chromatographs used packed columns with flow rates (20–30 mL min^{-1}) that overwhelmed the required low pressures of the mass spectrometer. One of the main obstacles to the technique of GC/MS was this incompatibility in pressure requirements. Today's instrumentation is faced with far fewer such problems because of the use of capillary columns with flow rates that are usually 1.5 mL min^{-1} or less, and much better pumping systems to maintain the vacuum required for the mass spectrometer.

As will be forever debated, the actual first attempt to interface the gas chromatograph and the mass spectrometer was accomplished by either Joseph C. Holmes and Francis A. Morrell at Philip Morris, Inc., in Richmond, Virginia, who published their work on the interfacing of a gas chromatograph with a Consolidated Engineering Corporation (CEC) Model 21-103B magnetic-sector mass spectrometer in 1957 [12] or Roland S. Gohlke and Fred McLafferty (both at Dow Chemical Company in Midland, Michigan, at that time) who presented their work on interfacing a gas chromatograph with a time-of-flight (TOF) mass spectrometer at the 129th National American Chemical Society (ACS) meeting in April of 1956 in a symposium on *Vapor Phase Chromatography* [13]. This work was first published in a paper authored only by Gohlke in the April 1959 issue of *Analytical Chemistry*, almost a year after it was received by the journal on May 31, 1958 [14], and almost 3 years after the 129th ACS meeting. The Holmes/Morrell work was first presented at the Fourth Annual Meeting of American Society for Testing and Materials (ASTM) Committee E-14 on *Mass Spectrometry and Allied Topics* in Cincinnati, Ohio, in May of 1956.

The GC-MS of today is a unique instrument. Gohlke/McLafferty and Holmes/Morrell treated their systems as a gas chromatograph being used as an inlet to a mass spectrometer; there are some who would treat the mass spectrometer as a detector for the gas chromatograph. Neither of these is true. The mass spectrometer is not a gas chromatograph detector, and the gas chromatograph is not an inlet for the mass spectrometer. It is important to remember that GC/MS is as different from either GC or mass spectrometry as GC and mass spectrometry are from one another. This is because an elevated pressure is required to separate the components of a mixture in a gas chromatograph and a greatly reduced pressure is required to separate the ions of various *mass-to-charge ratios* (*m/z* values) that characterize a pure component of that mixture.

Both the Gohlke/McLafferty and the Holmes/Morrell attempts at interfacing the gas chromatograph and the mass spectrometer involved splitting a small portion of the gas chromatograph eluate to the mass spectrometer, with the remainder being diverted to either a conventional gas chromatograph detector or the atmosphere. This was necessary to circumvent the conflicting high-/low-pressure needs of the two instruments. In the 1960s, devices were developed to enrich the eluate from packed gas chromatograph columns with respect to the analyte. These devices, for the most part, have now fallen into disuse because of the use of capillary gas chromatograph columns that produce eluates much richer in analyte concentration than packed columns and improved vacuum systems. Today's modern instrument has the exit end of the gas chromatograph column placed directly in the ion source of the mass spectrometer.

A very significant factor in the evolution of GC/MS was the development of data systems. When GC/MS was first being explored, it was readily seen that the potential for the volume of data was overwhelming. A 10-minute chromatographic run, acquiring spectra at the rate of one per second, would result in a total of 600 spectra. Extracting the spectra associated with various chromatographic peaks and then dealing with the presence of mass spectral peaks that were due to background associated with the sample or the gas chromatograph column was quite daunting.* It was not until the development and commercialization of the minicomputer (ca. 1965) that it was possible to bring the computer to the GC-MS. Before that time, the data had to be brought from the mass spectrometer to the computer and input manually. When Digital Equipment Corp. (acquired by Compaq Computer, which was then acquired by Hewlett–Packard) introduced the first commercial minicomputer, the PDP-8, one of its first uses was to acquire and process GC/MS data [15, 16]. Users quickly warmed to the abilities and speeds of these computers. As the minicomputer evolved, the speeds began to fade because of the overhead of the operating systems and software used to develop the GC/MS applications. As the speed of the second-generation individual computer (the microcomputer) continued to increase, the speed of the early 1970s GC/MS data systems was once again realized near the end of the 1990s.

Improvements in capillary column injectors [17–20], development of fused silica capillary columns [21], development of electronic flow and pressure control [22], and improvements in the bleed characteristics of open-tubular wall-coated columns [22] have led to easier-to-use GC/MS systems with lower and lower limits of detection. Improvements in gas chromatograph column stability, controlled rapid heating rates for fast chromatography, and reduced oven cool-down times have also partially contributed to the technique.

The GC/MS instrument of today allows for more flexibility in ionization, speed of data acquisition, and ease of use for less-skilled practitioners who are more interested in answers than the art and science of the technique, and, very importantly, much smaller size. The GC/MS floor-standing instrumentation of the 1970s and early 1980s would fill a 10-m × 10-m room. Some of today's laboratory instrumentation can be accommodated by 2–4 square feet of benchtop space.

---

*The term "peak" has already been used twice in this paragraph: once referring to the chromatographic peak and once referring to the mass spectral peak. This is an excellent example why it is always important to identify the type of peak being discussed when GC/MS is involved. In GC/MS, the word "peak" should never be used unless it is preceded by "chromatographic" or "mass spectral".

GC/MS is limited to analytes that are not only volatile and thermally labile but can also withstand the harsh partitioning conditions of the gas chromatograph.* Even with this limitation, there are many analytes that can only be separated from complex mixtures and identified by GC/MS. Compounds that exist only in the gas phase at temperatures below 100 °C cannot be separated and ionized using techniques other than GC/MS. Due to the ability to form stable, volatile derivatives of many compounds not suited for GC/MS in their natural forms, the number of possible analytes can be significantly expanded. Because of the extensive fragmentation experienced during electron ionization (EI), it is the most widely used GC/MS ionization technique; there are many compounds that produce unique patterns that can be used in conjunction with gas chromatographic retention-time data for an unambiguous identification.

Limits of detection can be lowered using special data acquisition techniques such as selected ion monitoring (SIM).** Three-ion ratios can be used for unambiguous identification because of extensive fragmentation when SIM is employed [23].*** Formation of electrophilic derivatives through the use of reagents such as perfluoropropionic anhydride allows limits of detection to be greatly reduced in the presence of complex matrices due to the ability to take advantage of techniques such as electron capture/resonance ionization.

Figure 1.2 is a conceptual illustration of the GC-MS. There are a number of variables associated with a GC/MS analysis. These variables fall broadly into two categories—instrumental variables and operational variables. For the most part, the instrumental variable must be decided before the actual GC-MS is acquired. The operational variables are those decided for each specific analysis.

---

*Compounds that can be ionized by atmospheric pressure chemical ionization (APCI) used in combination with LC/MS must be volatile and nonlabile-like compounds analyzed by using GC/MS; however, in LC/MS, the analytes do not have to withstand the rigors of the gas chromatographic partitioning process. Therefore, the range of analytes for APCI LC/MS can be expanded in the direction of lack of volatility and thermal lability.

**In many cases, spectral acquisition of GC/MS data involves a contiguous range of m/z values recording any ion abundances at any m/z value in the range. Instruments such as the transmission quadrupole mass spectrometer can be operated where each acquisition cycle of the instrument measures the ion current for a few (usually less than 8) m/z values by jumping from value to value and spending a specified "dwell time" on each ion to measure its ion current (abundance). This technique is called selected ion monitoring (SIM).

***Three-ion ratios can also be used with full-spectrum acquisition methods. Although it may not appear to be necessary to use these ratios when a complete spectrum is generated by full-spectrum acquisition, the three-ion ratio is required in some regulated methods, such as screening for drugs of abuse.

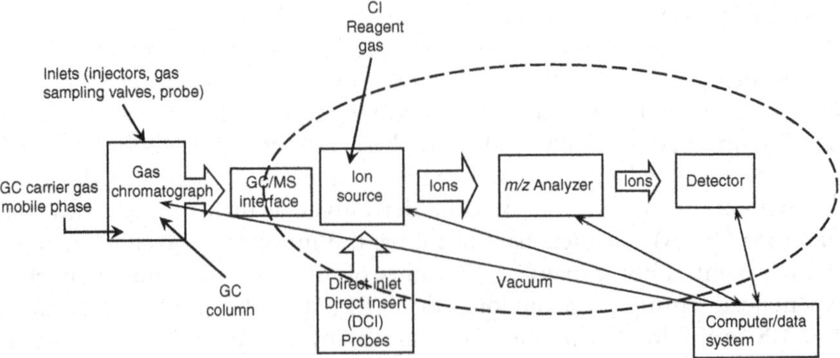

Figure 1.2 Conceptual illustration of the GC/MS system reveals the major components: the GC and its inlets, other detectors, the ion source, inlets other than the GC to the mass spectrometer, the *m/z* analyzer, the ion detector, and the data system. The components of the mass spectrometer must be maintained under vacuum, allowing ions to be independent of all other matter.

1.1. Instrumental Variables

- Automation (requiring some type of autosampler).
- Type of sample introduction(s)—e.g., type of gas chromatograph injector, direct introduction, pyrolysis, gas sampling valves, etc. If subambient cooling is required, will it be CO_2 or LN_2?
- Gas chromatograph oven-temperature requirements—e.g., subambient cooling (CO_2 or LN_2 option), rapid temperature ramping, minimal cool-down times, etc.
- Requirement (if any) for conventional gas chromatograph detection.
- Gas chromatographic mobile phase to be used.
- Type of mass spectral ionization—e.g., EI, chemical ionization (CI), field ionization (FI), electron capture negative ionization (ECNI), etc.
- Type of *m/z* analyzer—this is based on data requirements such as mass accuracy, spectral acquisition rate, lower limits of detection, linearity for quantitation, etc.
- Data system requirements—e.g., analysis reporting, database searching, automated quantitation, accurate mass from integer data, etc.

Some of these instrument variables can be added after the purchase and installation of the initial instrument, like software items; others such as analyzer and ionization types are not changeable. The same could be true for the sample introduction types and oven-temperature requirements. This is why a careful "needs analysis" should be performed before the instrument is purchased.

1.2. Operational Variables

- Gas chromatograph column to be used—e.g., length, diameter, stationary phase, thickness of stationary phase, etc.
- Injector settings—e.g., temperature, split ratios, split times, etc.
- Need for flow-rate adjustment to obtain proper linear velocity of mobile phase.
- Calibration of *m/z* scale.
- Gas chromatograph oven-temperature program rate(s)—e.g., initial hold time and temperature, temperature ramp(s), upper temperature and hold time, cool-down time, etc.
- Temperature of interface between gas chromatograph and mass spectrometer.
- Temperature of ion source and analyzer.
- Type of ionization—e.g., EI, CI, FI (may require hardware change and introduction of an auxiliary gas).
- *m/z* range to be acquired.
- Rate of spectral acquisition (may involve stating number of spectra to be averaged before storing data).
- Spectral acquisition type (centroid [default] or profile) based on instrument type.

The various gas chromatograph sample inlets, sample introduction methods, columns, GC detectors, and operating conditions are described in Chapter 2. The reason the gas chromatograph mobile phase, column size, and ionization type need to be known at the time of an instrument purchase is that this has bearing on the pumping system required for the *m/z* analyzer. This will be explained in more detail in Chapter 4.

GC involves a lot more than just selecting an appropriate column length, diameter, and stationary phase. With modern open-tubular columns, sample injection has become as significant as the column selection. As detailed in Chapter 2, there are several different types of injectors; the most widely used injector is the so-called split/splitless injector used for a spilt or a splitless injection. Analytes are usually present in a solvent. Selection of the solvent, again, is as important as the column selection. Even though GC conditions are provided for many types of individual compounds in Section II, there are many analyses that require the determination of multiple types of compounds at the same time. For this reason and others, GC/MS is an experimental science. More often than not, an analysis must be developed. The chromatographic and data acquisition must be determined empirically.

The most widely used ionization technique in GC/MS is EI. Probably greater than 90% of all GC/MS analyses are performed using EI. Many instruments require a physical change in the ion source when switching from one ionization technique to another. The internal ionization quadrupole ion trap (QIT) allows for EI and CI without any physical changes. Some manufacturers have combination EI/CI ion sources and others provide instructions for obtaining EI data from CI sources. FI is the least used GC/MS ionization technique and is only offered by two instrument manufacturers: Waters Corp. as a TOF and tandem quadrupole* instrument; and JEOL as a TOF GC/MS system. Another ionization technique that was available in the past for GC/MS is APCI. There are no commercial instruments offering this technique at this time. A paper was published with instructions on how to configure an atmospheric pressure ionization (API) instrument (mostly APCI and electrospray ionization (ESI) instruments used in liquid chromatography/mass spectrometry (LC/MS) to take the eluate from the gas chromatograph for APCI) [24].

ECNI, also known as resonance electron capture negative ionization, is another somewhat widely used GC/MS technique because of the low detection limits that can be obtained for electrophilic compounds such as halogenated pesticides or halogenated derivatives. Due to the specificity of this technique, matrix compounds will not be detected for extremely low limits of quantitation, especially when used in conjunction with an SIM analysis.

GC/MS has used many types of *m/z* analyzers to separate ions according to their *m/z* values. The device that is the most ubiquitous in GC/MS instrumentation is the transmission quadrupole analyzer also known as the quadrupole mass filter (QMF). The device that follows this in popularity is the QIT. Both of these devices are limited to producing data with an integer *m/z* value. Such data does not produce unambiguous elemental composition for ions. However, an accurate mass measurement, within ± 0.1 millimass units, does provide an unambiguous elemental composition. This has led to an increased popularity of the reflectron TOF analyzer, which will generate such data. The TOF provides lower detection limits than can be achieved with the QMF analyzer and does not produce skewed spectra due to spectral acquisitions occurring during rapid changes in the concentration of the analyte in the ion source due to the chromatographic peak width. The TOF analyzer will acquire spectra at a faster rate than can be achieved with any other type of analyzer. Software has been developed

*Tandem quadrupole is a term used to describe an instrument used for mass spectrometry/mass spectrometry (MS/MS). This instrument is sometimes referred to as a *tandem-in-space* instrument, which means that ions from an initial ionization isolated by one mass spectrometer are activated in such a way as to bring about their subsequent decomposition. These ions produced by this decomposition of a precursor ion are then separated according to their *m/z* values, using a second *m/z* analyzer. The instrument is called a triple quadrupole because the two tandem QMF instruments are separated by a third device which brings about the collisional activation of the precursor ion. In the original design of this instrument, this third device was also a quadrupole operated in such a way as to be used only for collisional activation and not ion separation. Today, this third device is usually not a quadrupole but the name *triple quadrupole* remains.

that allows for the assignment of accurate *m/z* values to QMF and QIT data through the use of mass spectral peaks of known purity and known exact *m/z* values, which has expanded the utility of these instruments.

Magnetic-/electric-sector instruments (the double-focusing mass spectrometer), along with magnetic-sector (single-focusing) instruments that dominated mass spectrometry until the commercialization of the QMF by Finnigan Corp. (now Thermo Scientific of Thermo Fisher), Hewlett–Packard (the analytical instrument division of HP is now part of Agilent Technologies), and Extranuclear Corporation (Extrel CMS, LLC) in the late 1960s and early 1970s, are still in limited use and commercially available from at least three different manufacturers.

GC/MS/MS (also known as tandem mass spectrometry) is another technique that is being increasingly employed for various types of analyses and will be discussed in more detail in Chapter 4. GC/MS/MS is commercially limited to the use of the triple quadrupole instrument (tandem-in-space) and to the QIT systems (tandem-in-time). Although commercially available, the Fourier transform ion cyclotron resonance (FTICR) mass spectrometer (the magnetic ion trap) has also been used for GC/MS. These instruments are capable of very accurate mass measurements; but due to their complex operation, need for a cryogenically cooled superconducting magnet, and high initial cost, they are rarely found in a GC/MS laboratory.

GC/MS presents a paradox (Figure 1.3). As GC has developed, the width of the chromatographic peak has continually been reduced; i.e., the elution time for a component has become less. Acquisition of a mass spectrum from a scanning beam-type instrument (the QMF and the double-focusing sector instrument) is obtained over a finite period. Narrow chromatographic peaks result in rapid changes of analyte concentration in the ion source during the acquisition of a single spectrum. This spectrum can exhibit a skew (relative intensities of peaks at different *m/z* values that are different from what would be observed if there was no change in concentration in the ion source during the spectral acquisition). This skew will not be that reproducible and will make searches against standard mass spectral databases difficult. Shorter acquisition time can reduce the spectral skewing; however, the quality of the spectrum can deteriorate because the time spent measuring the ion current for any single *m/z* is so short that there is such a poor signal-to-noise ratio that the spectrum is uninterpretable. Reducing the period of elution for the analyte from the gas chromatograph column would greatly reduce the chromatographic resolution (the peak capacity). This reduced elution period would also have a tendency to increase the analysis; and with the current interest in fast GC to shorten analysis times, this would not be acceptable. Instrument companies are offering QMF *m/z* analyzers with shorter "scan time" by reducing the noise, but the minimal signal will remain the same. This paradox, in part, is the reason that there is an increased interest in the pulsed instruments such as the QIT and TOF.

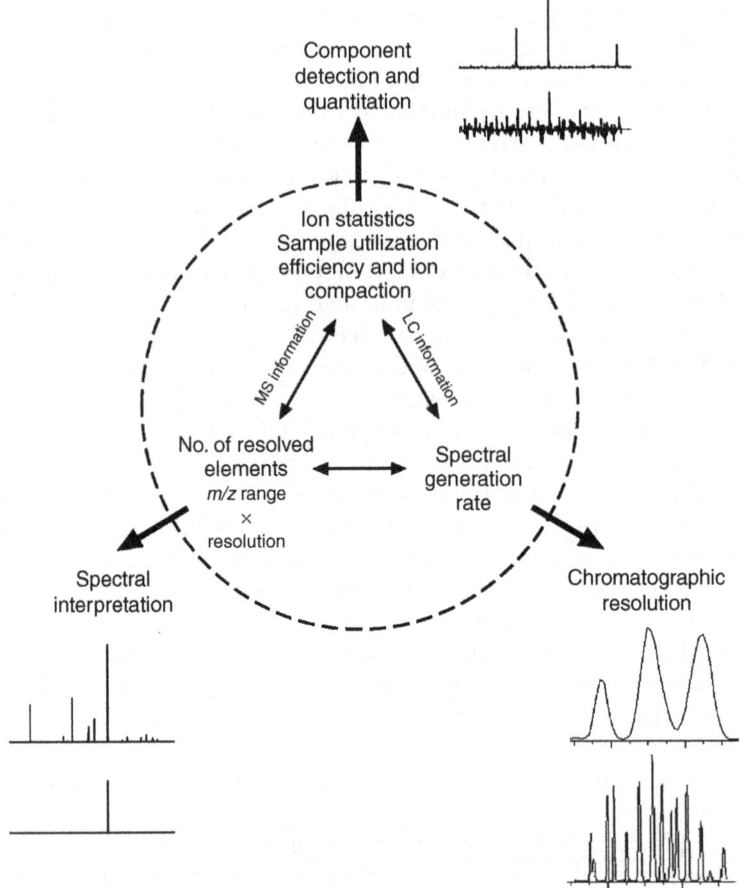

Figure 1.3 Conceptual illustration of the dilemma created by scan speed versus chromatographic peak width. Too slow of a spectral acquisition rate results in loss of chromatographic fidelity. Too fast of a scan speed results in poor spectral quality.

New practitioners enter the field of GC/MS; and, like all newcomers, they lack the experience needed to do many of the tricks that come with time spent in developing methods and interpreting data. This book should make that path to gaining the experience and knowledge easier and aid in resolving some of the various paradoxes that will be encountered.

REFERENCES

1. Abian, J. (1999). The coupling of gas and liquid chromatography with MS. *J. Mass Spectrom.*, 34, 157-68.
2. Thomson, J. J. (1899). On the masses of the ions in gas at low pressure. *Lond. Edinb. Dublin Philos. Mag.*, 48, 547-67.

3. Wien, W. (1898). Untersuchungen über die elektrische Entadung in verdünnten Gasen. *Annalen der Physik und Chemie*, 65, 440-52.
4. Tsvet, M. S. (Mswett—German Transcription of the name). (1903). A new category of adsorption phenome and its application to biochemical analysis. *Travl. Soc. Naturalistes Varisovic*, 14 (Russian).
5. Tsvet, M. S. (Mswett—German Transcription of the name). (1972). Tswett centenary issue. *J. Chromatogr.*, 73(2), 303.
6. Tsvet, M. S. (Mswett—German Transcription of the name). (1906). *Ber. Deut. Bot. Ges.*, 24, 316, 384.
7. Tswett, M. S. (1910). *Les Chromophless Dans Le Monde Végétal et Animal.* Varsovie (French), Paris.
8. Martin, A., Synge, R. A. (1941). New form of chromatography employing two liquid phases. I. A theory of chromatography. II. Applications to the microdetermination the higher monoamino acids in proteins. *Biochem. J.*, 35, 1358-68.
9. Martin, A., Synge, R. A. (1941). Separation of the higher monoamino acids by counter-current liquid-liquid extraction: the amino acid composition of wool. *Biochem. J.*, 35, 91-121.
10. James, A. T., Martin, A. J. P. (1951). Liquid-gas partition chromatography. *Biochem. J. Proc.*, 48, VII.
11. James, A. T., Martin, A. J. P. (1952). Gas liquid partition chromatography: a technique for the analysis of volatile material. *Analyst*, 77, 915-32.
12. Holmes, J., Morrell, F. (1957). Oscillographic mass spectrometric monitoring of gas chromatography. *Appl. Spectrosc.*, 11, 86, 87.
13. Gohlke, R. S., McLafferty, F. W. (1993). Early GC-MS. *J. Am. Soc. Mass Spectrom.*, 4, 367-71.
14. Gohlke, R. (1959). Time-of flight mass spectrometry and gas liquid partition chromatography. *Anal. Chem.*, 31, 535-41.
15. Reynolds, W. E., Bacon, V. A., Bridges, J. C., Coburn, T. C., Halpern, B., Lederberg, J., Levinthal, E. C., Steed, E., Tucker, R. B. (1970). A computer operated mass spectrometry system. *Anal. Chem.*, 42, 1122-9.
16. Sweeley, C. C., Ray, B. D., Wood, W. I., Holland, J. F. (1970). On-line digital computer system for high-speed single focusing mass spectrometry. *Anal. Chem.*, 42, 1505-16.
17. Desty, D. H., Goldrup, A., Whyman, B. A. F. (1959). The potentialities of coated capillary columns for gas chromatography in the petroleum industry. *J. Inst. Petro.*, 45, 287-98.
18. Schomburg, G. (1981). Sampling systems in capillary chromatography. In: *Proceedings of the Fourth International Symposium on Capillary Chromatography* (R. E. Kaiser, ed.), pp. 371-404. Huethig, Heidelburg, Germany.
19. Zlatkis, A., Walker, J. Q. (1963). Direct sample introduction for large bore capillary columns in gas chromatography. *J. Gas Chromatogr.*, 1(5), 9-11.
20. Grob, K., Grob, K., Jr. (1978). On-column injection onto glass capillary columns. *J. Chromatogr.* A., 151(21), 311-20.
21. Dandeneau, R. D., Zerenner, E. H. (1979). An investigation of glasses for capillary chromatography. *J. High. Res. Chromatogr.*, 2(6), 351-6.
22. Bartle, K. D., Myers, P. (2002). History of gas chromatography. *Trends Anal. Chem.*, 21(9.10), 547-57.
23. Sphon, J. A. (1978). Use of mass spectrometry for confirmation of animal drug residues. *J. Assoc. Off. Anal. Chem.*, 81(5), 1247-52.
24. McEwen, C. N., McKay, R. G. (2005). A combination atmospheric pressure LC/MS-GC/MS ion source: advantages of dual AP-LC/MS:GC/MS instrumentation. *J. Am. Soc. Mass Spectrom.*, 16(11), 1730-8.

CHAPTER 2

Gas Chromatography

2.1. Overview of a Gas Chromatograph

The technique of gas chromatography (GC) was introduced by James and Martin in 1952 [1]. The basic operating principle of GC involves volatilization of the sample in a heated inlet or injector of a gas chromatograph, followed by separation of the components of the mixture in a specially prepared column. Only those compounds that can be vaporized without decomposition are suitable for GC analysis. These compounds include most solvents and pesticides, numerous components in flavors, essential oils, hydrocarbon fuels, and many drugs. Acids, amino acids, amines, amides, nonvolatile drugs, saccharides, and steroids are among the compound classes that frequently require derivatization to increase their volatility. (See Appendix G for more information on derivatization of compounds for analysis by gas chromatography/mass spectrometry (GC/MS).

A carrier gas (sometimes referred to as the mobile phase), usually hydrogen or helium, is used to transfer the sample from the injector, through the column, and into a detector or mass spectrometer. The vast majority of columns used today are capillary tubes with a stationary phase coated on the inner wall. Separation of components is determined by the distribution (partitioning) of each component between the mobile phase (carrier gas) and the stationary phase. A component that spends little time in the stationary phase will elute quickly. After elution from the column, each component still in the carrier gas flows into a detector or a mass spectrometer. Detectors are dedicated tools designed specifically for use on a gas chromatograph; examples are thermal conductivity (TCD), flame ionization (FID), nitrogen–phosphorus (NPD), flame photometric (FPD), and electron capture (ECD). A mass spectrometer is a complex instrument that can be used with or without prior separation by a chromatographic instrument such as a gas chromatograph.

Key features of gas chromatographs are separate ovens that heat the individual injectors, the column, each detector, and the transfer line to the mass spectrometer. The column, and sometimes the injector ovens, allows the temperature to be increased at a regular rate (temperature programming) during the separation of the compounds in the sample. Figure 2.1 is a schematic representation of a gas chromatograph-mass spectrometer (GC-MS).

Gas Chromatography and Mass Spectrometry
DOI: 10.1016/B978-0-12-373628-4.00002-2

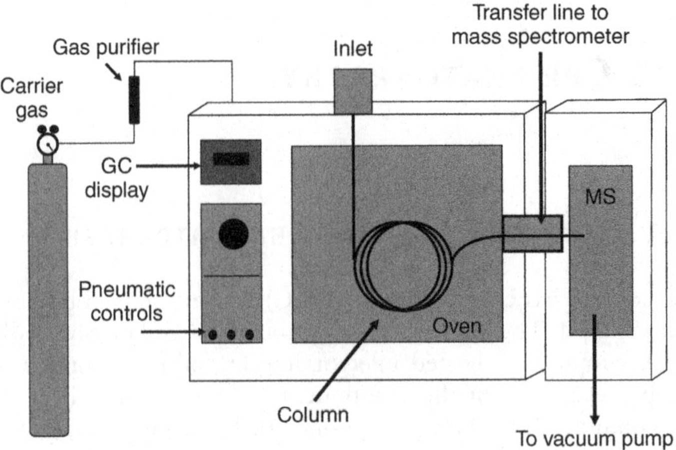

Figure 2.1 Schematic of a typical simple GC-MS. The configuration of the gas chromatograph can vary; for example, there can be two or more inlets or injectors, and several detectors can be installed on the gas chromatograph. GC/MS systems are controlled by a computer, which controls the physical parameters of the system such as the temperature zones and gas flows. The computer also handles the data generated during a run; i.e., compares mass spectra to a library to identify compounds and performs quantitation of peaks generated by the mass spectrometer and GC detectors. *Printed with permission of Restek Corporation.*

2.2. Sample Introduction

In the GC sample introduction process, the components in the mixture should be transferred to the column in as narrow a band as possible. The function of the column is to separate the components while minimizing the broadening of the band corresponding to each compound as it moves through the column. This results in the components in the mix eluting as sharp peaks, thus maximizing the signal to noise for each analyte.

Samples for GC/MS come in a wide variety of states, including gases containing organic compounds, liquid samples that may be organic solvents or water containing highly volatile to semivolatile compounds, and solids in which volatile or semivolatile compounds are embedded. Inlets for gas chromatographs are discussed, followed by a summary of recommended inlets for different types of samples.

2.2.1. Inlets for Liquid Samples

The majority of samples for GC analysis consist of a purified solvent containing the analytes of interest. This type of sample is introduced into a GC injector via a syringe. The sample is pulled into a syringe designed for use with gas chromatographs and typically about 1 μL is injected. There are several types of injectors that are designed for gas chromatographs. For samples to be analyzed by GC/MS, injectors that are designed for capillary columns (o.d. ≤0.53 mm) should be used. Injector hardware for capillary columns is described below; this is followed by sample types and injection parameters. Possible problems with each type of injection mode are also discussed.

Split/splitless injectors

The oldest and most commonly used injector for capillary columns is the heated split/splitless injector [2]. This injector can operate in two modes: split and splitless. Selection of either the split or the splitless mode depends on the concentration of the analytes in the sample. Figure 2.2 is a schematic representation of a typical split/splitless injector.

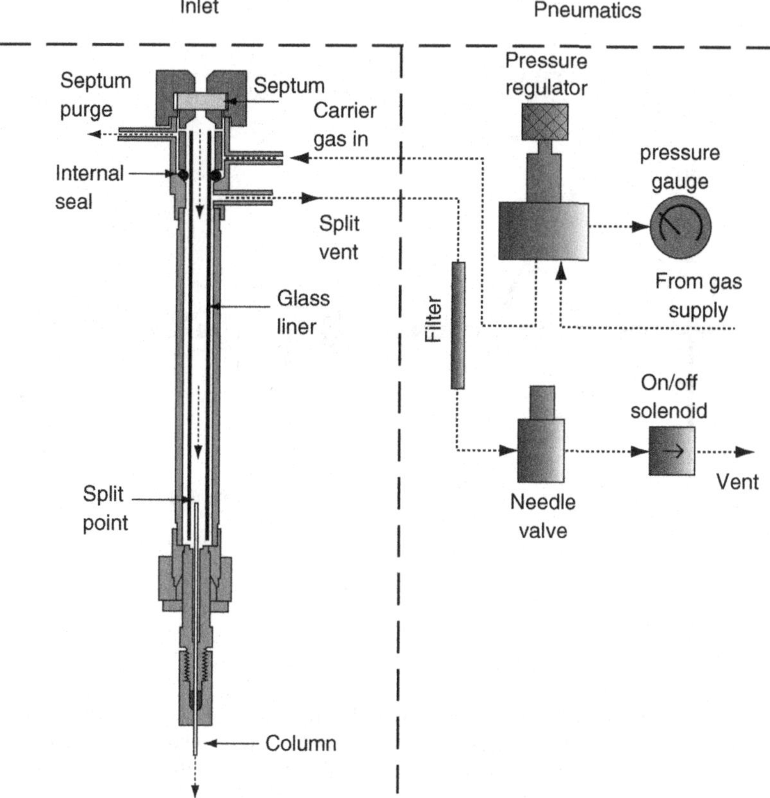

Figure 2.2 A cross section of a typical split/splitless injector. Features of these injectors include a glass liner where the length is fixed but the internal diameter and inner shape can be customized for the application. These injectors also typically include a septum purge line where a portion of the carrier gas flows past the septum and sweeps out impurities derived from the hot septum before they enter the column. The optimum position of the column in the injector is normally specified by the manufacturer.

Both split and splitless injection modes are hot isothermal injection techniques; that is, the injector is set at a temperature that is hot enough to vaporize the solvent and the analytes in the sample, and this temperature is constant throughout the GC run. Split injection is used for neat samples (not dissolved in a solvent) or samples where the analytes are dissolved in a solvent at relatively high concentrations. The splitless mode is used for samples containing analytes at trace levels.

Split mode

In the split mode, the injected sample is vaporized into the stream of carrier gas; and a small portion of the sample and solvent, if any, is directed onto the head of the GC column. The remainder of the sample is vented* (Figure 2.3). Typical split ratios range from 10:1 to 400:1 and can be calculated from the equation:

$$\text{Split ratio} = \frac{\text{Column flow} + \text{Vent flow}}{\text{Column flow}}$$

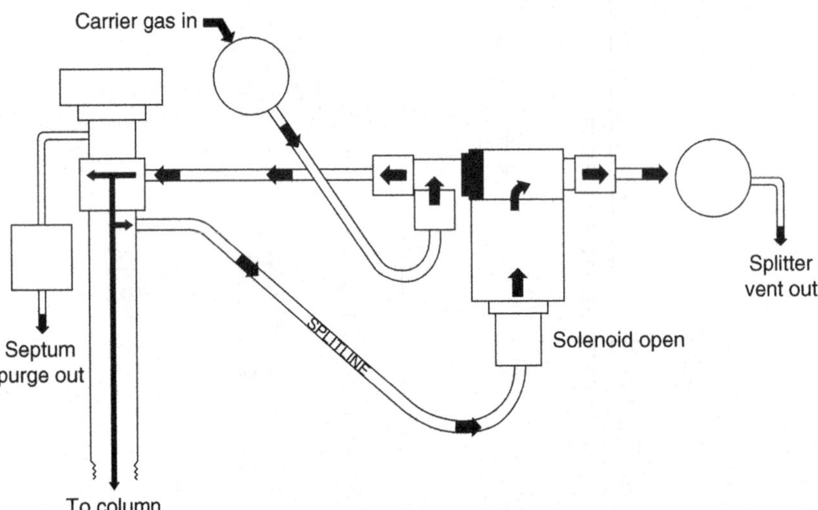

Figure 2.3 The flow path of the carrier gas in the split mode of a split/splitless injector. The sample (typically 1 μL) is injected into the hot glass insert where it is mixed, and a small part (normally between 5% and 0.5%) is directed onto the column while the remainder passes out of the split vent.

---

* All GC injectors will vent a portion of the vapors of the sample, solvent, or both to the atmosphere. These vapors should always be considered as hazardous and should be vented to a fume hood or trapped with a carbon filter. If a trap is used, it must be changed on a regular basis, and the trap material should be considered as hazardous waste.

where column flow refers to the flow of the carrier gas at the head of the column in $mL\,min^{-1}$ and vent flow refers to the flow of the carrier gas out of the splitter vent in $mL\,min^{-1}$.

In modern gas chromatographs, these flows are controlled electronically, and the split ratio can be set in the software. This will be discussed in more detail later in Section 2.3.3.

The split mode is used for samples that are fairly concentrated—generally, the compounds of interest are at levels greater than $100\,\mu g\,mL^{-1}$. The splitter vent is opened throughout the entire analysis. The injector temperature should be high enough to volatilize all of the analytes. The initial column temperature should be just below the temperature at which the first compound elutes.

There are many advantages of using the split injection mode:

- There are few parameters to optimize other than the split ratio.
- Analysis times are shorter than with other methods because there is no need for a low initial column temperature to focus the sample on the column.
- Problems with solvents that are not soluble in the stationary phase are avoided because very little solvent goes on the column.
- The technique is good for thermolabile samples because residence time in the hot injector is very short.

Injector liners for split injection are designed to aid mixing of the sample so that the split ratio is the same for all of the analytes in the sample. Sample size is usually 1 μL or less, and a fast injection speed is recommended.

Splitless mode

The splitless mode is another way to use a split/splitless injector. In this mode, the sample is injected with the splitter vent closed (Figure 2.4).

After a specific time (purge activation time), the splitter vent is opened to purge solvent from the injector. The analytes in the sample are deposited onto the head of the column and most of the volatile solvent is vented. For this reason, and because large amounts of sample can be injected, splitless injection is used for trace analysis. To prevent losses of the more volatile compounds of interest, the solvent boiling point should be at least 20 °C below the lowest boiling component of the sample.

With splitless injection, the injector temperature should be high enough to volatilize all of the analytes, and the initial column temperature should be 10–15 °C below the boiling point of the solvent. Under these conditions, the solvent and the analytes condense in a narrow band in the stationary phase at the head of the column. As the column temperature is increased, the analytes begin moving in the stationary phase and elute from the column as sharp peaks. The concentration of the solvent, followed by evaporation with increasing column temperature, is known as solvent

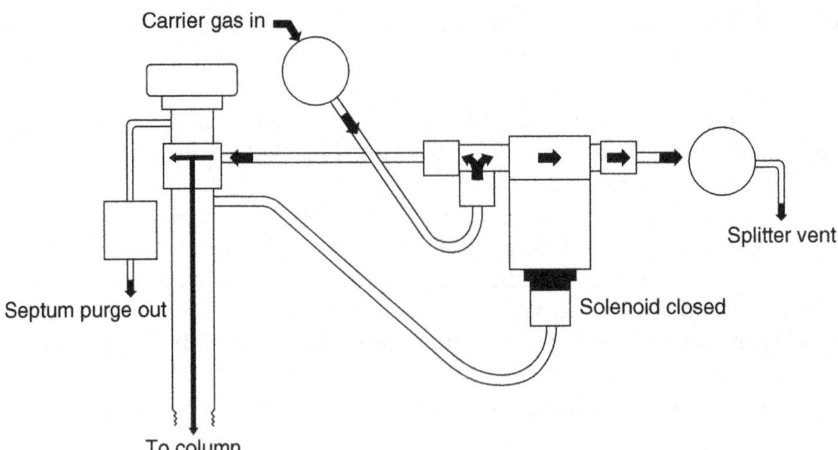

Figure 2.4 The flow path of the carrier gas in the splitless mode of a split/splitless injector. The sample (typically 1 μL) is injected into the hot glass insert with the split vent closed. Most of the sample is directed onto the column. After 1 minute or less (purge time), the split vent is opened and remaining solvent and a small portion of the sample are directed out of the split vent. The purge time should be optimized experimentally so that solvent peak tailing is eliminated, and losses of analytes are minimized.

focusing [3]. The opening of the splitter vent or purge time is usually about 0.5–1.0 minutes after the injection; however, this time should be optimized experimentally (Figure 2.5). If the splitter is opened too early, analytes are lost; if the splitter is opened too late, the solvent peak tails badly and obliterates the earlier eluting components (Figure 2.6).

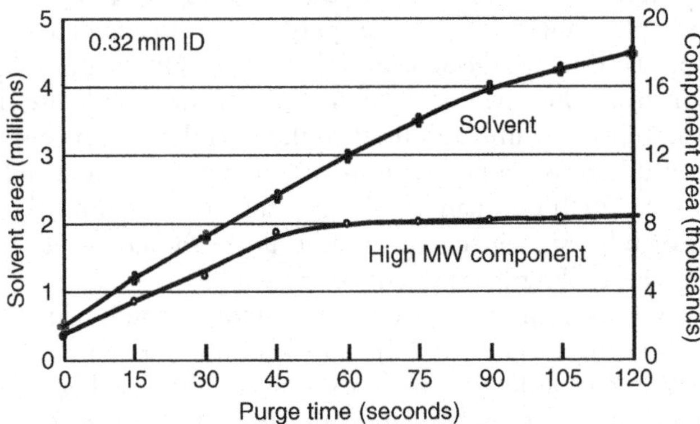

Figure 2.5 Optimization of purge time in the splitless mode. In this example, the optimum purge time is about 60 seconds where there is a good recovery of the high-boiling analytes in the sample and minimum solvent peak tailing. *Printed with permission of Restek Corporation.*

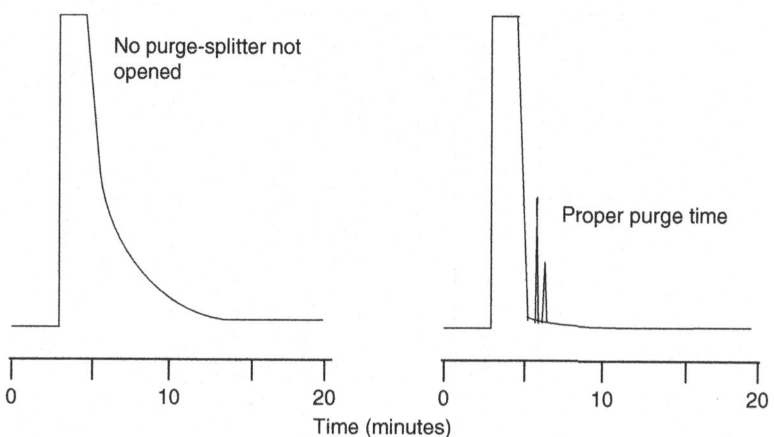

Figure 2.6 Comparison of a chromatogram that results from a splitless injection where the splitter is not opened to a splitless injection with a proper purge time. Note that the tailing solvent peak in the chromatogram on the left obscures the compounds that elute just after the solvent. *Printed with permission of Restek Corporation.*

One of the main problems with hot splitless injection is molecular weight discrimination. This means that if a sample contains compounds with a wide boiling-point range, lower quantities of the higher boiling compounds will be recovered, relative to the more volatile compounds that were in the sample. Another drawback of splitless injection is that residence time in the hot injector is relatively long, and thermally labile compounds tend to decompose. As a result of these disadvantages, splitless injection is recommended for trace analysis of samples that are in a fairly narrow boiling-point range and do not contain analytes that are thermally labile.

With splitless injection, a relatively large amount of solvent goes on the column and the solvent should be compatible with the column phase;* otherwise, solvent focusing will not occur. This will affect the early eluting analytes—they will be dispersed in the column and may elute as excessively broad or split peaks (Figure 2.7).

This must be avoided as sensitivity and precision will be adversely affected. Solvent compatibility with the column and initial column temperatures are important considerations with all injection modes except for the split mode.

* Samples with polar solvents such as methanol should be injected onto columns with polar stationary phases, such as waxes; suitable solvents for nonpolar columns are hexane, isooctane, and toluene. Ethyl acetate is compatible with most stationary phases.

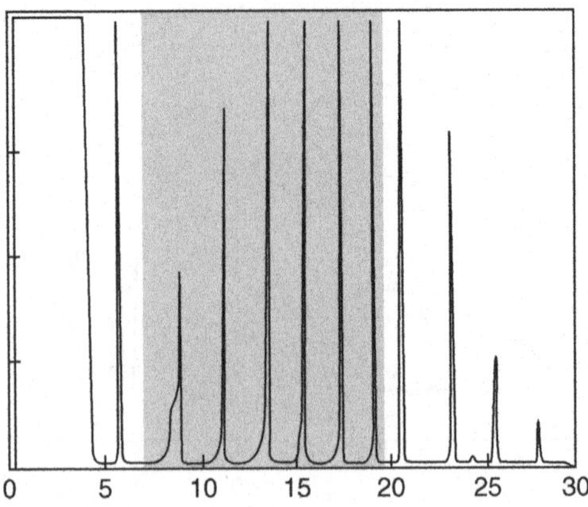

Figure 2.7 The peaks in the shaded area of this chromatogram were not focused at the head of the column after injection. The problem is seen with early eluting compounds that were introduced too rapidly into the column or into a column phase that was incompatible with the solvent.

Split/splitless injector liners

Split/splitless injectors can be used with several different glass inlet liners. Although the outer dimensions of these liners must be compatible with the requirements of a specific manufacturer's injector, liners with varied inner dimensions can be installed in a given injector. Some considerations that may influence the selection of a particular liner are as follows:

- Whether the injection mode is to be split or splitless
- Volume of sample to be injected
- Sample is dirty (contains nonvolatile components)
- Presence of thermally labile analytes in the sample

Liners for split injection are usually 4 mm i.d. and may contain a cup or glass frit to promote mixing of the sample, thereby assuring uniform splitting for all the components. The inserts may be packed with quartz wool to prevent nonvolatile material in the sample from entering the column. As stated above, injector residence time is short with split injection, and breakdown of active compounds in the liner is usually not a problem.

Although 4-mm packed liners are primarily used for split injection of dirty samples, these liners are sometimes used for splitless samples; however, liners for splitless injection are usually smaller in i.d. than liners for split injection (0.5–2 mm). The reason for the use of these smaller i.d. liners is that the residence time in the injector liner is relatively long with splitless injection, and choosing a liner with a smaller surface area reduces the likelihood of the breakdown of thermally labile compounds. These smaller i.d. liners require a slower injection speed.

In splitless injection, solvent volumes of 0.5 μL or greater are injected into a hot injector. Under these conditions, the solvent will expand to varying degrees and may overload the injector liner, resulting in "flashback" of the sample into the injector body. Severe tailing of the solvent and loss of analytes will result. Agilent Technologies has a free software program called "FlowCalc" that can be downloaded from their Web site (http://www.chem.agilent.com/cag/servsup/usersoft/files/GCFC.htm). This program allows the entry of the splitless injection parameters and calculates the maximum volume that can be safely injected (Figure 2.8) for a particular liner.

Most commercially available glass liners* have been deactivated to minimize decomposition of sensitive compounds and absorption of polar compounds on hot glass surfaces. Note that some users recycle liners by cleaning

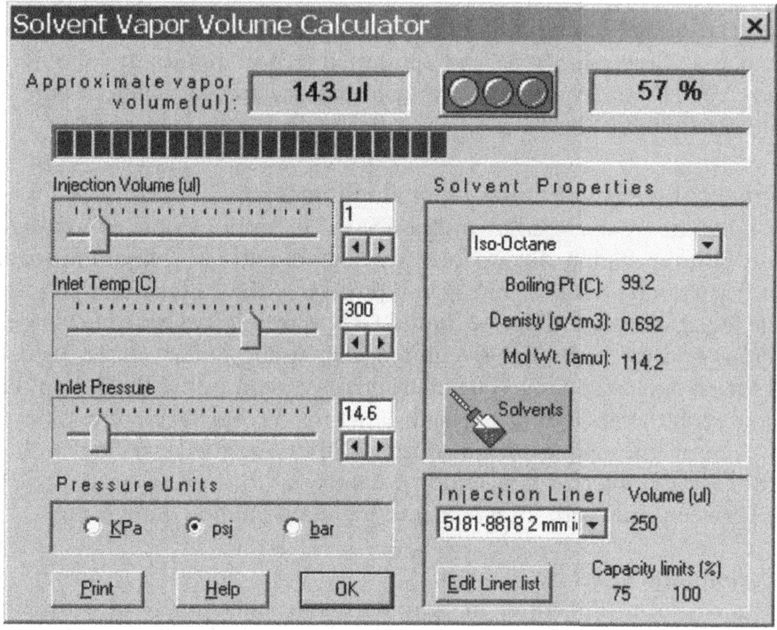

Figure 2.8 Screen capture of the "FlowCalc" program that can be downloaded for free from the Agilent web site. This useful software enables the user to enter the injection volume of a sample in a particular solvent at a given injector temperature, pressure, and liner volume. After these parameters are entered, the software calculates if the injector will be overloaded under the particular conditions.

* The deactivation procedures for inlet liners are proprietary and vary with the manufacturer. Liners are silane treated for general use. Base deactivated liners have been treated with amines and are suitable for trace analysis of basic compounds. Restek also sells Siltek™-treated liners, which they claim produces a highly inert glass surface that is suitable for compounds over a wide pH range and for chlorinated pesticides, and is stable at high temperatures.

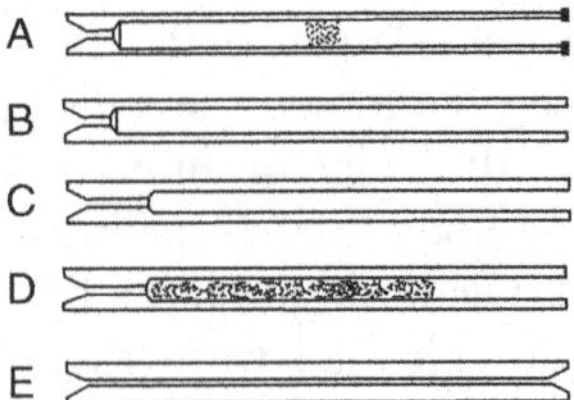

Figure 2.9 Cross sections of some commercially available liners for split/splitless injectors. Liner A contains a glass frit to promote mixing during split injection; B is an open insert that may be used as is or packed by the user usually for split injection; C is an open 2-mm i.d. insert for splitless injection; D is the same insert with packing; and E is an open 0.5-mm i.d. insert for splitless injection of labile compounds that require a liner with a minimum surface area.

and deactivating them with dimethyldichlorosilane or alcoholic KOH. Examples of liners for split and splitless injection modes are shown in Figure 2.9. The liners shown are only a small fraction of different types of commercially available liners.

Programmable-temperature vaporizing injectors

PTV injectors are enclosed in an injector oven that is capable of cooling as well as heating the injector. Coolants may be liquid CO_2, liquid nitrogen, or air (used rarely). If liquid nitrogen is piped into the laboratory, that might be the logical choice; otherwise, liquid CO_2 is readily available in cylinders. Different injector hardware is required, depending on which gas is to be used for cryogenic cooling. The appropriate version of the PTV should be ordered with the gas chromatograph. Note that cryogenic cooling is necessary even if the initial temperature of the injector is relatively high (50–80 °C); the injector cools slowly, compared to the GC oven, and the time between runs without a coolant installed would be quite long.

On-column injectors

With on-column injection, the sample is injected directly into the column. Unlike the conventional split/splitless injector, which has an injector oven that is set at a constant temperature throughout the GC analysis, the on-column injector works best with a programmed-temperature injector oven. The injector temperature is set below the boiling point of the solvent. After the syringe is withdrawn, the injector is rapidly heated (normally at a rate of at least 50 °C min^{-1}) to the maximum temperature, which should be high enough to volatilize all of the compounds of interest. As the sample components vaporize, they are transferred onto the head of the GC column, and the column is temperature programmed in the usual manner. This injection technique gives precise and complete recovery of most analytes and is especially valuable for samples with compounds that have a wide-boiling range or that contain thermally labile analytes.

In the past, syringes were used with very small o.d. fused silica or metal needles that could be inserted into fused silica columns with i.d.'s as small as 250 μm. These needles are still available, but they are very fragile; and now most on-column injection is accomplished using standard GC syringes and wide-bore columns with i.d.'s of 530 μm. In GC/MS systems, narrow-bore (250–320 μm) columns are preferred, and a short section of inert* 530-μm fused silica capillary tubing is often inserted into an injection port and attached to a narrower-bore capillary analytical column, using a glass connector such as the one in Figure 2.10.

Multifunction PTV injectors

As discussed in the previous section, true on-column injection with narrow-bore columns is not a rugged technique because the required small o.d. syringe needles are fragile. However, temperature-programmable injectors can be operated in a "direct injection" mode that approximates the

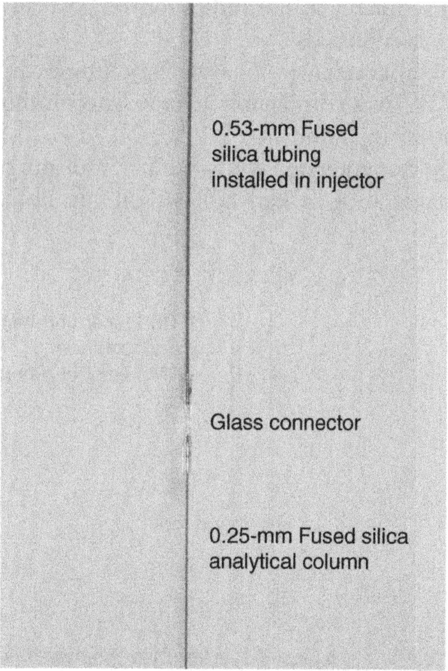

Figure 2.10 Photo of two columns of different o.d.'s connected with a glass connector.

* The 530-μm fused silica tubing may be uncoated deactivated tubing, or it may be coated with a thin layer of a nonpolar stationary phase.

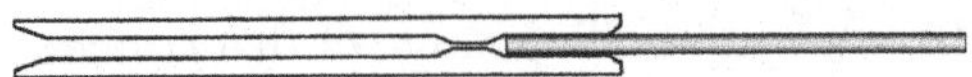

Figure 2.11 Glass insert with a column installed to form a butt connection. Under these conditions, performance approximates on-column injection.

performance of a true on-column injector. To operate in this mode, a special injection liner (Figure 2.11) is required. The narrow-bore analytical column is inserted into the lower portion of the liner to form a butt connection, and the injector needle delivers the sample into the upper portion of the liner. Injector and column-temperature conditions are the same as with true on-column injection.

Soon after its introduction at Varian, it was demonstrated [4] (Figure 2.12) that the temperature-programmable feature of a PTV injector, containing the insert shown in Figure 2.11, allowed better recovery of thermally labile compounds than hot isothermal injection. A test sample developed by Donike [5] was used to determine the injector's thermal stability. This sample consists of the two TMS esters which are thermally labile and *n*-triacontane which is stable; therefore, the relative chromatographic peak heights of the TMS esters to those of the *n*-triacontane are an indication of how an injector performs with thermally labile compounds.

Figure 2.13 shows that recovery of a wide-boiling hydrocarbon mix after injection into the PTV was comparable to true on-column injection and far superior to hot splitless injection.

A split/splitless injector equipped with a temperature-programmable injector oven is a very versatile tool. It can be used in the isothermal mode as a

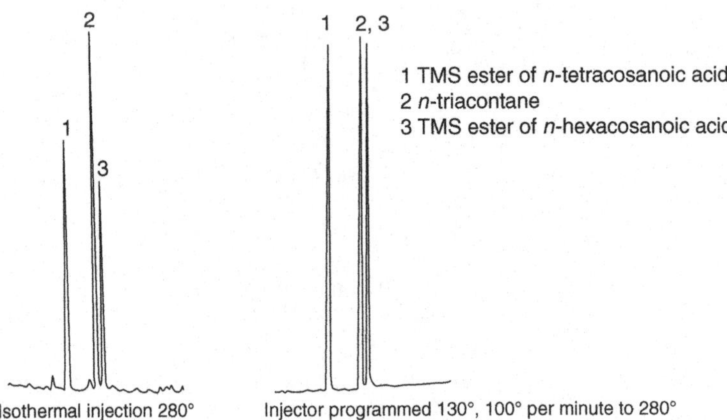

Figure 2.12 The chromatogram on the left shows losses of thermolabile compounds 1 and 3, relative to a stable hydrocarbon (compound 2), with hot isothermal injection. On the right, recovery of the thermolabile compounds was improved with temperature-programmed injection. The same injector hardware and column temperatures were used for both injections.

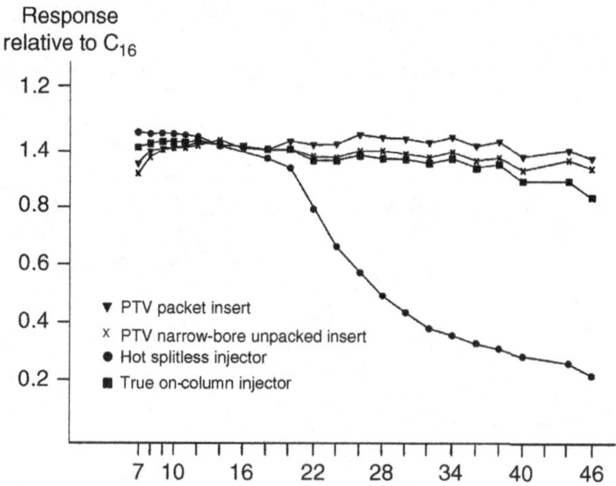

Figure 2.13 Responses relative to C_{16} for normal alkanes from C_7 to C_{46}. The experiments were conducted using two mixes: one containing 30–40 ng μL^{-1} of normal alkanes from C_7 to C_{16} in hexane and the other containing 30–40 ng μL^{-1} of normal alkanes from C_7 to C_{46} in isooctane. Injection volumes were 0.6 μL. Three injectors were used: conventional split/splitless, a PTV, and a true on-column injector. See the original reference (http://www.varianinc.com/media/sci/tech_scan/A01852.pdf) for all conditions.

conventional split/splitless injector; with temperature programming, it can operate in the direct injection mode; or it can be used in cold split and splitless modes.

Cold split and splitless injection

The cold split mode is used for samples that are fairly concentrated—generally greater than 100 μg mL^{-1} with a wide boiling-point range of analytes. Injector temperature conditions are the same as with on-column injection. As discussed in the section on *Splitless mode* under Section 2.2, the initial column temperature should be lower than with hot split injection in order to aid focusing of the analytes on the column.

Cold splitless injection is used for trace analysis of wide-boiling samples. Conditions are similar to hot splitless injection except that the injector is cooled during the injection of the sample and then heated rapidly (50–200 °C min^{-1}) as for the other cold injection techniques described above. Molecular weight discrimination is greatly reduced, but some degradation of thermally labile compounds may occur due to the long residence time of the analytes in the liner.

Large-volume injection

Vogt [6,7] first described the PTV for use as a large-volume injector (LVI). The LVI is a modification of the cold splitless mode. Sample volumes ranging from 5 μL to 100 μL are injected to improve the detection limits of analytes at very low

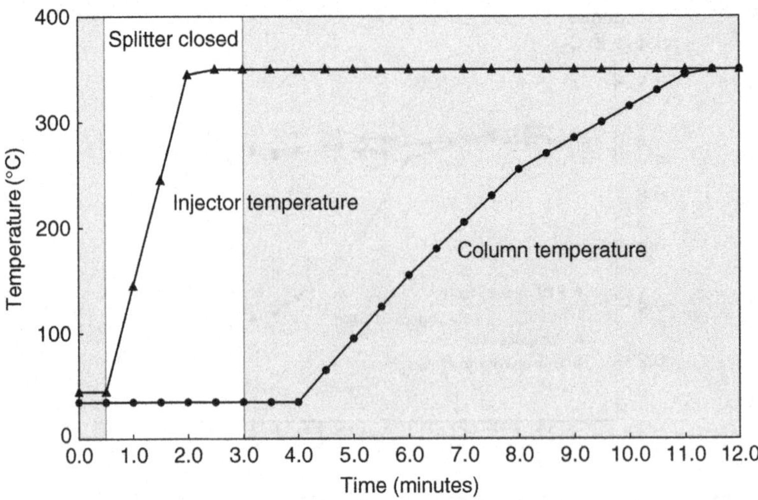

Figure 2.14 Conditions for large-volume injection.

concentrations. In this mode, the compounds of interest should have a boiling point substantially above that of the solvent to avoid losses when the solvent is vented. The sample is injected slowly, with the column and injector just below the boiling point of the solvent and the splitter open. Most of the solvent is vented, whereas the high-boiling analytes are retained in the liner. After an experimentally determined hold time (usually up to 1 minute), the splitter is closed and the injector rapidly heated, transferring the analytes to the column. After another few minutes, the splitter is opened and the column is programmed until all of the compounds in the sample elute (Figure 2.14). Normally, a liner packed with Carbofrit™ [8] or glass wool is used, but successful results have also been reported [9] with an unpacked liner. The technique requires that some time be spent determining the optimum conditions for each analysis.

Impurities in the solvent will be more apparent with this technique, but this should not be a problem with highly selective mass spectrometer detection.

Matching the injector to the sample

Some factors to consider when deciding what type of injector or injection mode is most suitable for a given liquid sample are concentration, boiling-point range, and thermal stability of the analytes.

The expected concentration of each analyte should be between the highest and lowest levels of a calibration curve that was constructed using standards at known concentrations. Generally, this would be in the range of picograms to nanograms passing into a GC detector or mass spectrometer. Samples where the concentration of analytes is lower than a few nanograms per microliter are considered to be *trace analysis samples.*

Table 2.1 Injection mode recommended for various liquid samples

Injection mode	Sample type
Split/splitless injector (hot isothermal injection) in split mode	Samples that are fairly concentrated—generally containing analytes at concentrations greater than 100 μg mL^{-1} with a narrow boiling-point range
Split/splitless injector (hot isothermal injection) in splitless mode	Trace analysis of compounds that are in a fairly narrow boiling-point range and are not thermally labile
Cold on-column	Trace analysis of wide-boiling samples and/or samples containing thermally labile compounds
PTV in direct injection mode	Same as cold on-column
PTV in cold split mode	Samples that are fairly concentrated—generally greater than 100 μg mL^{-1} with a wide boiling-point range of analytes
PTV in cold splitless mode	Trace analysis of compounds with a wide boiling-point range
PTV in large-volume mode (LVI)	Trace analysis of samples where the analytes are too low in concentration to be detected if only 1 μL or 2 μL is injected

Table 2.1 summarizes the type of sample recommended for each injection mode.

Based on the above discussion, it can be concluded that cold injection followed by temperature programming gives the best results as far as recovery of all analytes. This may bring into question why isothermal injection is so widely used. The answer is that isothermal injection is more convenient. An inconvenience with cold injection methods is the necessity of cooling the injector with liquid CO_2 or liquid nitrogen. A tank of liquid CO_2 or a Dewar filled with liquid nitrogen usually needs frequent attention. As the cooling gas is depleted, the time between GC analyses becomes longer. Many samples are successfully analyzed with hot split or splitless injection; and these methods are easier to optimize and cost less because there is no need to use coolants, and, especially with hot split methods, the run time is shorter.

Injection techniques

Filling the syringe

Injections of liquid samples into a gas chromatograph are usually made with a 10-μL syringe. In order to obtain reproducible manual injections when injecting into a hot injector, the "sandwich technique" is recommended.

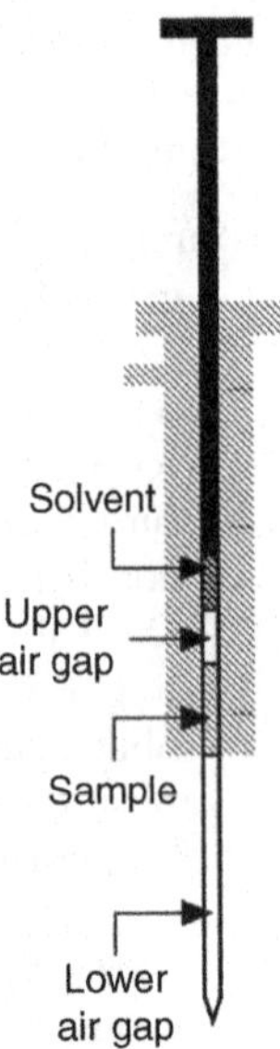

Figure 2.15 A syringe filled with sample and solvent prior to injection with the "sandwich technique" described in the text.

A small amount of solvent is drawn into the syringe barrel, followed by an air plug, next the sample, and finally another air plug, leaving the needle empty (Figure 2.15). With this technique, discrimination against both high- and low-molecular-weight compounds is reduced [10].

As stated earlier in the discussion of hot splitless injection technique, there is a limit as to how much can be injected into a hot injector without overloading it. The volume of the solvent plug must be considered when using the sandwich technique in hot splitless injection.

Injection speed

Injection speed has a significant effect on chromatographic results. For hot split injection, a very fast injection speed is recommended.

For hot splitless injection, the optimum injection speed depends on the capacity of the glass liner and the boiling-point range of the compounds in the sample.

1. For smaller i.d. liners, injection speed should be slower.
2. Fast injection speed will result in discrimination against the more volatile compounds.
3. Slow injection speed will result in discrimination against the less volatile compounds.

For cold injection, a slower speed is preferable.

Injection speed is an important parameter and must be optimized empirically. Many autosamplers allow the setting of different injection speeds, thus assuring not only the proper speed but also a reproducible speed from injection to injection—this is one of the many reasons that injection with an autosampler (as opposed to manual injection) is recommended.

Pressure-pulsed injection

With pressure-pulsed injection, the pressure of the carrier gas is increased by 30–50 psi just before the injection. Afterward, the pressure is lowered to maintain the desired flow through the column. This has the effect of compressing the sample to a much smaller volume and avoiding the problems of overloading the injector that occur with hot splitless injection.

Septum problems and how to avoid them

The GC septum is a silicone disk that seals the injector and maintains the elevated pressure that is within the injector body. The septum must be able to maintain this pressure even after it has been punctured dozens of times by the syringe needle. GC injectors are heated, as high as 400 °C, and the septum must be able to withstand high temperatures* without softening and melting into the injector body. If a septum becomes too perforated, it can be a major source of air** leaking into the GC column, which may shorten column life, especially in columns with a polar stationary phase. Another problem with septa is bleed that manifests itself as extraneous (ghost) chromatographic peaks during the temperature programming of a GC analysis and/or extraneous mass spectral peaks. Septa are also subject to "coring" by the syringe needle. When coring occurs, pieces of septum fall into the hot inlet where they can release compounds such as phthalates and organosiloxanes and cause ghost peaks in the chromatogram. With a mass spectrometer, spurious peaks due to septum bleed have *m/z* values of 355, 429, and 503. Ghost chromatographic and mass spectral peaks from the septum are especially problematic with trace analysis.

Pieces of the septum can also clog the GC syringe needle. It was stated earlier that automatic injection is preferable to manual injection because the injection speed is easily replicated. Another advantage with automatic injection is that the syringe needle tends to penetrate the same part of the septum with each injection, thereby prolonging septum life. Changing the septum at regular intervals is an important part of the maintenance of a gas

---

* The septum is somewhat protected by its position at the top of the injector, where it may be as much as 100 °C cooler than the inside of the injector.

** With a mass spectrometer, air can be detected by observing peaks at *m/z* 28 and 32. In some cases, it might be practical to start mass acquisitions at *m/z* 20 or lower to check for the presence of air.

chromatograph. The frequency depends on whether injection is manual or automatic, the polarity of the stationary phase, and observation of the chromatogram for ghost peaks. A reasonable procedure would be to cool down the GC ovens and shut down the mass spectrometer at the end of a week's work, change the septum and the GC injector inlet if necessary, and then turn on the system and let it bake out over the weekend.*

In order to minimize septum problems, GC syringes are available with several different types of tips (Figure 2.16); the most common are as follows:

Beveled point (the standard point style) is suitable for septum penetration for all chromatographic applications. With this point, the end of the needle can easily be deformed and then is liable to damage the septum. This occurs when the needle is dropped point down on a hard surface. The long end of the needle develops what is called in the trade a "burr" or spur.

Side-hole point prevents the needle from being plugged when septa are penetrated (coring). Although this needle is less likely to damage the injector septum (because the tip comes to a smooth conical point), it can also direct the sample onto a hot glass liner and exacerbate decomposition of sensitive compounds.

Conical point syringe is recommended for use with the Merlin™ Microseal and the Jade™ Inlet, which are described below.

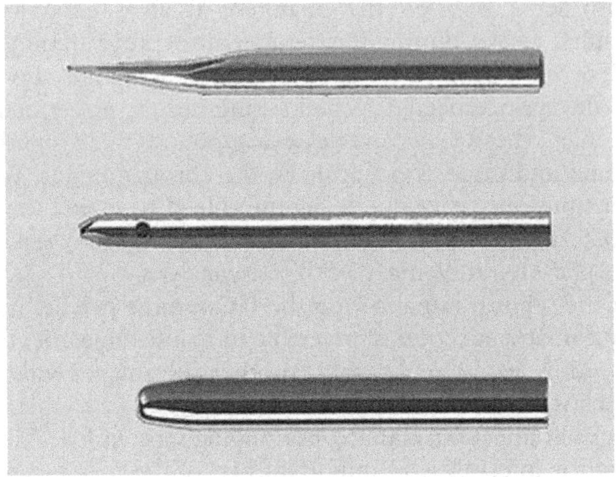

Figure 2.16 Needle-point styles for GC syringes: (top) beveled; (middle) side-hole; (bottom) conical.

* Another possibility is to count the number of injections and change the septum every 100 injections. (Many GC software packages allow the user to keep track of the number of injections and reset the count to 0 when the septum is changed.)

The needle on a GC syringe is normally 26 gauge (0.46 mm o.d.), which is thin enough to penetrate the injector septum without too much damage but is rugged enough to withstand many GC injections.

Merlin™ Microseal

The Merlin Microseal is a replacement for the septum and nut for GC injectors; the product is custom designed to fit specific injectors of major manufacturers of gas chromatographs. According to the manufacturer (Merlin Instrument Company), the Merlin Microseal will last for several thousand injections.

The seal is made of a fluorocarbon elastomer, which has less bleed than a conventional septum. Bleed is also significantly reduced because the needle penetrates an opening on top of the seal; and unlike a conventional septum, a fresh surface is not exposed each time an injection is made. On the bottom surface of the seal is a metal duckbill (Figure 2.17) that opens to admit the needle and then closes when the needle is withdrawn.* The Merlin Microseal must be used with a 23 gauge (0.64 mm o.d.) syringe needle with a conical opening.

Jade Inlet

Another septumless solution for capillary injectors is the Jade Inlet (Figure 2.18), which is available for most GC injectors. The manufacturer

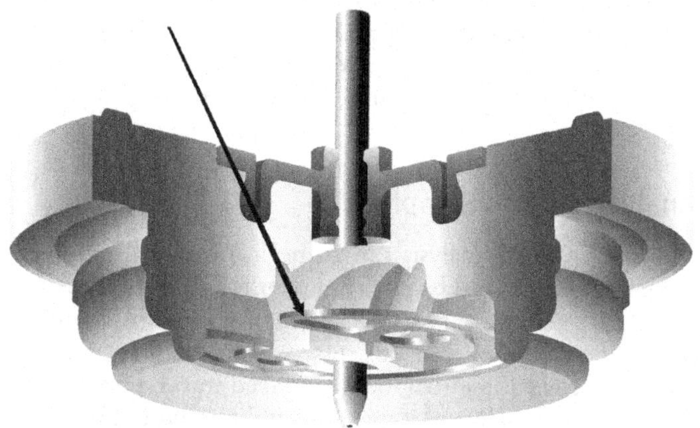

Figure 2.17 Schematic of the Merlin Microseal. The metal duckbill opens to admit the syringe needle as described in the text.

* There has been a report of air leaks associated with use of the Merlin Microseal. The author used this product extensively and did not experience leakage problems. Nevertheless, the user should be aware that air leaks may be a problem with any changes to the inlet system and should monitor for air leaks regularly.

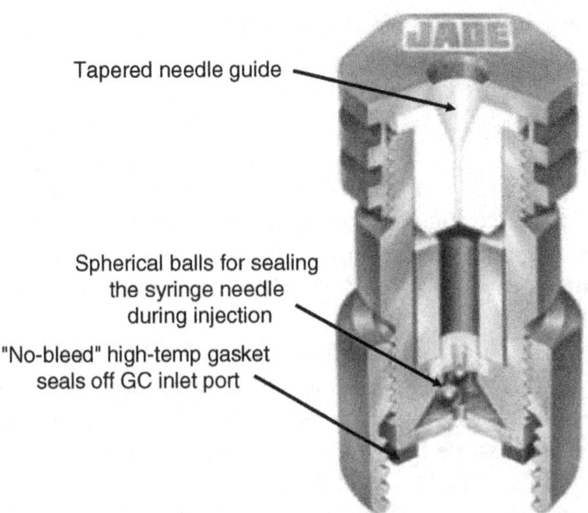

Figure 2.18 Schematic of the Jade™ Inlet.

(Alltech Associates, Inc.) claims that there is no bleed, the device can withstand inlet temperatures of up to 400 °C, and the only maintenance is an annual replacement of the needle guide. The syringe needle enters the needle guide and forms a seal. Then it encounters a ball and seat valve; the needle unseats the ball and enters the injection port. After the sample is injected, the needle is withdrawn from the ball valve and a magnet pulls the ball back into the seat, sealing the injector. The Jade Inlet is used with a syringe needle with a conical opening—either 23 or 26 gauge.

2.2.2. Analysis of Gas Samples

Gas samples for GC/MS analysis can come from several different sources including a gas tank, a special bag or container for collecting gas samples, or a headspace or purge-and-trap device. Sometimes the gas is introduced by injecting into one of the injectors discussed above with a gas-tight syringe. Alternatively, the gas sample may be introduced into the carrier gas stream through a gas sampling valve. A typical gas volume would be 1 mL.

Finally, the analytes in a gas sample may be adsorbed into a trapping material and then released by heating the trapping material and transferring the analytes into the gas chromatograph. Trapping the analytes in a gas sample yields a lower limit of detection as opposed to merely injecting 1–2 mL of the gas. The lower detection limit is due to the fact that a much larger volume of gas is sampled with trapping. Additionally, when the trap is heated, the analytes are concentrated into a narrow band, thus resulting in narrower chromatographic peaks with a higher signal to background. The hardware and some techniques for analysis of gas samples follow.

Gas-tight syringes

Gas-tight syringes may be used with any type of GC injector including split/splitless in either mode. The sample size usually varies from 100 μL to 2 mL or more. Injection with a gas-tight syringe can be automated—in fact, a 100-μL syringe can be installed in some autosamplers that were designed for liquid injection [11]. There are also some automated systems like the versatile *CTC CombiPal*™* that use gas-tight syringes that can be heated to prevent condensation of analytes in the gas sample.

Gas sampling valves

Valves are used in GC for many purposes and can add versatility and convenience to a GC/MS system. Some additional applications of valves will be discussed later in this chapter.

Valves are often used with a gas sample loop to introduce a sample into a column. Typically, a gas sampling valve is installed on top of the GC, usually in its own separately controlled heating oven. Even when the system is dedicated to the analysis of permanent gases, some heating of the valve is desirable to prevent condensation of water. The gas sample loop is available in different sizes; but for sample introduction into a gas chromatograph, 0.5 mL, 1 mL, and 2 mL are typical. Figure 2.19 is a photo of a valve and a gas sample loop.

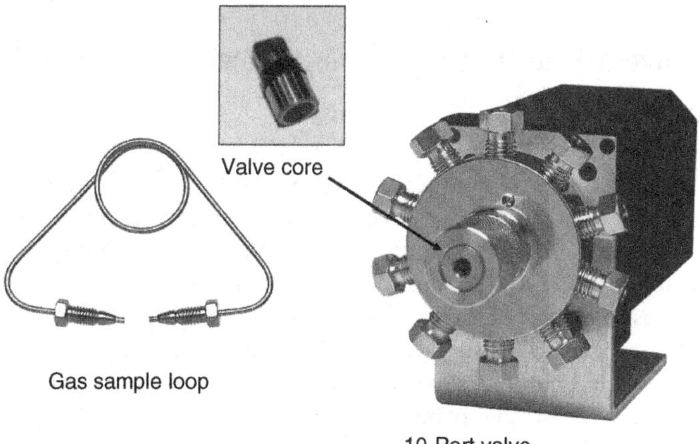

Figure 2.19 A 10-port valve. The core (shown in the inset) rotates to change the path of flow in the valve. On the left is a gas sample loop that can be installed on the valve to deliver a known volume of gas to the analytical column.

---

* CTC Analytic AG, www.CTC.ch; the CTC CombiPAL is marketed by several companies under different names.

Some points to note about the use of gas sampling valves are as follows:

- For most applications, valves with 4–10 ports are used.
- Gas sampling valves are usually heated, and they are designed to operate within a specified temperature range.
- They are available in a wide range of sizes for different size columns and tubing.
- They are available in materials with different degrees of inertness.
- High-vacuum valves are available for use with GC/MS.
- They operate either manually by turning a handle that rotates the valve core or automatically with an actuator that is timed to rotate the core during the GC analysis.
- They may be installed in a gas chromatograph and used instead of an injector for gas samples, or they may be part of an extraneous device such as a headspace or purge-and-trap system.

The series of schematics in Figure 2.20 illustrates the operation of a six-port gas sampling valve.

When a gas sampling valve is part of an extraneous system, the transfer line to the gas chromatograph can be equipped with a needle at the end that is inserted into a GC injector; or, alternatively, the transfer line can be connected to the carrier gas lines of the gas chromatograph directly, thereby eliminating the need for a GC injector. The transfer line would normally be enclosed in a heated jacket.

2.2.3. Solvent-Free Determination of Volatile Compounds in Liquids and Solids

Static headspace

Headspace GC is a sample preparation method for determining volatile compounds in solid and liquid samples. The technique has existed since the late 1950s [12] and is still actively used. With this technique, only the gas phase above the sample is introduced into the GC column. The popularity of headspace analysis is due to its simplicity and the fact that it is a very clean* method of introducing volatile analytes into a gas chromatograph; the injector system and column should require virtually no maintenance.

A liquid sample that is to be analyzed with the headspace technique is usually collected from the source and refrigerated in a container that is filled to the top so that no volatiles are lost from the sample. Just prior to analysis, the liquid or solid sample is placed in a vial at room temperature, leaving a significant amount of headspace over the sample; then the vial is sealed and

* With the headspace technique, only vapor above the sample is introduced; and there is no possibility of introducing nonvolatile material into the sample, the injection port, or column.

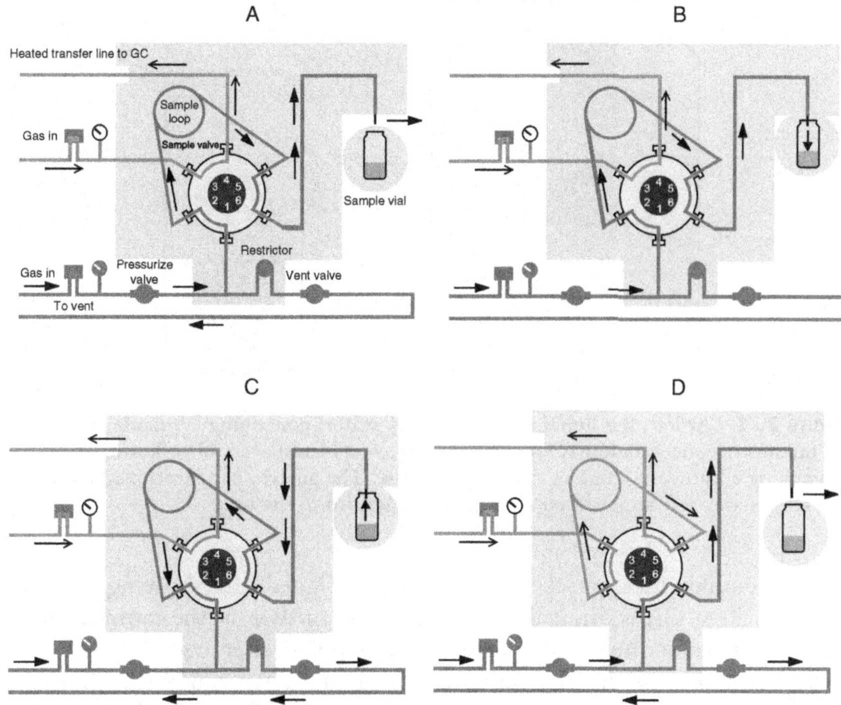

Figure 2.20 A schematic illustrating the delivery of a gas sample from a vial to a gas chromatograph. In this case, the vial contains a liquid or solid sample, and the gas phase over the sample is to be delivered to a gas chromatograph or a GC-MS. (A) standby position where the sample is heated; (B) the sample is pressurized with carrier gas; (C) the vent valve is opened and the pressurized gas in the vial flows into the gas sample loop; (D) the core of the valve is rotated and the gas sample from the vial is swept out to the gas sample loop and into the GC column.

heated. The analytes (volatile compounds) begin to move into the gas phase above the sample until a state of equilibrium is reached. At this point, the ratio of the analyte concentrations in the gas phase and in the liquid or solid phase is a constant. The constant is defined as the partition coefficient, *K*:

$$K = \frac{C_S}{C_G}$$

where, C_S is the concentration of the analyte in the sample after equilibrium (the term C_L may be used if the sample is a liquid) and C_G is the concentration of the analyte in the gas phase after equilibrium.

The partition coefficient is lowered as the temperature increases and also varies with changes in the matrix; for example, when organic analytes are dissolved in water, the addition of salts lowers *K*.

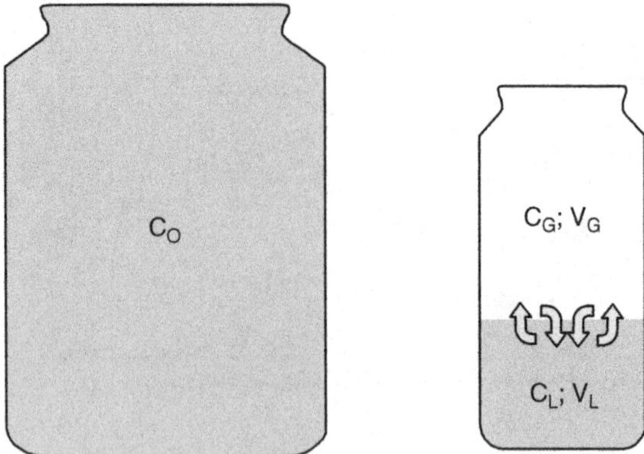

Figure 2.21 At left, the liquid sample (usually water) containing a volatile analyte at an unknown concentration (C_O). On the right is a headspace vial of known volume containing a known volume of the liquid sample. The analyte has partitioned between the two phases and achieved equilibrium, as described in the text.

The equilibration process is illustrated in Figure 2.21 showing a liquid sample where C_O is original concentration of analyte in the sample; C_L is concentration of the analyte in the liquid phase after equilibration; C_G is concentration of the analyte in the gas phase after equilibration; V_L is volume of the liquid phase in the headspace vial; and V_G is volume of the gas phase in the headspace vial.

After equilibrium is reached, an aliquot of the gas-phase sample is removed from either the headspace vial with a gas-tight syringe or a headspace instrument with a gas sample valve, and the sample is introduced into the gas chromatograph.

From the equation above, it can be seen that analytes with a very high partition coefficient tend to remain in the sample phase and those with a low partition coefficient tend to accumulate in the gas phase. Therefore, headspace detection limits for analytes in a compatible sample matrix (such as methanol in water) will be much greater (less sensitivity) than for analytes in an incompatible sample matrix (toluene in water). Another significant point is the importance of the phase ratio β, the ratio of the volume of the gas phase to the volume of the liquid phase in the headspace vial. It can be shown [13] that for analytes with a very high partition coefficient in the solvent, β is not significant in determining the concentration of the analyte in the headspace, and the relative volume of the two phases will not affect the headspace detection limits. When the partition coefficient is low, there will be lower detection limits if the volume of the liquid phase is increased relative to the gas phase in the vial.

Although headspace sampling can be done manually, better precision and lower detection limits are usually obtained with a dedicated headspace system. Several headspace systems are available; transfer to the gas chromatograph is with a gas sampling valve, a gas-tight syringe, or the pressure-balanced system that is unique to Perkin–Elmer [14]. Recently, commercial headspace samplers have become available with the capability of trapping the gas sample onto a sorbent instead of introducing it directly into the gas chromatograph.

With conventional headspace, the analytes are normally present at concentrations ranging from parts per billion to low percentage levels; with trapping, samples with much lower levels of analytes can be accurately analyzed.*

Common applications of the headspace technique are determination of blood alcohol [15], residual solvents in pharmaceuticals [16–18], and flavors in food [19,20]. For more information on optimization, quantitation, and troubleshooting in headspace, further reading is recommended [13,21].

Dynamic headspace (purge and trap)

The term "headspace GC" is commonly used to describe static headspace (described above) but is sometimes used to refer to dynamic headspace or what is also known as "purge and trap."

In the purge-and-trap technique (Figure 2.22), an inert gas is bubbled through the sample and the volatile analytes are transferred to an adsorbent

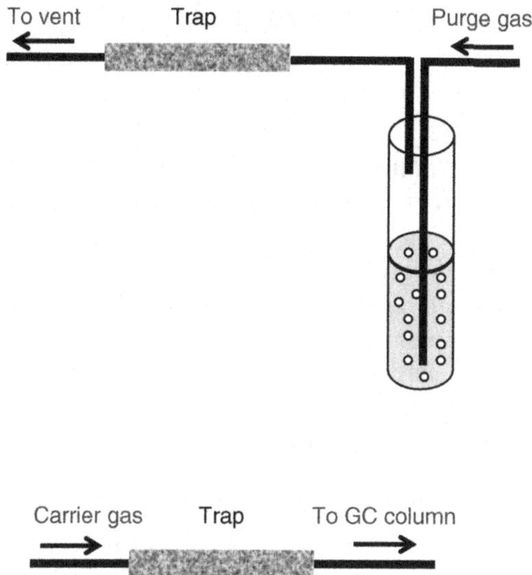

Figure 2.22 Purge gas bubbling through a sample containing volatile analytes that are directed to a trapping material. At the bottom of the schematic, the flow of gas is reversed using a valve (not shown) and directed to the GC column. The trap is heated and releases the analytes.

* The trapping capability on commercial headspace samplers is normally sold as an optional add-on since many applications do not require the additional sensitivity afforded by trapping.

trap [22–24]. The adsorptive material is usually Tenax®, a synthetic polymer that does not react with the analytes but efficiently binds them under ambient conditions and releases them at an elevated temperature without chemical modification; other adsorptive materials include silica gel, activated charcoal, and carbon molecular sieve. The trap is heated and the volatiles are released or desorbed and transferred to the gas chromatograph. Finally, the trap is heated to a higher temperature than was used during the desorption step to remove residual analytes and moisture, and the system is ready for another sampling. Application-specific packed traps for different purge-and-trap systems are commercially available.

Purge and trap generally allows for lower detection limits than static headspace because the former technique should theoretically remove all of the analyte from the matrix, whereas the latter is limited to removal of a single aliquot of the gas phase. With the new trapping capability of many headspace samplers, the difference in detection limits between the two techniques is reduced.

Purge-and-trap systems require more maintenance than headspace systems and are subject to problems such as foaming of the sample. In the United States, purge and trap is the standard method for determination of volatile organic compounds in water [23], whereas in many European countries, static headspace is used for this application. Other applications of the purge-and-trap technique include flavors in food [25,26] and toxic compounds in biological fluids [27,28].

Solid-phase microextraction

Solid-phase microextraction (SPME) was developed by Janusz Pawliszyn and his associates [29] at the University of Waterloo in Ontario, Canada. It is a solvent-free thermal desorption technique for determining volatiles in the headspace over aqueous and solid samples or determining semivolatiles directly in aqueous samples. Supelco Analytical (a division of Sigma-Aldrich Co.) is the licensee on Pawliszyn's patents and controls the distribution of the materials used for SPME.

The technique utilizes a short thin rod of fused silica (typically 1 cm long and 0.11 μm o.d.) coated with an absorbent polymer. The coated fused silica (SPME fiber) is attached to a metal rod. The coated fiber and metal rod comprise the SPME fiber assembly,* as shown in Figure 2.23.

Fiber assemblies are available with different polymers and different coating thicknesses. Some fibers are coated with a mixture of an absorbent polymer and an adsorbent compound related to charcoal. Most of the absorbent polymers are silicones related to GC column stationary phases.

* For greater fiber life with automated systems, Supelco has introduced a metal alloy fiber that is commercially available. However, not all phases are available on these fibers.

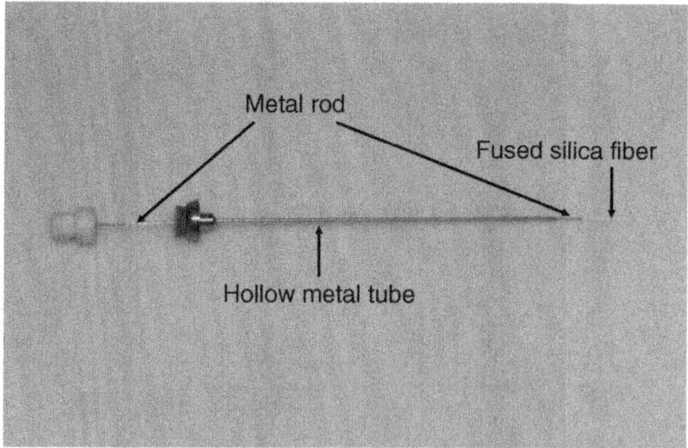

Figure 2.23 A photograph of an SPME fiber assembly. The plastic nut is color coded, according to the phase coated onto the fiber. The assembly shown is for a fiber holder to be used in an automated system; for manual fiber holders, a fiber assembly is used with a spring inserted under the nut.

The SPME fiber assembly is disposable and may be used for approximately 100 injections. The SPME fiber is installed in a modified syringe. In the standby position, the fiber is withdrawn into a protective hollow metal rod (sheath). For sampling, an aqueous sample containing organic compounds or a solid containing volatile organic compounds is placed into a vial,* and the vial is closed with a cap fitted with a septum. The sheath is pushed through the septum and the plunger is lowered, forcing the fiber into the vial where it is immersed directly into the aqueous sample or the headspace above the sample. Organic compounds from the sample are subsequently absorbed onto the fiber. After a time interval determined during the method development phase for the particular analysis, the fiber is drawn into the protective sheath and the sheath is pulled out of the sampling vial. Immediately after, the sheath is inserted through the septum of a GC injector, the plunger is pushed down, and the fiber is inserted into the injector insert where the analytes are thermally desorbed and transferred onto the GC column. The desorption step usually occurs over a period of 1–2 minutes; afterward, the fiber is withdrawn into the protective sheath, and the sheath is removed from the GC injector. Figure 2.24 is a schematic representation of this process.

Most capillary injectors are suitable for SPME sample introduction. The injector should be used with a straight glass liner with an i.d. of 0.75 mm. The chromatographic peaks tend to be narrow with SPME; band broadening is not a problem as there is no solvent or large volume of gas introduced into the injector. The protective sheath of the fiber tends

* Usually a standard headspace vial.

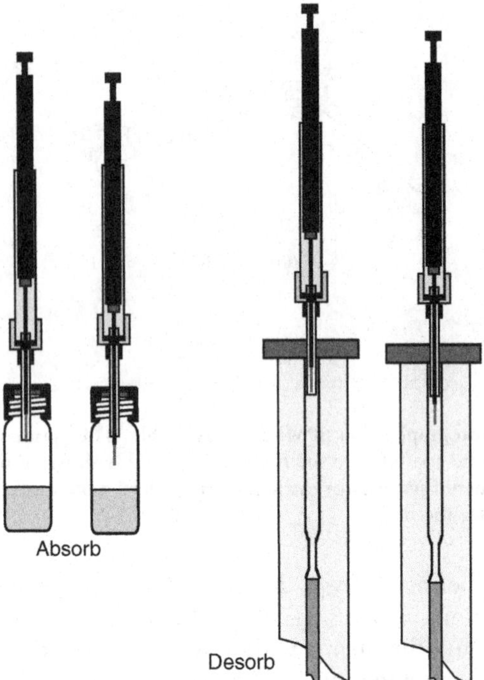

Figure 2.24 The SPME sampling procedure. On the left, the fiber holder is inserted into the sample vial followed by absorption of the analytes onto the fiber. On the right, the fiber holder is inserted into a hot GC injector followed by desorption of the analytes in the injector.

to core the GC septum; for this reason, the Merlin Microseal or the Jade Inlet is strongly recommended with SPME. SPME is a very popular technique—it is much less expensive to purchase a fiber holder and some fiber assemblies than to buy a headspace or purge-and-trap system. Many users use the technique in manual mode, but automation is also available with the CTC CombiPAL (available for nearly all commercial brands of gas chromatographs). Like static headspace, SPME is an equilibrium-based technique and extraction is not exhaustive; however, results are precise and consistent with other analytical methods, and detection limits as low as the part-per-trillion level have been reported [30]. Numerous applications are in the literature. Some of these applications are determination of flavors and impurities in food and beverages [31–33], pesticides and volatiles in environmental samples [34–36], and analysis of drugs and toxic compounds in biological fluids [37,38]. For a deeper understanding of the theory of SPME and method development, the book by Pawliszyn is recommended [39].

Stir-bar sorptive extraction

Stir-bar sorptive extraction (SBSE) was developed at the Research Institute for Chromatography in Belgium [40] and is marketed by GERSTEL (www.gerstel.com) under the name GERSTEL *Twister*™. The Twister device comprises a magnetic stir bar with a polydimethylsiloxane (PDMS) coating. The bar is available in two lengths (1 cm or 2 cm) and two phase thicknesses (0.5–1.0 mm). It is similar to SPME in that it is an equilibrium-based technique whereby analytes are absorbed from aqueous samples or the headspace into a coating of PDMS. Unlike SPME, which utilizes PDMS and several other phases, PDMS is the only phase currently available.

The Twister device is placed in a headspace vial with the sample and stirred for 30–90 minutes on a stir plate. The analytes are absorbed into the PDMS coating. The analytes can be desorbed from the PDMS coating with a solvent such as ethyl acetate or acetonitrile or by thermally desorbing into any split/splitless inlet or thermal desorption system in which the stir bar will fit. It was shown [41], however, that extraction efficiency is maximized when the stir bar is thermally desorbed into a proprietary (GERSTEL) thermal desorption unit (TDU) that is connected to the GERSTEL Cooled Injection System (CIS). The CIS serves as both a cryo-focusing trap and a PTV split/splitless injector. The TDU and the CIS can be installed on any modern gas chromatograph. The Twister bar is heated in a dedicated TDU liner, and the analytes are transferred to the CIS and finally to a GC column. SBSE desorption has been automated, using the GERSTEL MultiPurpose Sampler.* There is a greater volume of the absorptive phase on the stir bar than on a SPME fiber; therefore, the SBSE extraction technique offers the advantage of greater sensitivity and a wider linear capacity. The coated stir bar is also more rugged than an SPME fiber. However, SPME has the advantage of offering more selective coating fibers and of utilizing ordinary GC injectors for thermal desorption; therefore, SPME is more widely used.

Applications of SBSE have been published for analysis of food samples [42] and for environmental analysis of water [43]. A recent review [41] listed over 140 publications of SBSE application in the food, environmental, and life science areas.

2.2.4. Probes and Other Direct Sample Introduction Techniques for GC

Direct sample interface for liquid and solid samples

Aviv Amirav and his colleagues at Tel Aviv University developed the direct sample interface (DSI) [44]. A PTV GC injector is modified so that

* This is a modified version of the CTC CombiPal.

semivolatiles in dirty matrices such as food, soil, or blood can be determined without cleanup procedures.

The DSI can be used in two modes: (1) the analytes can be separated by conventional GC columns, and the analytes can be determined with a mass spectrometer or a GC detector; or (2) the analytes can be transferred rapidly into a mass spectrometer without separation. The first mode is recommended when many compounds that are amenable to a GC separation are to be determined; the second is for determination of one or two compounds, especially for compounds that are thermally labile and decompose under the conditions of long residence time in a GC column.

In both modes, a liquid or solid sample is placed in a small disposable glass microvial (Figure 2.25). The analysis begins with vaporization of the solvent (for liquid samples) at a relatively low injector temperature followed by rapid heating of the injector to the desired temperature required for sample compound vaporization. The nonvolatile matrix residues are retained, whereas the analytes are vaporized. With conventional GC separation, semivolatile compounds in the sample are focused on the early portion of the GC column and are analyzed by mass spectrometry or with GC detectors.

For direct sampling into a mass spectrometer, a short column is recommended. The column should be approximately 2 m in length and 0.1 mm i.d. with either no coating or a thin (0.1 μm) film thickness. The column in this case serves as a fast transfer line; therefore, the column temperature should be high enough so that the analytes are not retained on the column.

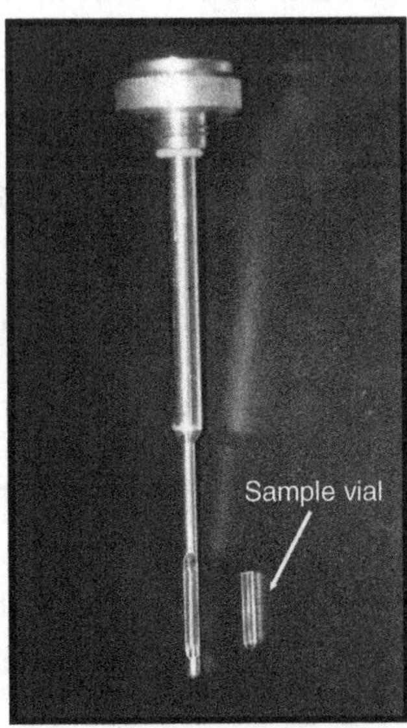

Figure 2.25 Photograph of a direct sample interface with a glass microvial that is used to contain the sample.

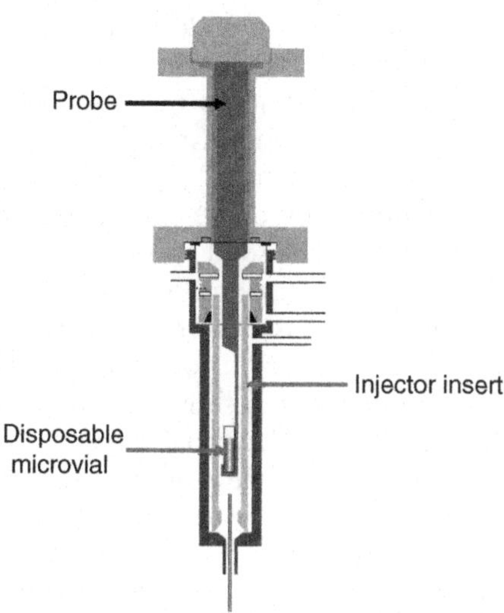

Figure 2.26 Cross section of the direct sample interface installed in a GC injector.

The DSI has a significant advantage over the direct insertion probes that are sometimes used with mass spectrometers to introduce solids and thermally labile samples into the ion source, in that any sputter of nonvolatile materials is retained in the transfer line and does not foul the ion source. The DSI is much cleaner with respect to ion source contamination than the direct insertion probe.

The DSI device is commercially available in several different configurations. It is available as the "DMI" (difficult matrix introduction) from ATAS GL International. The ATAS GL version can be installed in the glass liner of their Optic PTV injector. This injector can be installed on any gas chromatograph. The device is also available under the name "ChromatoProbe" for Varian and Agilent PTV injectors. The ChromatoProbe is available from Varian and Aviv Analytical. Figure 2.26 shows the ChromatoProbe installed in a GC injector.

The DSI has been used for the analysis of pesticides in food [45–47], biological samples [48], and aerosols [49]. A particularly interesting application was the determination of cocaine in a single hair [50].

Pyrolizers

In pyrolysis GC, nonvolatile solid and liquid samples are heated to temperatures that are generally higher than the temperatures in GC injectors. Under these conditions, specific chemical bonds are broken; and volatile decomposition products are released and transferred to a gas chromatograph or GC-MS for examination and possible identification.

In some applications, flame ionization detection is used; and the number of peaks, relative area counts, and retention times of the peaks is compared to chromatograms obtained with known materials. Comparison of chromatograms resulting from decomposition without identifying the product compounds is known as “fingerprinting.” Pyrolysis may also be used with mass spectrometers for identification of the chromatographic peaks resulting from the thermal breakdown of the solid sample.

In order to use the pyrolysis technique, a separate pyrolysis system must be purchased where the sample is decomposed and the decomposition products are transferred to the gas chromatograph. Figure 2.27 is a schematic of such a system. In the configuration shown in the figure, the trap is not used but is kept clean with a secondary flow of gas.

In another configuration, the sample is pyrolyzed and collected onto the trap; then desorbed from the trap to the gas chromatograph. Advantages are that the sample may be heated slowly or for long times without degrading the chromatography or pyrolyzed in air or oxygen to perform oxidation studies without introducing air to the GC-MS.

The pyrolysis temperature must be optimized for each sample type; the temperature should be high enough to produce a wide variety of products. At relatively low temperatures, the degradation may be too slow to be

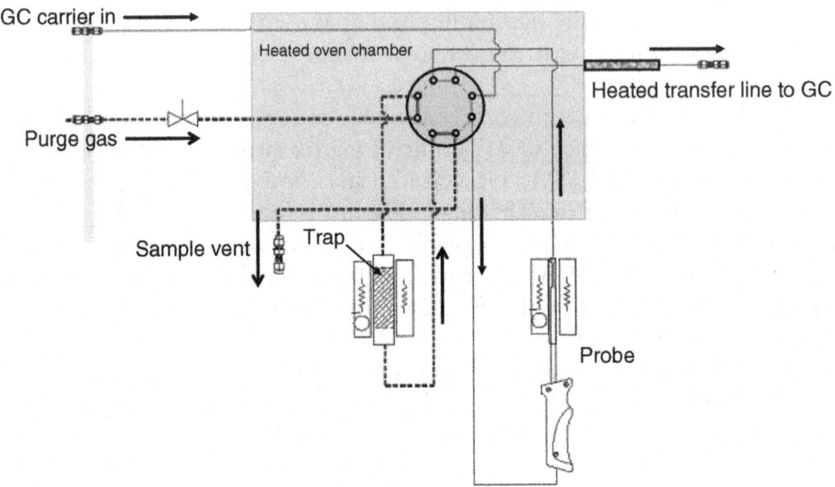

Figure 2.27 This is a schematic of the pneumatics for the CDS Pyroprobe 5200, which connects to the gas chromatograph via a heated transfer line. In this configuration, the Pyroprobe is directly online with the gas chromatograph for traditional pyrolysis GC/MS. The arrows show the flow of carrier gas through an eight-port valve in a heated chamber, then through the Pyroprobe and to the gas chromatograph. *Courtesy of Thomas Wampler, CDS Analytical, Inc., Oxford, PA.*

useful; very high temperatures may result in excessive degradation to only a very few small nonspecific products. Some pyrolizers can achieve temperatures up to 1400 °C, but most analytical work is conducted between 500 °C and 800 °C [51].

To achieve meaningful results when comparing multiple samples, the pyrolysis temperature must be tightly controlled and reproducible. This is accomplished among various commercial pyrolizers by various means:

- "Curie point pyrolizers" [52] where the sample is applied to a strip of a ferromagnetic alloy and is heated to its Curie point, the temperature where it loses its magnetism and does not absorb additional energy. At this point, the temperature ceases to increase. The specific composition of the alloy is chosen to attain a specific temperature.
- Microfurnaces [53] with a constantly heated isothermal zone into which samples are introduced into a little cup.
- Filament (usually platinum) pyrolyzers [54] can attain a wide temperature range and can be heated rapidly or at a controlled temperature rate to provide additional analytical information.

With commercial pyrolyzers, analyses can be automated. The analysis can be conducted utilizing different modes:

- Rapid heating to a single temperature, followed by the GC analysis.
- Programmed pyrolysis during the GC analysis.
- Multistep pyrolysis with a GC analysis after each step.
- "Double-shot mode" (Shimadzu) in which volatiles are released from the sample at 100–300 °C, transferred to the GC-MS, and then the sample is pyrolyzed at a higher temperature.

The major application of pyrolysis GC or GC/MS is the characterization of polymers [55,56]; another important application is the identification of the model of an automobile from a few milligrams of paint left at the scene of an accident [57]. Other applications include examination of art materials and paper [58], biological samples including microorganisms [59], and environmental samples [60].

2.2.5. Summary of GC Sample Introduction Methods

This section matches specific samples to the sample introduction methods already presented. For some sample types, the GC-MS or GC analytical procedure is determined by regulatory requirements and/or by many years of experience. For other sample types, several different sample introduction methods would give successful results; and the choice of technique depends on the experience of those performing the analysis, the number of samples of a given type to be analyzed, and the budget of the laboratory. For example, it is not worth investing in a static headspace or a purge-and-trap system or dedicating a GC-MS to analyses of volatiles if the laboratory has a large volume of samples for determination of semivolatiles and only an occasional need to identify volatiles. In this case, SPME might be the technique of choice for the volatiles, as the outlay for capital equipment would be lower. Table 2.2 lists sample types and suggests sample introduction techniques for these types.

Table 2.2 Recommended sample introduction methods for various applications

Application	Sample introduction technique
Volatile organic compounds in drinking water	Purge and trap in the United States; generally static headspace in Europe
Volatiles in blood and urine	Static headspace
Residual organic volatiles in pharmaceutical compounds	Static headspace
Flavors and compounds giving “off-flavors” in food and beverage samples; volatiles in aqueous matrices	Static headspace, purge and trap, and SPME The latter is especially useful if compounds of lower volatility are of interest
Neat samples such as gasoline and essential oils	Hot split injection
Organophophorus and chlorinated pesticides extracted from water	Hot splitless injection
Pesticides and drugs extracted from dirty matrices	Hot split or splitless injection depending on the concentration
Polyaromatic hydrocarbons (PAHs)	Temperature-programmed on-column or direct injection will give the best recovery of the later-eluting compounds
High-boiling hydrocarbon mixtures such as waxes	Temperature-programmed on-column or direct injection
Thermally labile compounds	Temperature-programmed on-column or direct injection Hot split injection with a short column may also give acceptable results because the time in the GC system will be minimized Direct sample interface (DSI) probe with a short column might be successful for the analysis if there are only one or two compounds to be determined
Identification of automobiles by paint-chip analysis	Pyrolysis
Analysis of polymers	Pyrolysis
Residual solvents in polymers	Static headspace, purge and trap

2.3. Separation of Components in the GC System

2.3.1. Columns for GC/MS

In a gas chromatograph, separation occurs within a heated hollow tube (the column) that contains a stationary phase. The components of the injected sample are carried onto the column by the carrier gas (the mobile phase) and selectively retarded by the stationary phase. The temperature of the oven in which the GC column resides is usually increased at 4–20 °C min^{-1} so that higher boiling and more strongly retained components are successively released from the stationary phase.

In the early decades of GC, columns were made of copper, aluminum, stainless steel, or glass, and were packed with small particles of an inert solid (support) that were coated with a thin layer of a stationary phase. The support consisted of cleaned diatomaceous earth particles that were of a uniform size,* and the stationary phase was a high-boiling viscous liquid polymer. These columns were used for separation of compounds that were liquids or solids at room temperature. Packed columns were also filled with small particles of a solid polymer or a molecular sieve** and were used for separation of permanent gases and very volatile liquids.

Golay [61] introduced wall-coated open-tubular (WCOT) columns in 1958; the early open-tubular columns were glass capillaries, and the fragility of these columns prevented them from being widely accepted. The problem of fragile glass was overcome with the development of the fused silica columns, invented by Dandeneau [62] in 1979. Fused silica columns consist of a highly purified silicon dioxide that is coated on the outer wall with a polyimide material to eliminate the brittleness that was originally a problem with these columns. The inner surfaces of WCOT columns are coated with a uniform thin layer of polymer. Several different polymers are available; these polymers are similar to the polymers that served as the stationary phase in the packed columns.

Another type of open-tubular column not normally used with a GC-MS is the porous-layer open-tubular (PLOT) column. PLOT columns have a layer of a solid polymer around the inner wall of the column and are used for separating permanent gases. A variation of the PLOT column is the support-coated open-tubular (SCOT) column in which the porous layer

---

*The diatomaceous particles were sized by passing through sieves.

**A molecular sieve is a material (often an alumino silicate) containing tiny pores of a precise and uniform size that is used as an adsorbent for gases and liquids. Molecules small enough to pass through the pores are adsorbed, whereas larger molecules are not.

consists of support particles coated with a liquid phase that was deposited from a suspension [63].

Since the early 1980s, the use of packed columns has dramatically declined to only a few specialized applications, and virtually all GC columns are fused silica capillary columns.

Grob reviews the theory of GC columns in detail [64].

The stationary phase

Selectivity and retention indices

Dimethylsiloxanes and 5% phenyl/95% dimethylsiloxane are good general purpose phases for many applications. These relatively nonpolar phases tend to separate compounds by boiling point differences. However, to separate compounds that are similar in boiling point but different in polarity, more selective phases are often required. In order to select the proper phase for a specific application, the user may consult the literature or use the recommendations of column manufacturers for specific classes of compounds. There is also software available to help the user choose the appropriate column and conditions for an analysis (see Appendix K).

However, the reader should also be aware of the extensive body of data that exists comparing the retention of numerous compounds on stationary phases of different polarities.

Kováts [65] devised a widely used system of retention indices that is useful for reporting the retention of a compound on a specific stationary phase under a wide range of chromatographic conditions. The actual retention time of a compound on a specific phase varies with the length of the column, thickness of the stationary phase, temperature conditions, and carrier gas flow. In the Kováts system, a series of *n*-alkanes is injected and the adjusted retention time* of each compound is noted. The Kováts retention index of each of the *n*-alkanes is, by definition, the number of carbons in the alkane multiplied by 100. Therefore, the retention index of *n*-octane would be 800; and that of *n*-decane would be 1000. To determine the Kováts retention index of a given compound on a given phase, the compound is injected onto the column under the same chromatographic conditions as the mix of *n*-alkanes. Next, a plot is made where the *x* axis is the retention index for the alkanes and the *y* axis is the log of the adjusted retention times of the *n*-alkanes. To determine the retention index for the compound of interest, a line is drawn from the *y* axis at the point representing the log of the adjusted retention time for that compound to the plot of

*The adjusted retention time t'_R for a compound is the actual retention time minus the retention time of an unretained compound such as methane.

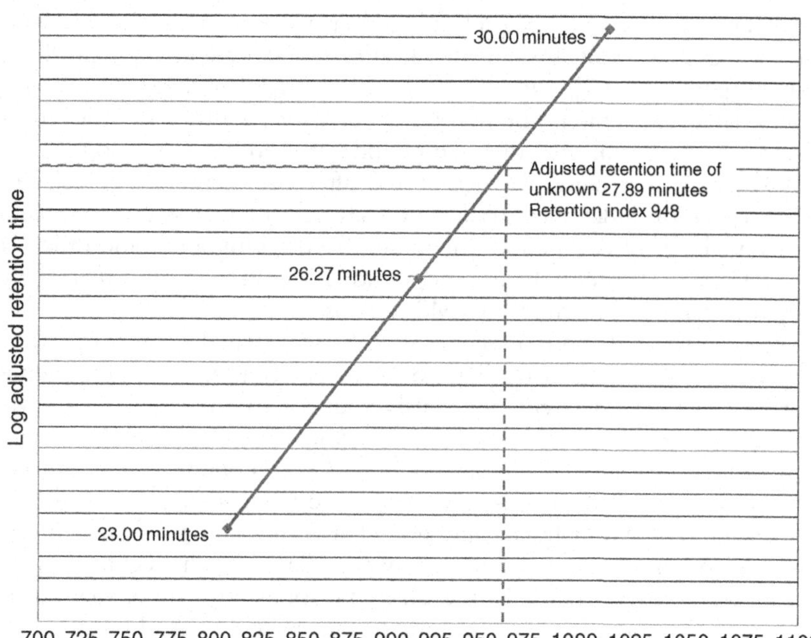

Figure 2.28 Determination of the Kováts retention index for an unknown compound by a graphical plot of the log of the retention time versus the retention index of normal alkanes (by definition, 100 times the carbon number). Three *n*-alkanes were injected—octane, nonane, and decane. The unknown eluted at an adjusted retention time of 27.89 minutes. By interpolating between 900 and 1000, the retention index of the unknown was determined to be 948. Isothermal column temperatures are used in this method of determining retention index.

the data for the alkanes. Next, a vertical line is drawn from the point where the horizontal line intersects the plot to the x axis where the line will intersect at the value for the retention index (Figure 2.28).

The Kováts retention index may also be calculated according to the equation:

$$I_x = 100\,Z + 100\left(\frac{\log t'_{R,x} - \log t'_{R,z}}{\log t'_{R,z+1} - \log t'_{R,x}}\right)$$

where I_x is the retention index of a given compound X; Z is the carbon number of the n-alkane that elutes before X; $t'_{R,x}$ is the adjusted retention time of the compound X; $t'_{R,z}$ is the adjusted retention time of the n-alkane that elutes before X; and $t'_{R,z+1}$ is the adjusted retention time of the n-alkane that elutes after X.

Kováts retention indices apply to isothermal conditions; for temperature-programmed runs, several modifications of the above equation have been proposed [66,67]; the values may be somewhat different than for indices determined for isothermal runs, but they are also based on interpolating the retention time of a compound between the retention times of two *n*-alkanes that elute just before and after the compound. Tables of Kováts retention indices are available on several different stationary phases for many classes of compounds, including flavor compounds and drugs of abuse. Their use in identifying compounds is diminished now that many laboratories utilize benchtop GC/MS systems.

McReynolds [68] and Rohrschneider [69] each devised a system to classify liquid stationary phases according to their interactions with compounds (probes) containing different functional groups. Each of these probes was assigned a symbol. The probes used by McReynolds are listed with the symbol in parenthesis: benzene (X′), *n*-butanol (Y′), 2-pentanone (Z′), nitropropane (U′), pyridine (S′), 2-methyl-2-pentanol (H′), iodobutane (J′), 2-octyne (K′), 1,4-dioxane (L′), and *cis*-hydrindane (M′). For each liquid phase studied, the Kováts retention index of a given probe was compared to the retention index of that probe on squalene under the same chromatographic conditions. Squalene was used as the basis of comparison to other liquid phases because it is a cycloparaffin and nonpolar in its interaction with other compounds. To use the McReynolds constants to select a stationary phase for separating a specific class of compounds, you would choose a phase with a high McReynolds constant for that type of compound. For example, to separate alcohols with similar boiling points, you would examine the Y′ value (corresponding to *n*-butanol). For dimethlysiloxane, the Y′ value is 53; for polyethylene glycol, the Y′ value is 536; therefore, the latter would be a more suitable choice. The structures of some common GC phases of varying polarities are shown in Figure 2.29.

Maintenance of the column phase

GC column phases have a recommended range of operating temperatures. Below the minimum temperature, the viscosity of the phase is high; and diffusion of analytes between the stationary and mobile phases will be impeded. As columns are heated, they begin to bleed; that is, some phase is lost from the column.* Above the maximum temperature, the bleed

*This can be loss of the intact phase or decomposition of the polymer chain that makes up the stationary phase. This results in highly volatile compounds composed of three, four, five, or more monomer units. These compounds are ionized in the mass spectrometer and result in spurious peaks in the mass spectrum. Sometimes, the deleterious effect of these peaks can be minimized with background subtraction, but they still pose a potential problem in library searches and interpretations of spectra. The presence of oxygen and water in the column (due to air leaks) can aggravate this decomposition; therefore, it is very important to be ever mindful of the possibility of air leaks.

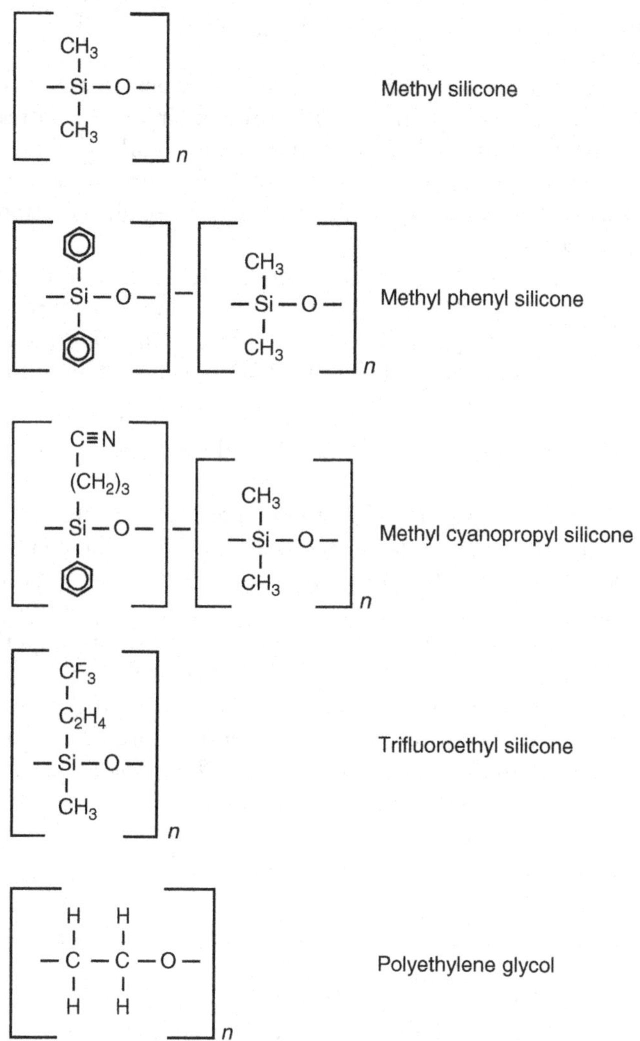

Figure 2.29 Structure of several common GC phases. The phases shown are of increasing polarity with the least polar at the top of the figure and the most polar at the bottom. *Printed with permission of Restek Corporation.*

becomes excessive, shortening the column life and contributing extraneous ions to the mass spectrum. Columns with polar stationary phases tend to bleed more than columns with nonpolar stationary phases. Column manufacturers have spent a great deal of effort in developing proprietary procedures for stabilizing phases to reduce bleed and extend column life, but some bleed is still seen at temperatures approaching 350 °C for nonpolar phases and lower for polar phases.

The thickness of the stationary phase is an important variable to consider. In general, a thin stationary phase (~0.2 μm) is desirable for working with mass spectrometry because there is less bleed. This works well when high-boiling compounds are to be separated; but for low boilers, a thicker stationary phase (1.0 μm) is necessary to provide enough retention for the compounds to separate.

New columns should be conditioned before use to remove excess volatiles from the manufacturing process and achieve a stable baseline. The following is a good general procedure for conditioning columns, but the specific recommendations of the manufacturer should be followed:

1. The column should be installed in the *cooled* injector but *not* in a mass spectrometer or other GC detector. The column temperature should be set at 40 °C.
2. Column phases, especially the more polar ones, are sensitive to oxygen. Therefore, after installation of a new column, the column should be flushed for approximately 30 minutes with a highly purified carrier gas.
3. At this point, the injector oven should be heated to the planned operating temperature or maximum planned operating temperature for a PTV, and the column oven should be heated at 5–10 °C min^{-1}.
4. The final temperature in the column temperature program should be 20 °C above the final planned operating temperature but not above the maximum isothermal temperature given by the manufacturer.
5. Maintain the final temperature for 30 minutes.

Physical dimensions and performance of the column

Earlier in Section 2.3.1, it was stated that capillary columns have supplanted packed columns. Table 2.3 compares the dimensions* of the two types of columns.

For a column with a given stationary phase, changing the physical column parameters can vastly improve GC separations. An important measure of the performance of a column is the number of theoretical plates N. This is a term borrowed from distillation theory and is expressed by the equation:

$$N = 5.55\left(\frac{t_R}{W_{1/2}}\right)$$

* Typical dimensions for analytical columns; occasionally, columns are used that deviate from these values.

where t_R is the retention time of the compound in the column (isothermal column temperature) and $W_{½}$ is the width of the peak at half height (Gaussian peak).

Table 2.3 Physical characteristics of capillary versus packed columns

	i.d. (mm)	Length (m)	Comments
Capillary columns	0.10–0.53	10–60	Open tube allows less resistance to flow through the column so longer lengths are possible without a high carrier gas pressure at the head of the column With mass spectrometry, the use of 0.53 mm or very short 0.25–0.32 mm columns is impractical because the vacuum makes it impossible to maintain the necessary flow rate of helium through the column Open-tubular structure allows easier transfer of the analytes between the stationary phase and the carrier gas
Packed columns	2–4	1–3	Requires a high back-pressure to force the carrier gas through the column because of the packing

Figure 2.30 compares the chromatograms resulting from injecting a given compound on two columns with the same stationary phase but with different plate numbers. Column efficiency depends on many factors including the temperature of the column, the type of carrier gas, the velocity of the carrier gas, the ease with which the analyte molecules diffuse in and out of the stationary phase, and the physical parameters of the column.

Table 2.4 lists some advantages and disadvantages of varying the physical parameters of the column.

In Table 2.4, it was pointed out that reducing the i.d. and the stationary-phase thickness reduces the capacity of the column, and reducing the column-phase thickness may also expose active sites. Frequently, column problems are manifested as frontal and rear tailing peaks (Figure 2.31). The peak on the left has a frontal tail due to column overload. The column

Table 2.4 Effects of changing parameters on column performance

Column parameter	Advantages	Disadvantages
>Length	Increases efficiency (first-order effect)*	Necessitates a higher back-pressure of carrier gas Lengthens run time
<i.d.	Efficiency is increased as the column i.d. is decreased (second-order effect) Lower flow of carrier gas is more suitable for operation of a mass spectrometer Less column phase in a smaller i.d. column reduces column bleed	Necessitates a higher back-pressure of carrier gas Reduces the amount of stationary phase per unit of column length and lowers the capacity of the column, thereby increasing the likelihood of overloading the column
<Thickness of stationary phase (d_F) (Typical phase thicknesses for capillary columns are 0.1–3.0 μm)	Efficiency is increased as d_F is decreased (second-order effect) Decreases column bleed Shortens run time A thin phase is necessary to elute very high boiling analytes such as polynuclear aromatic compounds	Reduces the ability of the column to retain highly volatile analytes, possibly to the extent that the column is useless for low boilers. For determining very low boilers with GC/MS, the column thickness should not exceed 1.0 μm to prevent excessive bleed. Reduces the amount of stationary phase per unit of column length and lowers the capacity of the column, thereby increasing the likelihood of overloading the column Active sites on the column may be exposed if the stationary phase is very thin, causing absorption of polar compounds

*The effect of column length, column i.d., and thickness of stationary phase on column efficiency will be described later in the discussion of the Van Deemter equation. As mentioned in Appendix M, the more efficient the column, the narrower the peak width at a given retention time.

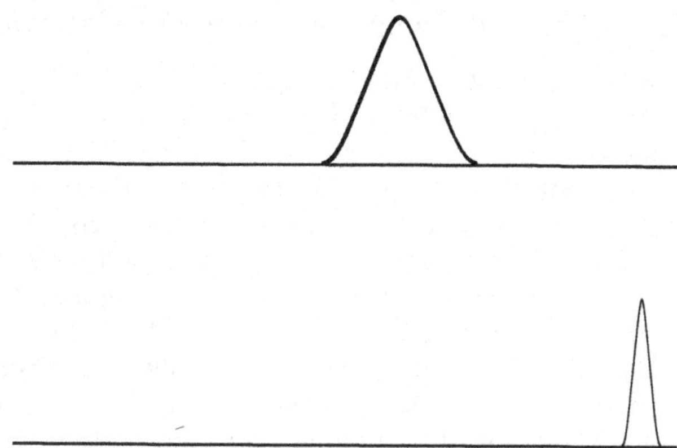

Figure 2.30 Two chromatograms resulting from injecting a compound on columns with different plate numbers. The top column has 152 plates and the bottom has 6,167 plates, allowing the compound to be retained longer and with less broadening.

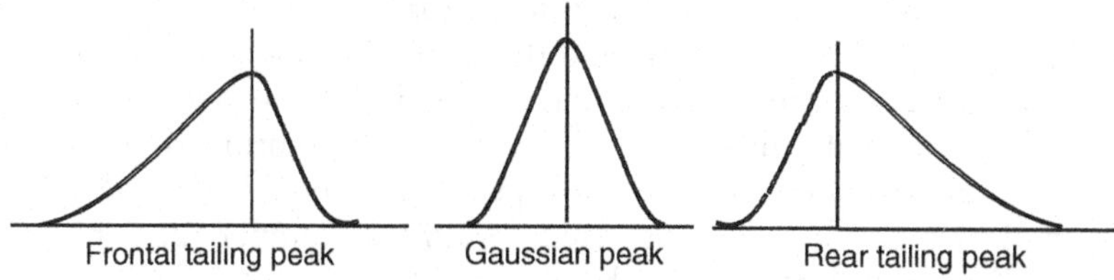

Figure 2.31 Tailing peaks in a chromatogram as compared to the ideal Gaussian peak. See the text for a discussion of the causes. *Printed with permission of Restek Corporation.*

cannot retain all of the analyte, and some of the excess elutes too early causing a peak that is skewed on the left side. The peak on the right has a rear tail caused by some of the analyte being retained too long when there are active sites in the column. This can be due to exposed fused silica in a column with a thin stationary phase.

As stated in the section on The stationary phase under Section 2.3, a column may be highly efficient; but in order to separate compounds with similar boiling points, but different polarities, it is necessary to consider the selectivity of the column. It has been said that capillary columns have such high efficiency that selectivity is not as important as it was when packed columns dominated; nevertheless, in order to separate similar compounds, a selective column may be desirable. With GC/MS, separation may not be important if the coeluting compounds have different mass spectra.

In the chapters on the analysis of different types of compounds comprising Section II, specific column dimensions and stationary phases are recommended.

2.3.2. Selecting Column Temperatures in a GC Separation

After column selection, the most critical step in developing a GC separation is selecting the column oven temperatures for the analysis. Possibilities include an isothermal run where the temperature remains constant, a temperature-programmed run where the temperature is increased at a constant rate, and a multilevel run where the temperature rate is increased at different rates at different times during the GC run. Finally, the temperature may be held constant for various periods of time—at the beginning, in the middle of a temperature program, or at the end of the analysis. Generally, isothermal conditions are used when only a few compounds are in the sample; and these compounds are similar in boiling point and polarity. When the sample contains many compounds of a wide boiling-point range, temperature programming is necessary to separate all of the components of the mixture.

In Figure 2.32, the top chromatogram represents an isothermal analysis at 100 °C. The early peaks are well separated, but C_{13} does not elute until 23.44 minutes into the run and it is broad. Under these conditions, the run time would be very long; and the final peak would probably be so wide (low signal to noise) that it would not be detected by the data system. With an isothermal analysis at 150 °C (middle chromatogram), the final peaks are separated; but the early peaks are merged into the solvent peak and would not be detected. With a temperature program (bottom chromatogram), all of the peaks are separated in a reasonable time; and they all have an excellent signal-to-noise ratio and would be easily measured by the data system. The object is to select column temperature parameters that will effect the best separation of the analytes in as short a time as possible while operating within the temperature constraints of the column phase.

The following are some general suggestions. In order to optimize the column temperatures, it is necessary to spend the time to determine the best conditions empirically. There are certain precautions to take when utilizing temperature programming:

- After the final run in a series, the column (and the injector if a PTV is being used) should be left at the final run temperature to minimize accumulation of impurities from the sample and the inlet system in the column and injector.
- The first run, after not using the system overnight, should always be a blank run where nothing is injected to verify that the system is clean and there are no ghost peaks.
- Temperature programming imposes a strain from repeated heating and cooling of the components of the injector and column oven, and leaks are more likely to develop than with isothermal analysis.

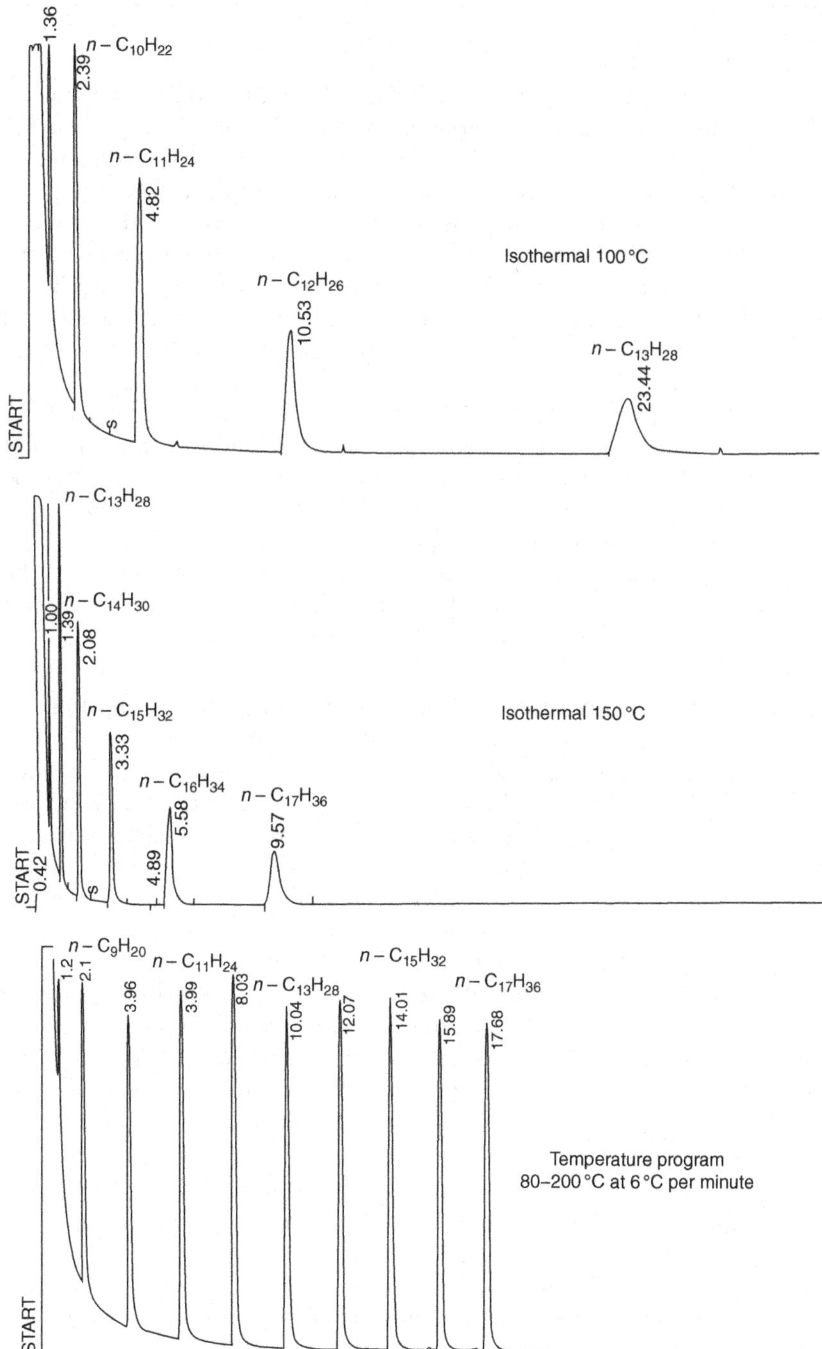

Figure 2.32 Separation of a mixture of *n*-alkanes from C_9 to C_{17} under three different temperature conditions: top chromatogram, isothermal analysis at 100 °C; middle chromatogram, isothermal analysis at 150 °C; bottom chromatogram, temperature-programmed separation.

Initial temperature

The initial temperature* of the analytical run depends on the temperature at which the first compound of interest elutes, the solvent used to dissolve the analytes, and the injection mode. In the section on *Splitless Mode* under Section 2.2, it was suggested that the solvent should have a boiling point about 20 °C below the boiling point of the first analyte; and the initial temperature should be about 10–15 °C below the boiling point of the solvent. It was explained that these conditions would be most favorable to affect solvent focusing, which would result in sharp initial peaks that are well separated from the solvent. These suggestions should also be followed for injection into a PTV.

In order to minimize the time of the analytical run, it is desirable to choose a solvent with a boiling point that is not too low compared to that of the first compound of interest; for example, rather than using *n*-hexane (BP = 69 °C), you might use *n*-nonane (BP = 151 °C) when analyzing compounds with boiling points greater than 170 °C.

For split injection where only a small quantity of solvent goes onto the column and there is no focusing, the initial column temperature can be just before the temperature at which the first compound elutes.

Temperature program rates and multilevel temperature programs

Temperature program rates are usually 4–20 °C min^{-1}**; the actual rate depends on the number of analytes and how closely they elute. In a multi-level temperature program, the temperature program should be relatively slow to separate compounds with similar boiling points that near one another; then the rate of programming should be increased when only a few compounds with widely different boiling points are expected to elute. With mass spectrometry, separation of all peaks is not critical if the compounds of interest have different quantitation ions.

When all of the compounds of interest have eluted, it is common practice to rapidly increase the temperature program rate to around 50 °C min^{-1} to just below the column limit. The purpose of this is to elute extraneous high-boiling compounds from the sample or from other sources such as the septum.

Final temperature

If all of the compounds elute at a temperature that is well below the column temperature limit and there are no high-boiling impurities in the sample or

---

* The optimum initial temperature may be 35 °C or lower; in that case, cryogenic cooling of the GC oven with liquid CO_2 or liquid nitrogen would be required. This option should be ordered when the GC is purchased.

** When the column oven is programmed at a rate greater than 20 °C min^{-1}, the retention time reproducibility may be compromised.

in the column, then the final temperature can be that at which the last analyte elutes; otherwise, the final temperature would be after the last extraneous compound elutes from the column (always keeping at or below the maximum column temperature).

Hold times

Frequently, temperature programming is halted for a certain time (hold time) during the analytical run. Two common reasons to halt the temperature program are as follows:

- To facilitate the separation of very closely eluting compounds
- To allow time at the end of a run for the elution of elute extraneous peaks when the maximum operating temperature of the column has been reached

2.3.3. Carrier Gas Considerations

The mobile phase or carrier gas is critical to GC/MS operation. The carrier gas must be inert and of a very high purity (99.999–99.9999%) in order to maximize the life of the column and assure a quiet baseline for most detectors. Helium or hydrogen is normally used in a capillary GC/MS system; in addition to these gases, nitrogen and argon are sometimes used as carrier gases in GC. Special heat-treated 1/8-inch copper tubing is used to connect the gases from their source* to the gas chromatograph.

The carrier gas should pass through a hydrocarbon trap, then a moisture trap, and finally an oxygen trap before it is connected to the gas chromatograph. There are many different types of traps recommended for GC and GC/MS use; some of these are multifunctional and can remove both moisture and oxygen. The user should be aware of and follow the manufacturer's recommendations for maintaining and changing the traps.

After the carrier gases and filters are connected to the gas chromatograph, it should be verified that the system is leak-free (see Appendix M).

The pneumatics must be capable of providing a stable and reproducible flow of carrier gas. Most modern gas chromatographs are equipped with electronic flow controllers. They are associated with software which requires that various parameters** be entered into the instrument's data

*The carrier gas source can be a gas cylinder located next to the instrument or from a manifold system installed in the building in which the laboratory is located. In the latter case, the user should be even more vigilant in monitoring cleanliness of the gas.

**Type of carrier gas, the inner diameter and length of the column, and whether or not the end of the column is under a vacuum (as is the case with a mass spectrometer).

system. The flow rate of the carrier gas and the split ratio for a split/splitless injector can then be set. In older GC systems, a constant carrier gas pressure was maintained at the head of the column; and in a typical temperature-programmed analysis, the flow rate through the column decreased with increasing column temperature. This was due to the increasing viscosity of the gas with rising temperature. In a system with electronic flow control, the pressure at the head of the column can be continually increased to maintain a constant flow with increasing temperature. Electronic flow control is also necessary for pressure-pulsed injection (discussed earlier).

An important parameter that is related to the carrier gas flow (f) is the linear velocity (u) of the carrier gas in $\mathrm{cm\,sec^{-1}}$. The linear velocity can be determined by injecting a compound such as methane or butane that is not retained by the column stationary phase and measuring the time from injection to detection. The linear velocity is the column length in centimeters divided by the retention time of the unretained gas in seconds. Linear velocity and carrier gas flow are related as follows:

$$f = 0.6\,u\pi r^2$$

where r is the radius of the column in millimeters.

The difference between flow and linear velocity is that column flow is dependent on the i.d. of the column; linear velocity is not. With an electronic flow-controlled system, it is not necessary to calculate linear velocity—the software calculates linear velocity and displays the value on the gas chromatograph's monitor or on the data system's display.

It was stated in the preceding section that N, the number of theoretical plates, is an important term in defining the efficiency of a column. A closely related parameter is the height equivalent to a theoretical plate (HETP). This is defined as the length of the column L divided by the number of plates, N:

$$\mathrm{HETP} = \frac{L}{N}$$

The Van Deemter [70] equation relates HETP to the linear velocity (u) of the carrier gas for packed columns:

$$\mathrm{HETP} = A + \frac{B}{u} + Cu$$

where A, B, and C are constants for a given column:

A refers to the size of particles in a packed column and the tortuosity of the path of the carrier gas through the particles and is not a factor in open-tubular columns.

B is a molecular diffusion term relating to the longitudinal diffusivity of the analytes in the stationary phase.

C is a resistance to a mass transfer term referring to the movement of the analytes in and out of the stationary phase. This term increases proportionally to the square of both the column radius and the film thickness of the stationary.

Figure 2.33 is a Van Deemter plot for hydrogen, helium, and nitrogen carrier gases. At the minimum point for each curve, the column efficiency is the greatest; this point corresponds to the optimum linear velocity. With nitrogen carrier gas, the curve reaches the lowest HETP value, but the optimum linear velocity is much lower than for helium and hydrogen. If the optimum carrier gas flow with nitrogen is increased, the column efficiency declines rapidly. This is not the case with hydrogen and helium where the curves are much flatter. Thus, with hydrogen and helium, a higher flow of carrier gas can be used, resulting in a reduction of retention time with better efficiency than with nitrogen. Hydrogen appears to be the best carrier gas, but it is not used as often as helium due to safety considerations and the fact that most mass spectrometers are designed to be used with helium. A typical helium flow rate for a 250-μm i.d. column is 1.0–1.2 cc min^{-1}; for a 320-μm i.d. column, the flow is 1.7–2.0 cc min^{-1}.

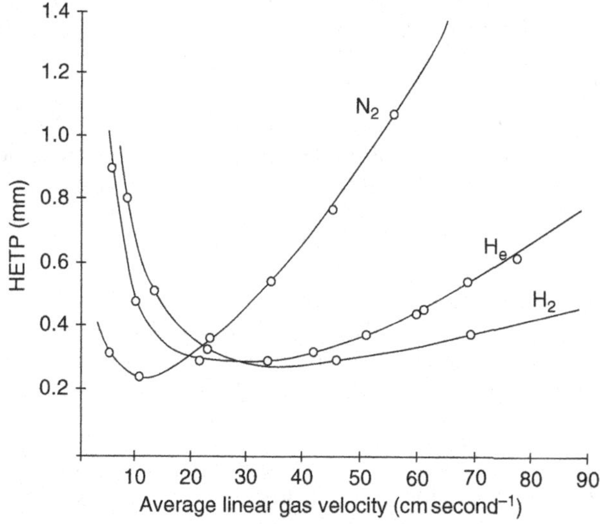

Figure 2.33 A Van Deemter plot (average linear gas velocity versus height equivalent to a theoretical plate) for three carrier gases. The minimum for each curve represents the carrier gas flow where the column efficiency is at a maximum.

2.3.4. Fast GC

With fast GC, the GC analysis time is substantially reduced from the typical analysis times of 10–60 minutes. In order for the reduction in time to provide useful results, the chromatographic peak width at half height must also be reduced. Consider the importance of extremely narrow peak widths in achieving baseline separation of several compounds that elute within a few seconds. Korytár et al. [71] defined three categories of fast GC according to analysis time and peak width at half height. Dagan and Amirav [72] defined a speed-enhancement factor using the same three categories. The speed-enhancement factor takes into consideration the increase in speed obtained by using a shorter column and a faster carrier linear velocity as compared to a conventional GC column with optimum carrier gas linear velocity. Table 2.5 summarizes the definitions of fast GC from the sources cited above.

Techniques of fast GC

In the above discussion of selecting column parameters and carrier gas flow to maximize GC efficiency, the fact that maximum efficiency is not always required to meet the practical aims of the analysis was not mentioned. It is not necessarily useful to separate compounds with baseline resolution* if their mass spectra are sufficiently different so that they can easily be resolved spectrally. With fast GC, the tendency is to sacrifice some efficiency to reduce the retention time of the compounds of interest while maintaining the requirements of the analysis—measuring the peaks of interest with sufficient accuracy. Some of the steps that are relatively simple to use [73] to achieve shorter retention times are listed below:

Shorter column length

As stated in Table 2.4, increasing column efficiency has only a first-order relationship with increasing column length; therefore, using a longer column increases the efficiency while increasing the retention time. The

Table 2.5 Definitions of fast GC

	Analysis time	Peak width at half height	Speed-enhancement factor
Fast GC	minutes	1–3 seconds	5–30
Very fast GC	seconds	30–200 ms	30–400
Ultrafast GC	$<$ seconds	5–30 ms	400–4000

*See Appendix A for the definition of resolution as it relates to gas chromatographic peaks.

column length can be reduced without a serious loss of efficiency at the same time the retention times are shortened.

Reducing column film thickness

It was also stated in Table 2.4 that column film thickness (d_F) has an inverse second-order relationship with column efficiency. Use of a thinner film thickness significantly shortens retention time without losing efficiency; the disadvantage is that the capacity of the column will be reduced. Therefore, a smaller sample size must be used and the limit of detection may be lowered.*

Increase the rate of temperature programming

Methods are often developed with temperature-programming rates that are far slower than required. With selective detectors or mass spectrometers, it is not always necessary to use a long temperature program to separate the compounds of interest. Additionally, prolonged periods of heating to clean out the column at the end of an analysis can be greatly reduced by using a backflush valve; this will be discussed in more detail later in this chapter.

Increase the carrier gas velocity while decreasing the column i.d.

In plots of HETP versus linear velocity where all parameters are identical except column i.d., it will be seen that efficiency falls off to a lesser degree as the linear velocity increases for smaller i.d. columns [74]. Note also from the Van Deemter plot (Figure 2.33) that the loss of efficiency is less with increasing linear velocity if hydrogen is the carrier gas. Therefore, with a narrow i.d. column and hydrogen carrier gas, a high linear velocity of carrier gas can be used without a severe loss of efficiency.

Practical considerations for using fast GC

Instrumental considerations

Matisová and Dömötörová [74] in their review article on fast GC include a very detailed discussion on instrumentation. They point out that modern conventional gas chromatographs can be used for conventional and fast GC and GC/MS, but dedicated gas chromatographs are often required for very fast and ultrafast GC. These instruments should be equipped with the ability to maintain high carrier gas pressures necessary for microcapillary columns, additional safety features to handle hydrogen carrier gas, injectors that can allow accurate control of very high split ratios, and ovens that can be accurately temperature programmed at faster rates than conventional gas chromatographs.

For the fastest separations, the spectral acquisition rates of quadrupole ion trap (QIT) and transmission quadrupole mass spectrometers are too low. Time-of-

* The peak width may decrease enough so that the increase in signal to noise will offset a loss in the detection limit.

flight (TOF) mass spectrometers [75] with the ability to provide 500 spectra per second are necessary. In addition, TOF mass spectrometers with narrow-bore columns offer the excellent detection limits [76] that are required for the very small samples that must be injected on columns suitable for ultrafast GC. At the present time, there is only one commercial provider of such a GC-MS—Leco Corporation (St Josephs, MI, USA; http://www.LECO.com).

The GC analysis time as part of the total analytic procedure

There are many factors affecting the speed of a GC analysis; among these are logging the sample into the laboratory, sample preparation time, the GC analysis time, cool-down time between analyses, analyzing the data, and preparing the analysis report. Sample preparation time is normally the limiting factor in determining throughput in the analytical laboratory; but in cases where the GC analysis time is significant in prolonging turnaround time, fast GC may be of use. Maštovská and Lehotay [73] in their review of fast GC/MS include a detailed discussion of how to determine if there is an economic benefit to modifying a GC/MS method so that GC analysis time is significantly reduced. For example, if the sample preparation of a batch of samples consumes an entire business day and the samples are put in an autosampler and analyzed during the night, it makes no difference to the productivity of the laboratory if the samples are analyzed in 1 or 16 hours.

2.4. Overview of GC Detectors

A great advantage of GC is the variety of detectors that are available. Several of the most common detectors are described below. These detectors are designed to work with a higher flow of gas than that delivered by a capillary column and usually require a makeup gas of helium or nitrogen.

2.4.1. Thermal Conductivity Detector

Unlike the other detectors to be discussed in this section, the TCD is truly a universal detector and can detect all compounds that can pass through a GC column. In the GC laboratory, the TCD is generally used for detecting permanent gases rather than organic compounds as the latter can be detected at lower levels with other GC detectors and mass spectrometers.

Figure 2.34 shows a typical TCD consisting of a cell with two channels for the flow of carrier gas. The eluate from the analytical column flows through one channel; the other channel serves as a reference and is connected to a source that provides pure carrier gas. Within the cell is a Wheatstone bridge circuit with four resistors or filaments. Gas from the column eluate passes over the two filaments on the sample side, and pure carrier gas passes over the

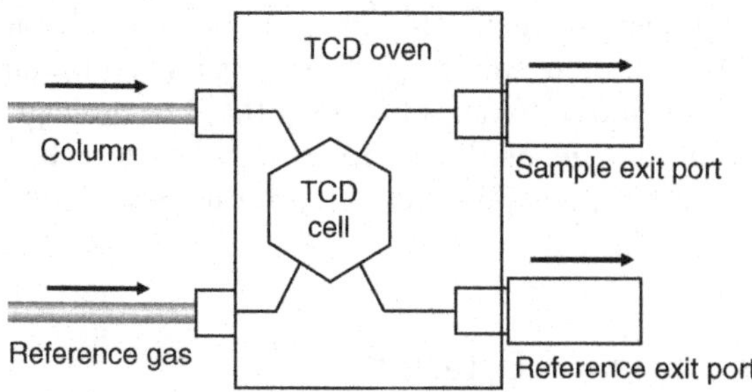

Figure 2.34 Cross section of a thermal conductivity detector (TCD).

filaments on the reference side. When pure carrier gas passes through both filament sections, the resistance of each section is the same and there is no signal. As compounds with different thermal conductivities elute and pass over filaments on the sample side, the different gas compositions cause heat to be conducted away from the filaments at different rates that, in turn, cause a change in the filament temperature and electrical resistance. Therefore, a signal is produced that is proportional to the difference in thermal conductivity between the analyte and the carrier gas and the quantity of the analyte.

The detection limit of analytes is related to the difference in thermal conductivity between the analytes and the carrier gas. Because helium has a higher thermal conductivity than most analytes (Table 2.6), helium is usually used as the carrier gas.

The conventional TCD has a large volume; and even with makeup gas, it may not be suitable for capillary columns. For this reason, a low-volume TCD is now available [77]. A micro-TCD is usually found on field-portable gas chromatographs as these detectors are very rugged and require no additional support gases.

The TCD is a nondestructive detector; therefore, the eluate can be connected with a transfer line to another detector for very low detection limits of various components in the sample, other than the permanent gases.

Table 2.6 Thermoconductivity ($\times 10^5$) at 0 °C of various compounds

Compound	cal cm^{-1} per second per °C
Hydrogen	41.6
Helium	34.8
Methane	7.2
Nitrogen	5.8
Pentane	3.1
Hexane	3.0

Source: Data from HM McNair and EJ Bonelli, Basic Gas Chromatography, 5th ed., Varian Instruments, 1969, p 93.

For example, hydrogen, oxygen, and nitrogen in the sample can be detected with the TCD; and then sulfur gases or organic compounds can pass through the TCD and into an FPD or FID. Flavor/fragrance analysts sometimes take advantage of the nondestructive feature to collect compounds or even have experts sniff them for critical components as they elute from the TCD exit port.

2.4.2. Flame Ionization Detector

The FID responds to compounds with a carbon–hydrogen bond. It is the most commonly used GC detector [78]. The detector requires hydrogen and air as support gases with a typical air:hydrogen ratio of 10:1. The gases are mixed and burn just above a flame jet (Figure 2.35).

Water is formed and it can be easily verified that the flame is lit by holding an object with a smooth surface over the detector and observing the water vapor condensing on the surface. A negative polarizing voltage is applied between the jet tip and a collector electrode. As analytes elute from the column, they pass through the flame and burn to produce ions. Electrons formed in the flame cause a current to flow in the gap between the jet tip and the electrode. By amplifying this current flow, a signal is produced. FIDs have a wide range of linearity ($\sim 10^7$).

2.4.3. Nitrogen–Phosphorus Detector

This detector (Figure 2.36) is also called the thermionic-specific detector (TSD). It responds to compounds in a sample that contain nitrogen and/or phosphorus. Nitrogen compounds generally give a good response if a C–N

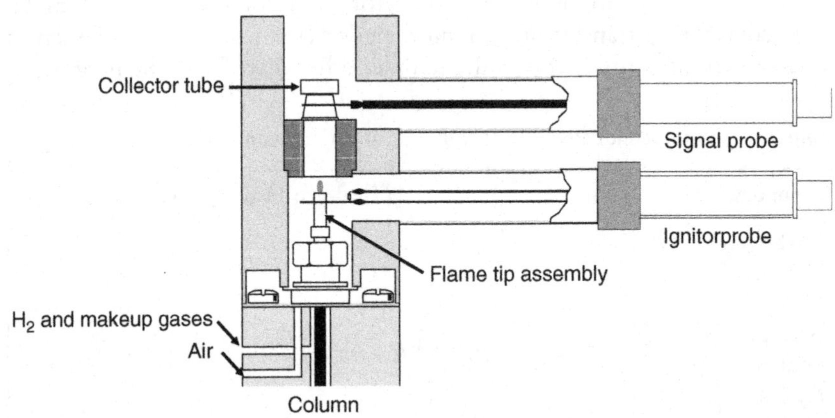

Figure 2.35 Cross section of a flame ionization detector (FID).

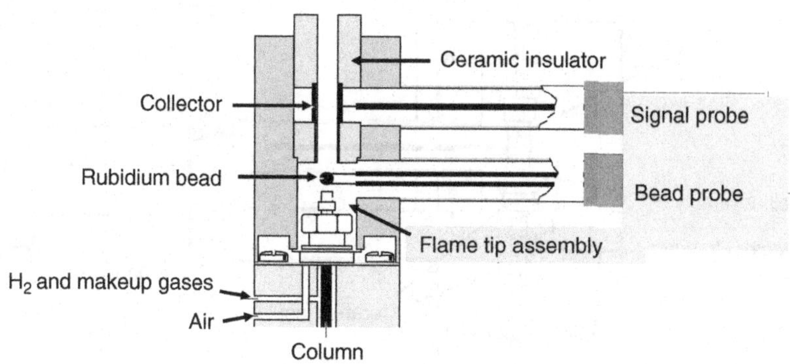

Figure 2.36 Cross section of a thermionic-specific detector (TSD or NPD). When the rubidium bead is heated and exposed to the reaction gases, compounds containing a carbon–nitrogen bond form a $CN^{-\bullet}$ ion which is detected. In compounds containing a phosphorus atom, HPO^- is formed and detected.

bond is present, and the detector is generally useful for all organic nitrogen compounds and for hydrogen cyanide (HCN). Most volatile organic and inorganic phosphorus compounds are responsive in this detector.

The NPD or TSD requires air and hydrogen as support gases; the air: hydrogen ratio is approximately 100:1, and there is no flame. The hydrogen flow is typically about 3–4 mL min^{-1}. The gases surround a heated rubidium sulfate bead. Nitrogen and phosphorus compounds increase the current in the plasma of vaporized rubidium ions. The reaction with phosphorus compounds is slower than with nitrogen compounds; this is manifested by some tailing with these compounds.

The detector was very popular for analyses of samples containing drugs and pesticides; but after the benchtop mass spectrometer became more affordable, mass spectrometry became more commonly used for these applications.

2.4.4. Electron Capture Detector

The ECD (Figure 2.37) has a very low detection limit for materials that readily capture electrons (electrophilic analytes) and gives a strong response to halogenated compounds as well as many other molecules such as N_2O, nitro-organics, diketones, and diacetal.

The ECD contains a radioactive foil, normally ^{63}Ni, which is a β-particle emitter and is used to ionize nitrogen or argon/methane carrier gas.*

* In a capillary system, the carrier gas is usually helium, and the detector makeup gas used with the ECD is nitrogen or argon/methane.

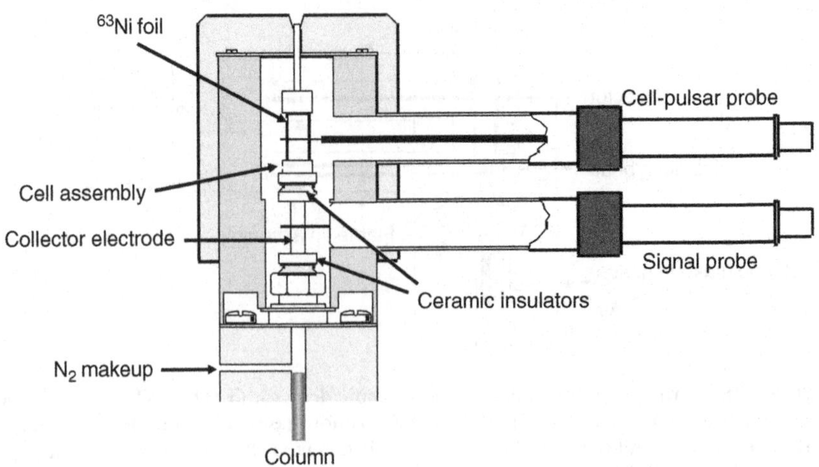

Figure 2.37 Cross section of an electron capture detector (ECD).

Electrons from the ionization migrate to the anode and produce a steady current. If the GC eluate contains a compound that can capture electrons, the current is reduced because the resulting negative ions move more slowly than electrons. The signal measured is the loss of electrical current. For many compounds such as multihalogenated organic molecules, the ECD has a lower limit of detection than any other GC detector, including the mass spectrometer.

Disadvantages of the ECD include a relatively short linear range (10^3–10^4) and a response that depends heavily on the number and type of electron-capturing functional groups in the molecule. This means that CCl_4 gives a much greater response than CH_3Cl. Among the halogens, the response is greater with increasing atomic weight; iodine produces the strongest response, and fluorine produces the least.

2.4.5. Flame Photometric Detector and Pulsed-Flame Photometric Detector

The FPD is used primarily to detect volatile organic and inorganic molecules containing sulfur and phosphorus. With the FPD, as with the FID, the GC eluate containing the analyte is burned in a hydrogen/air flame and chemiluminescent products are formed. By using optical filters to select wavelengths specific to the element of interest and a photomultiplier tube, compounds containing sulfur or phosphorus can be selectively detected. As opposed to the NPD, which responds to both nitrogen and phosphorus, and the ECD, which responds to many different groups, the FPD is highly selective and responds only to either sulfur or phosphorus, depending on which filter is installed.

When sulfur compounds burn in the flame, an excited disulfur molecule, S_2,* is formed; and the response is second order, rather than linear. Phosphorus gives a linear response. In the FPD, there are background products from the emission of hydrocarbons; and narrow bandpass filters must be used to eliminate these interfering emissions. These narrow bandpass filters adversely affect the detection limit.

In the early 1990s, Professor Aviv Amirav [79] invented the PFPD (Figure 2.38).

Rather than using a continuous flame as with the FPD, the flame in the PFPD is extinguished and reignited 3–4 times per second (Figure 2.39). This is accomplished by using a very low flow of hydrogen so that a continuous flame cannot be sustained. The combustion of interfering carbon and oxygen bonds produces emissions that last only about 1–3 milliseconds (ms), whereas the emissions of the various heteroatoms in the molecule are delayed (sulfur 6–26 ms and phosphorus 4–14 ms). The pulsed flame enables the emissions of the atoms of interest to be separated from the carbon interferences (Figure 2.40). The photomultiplier can be turned on and off (gated) to respond only to the emissions of the heteroatom of interest, and a wide bandpass filter is possible. Thus, the limits of detection are reduced.

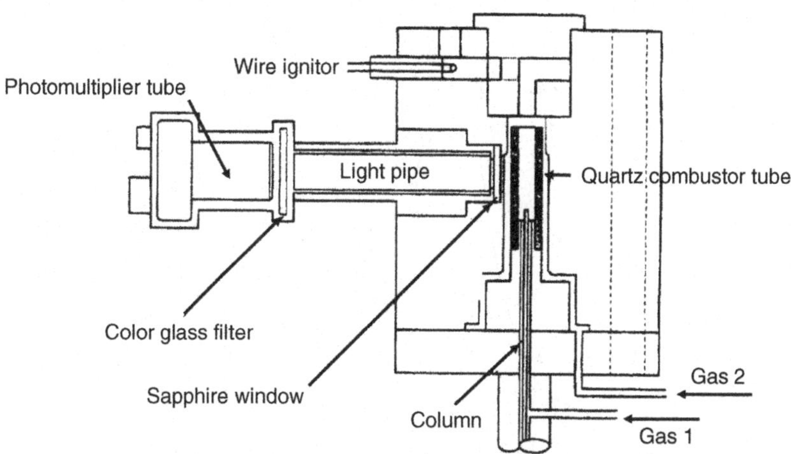

Figure 2.38 Cross section of a pulsed-flame photometric detector (PFPD). The PFPD body is mounted on the detector base of the gas chromatograph. A hydrogen-rich combustible gas mixture (gas 1) is fed through a gas tube around the GC column. A separately controlled second combustible gas mixture (gas 2) is fed around the combustor holder that supports the quartz combustor tube.

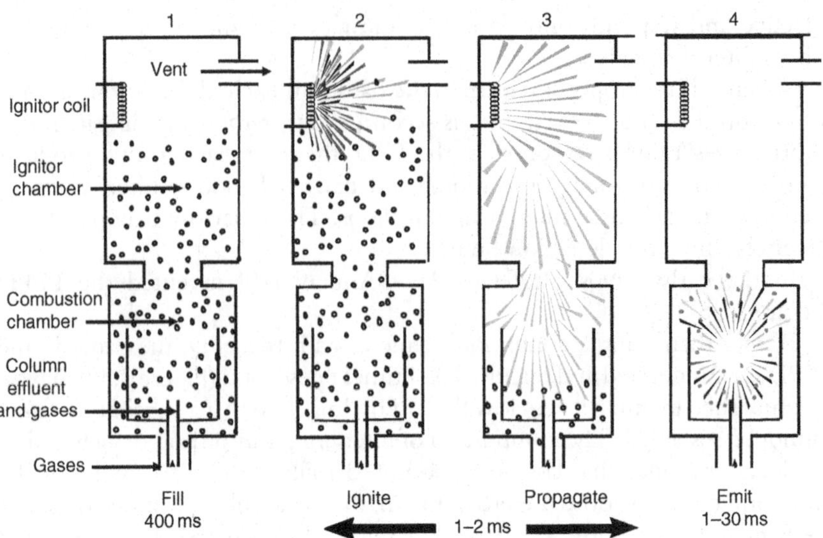

Figure 2.39 Steps in the operation of the PFPD: the carrier gas containing the analyte fills the chamber (1); the pulsed flame ignites on the heated wire igniter (2) and propagates through the igniter light-shield chamber to the combustor (3) and is self-terminated at the combustor holder; the pulsed flame emits light (4) which passes through a sapphire window, quartz light guide rod, and a color glass filter for detection by a photomultiplier.

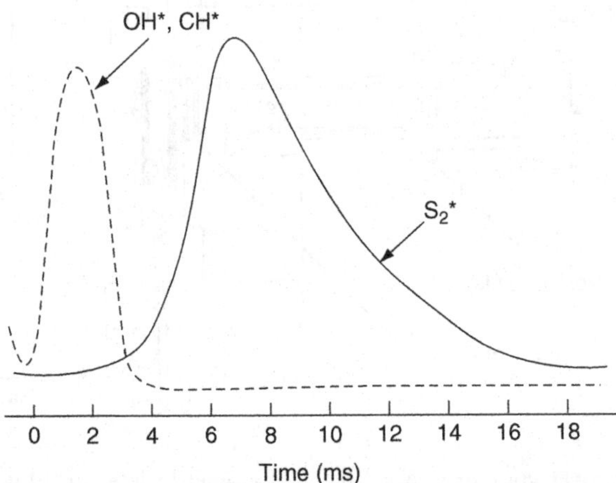

Figure 2.40 The hydrocarbon and sulfur emission profiles from the PFPD as a function of time in milliseconds. A gated amplifier is used to exclude the hydrocarbon signal and transmit a band of the sulfur signal from 5 ms to 10 ms.

Both the FPD and the PFPD are capable of responding to several additional elements besides sulfur and phosphorus if different filters are installed. Because the FPD is used primarily for phosphorus and sulfur, Jing and Amirav [80] conducted a systematic study of the PFPD for use as a heteroatom detector. Utilizing different filters and photomultiplier tubes, they were able to detect 28 elements, including carbon, nitrogen, sulfur, phosphorus, antimony, tin, germanium, lead, nickel, copper, chromium, manganese, and iron. The ranges of sensitivity and selectivity against carbon varied.

2.5. Adding Versatility to the GC/MS System with Valves, Splitters, and Thermal Modulators

In this section, the use of selective detectors with the mass spectrometer will be discussed as well as other techniques to increase the versatility of the GC/MS system.

The installation of a valve in the gas chromatograph can add to the functionality of a GC/MS system. Valves can be installed when a new instrument is ordered, or they can be added to an existing gas chromatograph. It is recommended that the above discussion on valves for GC be reviewed (Section 2.2.2). It has already been stated that a loop could be added to a valve for gas sampling; in addition, column switching and/or backflushing valves can provide the GC/MS user with the following advantages:

1. Supplemental information from the mass spectrometer with information from selective GC detectors.
2. Allow the use of completely different injection modes without changing columns.
3. Use columns of different polarity for different samples without shutting down the mass spectrometer.
4. Injectors can be serviced and columns can be changed without shutting down the mass spectrometer.
5. Improve separation by using two columns in tandem (GC^2).*
6. Shorten analysis time with backflushing and GC^2.

Splitters are also useful for obtaining additional information in GC/MS systems. The simplest splitters to use are glass "Y"-shaped connectors that are available from chromatography supply companies (Figure 2.41). These connectors can be used for dual-column or dual-detector analysis.

* Also referred to in the literature as two-dimensional GC.

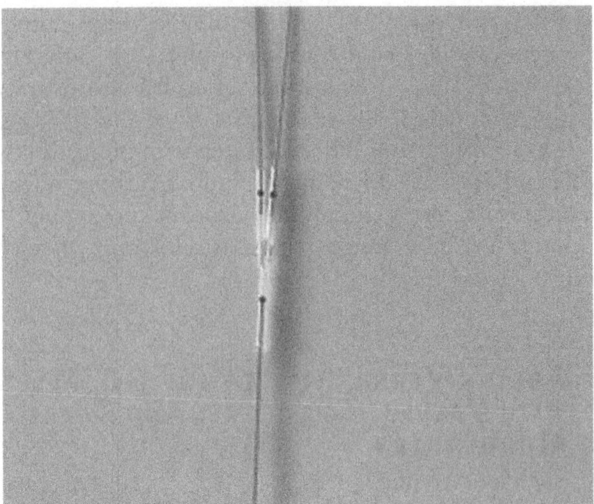

Figure 2.41 Photo of fused silica tubing installed in a "Y" connector.

2.5.1. Dual-Detector Analysis

The sample can be analyzed with two different GC detectors or with a GC detector and a mass spectrometer. Dual detection can be accomplished in two ways:

1. Connecting the end of the column to a splitter and directing a portion of the flow into the detector and the remaining portion into the mass spectrometer.
2. Using a valve to direct the flow from the column between a GC detector and a mass spectrometer as different compounds elute from the column.

When performing dual detection using a GC detector and a mass spectrometer with an eluate splitter, a potential problem exists due to the fact that the mass spectrometer operates under a vacuum and the GC detector operates at atmospheric pressure. The eluate will not split evenly between the GC detector and the mass spectrometer; most of the sample will go into the mass spectrometer unless conditions are adjusted to compensate for the pressure differences (Figure 2.42).

The fused silica tubing that is directed to the mass spectrometer side should have a smaller i.d. than the tubing that is directed to the GC detector (e.g., use 0.10-mm tubing on the mass spectrometer side and 0.25-mm tubing on the GC detector side). Further adjustments can be made by adjusting the relative lengths of the tubing. It is recommended that the fused silica tubing be coated with 0.1-mm methyl silicone to maximize inertness of the system. Amirav and Jing [81] used this technique for simultaneously detecting pesticides with a PFPD and a mass spectrometer.

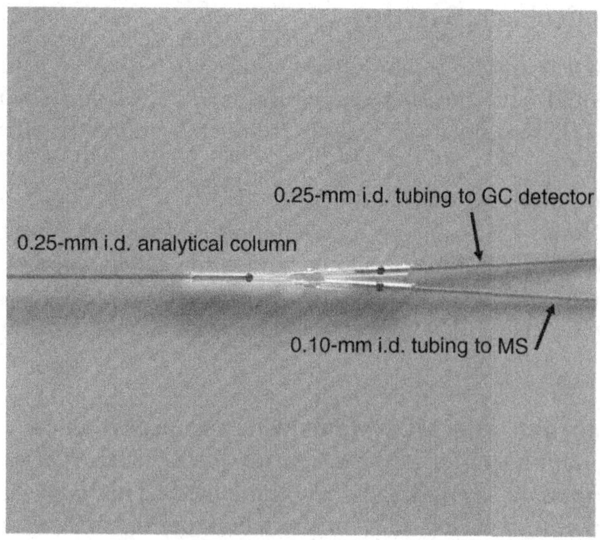

Figure 2.42 The end of an analytical column is installed in a "Y" connector for splitting the flow of effluent between a GC detector and a mass spectrometer. Note the connector tubing that connects the column to the mass spectrometer is of a narrower i.d. than the tubing connecting the column to the GC detector. This is to compensate for the vacuum in the mass spectrometer; if not for the difference in i.d. of the two tubing connectors, virtually all of the sample would flow to the mass spectrometer.

When a valve is used, *all* of the flow is directed into the detector or the mass spectrometer, and the detection method changes throughout the analysis as the valve is switched back and forth. Figure 2.43 is a schematic of the system. Note that an actuator should be used with the valve to assure automatic switching between the detector and the mass spectrometer at the correct time.

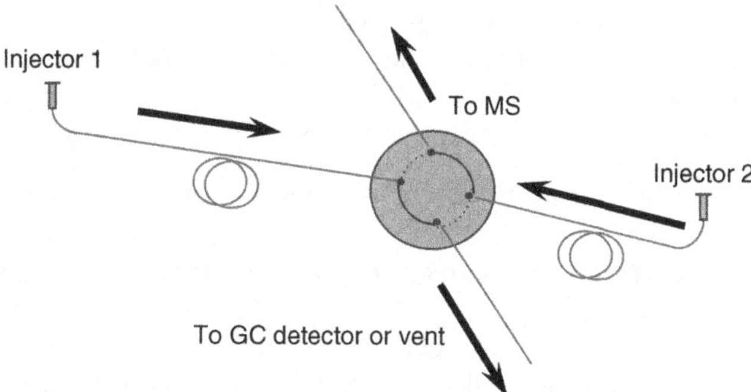

Figure 2.43 Schematic showing a gas chromatograph with two injectors and two columns connected to two ports of a four-port valve. The other two ports are connected to a mass spectrometer and either a GC detector or vent. This type of setup offers the user a great deal of versatility as described in the text.

2.5.2. Multicolumn Systems

In all cases where multiple columns are installed in a gas chromatograph, the columns should have similar temperature limits. Analyses of samples with columns of different polarities may be conducted in several different ways.

Dual-column analysis with a splitter

In this case, fused silica tubing is installed in an injector and is connected to a "Y" splitter. Two columns of different polarities are installed between the "Y" connector and two GC detectors. This application might be used, for example, with two FIDs where each column would separate different compounds in a sample.

Dual-column/dual-injector systems with a four-port valve

The configuration shown in the schematic for GC/MS systems with two injectors (Figure 2.41) is highly recommended. This type of system is extremely useful in saving time in the laboratory. It requires a four-port valve that can be operated manually or with an actuator. The main advantage of this configuration is that injector liners and columns can be changed without shutting down the mass spectrometer. When it is necessary to service an injector or change a column, the valve is rotated to isolate the injector and column from the mass spectrometer. The mass spectrometer vacuum is maintained while the carrier gas flow in the injector and/or column that is being serviced is directed to vent.* This system also allows a great deal of flexibility in analysis options:

1. Sequential analysis of a sample on two columns with different polarities with the same injector mode.
2. Sequential analysis of a sample on columns with the same polarity but completely different injector modes (e.g., split injection on one side and on-column injection on the other side).
3. Analysis of different samples with different columns and/or different injection modes (e.g., one side can have a conventional splitless injector and the other side can have sample introduction with a probe).

All of these options are immediately available by turning the valve.

Dual-column systems with valves for heart-cutting and backflushing

Two-dimensional GC systems can reduce analysis time and are often used in fast and ultrafast GC and GC/MS systems. Ong and Marriot [75] have reviewed the concepts and techniques of GC^2 and stated the role of valves in these systems.

---

* Varian, Inc., sells this configuration under the name "Quick-Switch Valve"; but the same configuration can be installed in GC/MS systems manufactured by other companies.

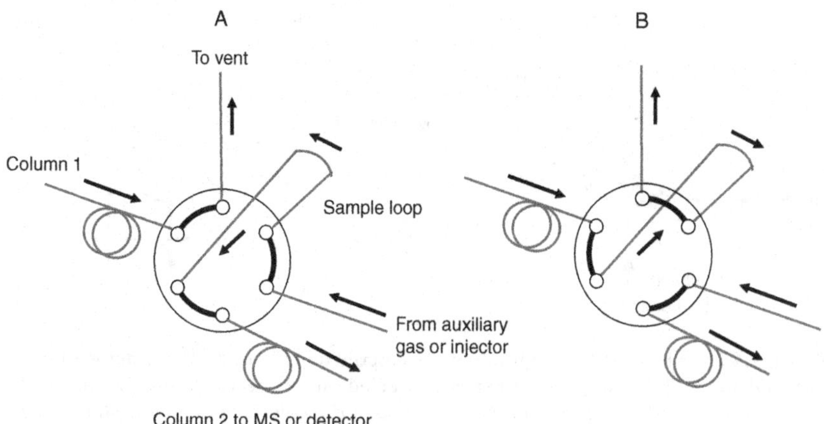

Figure 2.44 The configuration is set up for heart-cutting (isolating a small portion of the sample from a complex mixture). The various steps are described in the text.

In heart-cutting, a group of closely eluting compounds is transferred from one column to a second column that is capable of resolving them. Normally, the compounds of interest are part of a complex sample with many compounds eluting before and/or after them.

An example of an application using heart-cutting is shown in Figure 2.44. Two columns of different polarities are connected to a six-port valve that contains a trapping loop; the first column is nonpolar, and the second column was chosen for its ability to separate the compounds of interest. Three steps are involved as follows:

- Valve position A: The early eluting extraneous compounds rapidly reach the end of column one and are directed to a vent.
- Valve position B: The compounds of interest are then directed to the sample loop.
- Valve position A: The compounds of interest move from the sample loop to the second column where they are separated due to the greater selectivity of the second column. They then move into the mass spectrometer or a detector. Meanwhile, the later-eluting extraneous columns pass through column one and out of the vent.

A backflush valve (Figure 2.45) is used to direct high-boiling compounds that are not of analytical interest out of the system. After the more volatile compounds pass into the detector or mass spectrometer, the flow of the column is reversed so the compounds of low volatility are expelled from the front of the column. This eliminates the need to bake out the column and shortens run time.

Valves used for heart-cutting and backflushing need an actuator so the compounds in the sample can be directed to the proper columns at precise

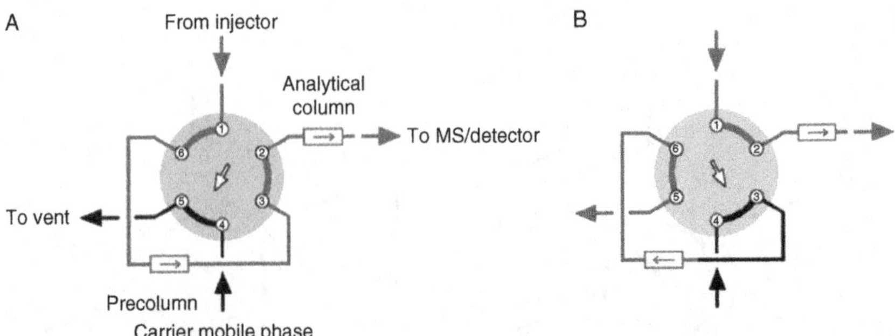

Figure 2.45 A backflush configuration. As described in the text, the carrier gas flow is reversed and high-boiling compounds are directed out of the vent. This eliminates the need to wait for these compounds to pass through the column and greatly shortens the run time.

times during the run. For more information on valve applications, the reader is referred to the Valco, Inc., Web site (http://www.vici.com/support/app/2p_japp.php) for a clear and interactive presentation of different valve configurations.

Dual-column systems with thermal modulators

A thermal modulator device as used in GC^2 is a short (1–200 mm) electrically conductive capillary column* that is connected to a cooling device. The cooling device can utilize a liquid or solid coolant. There is also a power supply for rapidly heating the short capillary. The thermal modulator device is inserted between two analytical capillary columns. Analytes from the first column are trapped in the cooled thermal modulator device in a narrow band; then the device is rapidly heated, and the narrow band of analytes is transferred to the second column where the individual compounds are separated. One of these devices is described in detail [82] in U.S. Patent No. 7,284,409.

Thermal modulators are useful for separating complex mixtures of compounds. In an example described by Crimi and Snow [83], thermal modulation was used to separate 58 solvents used in the purification and manufacture of pharmaceuticals that are regulated by the International Conference on Harmonization, the United States Pharmacopeia, and the European Pharmacopeia. Two columns were connected in series; the first column was relatively nonpolar and the second column was much shorter and more polar. The first column was programmed by the GC oven; the second column was heated isothermally in a separate oven within the GC

* An example is a Silco-Steel™ column available from Restek Corporation, Bellefonte, PA. The column can be coated with a suitable stationary phase.

oven. During the temperature program, groups of compounds with similar polarity and boiling point eluted from the first column into the second column. At this point, they were focused into a narrow band with a pulse of liquid nitrogen. A hot pulse on the second column then allowed the compounds to migrate and separate in the second column. Thus, compounds with the same retention time on the first column could be separated. Pursch et al. [84] reviewed GC^2 with special attention devoted to modulation devices.

REFERENCES

1. James, A. T., Martin, A. J. P. (1952). Gas-liquid partition chromatography: the separation and microestimation of volatile fatty acids from formic acid to dodecanoic acid. *Biochem. J.*, 50(5), 679–90.
2. Grob, K. (1988). *Classical Split and Splitless Injections in Capillary GC.* London: Huethig Publishing Ltd.
3. Snow, N. H. (2004). Inlet systems for gas chromatography. In: *Modern Practice of Gas Chromatography.* (R. L. Grob, E. F. Barry eds.) 4th ed. Hoboken, NJ: John Wiley & Sons; ISBN: 0471229830: 461–90.
4. Penton, Z. Determination of trace amounts of thermolabile compounds by capillary GC: comparison of a septum-equipped programmable injector (SPI) with a hot splitless injector. GC Application Note 11 (http://www.varianinc.com/media/sci/tech_scan/A01848.pdf).
5. Donike, M. (1973). Temperature-programmed analysis of fatty acid trimethyl silyl esters: a critical quality test for gas chromatographic columns. *Chromatographia.*, 6(4), 190–5.
6. Vogt, W., Jacob K., Obwexer H. W. (1979). Sampling method in capillary column gas–liquid chromatography allowing injections of up to 250 μL. *J. Chromatogr. A*, 174(2), 437–9.
7. Vogt, W., Jacob, K., Ohnesorge, A. B., Obwexer, H. W. (1979). Capillary gas chromatographic injection system for large sample volumes. *J. Chromatogr. A*, 186, 197–205.
8. Martinez Vidal, J. L., Arrebola F. J., Mateu-Sánchez M. (2002). Application of gas chromatography–tandem mass spectrometry to the analysis of pesticides in fruits and vegetables. *J. Chromatogr. A*, 959(1, 2), 203–13.
9. Jeannot, R., Sabik, H., Sauvard, E., Dagnac T., Dohrendorf K. (2002). Determination of endocrine-disrupting compounds in environmental samples using gas and liquid chromatography with mass spectrometry. *J. Chromatogr. A*, 974(1, 2), 143–159.
10. Penton, Z. (1991). Optimization of sample introduction parameters for determination of pesticides by capillary gas chromatography using a two column, two detector system. *J. Assoc. Off. Anal. Chem.*, 74(5), 872–5.
11. Penton, Z. (1994). Applications of a GC autosampler modified for headspace sampling. *J. High. Resolut. Chromatogr.*, 17, 647–50.
12. Loffe, B. V., Vitenberg, A. G. (1984). *Headspace Analysis and Related Methods in Gas Chromatography.* New York: Wiley-Interscience.
13. Penton, Z. (2002). Headspace gas chromatography. In: *Sampling and Sample Preparation for Field and Laboratory* (J. Pawliszyn, ed). 1st ed. Amsterdam, The Netherlands: Elsevier Science BV; ISBN: 0444505105: 279–96.
14. Closta, W., Klemm, H., Pospisil, P., Riegger, R., Siess, G., Kolb, B. (1983). *Chromatogr. Newslt.*, 11, 13–7.
15. Penton, Z. (1985). Headspace measurement of ethanol in blood with a modified autosampler. *Clin. Chem.*, 31(3), 439–41.

16. Li, Z., Han, Y. H., Martin, G. P. (2002). Static headspace gas chromatographic analysis of the residual solvents in gel extrusion module tablet formulations. *J. Pharmaceut. Biomed. Anal.*, 28(3, 4), 673–82.
17. Hong, L., Altorfer, H. (2001). A micro-sized headspace GC technique for determination of organic volatile impurities in water-insoluble pharmaceuticals. *Chromatographia.*, 53 (1, 2), 76–80.
18. George, R. B., Wright, P. D. (1997). Analysis of USP organic volatile impurities and thirteen other common residual solvents by static headspace analysis. *Anal. Chem.*, 69 (11), 2221–3.
19. Buecking, M., Steinhart, H. (2002). Headspace GC and sensory analysis characterization of the influence of different milk additives on the flavor release of coffee beverages. *J. Agric. Food Chem.*, 50(6), 1529–34.
20. Alonso, L., Fraga, M. J. (2001). Simple and rapid analysis for quantitation of the most important volatile flavor compounds in yogurt by headspace gas chromatography-mass spectrometry. *J. Chromatogr. Sci.*, 39(7), 297–300.
21. Kolb, B., Ettre, L. S. (2006). *Static Headspace-Gas Chromatography: Theory and Practice.* 2nd ed. Hoboken, NJ: John Wiley & Sons, Inc; ISBN: 0471749443.
22. Bianchi, A., Varney, M. S., Phillips, J. (1989). Modified analytical technique for trace organics in water using dynamic headspace and GC/MS. *J. Chromatogr. A*, 467, 111–28.
23. Huybrechts, T., Dewulf, J., Van Langenhove, H. (2003). State-of-the-art of GC-based methods for analysis of anthropogenic volatile organic compounds in estuarine waters, Illustrated with the river Scheldt as an example. *J. Chromatogr. A*, 1000(1, 2), 283–97.
24. Santos, F. J., Galceran, M. T. (2003). Modern developments in GC/MS-based environmental analysis. *J. Chromatogr. A*, 1000(1, 2), 125–51.
25. Fleming-Jones, M. E., Smith, R. E. (2003). Volatile organic compounds in foods: a five year study. *J. Agric. Food Chem.*, 51, 8120–27.
26. Narain, N., Galväo M. S., Madruga, M. S. (2007). Volatile compounds captured through purge and trap technique in Caja-Umbu (Spondias sp.) fruits during maturation. *Food Chem.*, 102(3), 726–31.
27. Brčić I., Skender, L. (2003). Determination of benzene, toluene, ethylbenzene and xylenes in urine by purge and trap GC. *J. Sep. Sci.*, 26(14), 1225–9.
28. Fabietti, F., Ambruzzi, A., Delise, M., Sprechini, M. R. (2004). Monitoring of the benzene and toluene contents in human milk. *Environ. Int.*, 30(3), 397–401.
29. Arthur, C. L., Pawliszyn, J. (1990). Solid phase microextraction with thermal desorption using fused silica optical fibers. *Anal. Chem.*, 62(19), 2145–8.
30. Alzaga, R., Ortiz, L., Sachez-Baeza, F., Marco, M. P., Bayone, J. M. (2003). Accurate determination of 2,4,6-trichloroanisole in wines at low parts per trillion by solid-phase microextraction followed by GC-ECD. *J. Agric. Food Chem.*, 51(12), 3509–14.
31. Rodriguez-Bencomo, J. J., Conde, J. E., Rodriguez-Delgado, M. A., Garcia-Montelongo, F., Perez-Trujillo, J. P. (2002). Determination of esters in dry and sweet white wines by headspace solid-phase microextraction and gas chromatography. *J. Chromatogr. A*, 963(1, 2), 213–23.
32. Sala, C, Mestres, M, Marti, M. P, Busto O, Guasch, J. (2000). Headspace solid-phase microextraction method for determining 3-alkyl-2-methoxypyrazines in musts by means of polydimethylsiloxane-divinylbenzene fibers. *J. Chromatogr. A*, 880(1, 2), 93–9.
33. Mestres, M., Sala, C., Marti M. P., Busto, O, Guasch, J. (1999). Headspace solid-phase microextraction of sulfides and disulfides using carboxen-polydimethlysiloxane fibers in the analysis of wine aroma. *J. Chromatogr. A*, 835(1, 2), 137–44.
34. Przyjazny, A, Kokosa, J. M. (2002). Analytical characteristics of the determination of benzene, toluene, ethylbenzene and xylenes in water by headspace solvent microextraction. *J Chromatogr A*, 977(2), 143–53.

35. Jiang, G. B., Liu, J. Y., Yang, K. W. (2000). Speciation analysis of butyltin compounds in chinese seawater by capillary gas chromatography with flame photometric detection using in-situ hydride derivatization followed by headspace solid-phase microextraction. *Anal. Chim. Acta.,* 421(1), 67–74.
36. Doong, R. A., Liao, P. L., (2001). Determination of organochlorine pesticides and their metabolites in soil samples using headspace solid-phase microextraction. *J. Chromatogr. A,* 918(1), 177–88.
37. Koide, I., Noguchi, O., Okada, K., et al. (1998). Determination of amphetamine and methamphetamine in human hair by headspace solid-phase microextraction and gas chromatography with nitrogen-phosphorus detection. *J. Chromatogr. B: Biomed. Sci. Appl.,* 707(1, 2), 99–104.
38. Namera, A., Watanabe, T., Yashiki, M., Iwasaki, Y., Kojima, T. (1999). Simple analysis of arylamide herbicides in serum using headspace-solid phase microextraction and GC/MS. *Forensic. Sci. Int.,* 103(3), 217–26.
39. Pawliszyn, J. (1997). *Solid Phase Microextraction: Theory and Practice.* New York: Wiley-VCH, Inc; ISBN: 0471190349.
40. Baltussen, E., Sandra, P., David, F., Cramers, C. (1999). Stir bar sorptive extraction (SBSE), a novel extraction technique for aqueous samples: theory and principles. *J. Microcol. Sep.,* 11(10), 737–47.
41. David, F., Sandra, P. (2007). Stir bar sorptive extraction for trace analysis. *J. Chromatogr. A,* 1152(1, 2), 54–69.
42. Bicchi, C., Iori, C., Rubiolo, P., Sandra, P. (2002). Headspace sorptive extraction (HSSE), stir bar sorptive extraction (SBSE), and solid phase microextraction (SPME) applied to the analysis of roasted Arabica coffee and coffee brew. *J. Agric. Food Chem.,* 50 (3), 449–59.
43. Loughrin, J. H. (2006). Comparison of solid-phase microextraction and stir bar sorptive extraction for the quantification of malodors in wastewater. *J. Agric. Food Chem.,* 54(9), 3237–41.
44. Amirav, A., Dagan, S. (1997). A direct sample introduction device for mass spectrometry studies and GC-MS analysis. *Eur. J. Mass Spectrom.,* 3(2), 105–11.
45. De Koning, S., Lach, G., Linkerhagner, M., Loscher, R., Tablack P. H., Brinkman, U. A. T. (2003). Trace-level determination of pesticides in food using difficult matrix introduction-gas chromatography-time-of-flight mass spectrometry. *J. Chromatogr. A,* 1008(2), 247–52.
46. Lehotay, S. J. (2000). Analysis of pesticide residues in mixed fruit and vegetable extracts by direct sample introduction/gas chromatography/tandem mass spectrometry. *J. AOAC. Int.,* 83(3), 680–97.
47. Lehotay, S. J., Lightfield, A. R., Herman-Fetch, J. A., Donoghue, D. J. (2001). Analysis of pesticide residue in eggs by direct sample introduction/gas chromatography/tandem mass spectrometry. *J. Agric. Food Chem.,* 49, 4589–95.
48. Kakimoto, S., Kitagawa, M., Hori, S., (2001). Rapid and simple method for the analysis of organophosphorus pesticides and these metabolites in the blood by applying GC-MS with chromatoprobe injector. *Japan J. Food Chem.,* 8(3).
49. Falkovich, A. H., Rudich, Y. (2001). Analysis of semivolatile organic compounds in atmospheric aerosols by direct sample introduction thermal desorption GC/MS. *Environ. Sci. Technol.,* 35(11), 2326–33.
50. Wainhaus, S. B., Tzanani, N., Dagan, S., Miller, M. L., Amirav, A. (1998). Fast analysis of drugs in a single hair. *J. Am. Soc. Mass Spectrom.,* 9(12), 1311–20.
51. Wampler, T. P. (1999). Review: introduction to pyrolysis–capillary gas chromatography. *J. Chromatogr. A,* 842(1, 2), 207–20.
52. Barnett, J., Montoya, B. M. (2002). Viability of applying curie point pyrolysis/gas chromatography techniques for characterization of ammonium perchlorate based propellants. SAND2002–1922 Report, Sandia National Laboratories, Albuquerque, NM.

53. White, R. L. (1991). Microfurnace pyrolysis injector for capillary gas chromatography. *J. Anal. Appl. Pyrolysis.*, 18(3, 4), 269–76.
54. Ericsson, I. (1985). Influence of pyrolysis parameters on results in pyrolysis gas chromatography. *J. Anal. Appl. Pyrolysis.*, 8, 73–86.
55. Wampler, T.P, ed. (2006). *Applied Pyrolysis Handbook.* 2nd ed. Boca Raton, FL: CRC Press, Taylor & Francis Group; ISBN:13:9781574446418.
56. Wang, F. C. Y. (2004). The microstructure exploration of thermoplastic copolymers by pyrolysis-gas chromatography. *J. Anal. Appl. Pyrolysis.*, 71(1), 83–106.
57. Challinor, J. M. (2001). Pyrolysis techniques for the characterization and discrimination of paint. In: *Forensic Examination of Glass and Paint: Analysis and Interpretation* (B. Caddy, ed.). New York: Taylor & Francis, Inc; 165–182, ISBN:0748405798.
58. Shedrinsky, A. M., Wampler, T. P., Baer, N. S. (1999). Application of analytical pyrolysis to problems in art and archaeology: a review. *J. Anal. Appl. Pyrolysis.*, 15, 393–412.
59. Smith, C. C., Morgan, S. L., Parks, C. D., Fox, A., Pritchard, D. G. (1987). Chemical marker for the differentiation of group A and group B streptococci by pyrolysis-gas chromatography-mass spectrometry. *Anal. Chem.*, 59(10), 1410–3.
60. Vorhees, K. J., Schulz, W. D., Currie, L. A., Klouda, G. (1988). An investigation of the insoluble carbonaceous material in airborne particulates from vehicular traffic. *J. Anal. Appl. Pyrolysis.*, 14(2, 3), 83–98.
61. Golay, M. J. E. (1957). In: *Gas Chromatography* 1958 (D. H. Desty, ed.). London: Butterworth Scientific Publications; 36–55.
62. Dandeneau, R. D., Zerenner, E. H. (1979). An investigation of glasses for capillary chromatography. *J. High. Resolut. Chromatogr.*, 2(6), 351–56.
63. IUPAC Compendium of chemical terminology, 2nd ed. (1997), http://www.iupac.org/goldbook/S06148.
64. Grob, R. L. (2004). Theory of as chromatography. In: *Modern Practice of Gas Chromatography* (R. L. Grob, E. F. Barry, eds.). 4th ed. Hoboken, NJ: John Wiley & Sons; 25–63, ISBN: 0471229830.
65. Kováts, E. (1958). Gas-chromatographische charakterisierung organischer verbindungen. Teil 1: Retentionsindices Aliphatischer Halogenide, Alkohole, Aldehyde und Ketone. *Helv. Chim. Acta.*, 41(7), 1915–32.
66. Van den Dool, H., Kratz, P. D. (1963). A generalization of the retention index system including linear temperature programmed gas-liquid partitioning. *J. Chromatogr. A,* 11, 463–71.
67. Barry, E. F. (2004). Columns: packed and capillary; column selection in gas chromatography. In: *Modern Practice of Gas Chromatography* (R. L. Grob, E. F. Barry, eds.). 4th ed. Hoboken, NJ: John Wiley & Sons; 65–191, ISBN: 0471229830,
68. McReynolds, W. O. (1970). Characterization of some liquid phases. *J. Chromatogr. Sci.*, 8(12), 685–91.
69. Rohrschneider, L. (1966). A method of characterization of the liquids used for separation in gas chromatography. *J. Chromatogr. A,* 22(1), 6–22.
70. McNair, H. M., Miller, J. M. (1998). *Basic Gas Chromatography.* New York: John Wiley & Sons; ISBN: 047117260X.
71. Korytár. P., Janssen, H. G., Matisová E., Brinkman U. T. (2002). Practical fast gas chromatography: methods, instrumentation and applications. *Trends. Anal. Chem.*, 21 (9, 10)558–72.
72. Dagan, S., Amirav, A. (1996). Fast, very fast, and ultra-fast gas chromatography-mass spectrometry of thermally labile steroids, carbamates, and drugs in supersonic molecular beams. *J. Am. Soc. Mass Spectrom.*, 7(8), 737–52.
73. Maštovská K., Lehotay, S. L. (2003). Practical approaches to fast gas chromatography-mass spectroscopy. *J. Chromatogr. A,* 1000(1, 2), 153–80.

74. Matisová, E., Dömötörová, M. (2003). Fast gas chromatography and its use in trace analysis. *J. Chromatogr. A,* 1000(1, 2), 199–221.
75. Ong, R. C. Y., Marriott, P. J. (2002). A review of basic concepts in comprehensive two-dimensional gas chromatography. *J. Chromatogr. Sci.,* 40(5), 276–91.
76. Cochran, J. W. (2002). Fast gas chromatography—time-of-flight mass spectrometry of polychlorinated biphenyls and other environmental contaminants. *J, Chromatogr, Sci.,* 40 (5), 254–68.
77. Hinshaw, J. V. (2006). The thermal conductivity detector. *LC-GC,* 24(1), 38–45.
78. Hinshaw, J. V. (2005). The flame Ionization detector. *LC-GC,* 23(12), 1262–72.
79. Amirav, A., Jing, H. (1995). Pulsed flame photometer detector for gas chromatography. *Anal. Chem.,* 67(18), 3305–3318.
80. Jing, H, Amirav, A. (1998). Pulsed flame photometric detector—a step forward towards universal heteroatom selective detection. *J. Chromatogr. A,* 805(1, 2), 177–215.
81. Amirav, A., Jing, H. (1998). Simultaneous pulsed flame photometric and mass spectrometric detection for enhanced pesticide analysis capabilities. *J. Chromatogr. A,* 814(1, 2), 133–50.
82. Hasselbrink, E. F, Libardoni M, Sacks R. D. (2007). Thermal modulation for gas chromatography, U. S. Patent No: 7,284,409 B2, Oct. 23, (www.freepatentsonline.com).
83. Crimi, C. M., Snow, N. H. (2008). Analysis of pharmaceutical residual solvents using comprehensive two-dimensional gas chromatography. *LC-GC,* 26(1), 62–70.
84. Pursch, M., Sun, K., Winniford, B., et al., (2002). Modulation techniques and applications in comprehensive two-dimensional gas chromatography (GC×GC). *J. Analyt. Bioanalyt. Chem.,* 373(6), 356–67.

CHAPTER 3

The GC/MS Interface

The GC/MS interface is the section of the instrument starting at the column exit in the gas chromatograph and extending to the entrance to the ion source of the mass spectrometer. For most applications of modern GC/MS instrumentation, the GC/MS interface is a heated metal tube equipped with a temperature controller. Even through the gas exiting the column is at a pressure between 200 kPa* (30 psi) and 700 kPa (100 psi) and the pressure in an electron ionization (EI) source is about 10^{-1} Pa, the flow of about 1–2 mL min^{-1} from a 250–320-μm i.d. column can be accommodated by a 50- L sec^{-1} turbomolecular or oil diffusion pump. If the system is capable of chemical ionization, it will have a 250-L sec^{-1} pump, which is more than adequate for such a gas load. Therefore, the GC column is inserted directly into the ion source from the gas chromatograph or the eluate from the GC column passes through a piece of open-fused silica tubing that passes through the heated interface. In such an arrangement, the GC column is connected to the transfer line using a Press-Fit™ connector, an SGE SilGuard™ connector, a Swagelok-type mini-union, or the Agilent Ultimate Union Kit (Figure 3.1).

Use of larger diameter columns (530–750 μm) requires the use of some type of device to manage the larger flow volumes. The two devices in use today are the *open-split interface* and the *jet separator.*

3.1. Open-Split Interface

The open-split interface [1,2] conceptually illustrated in Figure 3.2 offers a convenient connection between the GC column and the mass spectrometer. The flow rate into the ion source is fixed by the dimensions of the capillary tube leading to the mass spectrometer. The function of this device

* The unit of pressure used in this book is the pascal (Pa). The pascal is the SI (Système International d'unités) pressure unit. For many years, the pressure unit in the United States vacuum industry was the Torr. The Torr is equal to 133 Pa. The pascal is more appropriate because it is a direct measure of a force-per-unit area. A good approximation for the conversion of Torr to pascal is 1 Torr = 100 Pa. Most pressures in mass spectrometry are within an order of magnitude; therefore, this crude approximation should suffice.

Gas Chromatography and Mass Spectrometry
DOI: 10.1016/B978-0-12-373628-4.00003-4

Figure 3.1 The Agilent Ultimate column union.

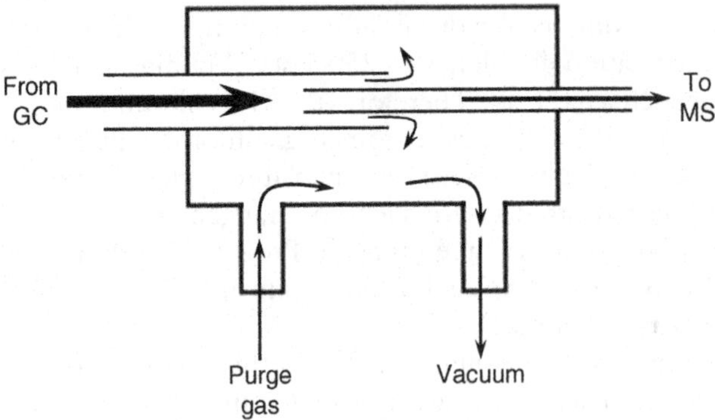

Figure 3.2 Representation of the open–split interface, illustrating the function of the purge gas.

is dependant on the use of a purge gas entering near the exit of the column (Figure 3.2). All or only a fraction of the GC column eluate enters the ion source, depending on the flow rates of the GC column and the purge gas. The open-split interface permits most of the solvent and other eluates to be shunted away from the ion source merely by increasing the flow rate of purge gas. It also permits the GC column to be changed without venting the mass spectrometer or using an isolation valve; the open-split interface ensures normal operation of the capillary column with the exit of the column at atmospheric pressure.

In the open-split interface, the column end is connected to an open sleeve, which has a flow of carrier gas (usually hydrogen or helium) through it acting as a purge gas. The column end is butted against a restrictor line going into the ion source. This restrictor allows a fixed amount of the column eluate to enter the ion source. The restrictor typically allows

1 mL min^{-1} to enter the ion source. If a 530-μm i.d. column is connected to the interface at a flow rate of 5 mL min^{-1}, only 20% of the eluate and therefore only 20% of the analyte would be introduced to the ion source. This limiting of the amount of analyte reaching the ion source should be considered as an issue when the limit of detection is important.

When using the open-split interface, the flow rate of the column can be changed without affecting the conditions in the ion source. Only a fixed amount of sample enters the mass spectrometer as established by the conductance of the smaller capillary (restrictor) shown in Figure 3.2. This means that the same fluid dynamic conditions for ionization will exist even if GC column flow rates change during temperature programming.

The open-split interface gives the better chromatographic integrity of the two types of interfaces. The chromatographic integrity is even better than that which can be obtained with the direct introduction of the column into the ion source. Its yield is dependent on the flow rate and the size of the restrictor. There is no enrichment of the analyte. The open-split interface has been found to discriminate against higher-mass and higher-boiling compounds, a feature that will limit its usability.

3.2. Jet Separator

The most popular of the various enrichment devices used with packed columns was the jet-orifice separator invented, and later patented, by Einar Stenhagen (Swedish medical scientist) and perfected by Ragnar Ryhage, also of Sweden. The jet-orifice separator continues to be used in modern instrumentation with capillary columns. The operation of this separator is based on a diffusion principle [3,4]. As shown in Figure 3.3, the eluate from the GC comes in from the left, expands through a nozzle at position A, and shoots toward an orifice in the wall of an adjoining chamber on its way to the ion source. Across the gap between points A and B, there is a tremendous expansion of the gases. Compounds that have high diffusivity

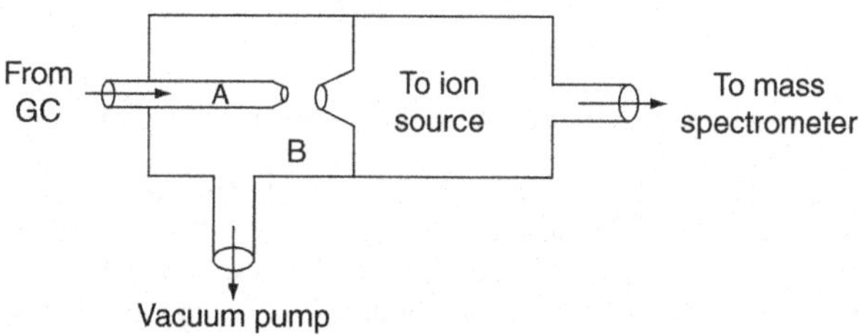

Figure 3.3 Representation of the jet separator showing the exit orifice from the GC eluate (A) and the inlet to the ion source for enriched gas stream (B).

will diffuse at right angles (a process known as effusion) much more than those having a lower diffusivity. Therefore, if a mixture of organic molecules of ~400 Da in helium, which has a mass of 4 Da, is forced through the nozzle at position A, the helium will have a greater tendency to diffuse at right angles into chamber B than will the molecules with a mass of 400 Da. This means a greater number of organic molecules of mass 400 Da will remain in the beam to enter the ion source than atoms of helium. This action imposes a discrimination against the carrier gas, which means that most of the carrier gas diffuses into chamber B and is removed by a vacuum pump. Most of the organic material is preferentially shot into the ion source, thereby achieving an enrichment of the analyte molecules among residual molecules of the carrier gas. The jet separator works in the so-called viscous-flow pressure region. A single-stage separator can accommodate flow rates from 5–30 mL min^{-1}; a double-stage separator is usually required for flow rates above 30 mL min^{-1}. The single-stage jet separator has an enrichment factor of ~50% and an efficiency of 30–40% [5]. The jet separator tends to discriminate against low-mass components in the GC eluate. In viscous flow, the flow rate through a pore or orifice is proportional to the square of the total pressure. The jet separators are prone not only to occasional clogging most often with condensed stationary phase that has bled out from the column but also with deposits of organic material that can condense on the orifice. Although the separator is heated, the tip of orifice A can be cold due to adiabatic gas expansion.

One important disadvantage to the jet separator is that it must have a flow rate of ~20 mL min^{-1} to perform properly. The largest of capillary columns (720 μm i.d.) have optimum flow rates of less than 15 mL min^{-1}, which means that for this device to function properly, a makeup gas must be added at the exit of the GC column. This will further dilute the analyte in the gas stream going to the ion source and have a deleterious effect on the limit of detection. With this added complication for the use of the jet separator, the results obtained in terms of detection limits are not that much better than those that can be achieved with the open-split interface.

REFERENCES

1. Bourne, S., Croasmun, W. (1988). Cross-bore open-split interface for GC-MS. *Anal. Chem.*, 60, 2172–2174.
2. Ligon, W. V. Jr., Grade, H. (1991). Adjustable open-split interface for GC/MS providing solvent diversion and invariant ion source pressure. *Anal. Chem.*, 63, 2386–2390.
3. Ryhage, R. (1964). MS as a detector for GC. *Anal. Chem.*, 36, 759–764.
4. Reis, V. H., Fenn, J. B. (1963). Gas separation in supersonic jets. *J. Chem. Phys.*, 39, 3240–3250.
5. Ryhage, R. (1967). Efficiency of separators in GC-MS. *Arkiv. Kemi.*, 26, 305–316.

CHAPTER 4

Mass Spectrometry Instrumentation

4.1. Overview of Mass Spectrometers

In GC/MS, the mass spectrometer ionizes the gas-phase eluate from the GC column as it enters the mass spectrometer. It is important to remember that this gas-phase eluate contains mobile phase (usually hydrogen molecules or helium atoms), analyte molecules, volatile matrix components that elute with the analyte, and molecules that have bled off of the stationary phase of the GC column (usually cyclic siloxanes) formed by decomposition of the stationary phase. Depending on the ionization technique, these ions representing the intact molecule may have sufficient energy to undergo fragmentation to ions that have a smaller mass. In GC/MS, the vast majority of the ions formed have only a single charge. Primarily, only ions of aromatic hydrocarbons subjected to electron ionization (EI) will form double-charge ions; and these ions are of low abundances compared to single-charge ions of the same mass. Mass spectrometers separate ions according to their mass-to-charge ratio (*m/z*) values; therefore, because virtually all of the ions have a single charge, the *m/z* value of ions formed in GC/MS is considered to be also the mass of the ion.

After the initial formation of ions representing the intact molecule and their subsequent fragmentation (ions of the intact molecule that are going to fragment will do so within less than a microsecond of their formation), they are accelerated out of the ion source with constant energy into the *m/z* analyzer. GC/MS uses EI, chemical ionization (CI), electron capture negative ionization (ECNI), field ionization (FI), and, to a much lesser extent, atmospheric pressure chemical ionization (APCI)—all defined in detail later in this chapter. No commercial instruments currently available offer this APCI technique. EI and FI form positive-charge molecular ions ($M^{+\bullet}$), and ECNI forms negative-charge molecular ions ($M^{-\bullet}$). These are ions that have the same elemental composition as the molecule from which they were formed but have one electron more or less than the original molecule. CI and APCI form protonated molecules (MH^+), deprotonated ($[M-H]^-$) ions, or adduct ions ($[M+C_2H_5]^+$ or $[M+Cl]^-$). All of these ions formed

Gas Chromatography and Mass Spectrometry
DOI: 10.1016/B978-0-12-373628-4.00004-6

by CI and APCI have different elemental compositions and therefore different masses than the mass of the original molecule.

CI, ECNI, FI, and APCI do not transfer much energy to the molecule during the ionization process and are called *soft ionization* techniques. EI produces very energetic molecular ions with only a positive charge ($M^{+\bullet}$). Depending on the structure of these molecular ions, a significant number will undergo fragmentation. Different molecular ions can produce fragment ions with different *m/z* values and different elemental compositions. The formation and subsequent detection of these fragment ions and their abundances produce the characteristic EI mass spectrum, sometimes referred to as a characteristic "EI fingerprint" for the compound. Molecular ions formed by EI are sometimes so energetic that their mass spectra do not exhibit a $M^{+\bullet}$ peak. This is why the soft ionization techniques can be considered complementary to EI because they usually provide the molecular mass of the analyte. The fragmentation of the ions representing the intact molecule is used to determine the structure of an analyte. Both the *m/z* values of the fragments and the *dark matter* (the elemental compositions implied by the difference in the *m/z* values of two mass spectral peaks) represented by these *m/z* values and that of the molecular ion are crucial in structure determinations. The fragments produced by EI are what are necessary to determine the structure of the original molecule. Soft ionization techniques have limited use in deterring the identity of the analyte because of the lack of fragment-ion formation.

The data resulting from a GC/MS analysis are known as "mass spectra." Mass spectra are acquired one after another at a consistent rate. The coordinates for each mass spectral peak represent the *m/z* value of an ion and the abundance of the ion with that *m/z* value. More often than not in GC/MS, these coordinates are presented by dropping a vertical line from their position on a two-dimensional Cartesian coordinate system to the *x* axis (the value of *x* represents the *m/z* value of the ion and the value for *y* represents the abundance of the ion, usually a relative intensity value). The data are displayed in this presentation because most of the mass spectrometers used for GC/MS are only capable of separating mass spectral peaks that differ by an integer *m/z* unit (Figure 4.1).

In some cases, the data are obtained by acquiring intensity information every 1/20 of an *m/z* unit and then summing these intensity values and recording the sum and the actual 1/20 of an *m/z* unit where the intensity is the greatest. This is called *centroid data*. Another form of centroid data can be obtained by recording only the intensity at the *m/z* value in the 20-step range where the maximum occurs. Centroid data can also be obtained by jumping 20 steps at a time to a position assumed to correspond to the maximum intensity of the peak and recording the observed intensity. Alternatively, the intensity at each of the 20 steps can be stored, and these

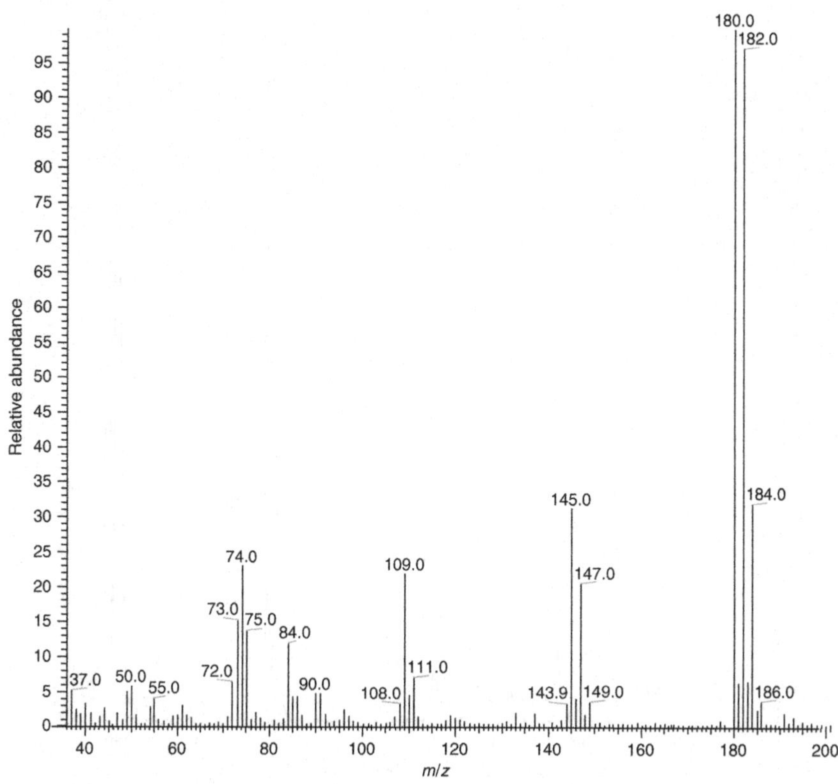

Figure 4.1 A typical mass spectrum acquired using EI on a GC-MS that is capable of separating ions that differ by only 1 *m/z* unit. This is the most common presentation of mass spectra in GC/MS. This could be the centroid presentation of data acquired in the profile mode and seen in Figure 4.2.

data then can be displayed as *profile data* (Figure 4.2). Spectra can also be displayed in a tabular format with pairs of numbers representing the *m/z* values and the corresponding intensities (as raw values or normalized to the maximum intensity value).

Either a single mass spectrum in a data file can be displayed or a series of spectra averaged together can be displayed as a single spectrum. A single mass spectrum or an averaged mass spectrum can be displayed with another spectrum or another averaged series of spectra subtracted as a background. A consecutive series of spectra can be displayed as the sum of all the intensities in each spectrum versus the spectrum number. The x axis of such a display (spectrum number) relates to time because the spectra are acquired at a constant rate. The y axis is related to analyte concentration in the ion source because the abundance of each ion in the mass spectrum is a function of the amount of analyte present at the time the spectrum is

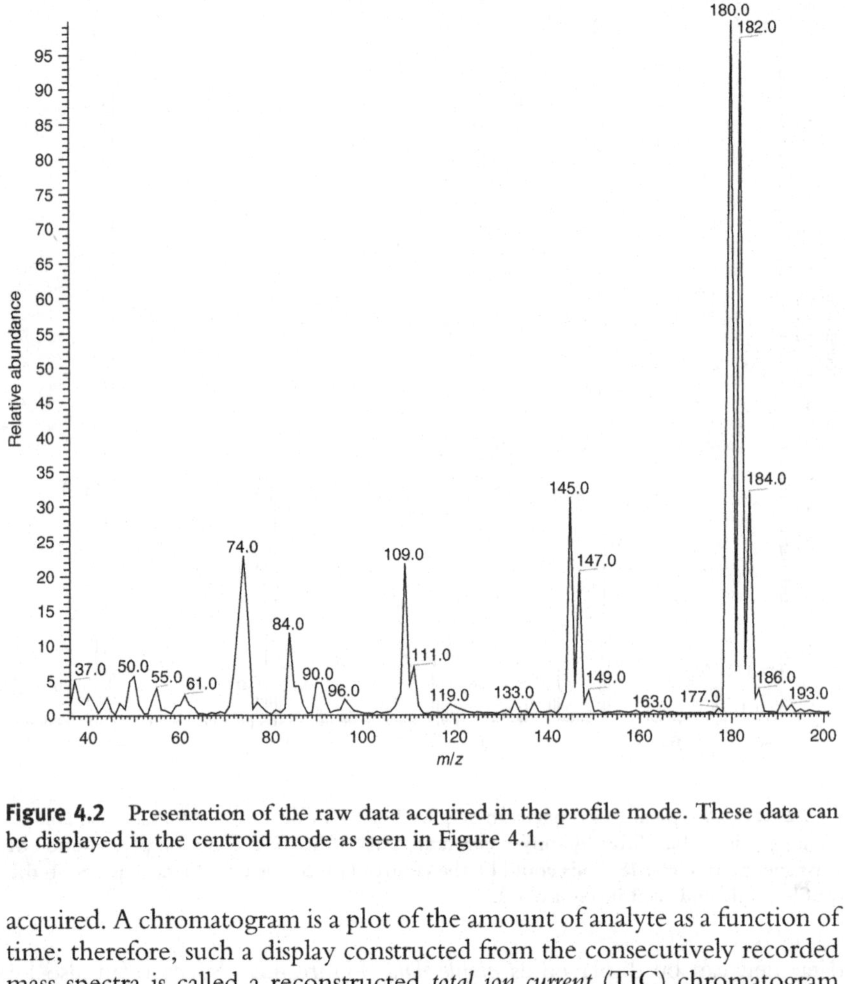

Figure 4.2 Presentation of the raw data acquired in the profile mode. These data can be displayed in the centroid mode as seen in Figure 4.1.

acquired. A chromatogram is a plot of the amount of analyte as a function of time; therefore, such a display constructed from the consecutively recorded mass spectra is called a reconstructed *total ion current* (TIC) chromatogram (Figure 4.3). The mass spectral data can also be displayed by plotting the ion current for a single *m/z* value, a small range of *m/z* values, or the sum of several *m/z* values vs time. These plots are called *mass chromatograms* or *extracted ion current* (Figures 4.4 and 4.5) chromatograms.

4.2. Resolution, Resolving Power, and Mass Accuracy

Not generally used that much in GC/MS, but of significance in all mass spectrometry, are the terms *resolution* (Δm), *resolving power* (R), and *mass accuracy* (sometimes symbolized as *parts per million*, ppm). These terms are

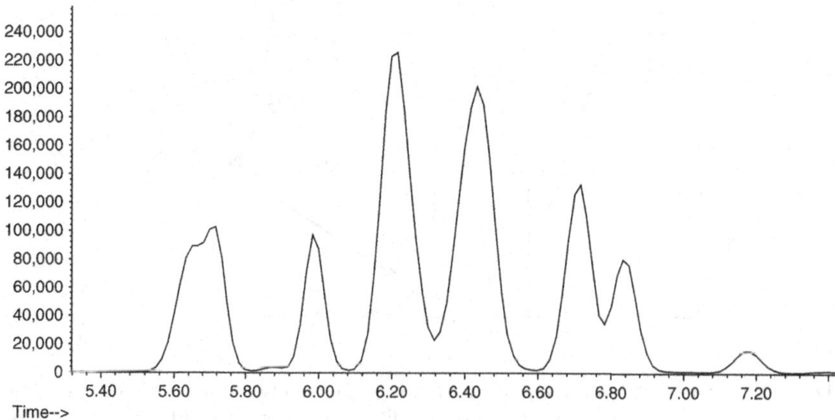

Figure 4.3 The display obtained by plotting the sum of the ion current represented by each of the peaks in individual spectra numbers (reconstructed total ion current (RTIC) chromatogram or an RTIC profile).

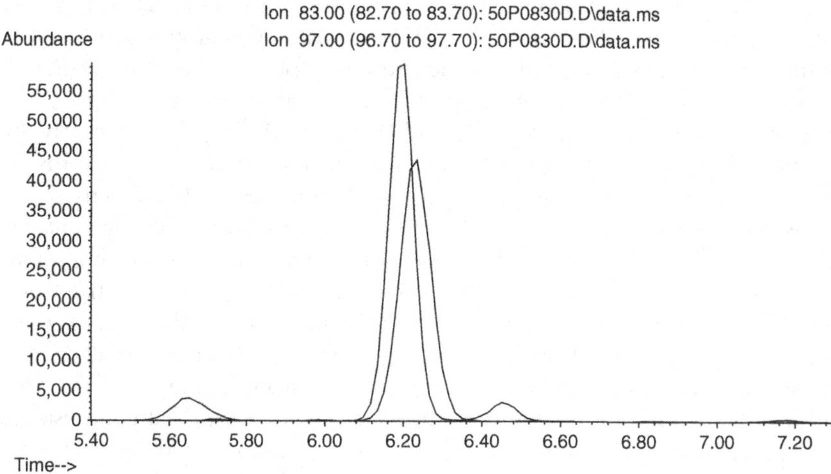

Figure 4.4 Mass chromatograms for two different integer *m/z* values. The display shows that each *m/z* value represents different analytes that are closely eluting from a chromatographic column.

often incorrectly used interchangeably; therefore, a good understanding of their definitions will facilitate an understanding of mass spectrometry.

First consider what is represented by a mass spectral peak of a pure compound. As an example, consider a compound that has an elemental composition of $C_{21}H_{25}F_3N_2O_4$. The lowest integer *m/z* value peak in a mass spectrum of the molecular ion of this compound is at *m/z* 426. This ion is called the

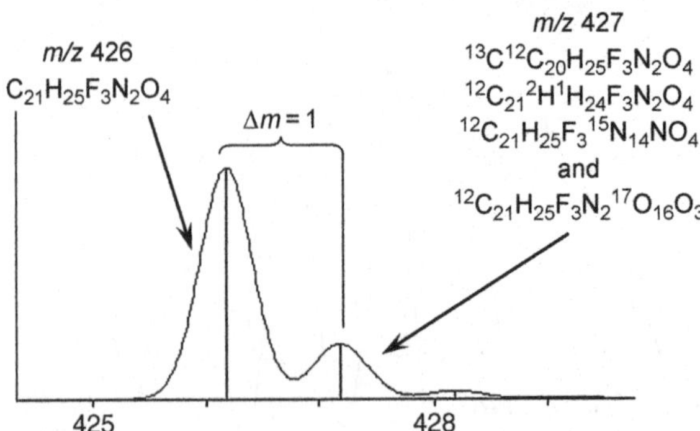

Figure 4.5 The peak on the left at *m/z* 426 represents only a single ion. The peak on the right at *m/z* 427 represents ions with four different masses. The differences are very small, and the mass spectrometer used for these data was not capable of separating ions with such small differences in *m/z* values.

nominal mass ion (the peak at an integer *m/z* value representing the molecular ion is called the *nominal mass molecular ion peak*) because the mass is expressed as an integer, which is the sum of the integer masses of all atoms of the most abundant naturally occurring stable isotopes for each of the elements present. The ion contains 21 atoms of ^{12}C, 46 atoms of protonium (^{1}H), 3 atoms of ^{19}F, 2 atoms of ^{14}N, and 4 atoms of ^{16}O. Another way of making this same statement is to say that the ion contains only atoms of the most abundant naturally occurring stable isotopes of its elements.

A third way to make this same statement is to say that the ion is *monoisotopic*. Carbon, hydrogen, and nitrogen have two naturally occurring stable isotopes (^{12}C and ^{13}C; ^{1}H and ^{2}H; and ^{14}N and ^{15}N). Oxygen has three naturally occurring stable isotopes (^{16}O, ^{17}O, and ^{18}O) and fluorine has only one naturally occurring stable isotope (^{19}F). A certain percentage of all carbon atoms are ^{12}C and a certain percentage of all carbon atoms are ^{13}C. This means that in addition to the monoisotopic ion, this elemental composition will be represented by ions that contain other isotopes of these elements. These other isotopic compositions will be represented by peaks at higher *m/z* values. In the case of this example with a monoisotopic peak (also called a nominal mass peak because it is expressed as the nominal mass of the ion) at *m/z* 426, there will be peaks at *m/z* 427, 428, 429, 430, and maybe higher. Each one of these integer *m/z* value peaks higher than *m/z* 426 will represent the same elemental composition, but they will represent different isotopic compositions; i.e., the peak at *m/z* 427 represents ions with the isotopic compositions of $^{13}C^{12}C_{20}H_{25}F_3N_2O_4$, $^{12}C_{21}^{2}H^{1}H_{24}F_3N_2O_4$, $^{12}C_{21}H_{25}F_3^{15}N^{14}NO_4$, and $^{12}C_{21}H_{25}F_3N_2^{17}O^{16}O_3$. There will not be individual peaks for each of these ions because the resolution (Δm) needed to separate them would have to be a very small number.

If the compound with an elemental composition of $C_{21}H_{25}F_3N_2O_4$ is present in the ion source with a compound that has the elemental composition $C_{21}H_{38}N_4O_5$, then the peak at *m/z* 426 would represent the integer mass of both ions. That is to say, the ion representing each compound will have the same nominal mass even though they have different elemental compositions. The data are presented at unit resolution; therefore, the peak at *m/z* 426 represents both ions.

A mass spectral peak representing an ion of a single elemental composition and a single isotopic composition (a monoisotopic peak) will have a width as well as an intensity. The intensity represents the abundance of the ion. The width (usually presented in terms of the width at half of the full height of the peak) represents the energy dispersion of the ions that have that *m/z* value. When the peak represents the most abundant isotopes of the elements of two or more elemental compositions, in addition to the energy dispersion, the peak width represents the range of masses that define the two ions. When data are reported to the nearest integer *m/z* value, these two ions are represented by a single mass spectral peak. As the *resolution* (defined as the difference between two adjacent *m/z* values that can be separated, Δm) of the data becomes less (improved data), the two ions will begin to be represented by two different peaks. In the example of ions with the elemental composition of $C_{21}H_{25}F_3N_2O_4$ and $C_{21}H_{38}N_4O_5$, the resolution must be at least 0.15 if two partial resolved peaks are to be observed in the mass spectrum. A resolution of 0.1 would provide a baseline separation of these two ions that have monoisotopic masses of 426.176642 Da ($C_{21}H_{25}F_3N_2O_4$) and 426.28421 Da ($C_{21}H_{38}N_4O_5$) or a difference of 0.107579 Da (Figure 4.6).

Not only must the data exhibit this type of resolution, but the mass spectrometer must also be capable of making an *accurate mass* assignment for these two ions. If ions are not separated so that the overlap of the two peaks occurs at a point no more than 50% of the peak height,

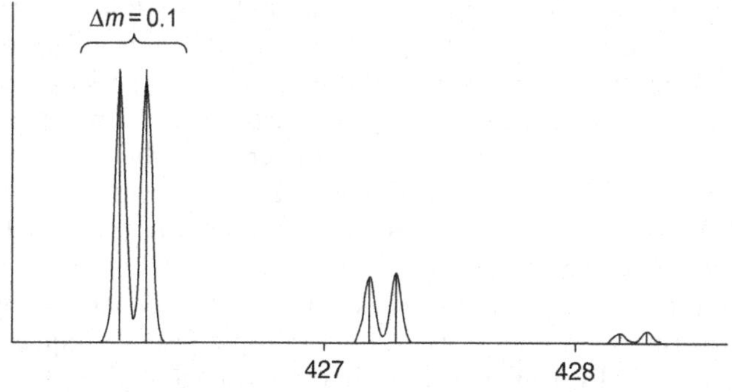

Figure 4.6 Ions of two different elemental compositions ($C_{21}H_{25}F_3N_2O_4$ and $C_{21}H_{38}N_4O_5$) having respective exact masses of 426.17664 Da and 426.284221 Da require a resolution (Δm) of 0.1 *m/z* units to be separated.

then there cannot be an accurate mass assignment for the purposes of this discussion.* However, even if the two mass spectral peaks are resolved, the instrument must still be capable of an accurate mass assignment to identify the unique elemental composition of each ion. Every ion that has a unique elemental composition has a unique monoisotopic mass. That monoisotopic mass may differ from the monoisotopic mass of another elemental composition by as little as a small fraction of a millimass unit; but if the mass spectrometer has sufficient resolving power (a quality of the mass spectrometer defined as the resolution at a specified *m/z* value and symbolized by $R = \Delta m/m$) to separate the two ions, it may be possible to assign accurate masses to each of the two ions.

It is important to understand what is meant by terms such as *high resolution* and *high resolving power.* Both of these terms have to do with the ability to separate ions that have smaller and smaller differences in *m/z* values. Higher resolving power is a property of the mass spectrometer; and, based on the definition above, it is clear that as the value for Δm gets smaller and smaller, the value for R is larger and larger. However, the term *high resolution* is a misnomer in that it is a self-contradiction. The term *high resolution* refers to the separation of peaks with *small differences* in *m/z* values in a mass spectrum, which means that the value for Δm is small. The term *low resolution* means that the value for Δm is large.

What defines whether an accurate mass can be assigned and how does resolution play a role in this assignment? The better the resolution, the narrower (sharper) the mass spectral peak will be. The narrower the mass spectral peak, the sharper the peak top. Different instruments produce mass spectral peaks of different shapes. A true Gaussian-shape peak will have a sharp peak top. Instruments that do produce Gaussian peaks are more likely to allow for accurate mass assignments than those that do not. Double-focusing (magnetic and electric sectors) and time-of-flight (TOF) mass spectrometers have Gaussian or triangular-shaped peaks. The mass spectral peak shape produced by transmission quadrupole (QMF (quadrupole mass filter)) and 3D quadrupole (QIT (quadrupole ion trap)) mass spectrometers is often referred to as a de-Gaussed shaped; therefore, peaks produced by these instruments are not as suitable for accurate mass assignments. All four of these different types of mass spectrometers and their relationship to GC/MS will be explained later in this chapter.

As stated above, the resolving power of a mass spectrometer is defined as the resolution at a given *m/z* value and symbolized by $R = \Delta m/m$ where *m* is the *m/z* value and Δm is measured. The value for Δm is either the difference in the *m/z* values of two monoisotopic peaks of equal intensity where the two peaks overlap at their base to a height that is 10% of their

* There is software that can assign an accurate mass to a monoisotopic mass peak that will be discussed later in this chapter.

height above the baseline (this is called the *10% valley* definition of resolution) or the *m/z* width of a monoisotopic peak at a position determined at half of the peak's maximum intensity (this is called the *full-width-at-half-maximum* (FWHM) definition for resolution). The value for Δm can also be obtained by determining the difference in the *m/z* values of two monoisotopic peaks that are separated in such a way that their bases overlap at a height equal to 50% of the peak height. This is the *50% valley* definition. The 50% valley definition results in a value for Δm equal to that obtained using the FWHM method with a single peak. Figure 4.7 is an illustration of the different definitions of resolving power and resolution.

On close examination of Figure 4.7, it will be noted that there are three labels for Δm. These have to do with different definitions used for resolution. When two monoisotopic peaks are separated so that their overlap is 10% of their height, the difference in the two *m/z* values is Δm. This is the so-called 10% valley definition. When there is only one monoisotopic peak, Δm is the width of the peak at half height. This is the FWHM definition of Δm. The value for Δm using the FWHM definition is half the value that would be obtained if the 10% valley definition was used. Using the

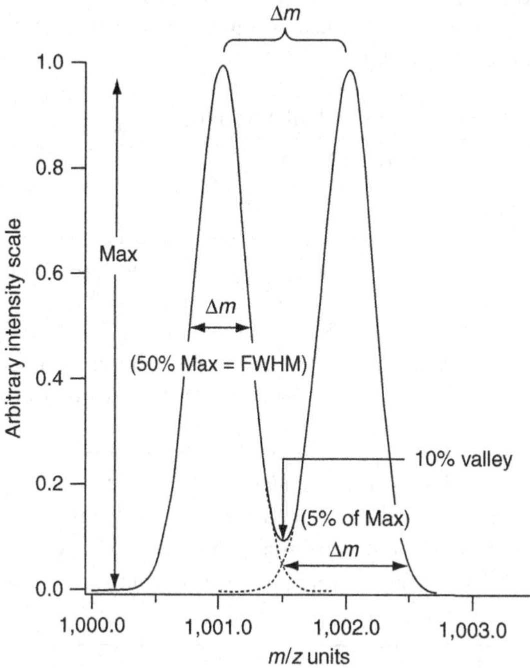

Figure 4.7 Two mass spectral peaks, each representing a unique monoisotopic composition that is about 1,000 Da. The two peaks are separated by 1 *m/z* unit. The FWHM is 0.5 *m/z* units in this example.

width of the mass spectral peak at 5% of its height is approximately equal to the Δm value determined using the 10% valley method. The 5% width method is seldom, if ever, used because of the problem with electrical and chemical noise at this signal level.

Double-focusing and TOF mass spectrometers have a constant resolving power performance, which means that the resolution decreases (the differences in *m/z* values that can be separated becomes greater) with increasing *m/z* values. The QMF (transmission quadrupole) and QIT instruments operate at a constant resolution, which means that the resolving power increases as the *m/z* value increases; however, *m/z* 100 can be separated from *m/z* 101 to the same degree that *m/z* 500 can be separated from *m/z* 501.

Accurate mass can be determined from low-resolution data; however, the peak used for the mass assignment must be a monoisotopic peak of a single elemental composition. High resolution is used mainly to assure that peaks of the same integer mass are separated so that the accurate mass being assigned is guaranteed to represent only a single elemental and isotopic composition. This assignment is usually accomplished by comparing the position of the peak being assigned an accurate value, with the position of a peak that represents a known composition and has a known calculated mass.* The accuracy of the mass assignment is sometimes reported in ppm. This value is obtained by

$$\frac{\text{assigned mass} - \text{calculated mass}}{\text{calculated mass}} \times 10^6 = \text{ppm}$$

This value represents how close the measured mass of an ion is to its calculated mass. The topics of mass accuracy, resolving power, and resolution are described in more detail in Watson and Sparkman [1] and de Hoffmann and Stroobant [2]. It should also be noted that the intensity of the isotope peaks relative to the intensity of the monoisotopic peak is another factor important in arriving at an elemental composition. This will be discussed further in the section on Data Interpretation.

> Cerno Bioscience has software, *MassWorks*, which allows the assignment of accurate mass values to ions represented by mass spectral peaks in data acquired with a conventional QMF GC/MS system. This software can deconvolute peaks representing doublets and triplets and has proved to be very reproducible.

* The calculated mass is derived by summing the number of each element multiplied by its published monoisotopic mass as assigned by NIST.

4.3. Vacuum System

As the ions travel from the ion source through the *m/z* analyzer into the detector they are in free motion. Their path and direction of travel must be determined only by the electrical or magnetic fields used to separate them according to their individual *m/z* values. This is why the mass spectrometer must be operated at a very low pressure, i.e., a vacuum. The ions must not collide with any other matter which could cause their direction to change or cause them to break apart; therefore, the mass spectrometer is operated at very low pressures on the order of 10^{-3} Pa or less. The GC-MS is a dynamic system with a continuous introduction of gas (the GC mobile phase). In order to maintain the required operating pressure, the introduced gas must be continually removed. The function of the vacuum system is to remove the entering gas at the same time the pressure is being maintained at the desired operating level.

The vacuum system is primarily comprised of two separate components: the low-vacuum (~10^{-1} Pa) component known as the *fore pump* or the roughing pump and the high-vacuum system that brings the pressure down to ~10^{-3} or less. In all mass spectrometers, except for the QIT, the rule is the lower the pressure, the better the performance. The TOF mass spectrometer requires the lowest operating pressure. In GC/MS units that have some type of readout for pressure, the pressure is that of the analyzer section or the ion-source housing (the part of the mass spectrometer where the ion source is located). Sometimes the pressure reading is said to be that of the ion source. This is rarely true. The pressure in the ion source itself is ~10^{-1} Pa for EI and FI and ~10–100 Pa for CI and ECNI.

4.3.1. Low-Vacuum Component

The low-vacuum component for most GC/MS systems is an in-line rotary vane pump, schematically illustrated in Figure 4.8. The gas flows from the area being pumped (usually the analyzer, but sometimes both the analyzer and the ion-source housing) to the pump where it is compressed and expelled through the oil in the pump. The gas then exits the pump. This gas will contain un-ionized material that has entered the mass spectrometer and should be considered as hazardous. It can be very rich in the solvent used to introduce the sample into the GC. It is important that the exhaust of the mechanical pump be sent to a chemical fume hood or at least filtered using a device such as that shown in Figure 4.9.

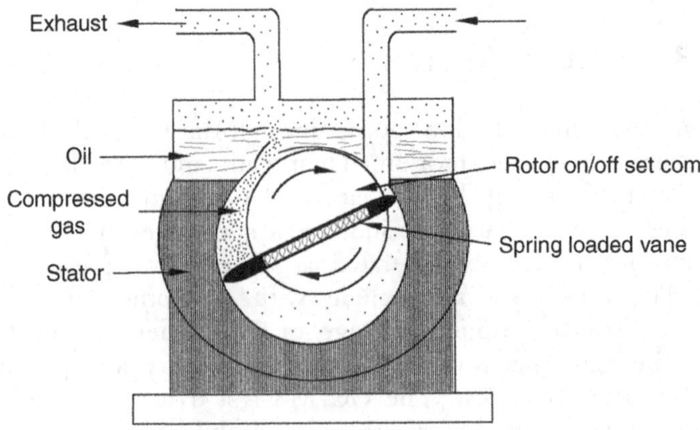

Figure 4.8 A graphic illustration of an oil rotary pump used as the low-vacuum component in most GC/MS systems.

Figure 4.9 Components required for the low-vacuum pump of a GC/MS system if it is not possible to exhaust it to a fume hoop or external vent with a fan.

> ***Warning!!!***
>
> The exhaust of the mechanical pump as well as the split vent of the gas chromatograph must be filtered (a carbon absorption filter) or vented to a chemical hood. Most of the sample introduced into a mass spectrometer ends up in the mechanical pump exhaust or the pump oil. The pump oil should be handled extremely carefully. This is a major hazardous waste.

The oil in these pumps should be treated as a hazardous waste and should not come in contact with anything other than its containment vessel. This oil can be the source of chemical background and needs to be changed on a regular basis. Changing intervals depend on what is being done with the mass spectrometer. Systems used 24/7 for the analysis of samples from the United States Environmental Protection Agency (U.S. EPA) Super Fund site will require pump-oil changes more often than systems used once a day for the analysis of volatile organic compounds in drinking water according to U.S. EPA Method 524. It is important to monitor the color and turbidity of the oil to determine the proper changing interval. In very recent years, the use of scroll pumps and root pumps have begun to replace the in-line rotary vane pump. However, they have not been showing up on GC/MS systems because of their higher cost. These new pumps are usually capable of higher gas flows than those associated with GC/MS; therefore, the increased cost is not warranted. However, these pumps have the distinct advantage that they do not use oil, which simplifies the maintenance and reduces the possibility of chemical background. It should be noted that these pumps require the same venting of exhaust as does the in-line rotary vane pump. Diaphragm pumps are sometimes used as roughing pumps on field-portable systems where the gas loads are lower than those associated with laboratory GC/MS systems.

4.3.2. High-Vacuum Component

The high-vacuum component is either a turbomolecular pump (Figure 4.10) or an oil diffusion pump (Figure 4.11). These pumps are, respectively, called the *turbo pump* and the *diff pump*. Both of these pumps must be evacuated to a pressure of ~10^{-1} Pa before they begin to operate properly; this is another function of the fore pump. The diff pump is based on older technology than is the turbo pump, and it is only used on QMF and internal ionization QIT GC/MS systems that are limited to EI. The diff pump is not used on

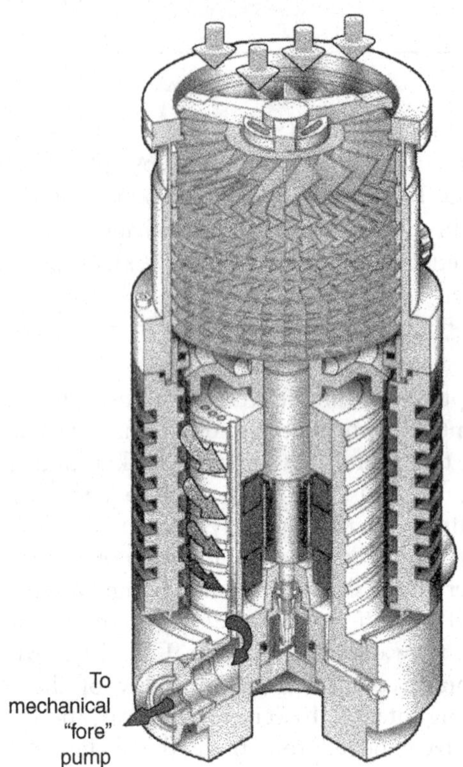

Figure 4.10 Schematic representation of a turbo molecular pump. These pumps are very popular in GC/MS. This configuration is widely used by all manufacturers.

mass spectrometers designed for purposes other than EI GC/MS. Diff pumps are less expensive and cheaper to maintain than the turbo pump. They have no mechanical parts and function by using molecular beams of hot oil vapor to physically force ambient gas molecules to move toward the throat of the fore pump. Sudden loss of power to the mechanical pump can cause the oil in the diff pump to be sucked into the areas being evacuated, creating a significant mess requiring a long downtime period to clean. Diff pumps require long startup and shutdown times due to the necessity for heating or cooling the oil in these respective processes. The speed of the diff pump used with EI-only GC/MS systems is usually about 35–50 L per second.

In EI-only QMF and internal ionization QIT GC/MS systems fitted with a turbo pump, the speed of the pump is about the same as that of an oil diffusion pump. GC/MS systems used for CI and/or ECNI have much larger gas loads on the ion source. These instruments may require differential pumping (two stages of high-vacuum pumping; one on the

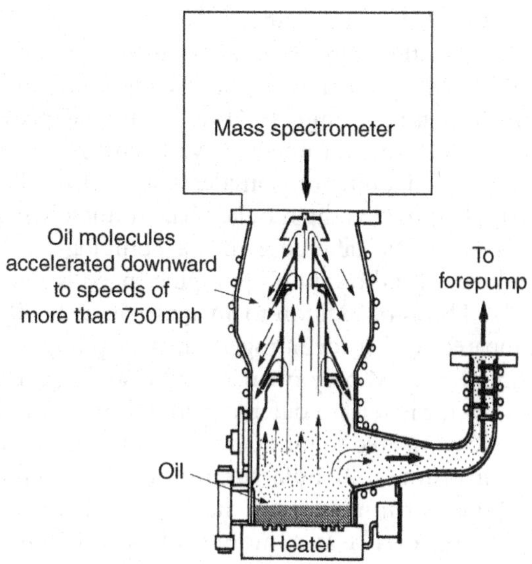

Figure 4.11 Schematic illustration of an oil diffusion pump, which provides a suitable vacuum system for EI-only QMF and QIT GC/MS systems.

ion source and one on the analyzer). In CI instruments that do not employ differential pumping, the single turbo pump has a speed between 250 L sec^{-1} and 300 L sec^{-1}. Differential pumping can be accomplished with two pumps, usually a 200 L sec^{-1} or higher speed on the ion-source housing and a 50 L sec^{-1} pump on the analyzer. Differential pumping can also be accomplished using a single *split-flow* turbo pump. This type of pump has separate pumping ports (two or more) so that the sum of the pumping speed from the multiple ports is greater than the speed obtainable by a single port from the same size pump. The turbo pump can be started and shut down quickly. It is turned off before the fore pump and is started after the fore pump has achieved its maximum vacuum. These pumps are essentially jet engines with a series of stators and rotor blades. Any foreign particulate material introduced into the pump rotors can cause a catastrophic failure when the turbine is turning at high rpm. These pumps have a finite life. They are very expensive. Most pump manufacturers have an exchange program that provides rebuilt pumps for about half the price of a new pump; however, the exchanged pump must be working at the time of the exchange. The pump should be in the turned-on state even when the GC-MS is not in use because many of these pumps have bearings that will deform and cause the pump to fail when started after a long (2–3 weeks) period of being idle.

Another important consideration regarding the high-vacuum pump on a GC-MS is the fact that the pumping speeds quoted by the GC-MS manufacturer are the speeds quoted by the pump manufacturer, which is usually different from the GC-MS manufacturer.* The pump manufacturer's speed specification is based on either nitrogen or air (which is 80% nitrogen). The gas being pumped in GC/MS is either He or H_2. With He becoming scarcer and more costly, more analysts are turning to H_2 as the mobile phase for both GC and GC/MS. Most GC/MS instruments have been optimized for He. It is possible that there are even ionization considerations that will have to be taken into account when H_2 is used as the mobile phase in these systems. However, a more acute problem is the difference in the pumping efficiency of turbo pumps for H_2 and He. A turbo pump's pumping speed is about 15–20% less for H_2 than for He. This is in addition to an approximately 30% reduction in speed for He compared to that for nitrogen/air. Diff pumps do not have the same reduced pumping-speed performance for H_2 as compared to He as do turbo pumps; however, there is no difference in diff pumps and turbo pumps when it comes to He compared to nitrogen/air. If H_2 is going to be used as a mobile phase, the instrument manufacturer should be carefully questioned about this before the instrument is purchased. There are also safety considerations for the use of H_2 as a mobile phase; like the necessity for venting the mechanical pump, an explosion-proof mechanical pump is necessary. And, what happens if the GC column breaks?

4.4. Ionization Types

EI is the most widely used type of ionization in GC/MS. Many instruments are sold to just perform this type of ionization. As pointed out above, EI is a very energetic process resulting in molecular ions ($M^{+\bullet}$) that have a great deal of internal energy. These ions want to get to a lower energy state and will fragment resulting in the multiple-peak mass spectra commonly associated with mass spectrometry. The other ionization techniques currently commercially available for GC/MS are called *soft ionization* techniques because they result primarily in ions that represent the intact molecule which does not produce many fragments. These soft ionization techniques are resonance ECNI, CI, and FI. ECNI and CI are much more widely used than FI. Waters and JEOL are the only companies that offer FI. CI is considered a complementary technique because it can provide the molecular weight information that may or may not be in the EI mass spectrum. Approximately 20% of the mass spectra of the 191,436 compounds in the NIST08 Mass Spectral Database do not exhibit a $M^{+\bullet}$ peak. ECNI is primarily used for quantitative

*The one exception to the pump manufacturer and the GC-MS manufacturer being different companies is Varian. Varian has both a vacuum division and an analytical instrument division. However, the pumping-speed specification quoted by the analytical instrument division for the high-vacuum system is the value quoted by the vacuum division.

purposes of electrophilic analytes in complex matrices. Due to the extreme specific for electrophilic compounds (especially halogenated compounds), this technique gives the lowest limits of quantitation (attogram μL^{-1} range) of any mass spectral technique. Still considered the ionization technique providing the lowest detection limit for the widest variety of compounds is APCI. As was pointed out above, this technique is not currently commercially available, but the interfacing of a GC to an atmospheric pressure ionization (API) interface on an LC/MS system has been described [3].

One very important consideration about different ionization techniques is that switching between them may require that the vacuum be broken (the instrument needs to be opened to the atmosphere). There are components in the mass spectrometer that require heating. Before the evacuated areas are opened to the atmosphere, the heating of these internal components should be terminated and the instrument cooled to room temperature. Care should be taken to make sure that the eluate from the GC column is not flowing into the mass spectrometer during this time because of the greater potential for condensation on the cooled metal parts that can result in ion-source and ion-optics contamination.

4.4.1. Electron Ionization

EI involves the interaction of a low-pressure (~10^{-1} Pa) gas cloud with electrons that have been accelerated through a 70-V electric field. The use of 70-eV electrons in EI is a convention. Most organic molecules have an ionization potential of about 10 eV. Maximum fragmentation results around 30 eV. By using 70-eV electrons, it is assured that the EI spectrum of a compound will be the same when obtained on different instruments. This reproducibility is what allows for the creation and use of mass spectral databases. Figure 4.12 is a schematic illustration of an EI source in its ion-source housing.

The EI source has an *ion volume* that is about 1 cm^3. The source is very *open,* which means that the conductance of gas out of it is high. This is one of the first ionization techniques for mass spectrometry developed in the 1930s by Arthur Jeffery Dempster, a Canadian–American physicist at the University of Chicago [4]. An electron beam is produced by accelerated electrons moving toward a trap with a positive charge on the opposite side of the ion volume. A permanent magnet is positioned on the source to cause the electron beam to travel a helical path. This improves the probability of an electron coming in proximity with the energy cloud that comprises most of the molecular volume of an analyte. This technique was originally called *electron impact.* The electron impact term implied that the electron struck the analyte molecule like a cue ball strikes a billiard ball. This is not the case. Molecules are made up mostly of open space with an orderly dispersion of atomic nuclei and electrons. The electrons in the molecule are in motion

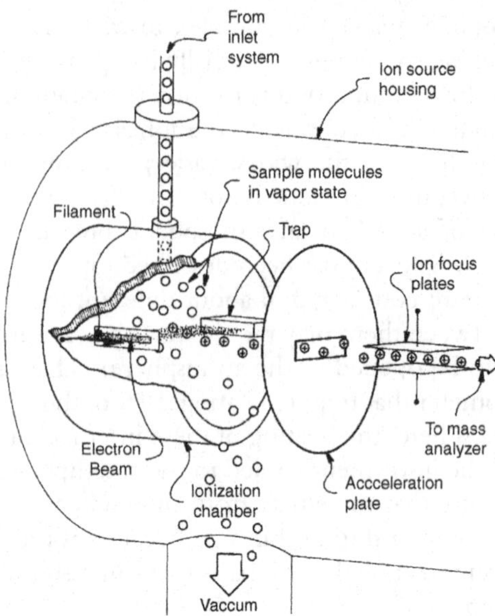

Figure 4.12 Schematic illustration of an EI source.

within certain limits. Even if by chance an ionizing electron came in contact with a nucleus of an atom or one of the molecule's electrons, the mass of the ionizing electron is so small that very little kinetic energy would be transmitted to the analyte [5]. The ionizing electron energizes the analyte molecule by being in the proximity of the molecule. The energized molecule wanting to achieve a lower energy state will expel an electron procuring the positive-charge molecular ion (positive charge because the resulting species now has one less electron than protons). The electron expelled by the analyte will suffer the same fate as the ionizing electron in that it will be attracted to and be neutralized by the positively charged trap. It is possible that the ionizing electron may interact with more than a single analyte molecule.

Even with all of these possibilities, the efficiency of EI is between 0.01% and 0.001%. It almost appears that EI is a technique that cannot work. However, it should be noted that based on manufacturers' specifications, as little as 10 pg of hexachlorobenzene can be injected onto a GC column and a spectrum representing the top of a reconstructed TIC chromatographic peak can be matched against the *NIST08 Mass Spectral Database* or the *Wiley Registry*, 8th ed., and a mass chromatogram of *m/z* 284 (the X+2 peak for the $M^{+\bullet}$) will have a signal-to-background of >10:1. This is the reason that it has been said that mass spectrometry produces more information from less sample than any other analytical technique. Another factor that adds to the power of EI is the large commercially available databases of EI spectra. *The NIST08 Mass Spectral Database*, which contains 220,435 spectra of 192,262 compounds, and the *Wiley Registry*, 9th ed., which contains 668,092 spectra, are the two major searchable commercially available databases. There are several other much smaller databases directed toward specific fields such as flavor and fragrances or forensics that are also commercially available. These databases are listed in Appendix I along with their number of spectra/ compounds and the publishers.

EI, like all the ionization techniques for GC/MS, except APCI, takes place in the evacuated region of the mass spectrometer. This means that the ion source is subjected to everything that comes from the GC column. To avoid contamination from less volatile column eluates, it is best to heat the ion source and the ion optics that precede the actual analyzer. Even the analytes that are in the gas phase can contaminate the ion source and ion optics. The gas is expanding into a vacuum, and the effects of adiabatic cooling should not be underestimated. The ion source and ion optics must be cleaned from time to time to maintain performance. The regularity with which cleaning is necessary will be a function of the types of samples and the analytical matrix. These cleaning requirements also apply to internal ionization QIT mass spectrometers. The three electrodes and spacers that electrically isolate them from one another must be cleaned on a periodic basis. Unlike the quadrupole elements of the QMF, these surfaces can be cleaned. In the event it is necessary to clean the QMF quadrupole surfaces, this should only be done by a trained instrument technician. Some QMF instruments have what is called a *prefilter* on the front of the quadrupole. This prefilter is safe to clean. QMF mass spectrometers using the quartz mandrel design for the quadrupole cannot have the actual quadrupole cleaned.

4.4.2. Chemical Ionization

CI was developed in the mid-1960s by Burnaby Munson and Frank Field [6,7]. This technique uses an ion source with about the same 1-cm^3 volume of the EI source, but it is more closed limiting the conductance so that the source pressure is very high (10–100 Pa). The pressure is maintained with a constant introduction of a *reagent gas*. The reagent gas is ionized by an electron beam produced by the acceleration of electrons from a filament, just as in EI. The potential used to accelerate the electrons in CI is much higher than that used in EI (~200 V). This is to assure that the electrons pass through the dense cloud of reagent gas. The high concentration of reagent gas molecules in comparison to analyte gas molecules means that there is little competition for ionizing electrons by the analyte molecules. The molecular ions resulting from the ionization of the reagent gas molecules will physically collide with residual reagent gas molecules in what is called an *ion/molecule reaction*. The products of this binary reaction are *reagent ions* that can collide with an analyte molecule to bring about another ion/molecule reaction resulting in an ionized species of the analyte molecule. The most often-used reaction to ionize analytes is a protonation reaction resulting in a protonated molecule (MH^+ or $[M+H]^+$). It is possible to produce deprotonated molecules ($[M-H]^-$) through reactions with negative-charge reagent ions, ions resulting from a hydride abstraction ($[M-H]^+$), or positive or negative ions resulting from the formation of collisionally stabilized complexes of reagent ions and molecules. Reactions in Scheme 4.1 shows the formation of reagent ions when methane is used as a reagent gas. Table 4.1 shows the various reactions that can occur in CI.

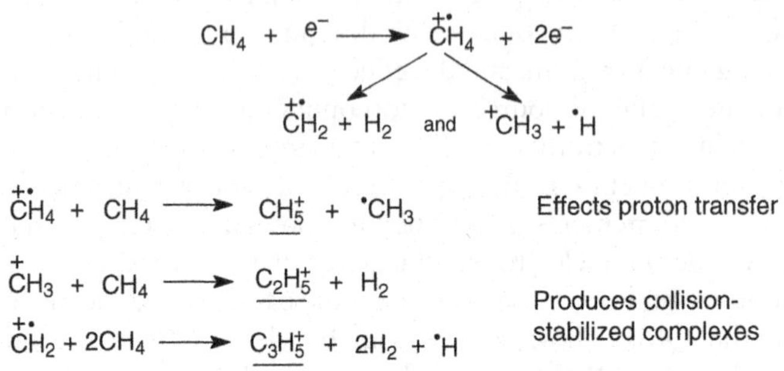

$$CH_4 + e^- \longrightarrow CH_4^{+\bullet} + 2e^-$$

$$CH_4^{+\bullet} \longrightarrow CH_2^{+\bullet} + H_2 \quad \text{and} \quad {}^+CH_3 + {}^\bullet H$$

$$CH_4^{+\bullet} + CH_4 \longrightarrow \underline{CH_5^+} + {}^\bullet CH_3 \qquad \text{Effects proton transfer}$$

$$CH_3^+ + CH_4 \longrightarrow \underline{C_2H_5^+} + H_2$$

$$CH_2^{+\bullet} + 2CH_4 \longrightarrow \underline{C_3H_5^+} + 2H_2 + {}^\bullet H$$

Produces collision-stabilized complexes

Scheme 4.1 Reaction of ions formed from methane by EI in the formation of CI reagent ions.

Table 4.1 Examples of CI reactions

Charge transfer (The reagent ion, missing an e^-, takes an e^- from the analyte.)

$CH_4^{+\bullet} + RH \rightarrow RH^{+\bullet} + CH_4$ $\quad M^{+\bullet}$

Proton transfer (Most common when the analyte molecule has a higher proton affinity than the reagent gas. The analyte takes H^+ from the reagent ion.)

$CH_5^+ + RH \rightarrow RH_2^+ + CH_4$ $\quad [M+1]^+$

$C_2H_5^+ + RH \rightarrow RH_2^+ + C_2H_4$

Hydride abstraction (The reagent ion has a high hydride affinity, the ability to remove a H^- from the analyte molecule.)

$CF_3^+ + RH \rightarrow R^+ + CF_3H$ $\quad [M-1]^+$

$C_2H_5^+ + RH \rightarrow R^+ + C_2H_6$

Collision-stabilized complexes (Occurs when the PA of the analyte and reagent gas are comparable. The reagent ion becomes attached to the analyte. When methane is used, this series of $[M+1]^+$, $[M+29]^+$, and $[M+41]^+$ is a very good confirmation of molecular weight.)

$C_2H_5^+ + RH \rightarrow (C_2H_5{:}RH)^+$ $\quad [M+29]^+$

$C_3H_5^+ + RH \rightarrow (C_3H_5{:}RH)^+$ $\quad [M+41]^+$

By far, the most common use of CI is to produce protonated molecules of the analytes. For this to happen, the analyte molecule must have a higher proton affinity (PA) than the proton affinity of the reagent gas. The difference in the proton affinity of the reagent gas and the analyte determines the amount of energy that the protonated molecule has. This internal energy of the protonated molecule is what determines the degree of fragmentation exhibited by the protonated molecule. Protonated molecules produced by the CH_5^+ reagent ion will exhibit more fragmentation than protonated molecules produced by NH_4^+ reagent ions because the proton affinity of methane is lower than the proton affinity of ammonia, which means that the difference in the proton affinity between an analyte molecule (which must be higher than that of the reagent) and methane is greater than the difference between the proton affinity of ammonia and the analyte molecule.

Table 4.2 lists some of the compounds used as reagent gases and their proton affinities. In addition to obtaining molecular weight information, CI is used to improve limits of quantitation by selecting a reagent gas that has a good proton affinity match for analyte but that will not result in ions from the matrix.

When used with the QMF, TOF, double-focusing or external ionization QIT mass spectrometer, CI is a much dirtier ionization technique than

Table 4.2 Characteristics of reagent gases for CI

Reagent gas	Predominant reactant ions	Proton affinity[a] (kJ mol^{-1}) of reagent gas[a]	Hydride affinity (kJ mol^{-1}) of reagent ion
He/H_2	HeH^+	176	–
H_2	H_3^+	424	1,256
CH_4	CH_5^+	551	1,126
	$C_2H_5^+$	666[b]	1,135
H_2O	H_3O^+	697	–
$CH_3CH_2CH_3$	$C_3H_7^+$	762[c]	1,130
CH_3OH	$CH_3OH_2^+$	762[d]	–
$(CH_3)_3CH$ (isobutane)	$C_4H_9^+$	821[e]	1,114
NH_3	NH_4^+, $(NH_3)_2H^+$, $(NH_3)_3H^+$	854	–
$(CH_3)_2NH$	$(CH_3)_2NH_2^+$, $(CH_3)_2H^+$, $C_3H_8N^+$	921	–
$(CH_3)_3N$	$(CH_3)_3NH^+$	943	–

[a] Lias SG, Bartmess JE, Liebman JF, Holmes JL, Levin RD, and Mallard WG. *J Phys Chem Ref Data.* 1988; Suppl. 1:17. All values converted from kcal to kJ.
[b] Of ethene.
[c] Of propene.
[d] From reference [1].
[e] Proton affinity of isobutene, which is the conjugate base of isobutane.

EI or FI. The high concentration of reagent gas contributes significantly to ion-source and ion-optics fouling. This means that cleaning of ion source/ion optics must be performed more frequently. Isobutane is one of dirtiest reagent gases. The use of this gas can produce a coating of polyisobutylene on all metal surfaces in the ion source and ion optics. This coating cannot be removed with common solvents, meaning that these surfaces must be cleaned by abrasion. Methane and ammonia are much cleaner reagent gases. Ammonia can cause a significant odor problem; however, 10% ammonia in methane will give the same results and does not produce the same unpleasant environment. The contamination resulting from a reagent gas is not experienced when CI is carried out using the internal ionization QIT mass spectrometer. This is because a time domain is used where the partial pressure of the reagent gas is about 10^{-3} Pa. The most important reference to understanding CI is the book *Chemical Ionization Mass Spectrometry*, 2nd ed., by Alex G. Harrison, Department of Chemistry, University of Toronto (CRC Press: Boca Raton, FL, 1992). The first edition appeared in 1983 and went through five printings. Even though this book is now more than 15 years old, the information it contains is invaluable for those wanting to do CI, especially in GC/MS.

Whereas EI uses an open source and CI uses a closed source, a physical source change is required to switch between the two techniques unless the laboratory has an instrument dedicated to each ionization method. In some cases, an actual source change where the vacuum on the ion-source housing has to be broken is not required; a probe is used to change the ion volume from that which is suitable for EI to that which is suitable for CI. Some manufacturers provide a combination EI/CI source; however, the performance is less than that of dedicated EI and CI sources also provide. Agilent has published an applications note on how to tune an instrument to obtain EI spectra when using their CI source. LECO provides a combination EI/CI source with their instrument, which they claim the same performance in EI as that which is obtained with the EI-only source. The internal ionization QIT does not require any hardware modifications to go from EI to CI because of the use of a time domain for CI. Requirements of a source change to switch between ionization methods will require downtime for the instruments. Even though instruments can easily be turned off and pumped down very fast using turbo pumps, it takes time to reach the desired pumping conditions due to the absorption of water on metal surfaces when the instrument was opened to the atmosphere.

4.4.3. Electron Capture Negative Ionization

ECNI was developed in the mid-1960s by George Stafford and Don Hunt [8]. The technique was popularized in environmental chemistry by Ron Hites [9].

ECNI is the least understood of the three main ionization techniques used by GC/MS practitioners. Ionization takes place when an electrophilic analyte captures a thermal-energy electron (0.1 to ~10 eV). This results in the formation of a negative-charge molecular ion ($M^{-\bullet}$). The source of these thermal electrons is the formation of methane molecular ions by EI in the same ion source and under the same pressure conditions as used for methane CI. This is one of the reasons that this technique is erroneously referred to as *negative chemical ion* (NCI). Even instrument manufacturers make this mistake. CI (regardless of the sign of the charge of the reagent ion or the product) is an ion/molecule reaction. ECNI is a reaction occurring between an electron and a gas-phase analyte molecule. The facts are that both EI and ECNI involve the gas-phase analyte and an electron; however, the energy of the electron is very different in EI and ECNI. CI (using methane as a reagent gas) and ECNI both involve the EI of methane present in a closed ion source at 1–100 Pa; however, methane CI uses the positive ions CH_5^+, $C_2H_5^+$, and $C_3H_5^+$ (reagent ions) to produce positive ions of the analyte, and ECNI uses the low-energy electrons expelled from the methane molecules in the formation of methane molecular ions to form molecular ions of the analyte that have a negative charge. The charge on the ion source is set to favor the existence of the electrons and neutralize the methane molecular ions in ECNI, whereas the ion source is set to favor the existence of positive-charge reagent ions in methane CI. Table 4.3 is a list of the various reactions that can occur in ECNI. The one most often encountered is the formation of the $M^{-\bullet}$.

ECNI is not available on the internal ionization QIT mass spectrometer. It would be necessary to store electrons like reagent ions are stored when using the QIT mass spectrometer for CI. The mass of an electron is so small as to make its storage impractical. The QIT fitted with an external ion source has been used for ECNI; however, these instruments do not exhibit the same performance as the QMF, TOF, or double-focusing mass spectrometers. Ion-source and ion-optics cleaning considerations are the same with

Table 4.3 Electron capture/negative-ion reactions

Resonance electron capture
$AB + e^-$ (~0.1 eV) $\rightarrow AB^{-\bullet}$
Dissociative electron capture
$AB + e^-$ (0–15 eV) $\rightarrow A^\bullet + B^-$
Ion-pair formation
$AB + e^-$ (>10 eV) $\rightarrow A^- + B^+ + e^-$
Negative ion/molecule reaction
$AB + C^- \rightarrow ABC^-$ or $(AB-H)^- + HC$

ECNI as they are with methane CI. Ion-source and ion-optic fouling from condensation of nonvolatile components in the eluate coming off the GC column are more deleterious to ECNI than they are to either EI or CI.

4.4.4. Field Ionization

FI is available on the Waters triple quadrupole GC/MS/MS (Quattro micro-GC) instrument and their GC TOF (GCT Premier) instruments and on the JEOL AccuTOF™ GC, a TOF mass spectrometer. FI of gas-phase analytes was an outgrowth of the development and study of field desorption, the first of the desorption/ionization techniques for nonvolatile and thermally labile analytes. FI is a technique rivaling CI, which results in a high abundance of ions representing the intact molecule. Unlike CI, which produces even-electron protonated molecules (MH^+), FI produces odd-electron molecular ions ($M^{+\bullet}$). In FI, ionization of an analyte molecule in the gas phase takes place in an electric field (107–108 V cm^{-1}) maintained between two sharp points or edges of two electrodes. The technique of FI was developed by H. D. Beckey, Institut für Physikalische Chemie der Universität Bonn (Bonn, Germany) in 1957 [10].

Like ECNI, this technique requires the use of an ion source separated from the *m/z* analyzer and is therefore not applicable to the internal ionization QIT. The technique is not as dirty as that of CI and maintenance requirements are more like that of EI. The products offered both by JEOL and Waters require that the ion source be changed to switch between EI or CI and FI.

4.4.5. Atmospheric Pressure Chemical Ionization

Although currently not commercially available, the technique of APCI should be considered as a viable alternative soft ionization technique for GC/MS. The ionization occurs outside of the vacuum of the mass spectrometer and therefore has less chance of fouling the ion optics just before the *m/z* analyzer. Used with GC/MS, this technique was developed in the mid-1970s by Evan Horning's group at the Baylor College of Medicine (Houston, TX). (This work was pioneered by Evan Horning's group at Baylor University, Houston, Texas [11–16]). Horning called the technique API. Today, the term API is used for any of the ionization techniques that occur at atmospheric pressure (electrospray ionization, atmospheric pressure photoionization, and APCI). APCI is the only API technique that is applicable to GC/MS. Ionized air (N_2, O_2, and H_2O) undergo ion/molecule reactions with gas-phase analytes at atmospheric pressure to produce protonated molecules, deprotonated molecules (negative-charge even-electron ions), and $[M-H]^+$ formed by hydride abstraction and various adduct ions, especially in the presence of a dopant. ECNI can also take place in an APCI system. As stated above, APCI is considered to be the mass spectrometry technique that will produce the lowest detection limit for the widest variety of analytes.

4.5. *m/z* Analyzer Types

The four types of *m/z* analyzers used in GC/MS are the QMF (most widely used), QIT, TOF (rapidly gaining in popularity), and the double-focusing mass spectrometer (rarely used in most laboratories today). The general operating principles and reasons that one might be chosen over another for GC/MS are described below. The *m/z* analyzer is unique with respect to other types of mass spectrometers that are generally calibrated at the time of manufacture. The *m/z* scale of the mass spectrometer must be calibrated on a regular basis. A compound that produces ions of known *m/z* values and abundances is used for this purpose. For work involving ions that are only of integer *m/z* values, this calibration can be performed once a day or less (should be checked every eight continuous hours of operation). When measurements are being made to a fraction of an *m/z* unit, it may be necessary to obtain data for a mass spectral peak representing an ion of a known elemental composition whose integer *m/z* value is close to that of the analyte being measured. It may also be necessary to obtain such data at the same time the data are obtained for the ion being measured.

4.5.1. Transmission Quadrupole GC-MS (a.k.a. Quadrupole Mass Filter GC-MS)

Quadrupole technology (which includes the transmission quadrupole and the QIT) was first explored by Wolfgang Paul and colleagues at the University of Bonn (Bonn, Germany) in the early 1950s [17]. The initial paper on the quadrupole technology was followed by a detailed description of the theory [18]. Paul shared half of the 1989 Nobel Prize in physics with the German-born American physicist, Hans Georg Dehmelt, the developer of the Penning (magnetic) trap, "for the developments of ion-trap techniques." The QMF was an outgrowth of his study of quadrupole fields. The seminal reference on the quadrupole technology is Peter Dawson's *Quadrupole Mass Spectrometry and its Applications* (Elsevier, Amsterdam, 1976; reprinted by the American Institute of Physics, Woodbury, NY, 1995). Almost from the beginning of its commercialization in the late 1960s, the QMF became the instrument of choice for GC/MS. The QMF instruments of that era were capable of faster spectral acquisition than the magnetic instruments. The QMF had constant resolution throughout their *m/z* scale, which gave it an advantage over the TOF and magnetic instruments of that time; and the QMF was very inexpensive compared to the magnetic-based instrument. The linearity of the *m/z* scale made them ideal for computerization using the newly developed minicomputers of the day. Once the QMF took control of the GC/MS market, it held on tightly. Today, in addition to both Agilent and Thermo still being major suppliers of the instrument, laboratory QMF GC/MS instruments are also available from Perkin Elmer, Varian, Shimadzu, and JEOL (in Japan and countries other than the United States). There is also a number of field-portable and process-control GC/MS instruments based on the QMF. In addition to the availability of new instruments, there are a large number of previously owned instruments that are available from several third-party suppliers, especially various models of the Agilent QMF GC-MS. The initial price of a new QMF GC-MS is low, usually less than \$100K, and the price for refurbished instruments is even lower.

The QMF uses dc (direct current) and ac (alternating current) in the form of an rf (radio frequency) electric field to separate ions according to their *m/z* values from a beam of ions of all *m/z* values that is continuously flowing from the ion source to the ion detector. The QMF, illustrated in Figure 4.13, is composed of four electrical poles (conducting surfaces).

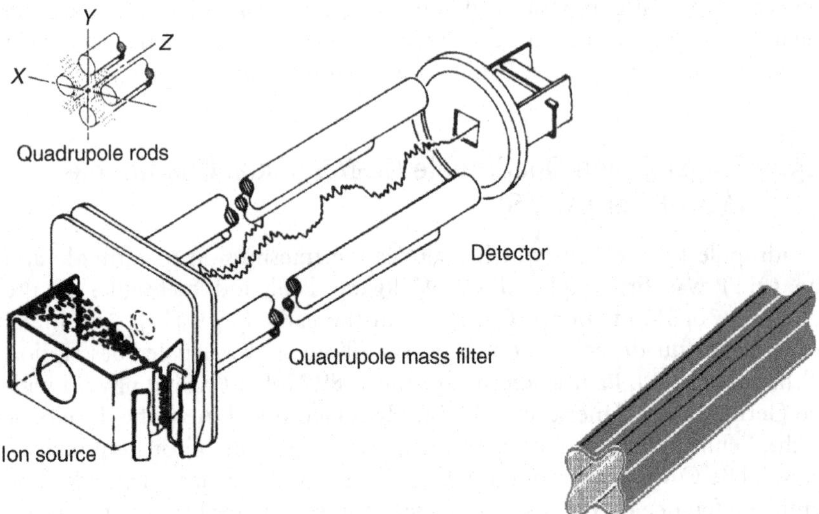

Figure 4.13 There have been almost as many geometries of the quadrupole mass filter as there have been manufacturers. When the quadrupole is envisioned, it is imaged as the positioning of four round rods as shown in the upper left. In addition to this geometry, poles machined to have a hyperbolic shape have been used as well as the quartz mandrel with the deposition of a conductive material such as gold, as shown on the lower right.

These poles are arranged so that two are across from each other on an x axis, and two are across from each other on a y axis. The area between the two sets of poles can be described as an x–y plot of two hyperbolic functions that are orthogonal to one another. By applying a positive dc potential to one set of poles and a negative dc potential to the other set, while at the same time applying the fixed-frequency rf voltage to all the poles, an alternating field that will push and pull ions as they travel along z axis of the device is created. Assuming a specific ratio of the rf amplitude relative to the dc amplitude, ions of only one m/z value will remain in the ion beam. Ions of other m/z values will be pushed or pulled to an extent that they will be filtered out of the beam. A mass spectrum is obtained by holding the dc:rf amplitude ratio constant and increasing or decreasing the two amplitudes together. The required rf is a function of the size of the poles, and their distance from one another is determined by the design of the instrument. The operator has no control over this parameter; and if the frequency of the rf generator changes, the instrument will no longer function. The dc:rf amplitude ratio is an operator-settable parameter and is set when the instrument is tuned for various functions.

The dc:rf amplitudes are changed digitally. In a scan of the m/z scale, the steps are usually 1/20 of an m/z unit. This exceeds the minimum differences in m/z values that can be separated using quadrupole technology. The resolution (see above for the definition of resolution) of the QMF is 0.3 m/z units. This means that ions that differ by less than 0.3 m/z units cannot be separated from one another using a QMF. A specific instrument may report m/z values to the nearest 0.05 m/z unit, but this is not an accurate mass to that value. It is only the 1/20 step within an m/z-unit interval where the signal reached a maximum. In some cases, in order to reduce the spectral acquisition time, instead of 20 steps per m/z unit, the acquisition algorithm shifts the m/z axis abruptly from one m/z unit to the next. Even with the use of this algorithm for data acquisition, only $1/x$ (where x is the number of integer m/z units in the acquisition range) of the ions of any individual m/z value is measured during a single spectral acquisition. The ion currents for the rest of the ions of that m/z value are ignored because they are not being measured. For instruments where a continuous beam of ions is being separated (the QMF and magnetic-sector-based instruments), the scan rate is very significant for establishing the detection limit and for problems of spectral skewing (the spectrum exhibits different intensities for individual m/z values because of changes in analyte concentration in the ion source during the acquisition of a single spectrum); the more narrow the chromatographic peaks, the more important the scan rate. More on this subject appears in the section on Selected Ion Monitoring vs Full-Spectrum Acquisition (Section 4.9).

QMF GC/MS instruments have lower acquisition limits of m/z 1 to m/z 10 and upper limits of m/z 650 to m/z 1,200.

4.5.2. Quadrupole Ion Trap GC-MS

Although QIT technologies were developed at the same time as the technology needed to bring about the QMF GC-MS, it took nearly another 20 years to develop the QIT GC-MS. The QIT GC-MS was developed by Finnigan Corp. (now Thermo Scientific of Thermo Fisher) who discovered that in order to store ions of different *m/z* values in concentric three-dimensional (3D) orbitals, it was necessary to collisionally cool the ions. Until this time, the paradigm for mass spectrometry had always been "the lower the pressure, the better the performance" with respect to separation of ions of different *m/z* values. Serendipitously, George Stafford and coworkers at Finnigan found that by using a low-molecular-weight buffer gas such as helium to maintain the trap at about 10^{-1} Pa produced excellent unit resolution and improved sensitivity [19]. This late 1982 discovery coupled with Stafford's earlier announcement of the mass-selective instability scan allowed the commercialization of the internal ionization QIT GC-MS. Ray March, in the preface to *Particle Aspects of Ion Trap Mass Spectrometry, Vol. 1: Fundamentals of Ion Trap Mass Spectrometry* (CRC, Boca Raton, FL, 1995), states that Stafford's mass-selective instability scan is as significant as Paul's work in the development of the trap. After Finnigan developed the second-generation QIT, which addressed a number of problems associated with the first instruments, the technology for use in GC/MS was licensed to Varian who, after selling the Finnigan second-generation internal ionization QIT GC-MS and through manufacturing refinements along with their own innovative technology, went on to develop a very competitive product. Thermo Scientific has now abandoned the internal ionization QIT GC-MS and only markets an external ionization product, leaving Varian as the only commercial manufacturer of the internal ionization QIT GC-MS. One of the significant advantages of the QIT GC-MS is that it is capable of using a temporal domain to perform MS/MS, reducing greatly the cost of the equipment needed for such an analysis.

Both the internal and the external ionization QIT are schematically illustrated in Figure 4.14. In the QIT GC-MS, ions are produced externally in a conventional ion source or internally by EI or CI. In the external ionization QIT, ions flow into the trap for a fixed period. In the internal ionization QIT, ions are formed in the trap for a fixed time. This fixed time for both systems is called the *ion time*. The ion time for ions formed in or that flow into the trap is controlled by an algorithm that is based on the concentration of analyte present called *automatic gain control* (AGC). As ions are formed in (or enter) the trap, they are stored in 3D concentric orbitals according to their individual *m/z* value with ions of the highest *m/z* values being closest to the center. The appearance of this "ball" of ions is analogous to the layers of an onion. The lowest *m/z* value stored is determined by the amplitude of a fixed frequency rf applied to the ring electrode while holding

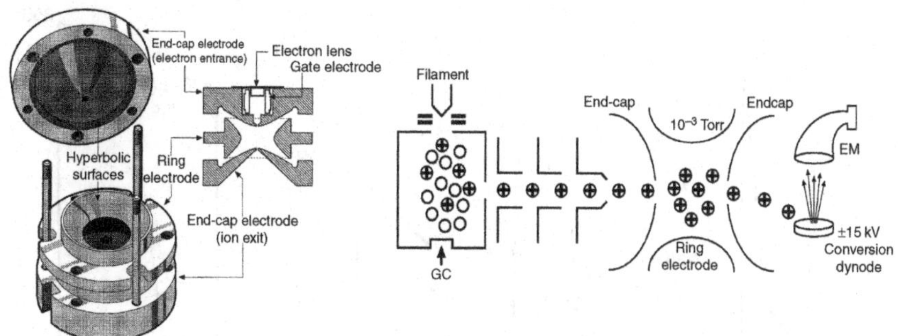

Figure 4.14 Internal ionization (left) and external ionization quadrupole ion trap as used in commercial GC/MS instruments.

the two electrically isolated end-cap electrodes at ground. The mass spectrum is then obtained by increasing the amplitude of the rf on the ring electrode, which results in ions of individual *m/z* values being destabilized one *m/z* value at a time. As ions become unstable, they are ejected towards the two end caps. Those ions that reach the detector end cap will exit and impact a high-energy dynode where they are accelerated into the ion detector.

The ions of all *m/z* values are scanned out of the trap at the maximum possible speed (usually at 5,600 *m/z* units per second). This pulsed process of obtaining a mass spectrum means that all of the ions of all *m/z* values formed during a spectral acquisition period are detected; therefore, there is no spectral skewing problem associated with the QIT as there is with the QMF. With all ions of all *m/z* values being detected, the detection limit for a full-spectrum acquisition is much lower for the QIT GC-MS than it is for the QMF GC-MS. Data acquisition times are slightly shorter than those experienced with the QMF. This increase in spectral acquisition speed allows two or more spectra to be averaged and saved as the spectrum to be displayed. This averaging of spectra can produce higher quality spectra for matching against databases and for use in data interpretation.

The *m/z* range of ions analyzed affects how many spectra can be obtained in a single time interval (sometimes referred to as a *duty cycle*). When the maximum *m/z* value of a desired range is reached, all the remaining ions are dumped from the trap and the process of acquiring the next spectrum is begun. Some instruments have a single-segment spectral function for the entire range of acquired *m/z* units; other instruments have a segmented spectral acquisition function as illustrated in Figure 4.15. Only the segments that contain the lower and upper *m/z* limits of the acquired data are involved in a single spectral acquisition. Segments that do not contain ions in the acquisition range are not used. A separate ionization takes place for each

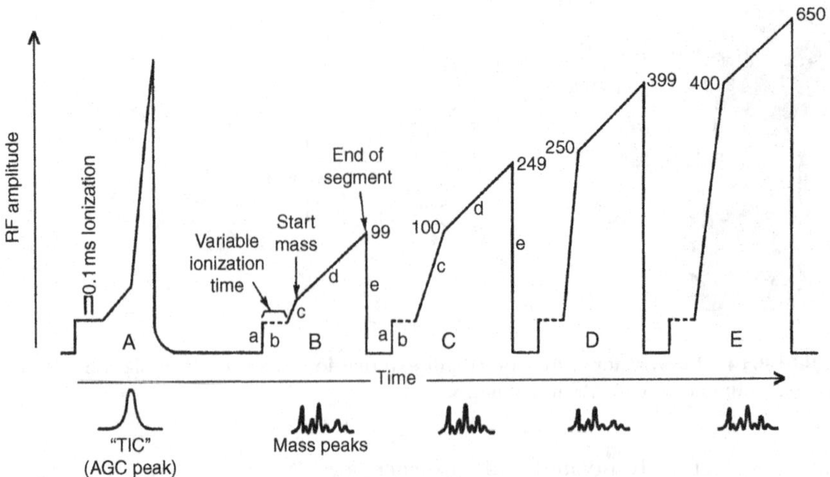

Figure 4.15 Illustration of multiple scan segments used in some commercial QIT GC/MS instruments. The 0.1-ms ionization at the beginning of the scan function is part of what is called *automatic gain control* (AGC). This determines the ionization time used in each of the up to four segments that can be used to acquire a spectrum.

segment of acquisition. Ionization times are controlled by the concentration of analyte present therefore avoiding the possibility of spectral skewing. If less time is required to obtain a spectrum than is set by the operator, multiple spectra are acquired, averaged together, and stored as the *analytical spectrum*.

Even though the ions are sequentially scanned out of the trap one *m/z* unit at a time, the spectral acquisition should be envisioned as a pulsed process rather than a scan process.

A significant disadvantage of the QIT is that only a maximum number of ions can be stored without loss of separation into individual orbitals. When this maximum number of ions is exceeded, there will be a *space-charge* effect, which results in unmatchable and uninterpretable spectra. If the number of ions formed during minimum ion time (10 μsec) exceeds the maximum storage of the trap, space charge will occur. Therefore, it is necessary to control the amount of sample that enters the ion source or the trap. It is easy to overload the QIT GC-MS, and adjustments in analytical procedures used with the QMF GC-MS may have to be made. Part of the acquisition faction is to perform a microscan to determine the amount of ionizable material that is in the trap to avoid overloading the trap when the GC analyte amount is high and to maximize the signal when the GC analyte is low. This process is called AGC. The ionization is experimental set to between 10 μsec and 25,000 μsec. The AGC limits the lower *m/z* value of an acquisition to 20. Although most QIT GC/MS systems can acquire data from *m/z* 10, AGC only functions when the starting *m/z* value is 20 or above.

This maximum number of ions allowed in the trap can also have an impact on quantitation. With respect to quantitation, the linear dynamic range of the QIT GC-MS is less than that obtainable with the QMF system. This number has been increased since the first QIT GC/MS instruments were introduced by the implementation of a feature called *axial modulation* (AM). When AM is being used, an rf potential is applied to the end caps that is half the frequency applied to the ring electrode and is the same amplitude as that applied to the ring electrode. AM increases the number of ions that can be accommodated in the finite volume of the ion trap.

In order to obtain the best limits of detection, the software with the ion trap will set an ion storage time that allows for the optimum number of ions to give the lowest limit of detection. This means that ions can be in the trap for up to 25,000 μsec before they are detected as compared to 10–15 μsec between formation and detection experienced in other types of mass spectrometers used in GC/MS. Not only is there the problem that ions have a finite stability and can over time decompose resulting in spectral differences between the QIT and other instruments, but the ions are also banging into He atoms as they continually move in their orbitals. Remember, the QIT uses He as a buffer gas and operates at a pressure of $\sim 10^{-1}$ Pa, which is at least two orders of magnitude higher than the operating pressure of other instrument types. Both time and potential collisional activation of ions can lead to spectral differences.

Although not thoroughly understood, ions that are believed to be products of ion/molecule reactions between analyte fragment ions or even analyte molecular ions and neutral analyte molecules are sometimes observed in mass spectra obtained with the QIT. These products usually manifest themselves in abnormally high $[M+1]^+$ peak intensities and reduced intensities at *m/z* values representing the fragment ion believed to react with the analyte molecules. These fragment ions that react with analyte molecules are sometimes odd-electron ions that are very acidic; however, even-electron ions can also be significant proton donors. This ion/molecule phenomenon is primarily seen with acids, aldehydes, ketones, alcohols, and amines. The phenomenon has been reported for both internal and external ionization QIT instruments, although it appears to be much more pronounced with the internal ionization instruments. This phenomenon in the internal QIT can be somewhat explained by why the instrument is so good for CI. Making sure that only low amounts of analyte are used will help to minimize the production of these protonated molecules of the analyte.

These ion/molecule reactions can be a benefit in cases where spectra do not exhibit much of a molecular ion peak; however, these ion/molecule reactions can not only cause problems with appearance of the spectrum in the molecular ion peak region, but they can also result in completely different fragmentation patterns. Figure 4.16 shows the spectra of amphetamine and a derivative of amphetamine prepared using carbon disulfide that were believed to have been obtained from a GC/MS analysis using a QIT (bottom two spectra). Based on assumptions made from information provided with the source of these two spectra, the amount injected onto the GC column was between 20 ng and 100 ng. The top two spectra in Figure 4.16 are spectra for these same compounds taken from the *NIST08 Mass Spectral Database*.

The lack of a discernable molecular ion peak in the *NIST08* mass spectrum of amphetamine is consistent with what is known about the mass spectra of aliphatic primary amines. The nominal mass of amphetamine ($C_9H_{13}N$) is 135 Da, consistent with the *Nitrogen Rule* (see Chapter 5). The appearance of a peak one *m/z* unit higher than that predicted for the

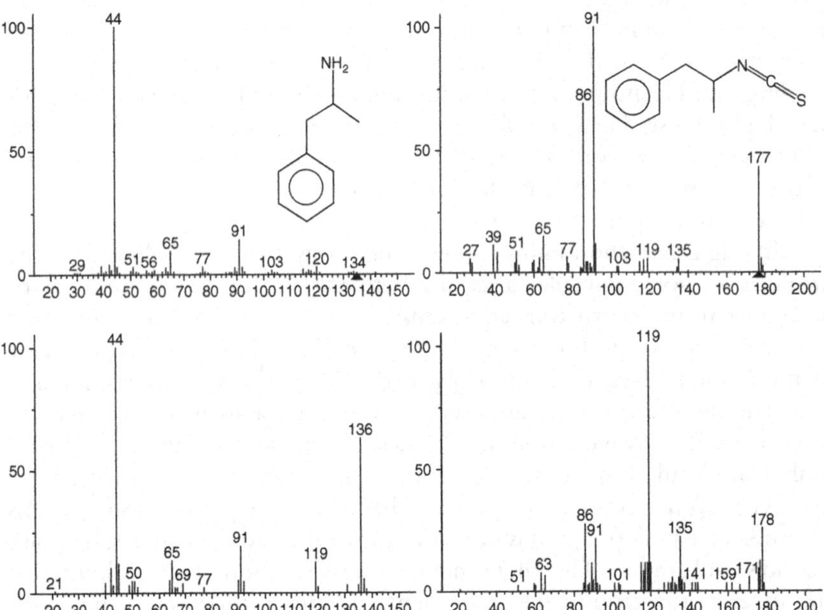

Figure 4.16 (Top) EI mass spectra of amphetamine (upper left) and a derivative of amphetamine (upper right) prepared using carbon disulfide as they appear in the *NIST08 Mass Spectral Database*. (Bottom) Mass spectra reported to be of the same two compounds (amphetamine, lower left; derivative, lower right) from the amphetamine chapter of *The Analysis of Controlled Substances* [29]. Based on anecdotal evidence gained by a careful reading of the entire book, it appears that these spectra were acquired using an internal ionization QIT mass spectrometer.

molecular ion (*m/z* 136 instead of *m/z* 135) in the mass spectrum of amphetamine obtained on the QIT mass spectrum is more than obvious. This peak's intensity is more than 50% of the intensity of the base peak. This peak probably represents protonated molecules of the amphetamine brought about by ion/molecule reactions between amphetamine molecules and methyl-substituted immonium ions formed by the fragmentation of amphetamine molecular ions. In this particular example, the proton donor is not an odd-electron ion, but an even-electron ion ($H_3C–HC=N^+H_2$, *m/z* 44).

The differences in the two mass spectra of amphetamine are somewhat defensible. There are 247 compounds in the *NIST08 Mass Spectral Database* that have a base peak at *m/z* 44, no molecular ion peak, and containing only a single atom of nitrogen and that contain only C, H, and N or C, H, N, and O. Being able to have an indicator of such a compound's molecular weight could be very beneficial to confirmation of an identification. On the other hand, the explanation of the differences between the two spectra of the carbon disulfide derivative would be much more challenging. Peaks with the same *m/z* values are present in both spectra, with the exception of the peak representing the intact molecule. The *NIST08* spectrum exhibits a molecular ion peak at *m/z* 177; again, consistent with the *Nitrogen Rule* because the analyte's elemental composition is $C_{10}H_{11}NS$—a single atom of nitrogen meaning an odd nominal mass. A peak is observable at *m/z* 177 in the mass spectrum of the derivative obtained with the QIT; however, there is a more intense peak at *m/z* 178 (probably representing the protonated analyte, which is obviously absent in the *NIST08* spectrum. The base peak in the *NIST08* spectrum is at *m/z* 91, with the peak at *m/z* 86 having a relative intensity of about 75%. The peak at *m/z* 119 has a relative intensity of <10%. In the spectrum obtained using the QIT, the base peak is at *m/z* 119; and the peaks at *m/z* 86 and 91 have relative intensities of ~25%, with the *m/z* 86 peak being somewhat more intense than the *m/z* 91 peak. Defense of such difference in the data obtained for the same compound from two different types of instruments would be a nightmare in a court of law.

In addition to limiting the amount of analyte used,* manipulation of the acquisition parameters can be made to not store the ions responsible for protonation of the analyte during the ionization period for the segment in which the data for the molecular ion is being acquired; however, this will increase the limit of detection. Refer to the manufacturer's instrumental manual for details on this technique.

* A good rule of thumb for the QIT mass spectrometer is to reduce the amount of analyte by an order of magnitude compared to what would normally be used for the QMF mass spectrometer.

Authors' Note: *With this somewhat detailed discussion of the potential problems with the QIT GC-MS, some may be dissuaded with respect to its use. It is not that these instruments should not be used because of their potential challenges; but rather that they have, perhaps, a few more considerations when they are used than do other instruments. GC/MS is not a black-box technique where a sample is introduced and results appear on the screen. Samples, usually in solution, are injected into a hot region where they are volatilized and reach temperatures far above their boiling point. This sudden conversion from the condensed state to the vapor phase in an area of flowing gas results in subjecting the mixture of analytes and solvent in the vapor phase to not only elevated temperature but also to pressure. Under these conditions, chemical reactions can and do occur.*

As an example of the types of problems that can come about from such reactions, a splitless injection of an aliphatic amine with an aromatic moiety in a methanol solution was made into a QMF GC-MS with an injection port temperature of 290 °C. The result was a very symmetrical reconstructed total ion current (RTIC) chromatographic peak made up of a series of identical spectra that looked nothing like the spectrum that should have been produced by the analyte. It turned out, after some investigation, that the supplier had not provided the wrong material; but the conditions used had caused the analyte to react with the methanol in the injection port. In another case, a fourth component was appearing in the chromatogram of a three-component mixture; again, after some investigation, it turned out that one of the analytes was reacting with the solvent that was chosen for the analysis before the mixture was injected. The point of these two examples is to emphasize the fact that when doing GC/MS, chemistry is being done. All possibilities exist, and it is important to be aware of the results and their meanings.

The QIT GC-MS continues to gain market position. It still lacks the success of the QMF GC-MS; but because of the lack of spectral skewing, the low detection limits for the full-spectra acquisitions, and the ability to perform MS/MS analyses (and the convenience of performing CI when using the internal ionization instrument) in a low-cost instrument, it continues to gain popularity. The QIT GC-MS, like the QMF GC-MS, is available at very low initial cost. The availability of previously owned instruments is not as great as with the QMF GC-MS, but that market is also growing. One of the reasons that the QIT GC-MS is not more popular is the limited number of manufacturers due to patent considerations. Thermo Scientific and Varian are the only manufacturers and marketers of an external ionization QIT GC-MS in the United States. Neither company's instruments are marketed by any other company either as an Original Equipment Manager (OEM) or branded product. Varian is the only company that is currently manufacturing and marketing the internal ionization QIT GC-MS in the United States, and it also has no OEMs.

Like the QMF mass spectrometers, the QIT instruments have a maximum resolution capability of about 0.3 *m/z* units. The QIT instruments, like the QMF mass spectrometers, often report data to the nearest 0.05 *m/z* units; but these values are meaningless.

Another factor that must be taken into consideration about the QIT GC-MS is that because of the necessity to determine the number of ions that will be produced based on the amount of sample present (the feature called AGC), there is a lower limit of *m/z* 20 for spectral acquisitions. Instrument designs prevent acquisitions below *m/z* 10 even when AGC is not used. The upper acquisition limit is similar to that for the QMF; i.e., *m/z* 65 to *m/z* 1,000.

4.5.3. Time-of-Flight GC-MS

There are two concepts associated with the currently marketed TOF GC-MS—rapid spectral acquisition and accurate mass measurement—although the accurate mass measurement instruments are capable of acquiring spectra at 25–50 Hz. The TOF mass spectrometer was first commercialized in the 1950s by the Bendix Aviation Corporation. Several notable names in the history of mass spectrometry had published sparingly on the TOF concept from as early as 1932. The seminal paper on TOF mass spectrometry *Time-of-Flight Mass Spectrometer with Improved Resolution* was published in 1955 by two researchers, William C. Wiley and Ian H. McLaren [20], working for Bendix Aviation Corporation Research Laboratories in Detroit, MI. This paper provides a detailed treatment of ion focusing in the TOF mass spectrometer. The instruments that became very popular in the 1950s lasting until the development of the QMF GC-MS are now called *linear TOF* instruments. Their operation is very simple. Ions are formed in an ion source. An acceleration pulse on the order of several hundred to a few thousand volts pushes a packet of ions of all *m/z* values into a field-free region called a *drift tube* or *drift region*. All of the ions of different *m/z* values have the same kinetic energy; therefore, ions of different *m/z* values are going to have different velocities. The ions of lower *m/z* values will reach the detector before those of higher values. The mass spectrum is obtained by recording arrival times of ions at the detector. This process had two problems when first commercialized in the 1950s. The first problem was the energy dispersion of ions of the same *m/z* value. Before ions were accelerated out of the ion source, they were moving in all directions; therefore, there was a distribution of initial kinetic energies. This resulted in somewhat broad mass spectral peaks. As can be seen from Eqn. (1), the mass spectral peaks get closer together as *m/z* values increase.

$$\mathrm{TOF} = \frac{L}{v} = L\left(\frac{m}{2zeV}\right)^{1/2} \tag{1}$$

This means that ions of successively larger *m/z* values will have larger differences in *m/z* values when separated. This difference in *m/z* values that can be separated is called *resolution*. The resolution in the data from a TOF mass spectrometer becomes poorer as the *m/z* value increases.

The second problem had to do with the speed and stability of electronics available at that time. The only way to observe a high-speed analog event was to use an oscilloscope. The only way to record the images from an oscilloscope was through the use of a Polaroid camera. This was not a practical way to record GC/MS data. If spectra were output to an analog recorder, then a separate pulse of ions would have to be accelerated for each *m/z* value in the spectrum range, and the detector would be turned on at different times to record the signal for each value. These problems led to the commercial discontinuance of the TOF mass spectrometer by the early 1980s because the same advantages of a low-cost instrument could be obtained with the QMF, and the QMF did not have the disadvantages associated with the TOF mass spectrometer. During this period, GC/MS was the predominant use of the mass spectrometer.

The one big advantage that the TOF instrument had over other instruments was that there was no upper limit to the *m/z* scale. Unlike mass spectrometers based on quadrupole technology that had an *m/z* upper limit imposed by the maximum amplitude that could be obtained for an rf voltage without disintegration of the sine wave or a magnetic instrument whose upper *m/z* limit was dependent on how high of a magnetic field could be achieved, the TOF mass spectrometer was only limited by the time it took a high-mass ion to reach the detector. The development of the desorption/ionization technique MALDI, matrix-assisted laser desorption/ionization (which produced single-charge ions with *m/z* values that exceeded those obtainable by either quadrupole or magnetic technology), in the mid-to-late 1980s for mass spectral determinations of nonvolatile thermally labile substances such as peptides and proteins resulted in renewed interest in TOF mass spectrometry. At about the same time, the former Soviet Union collapsed, allowing a great deal of research that had been taking place behind the Iron Curtain on TOF mass spectrometry to be released to the Western World. Through these two events emerged the *reflectron* TOF mass spectrometer, schematically illustrated in Figure 4.17. The reflectron TOF mass spectrometer sparked resurgences in the interest for these instruments in all areas of mass spectrometry including GC/MS.

The reflectron TOF mass spectrometer uses an electric field to act as an ion mirror. As ions of a single *m/z* value enter the ion mirror, their velocity is slowed to near zero. Ions of a given mass of higher energy, thus higher velocities, will penetrate the mirror deeper than those of lower kinetic

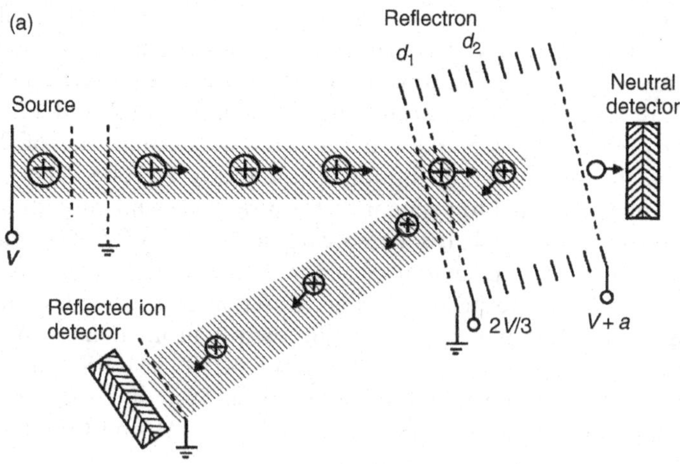

Figure 4.17 Ions of the same *m/z* value will have different velocities because of energy dispersions in the ion source prior to acceleration. Ions with higher initial energy will go into the reflectron first and stay longer. They will exit last. All the ions exit with the same velocity as they had when they entered. The faster ions will now be behind the slower ones and will catch up with the slower ions just before all the ions reach the detector. This gives the ions of that *m/z* value a narrow velocity dispersion, greatly improving the resolution.

energy. The ions are accelerated out of the mirror so that their exit velocities are the same as their entrance velocities; however, the lower velocity ions are now ahead of the higher velocity ions. The higher velocity ions will catch up with the lower velocity ions so that all of the ions of the same *m/z* value, regardless of their individual kinetic energies (velocities), reach the detector at the same time. This narrows the width of the mass spectral peak, allowing for very high mass accuracy measurements through the separation of ions of very small differences in *m/z* values.

Modern TOF mass spectrometers have also been able to take advantage of modern fast electronics and computer storage of data. This means that a complete mass spectrum can be obtained from a single pulse of ions.

Two tracks have been taken by the reflectron TOF in GC/MS. One has to do with rapid data acquisitions (500 spectra per second) at unit *m/z* value accuracy. LECO Corp. (St Joseph, MI, USA) is the only manufacturer of such an instrument, the EI-only Pegasus® HT, a floor-standing instrument. LECO also offers a unit *m/z* value accuracy benchtop EI/CI (not capable of negative-ion detection [ECNI]) instrument with an 80-Hz spectral acquisition rate, the TruTOF® HT. In order to take advantage of the rapid spectral acquisition rate to compress analysis time, LECO has a powerful chromatographic peak deconvolution feature built into their ChromaTOF® software, called *True Signal Deconvolution*®.

The other track is accurate mass measurements (±0.002 millimass units or better). There are two manufacturers of these instruments, JEOL and Waters. The JEOL instrument, the floor-standing AccuTOF™ GC with EI as a standard feature and the availability of CI and FI, has a resolving power (R) of 5,000 and a spectral acquisition rate of 25 Hz. The JEOL instrument has an optional heated direct probe. The Waters GCT Premier is a benchtop system that is provided standard with EI and has CI and FI available. It has a resolving power (R) of 7,000, claiming the same ±0.002 millimass unit accuracy as the JEOL instrument and a spectral acquisition rate of 20 Hz. The Waters system provides chromatographic peak deconvolution through their ChromaLynx® software. JEOL accomplishes the same thing using the NIST AMDIS (Automated Mass spectra Deconvolution and Identification System) to deconvolute chromatographic peaks. AMDIS is also used by QMF and QIT GC-MS manufacturers with their data systems.

All three TOF GC-MS manufacturers claim different linear dynamic ranges, which is important in quantitative applications. These instruments should be evaluated based on demonstrated performance, using samples that will be quantitatively analyzed rather than manufacturers' quoted specifications. All the TOF GC/MS systems have upper *m/z* limits of 1,000.

4.5.4. Double-Focusing GC-MS

Today's double-focusing GC/MS systems are based on the instruments first described by Joseph John Thomson [21] and Francis W. Aston [22]. The instrument's design was independently pursued in the United States by Arthur Jeffery Dempster [23]. Both of these mass spectrometers have a magnetic sector, which momentum focuses the ions, and an electric sector, which energy focuses the ions. The original instruments were configured with the electric sector preceding the magnetic sector. This configuration is called *forward geometry*. All the commercially manufactured instruments of today have a magnetic sector following the ion source and the electric sector preceding the ion detector. These instruments are described as having *reverse geometry*. Figure 4.18 is an illustration of the JEOL GCmate's ion optics showing a reverse geometry configuration with the magnetic sector preceding the electric sector. The advantage that a reverse geometry has over a forward geometry instrument is that it is easy to use the instrument for MS/MS. As ions are accelerated out of the ion source, they can enter a collision cell. Product ions of ions of all *m/z* values are formed in the collision cell. These product ions will have the same velocity as their individual precursor ions. The magnet can be set to a fixed field strength allowing ions of only a single velocity to reach the detector. The electric

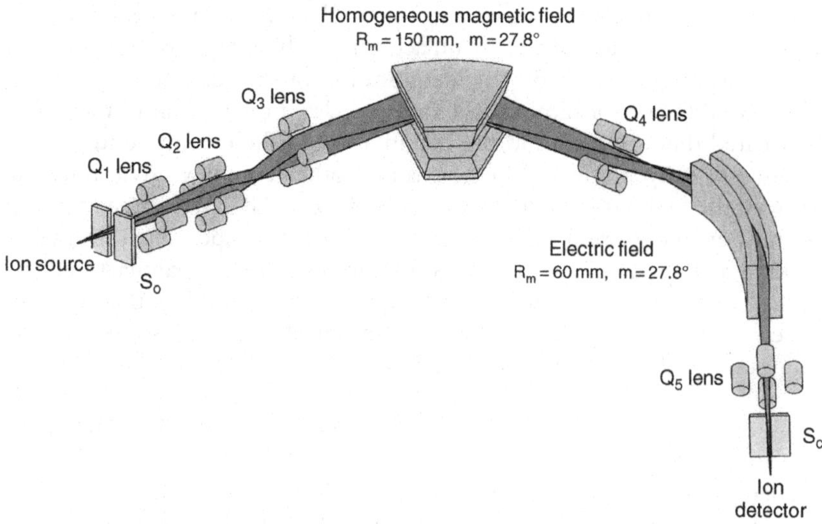

Figure 4.18 Ion optic system for a double-focusing mass spectrometer used in obtaining accurate mass measurements or acquire accurate mass SIM data in GC/MS. This is called a *reverse geometry* system because the magnetic sector precedes the electric sector. When double-focusing instruments were first developed, the electric sector preceded the magnetic sector and those instruments are said to have *forward geometry*.

sector is then scanned to obtain a product-ion spectrum or set to a specific field strength to allow only products of a specific *m/z* value to pass and reach the detector in an SRM experiment (see below under MS/MS for an explanation of SRM).

Unlike the TOF instrument, the double-focusing mass spectrometer can be operated at variable settings of resolving power. In the modern double-focusing instruments, ions are separated according to their *m/z* value in the magnetic sector based on their momentum. Ions of individual momentum will enter the electric sector. The electric sector can be tuned to limit the energy range of ions that pass to the detector. This process is known as setting the slit-widths. The narrower this setting, the higher the resolving power (R) of the instrument. This means that the differences in the *m/z* values that can be separated becomes smaller and smaller as the resolving power is increased. The problem is that the number of ions of a given *m/z* value also becomes smaller, resulting in an increase in the detection limit. When the detection limit is the driving force of the analysis, the instrument will be operated at a low resolving power. When mass accuracy and ion separation are the driving forces for the analysis, the instrument is operated at a high resolving power.

In addition to the mainstream mass spectrometry manufacturers, there are a couple of companies in Europe that make only a single model high resolving power magnetic-sector instrument. JEOL, Thermo Scientific, and Waters Corp. are the primary manufacturers of these types of instruments. Because of the requirements for identification of various isomers of polychlorinated dibenzo-*p*-dioxins and a related class of compounds, the polychlorinated dibenzfurans, and the quantitation of these compounds in the presence of polychlorinated biphenyls and other complex environmental matrices, the high resolving power magnetic-sector GC-MS operated in the selected-ion-monitoring (SIM) mode is still a very important instrument. Looking at the partial specifications for the instruments manufactured by JEOL, Thermo Scientific, and Waters Corp. primarily for dioxin analyses, it is seen that at least two of them use a specific dioxin isomer as the compound to demonstrate detection limits.

> Waters AutoSpec Premier R = 80,000 transmission at R = 10,000 relative to R = 1,000 is 10%.
> Thermo Scientific DFS High Resolution GC/MS EI/CI/ECNI 5 kV accelerating voltage optimization, R > 60,000 (10% valley), Scan rates 0.1–10,000 seconds per decade (continuously variable), Mass accuracy <2 ppm, Sensitivity EI GC/MS S/N >800:1 for 100 fg 2,3,7,8-TCDD (2,3,7,8-tetrachlorodibenzo-*p*-dioxins) at nominal *m/z* 322, R = 10,000, *m/z* range 2–6,000; *m/z* 2–1,200 at full accelerating voltage.
> JEOL MStation™ R = 60,000 at 10% valley, *m/z* range to 2,400 at 10 kV, Sensitivity 200 pg methyl stearate in EI gives a molecular ion peak with an intensity equal to 400:1 S/N or better; scan speed 0.1 second per decade.

A good example of what is involved with a dioxin analysis is that a resolving power of 13,000 ($\Delta m = \sim 0.025$ Da) is required for the separation of the $[M-Cl_2]^+$ ion of hexachlorobiphenyl from the ^{37}Cl ion of the molecular ion of 2,3,7,8-TCDD. These types of analyses are usually done using SIM (see below) and very high resolution GC.

In addition to the MStation, JEOL has a benchtop GC-MS with a resolving power of 5,000 (10% valley definition). Unlike the instruments used for dioxin analyses, which are priced in the hundreds of thousands of dollars, the GCmate is priced at about $125K, which includes EI, CI, and a direct inlet probe.

4.6. Ion Detection

Once the ions exit the *m/z* analyzer, they strike a detector which generates a cascade of electrons or photons to produce a signal. The intensity of this signal is directly proportional to the abundance of the ions with a specific *m/z* value. These signals are produced by ions striking a surface to initiate the first release of electrons or photons. This initial release is a function of the ions' kinetic energy ($E_k = mv^2/2$). In the case of the QMF mass spectrometer, this kinetic energy is ~15 eV or less, which results in a wide range of responses when ions of different *m/z* values strike the detector. Ion of higher *m/z* values produced lower responses than those of lower *m/z* value. This is one of the reasons for high *mass discrimination* (signal strength for the same number of ions of one *m/z* value is different than that for ions of a different *m/z* value) in the transmission quadrupole as compared to the mass spectrometers having ions of much higher kinetic energy such as the TOF and double-focusing instruments. For example, the kinetic energy of the ions in the double-focusing mass spectrometer is in the range of 1–5 keV. At this level, ions of all *m/z* values give approximately the same response when they strike the detector. To correct for this high mass discrimination, a post-acceleration dynode is placed at the exit of the *m/z* analyzers to elevate the kinetic energy of ions of all *m/z* values before they strike the detector. This postacceleration dynode can also be used as a conversion dynode as illustrated in Figure 4.19. The detectors only respond to positive ions, and the negative ions that can be detected in GC/MS must be converted to positive ions before striking the detector.

Initiation of photons is not as dependent on the kinetic energy of ions as is the initiation of electrons but, like the initiation of electrons, requires positive ions. Figure 4.20 is an illustration of a photomultiplier (PM)

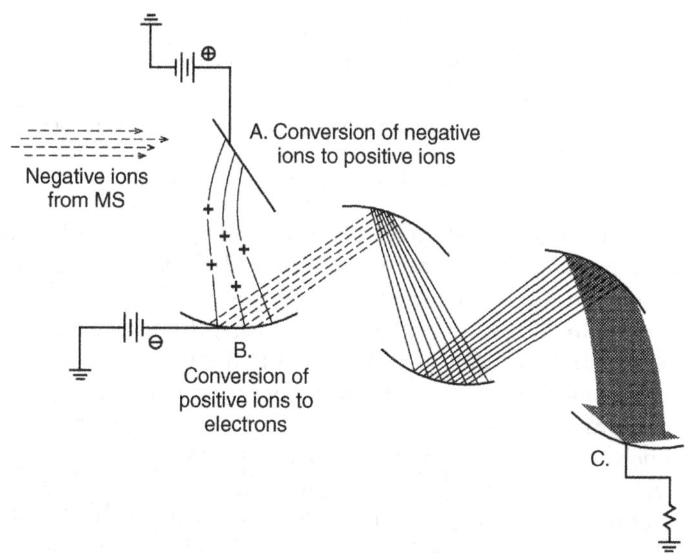

Figure 4.19 Illustration of a dynode that can be used to convert positive ions to negative ions and can also be used as a postacceleration device.

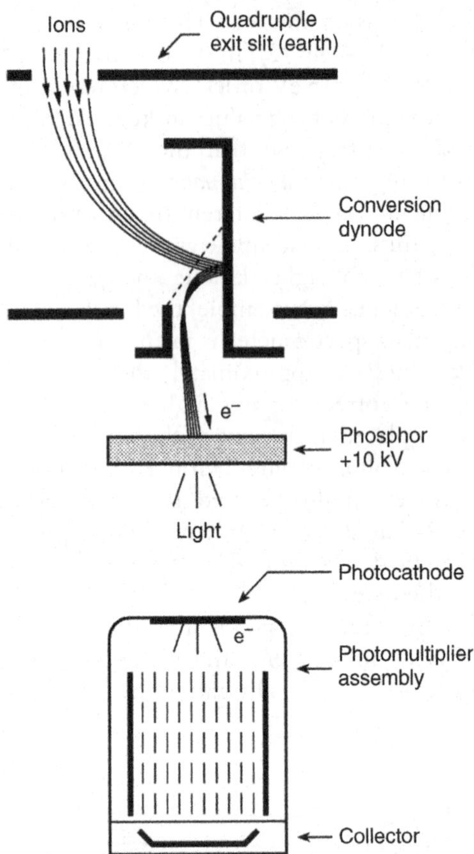

Figure 4.20 Illustration of a photomultiplier ion detector as used in some GC/MS instruments.

detector with a conversion dynode. The PM used as a detector for the ions has an advantage over the electron multiplier (EM) in that the detector does not deteriorate with time and therefore does not require replacement like the electron multiplier. Even with this advantage, the PM ion detector in GC/MS does not seem to be as widely used as the EM.

The EM produces about 10^5 electrons per ion. This is considered the optimal performance. Throughout its life, the voltage applied to the EM must be increased to maintain this level of *gain*. A voltage is eventually reached where the *noise* produced is so great that the signal at low levels is no longer distinguishable from the noise. This is when the EM has reached the end of its life. As the EM gets closer to the end of its life, the increases in voltages to maintain the 10^5 gain become greater. There are two types of electron multipliers: the continuous dynode and the discrete dynode. Figure

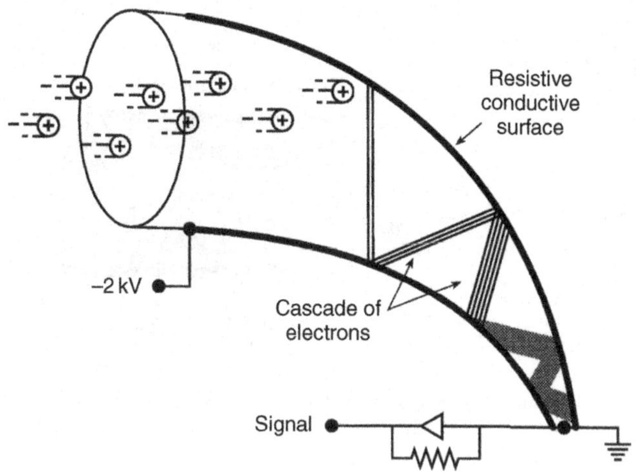

Figure 4.21 Representation of a continuous-dynode electron multiplier used in GC/MS.

4.19 is an illustration of a discrete-dynode EM. Figure 4.21 is an illustration of a continuous-dynode EM. The electron emitting surface from both of these devices is a lead-doped glass.

One of the most tedious aspects of GC/MS is the correct positioning of the EM when it needs to be replaced. In some cases, the EM is connected by flexible copper wires that make placing the new EM in the exact position as the old one difficult. Some manufacturers now use self-aligning devices such as that shown in Figure 4.22 to assure a reproducible positioning of the

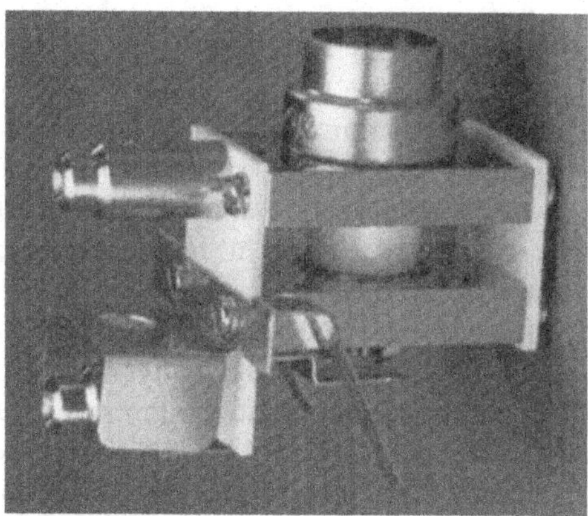

Figure 4.22 A self-aligning continuous-dynode electron multiplier replacement available from ITT.

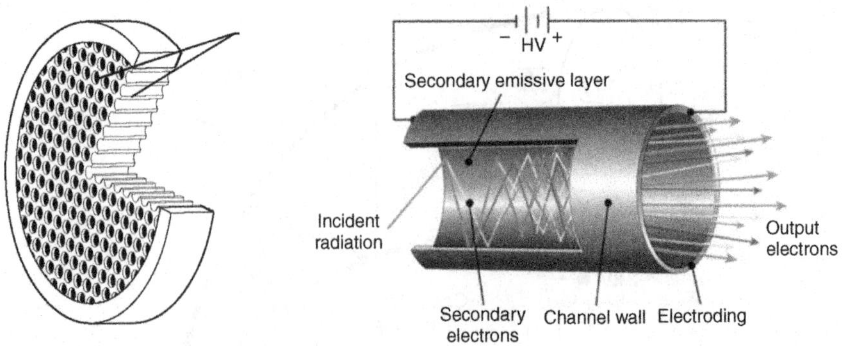

Figure 4.23 Illustration of how a microchannel plate electron multiplier functions.

EM. The inclusion of such a device should be an important factor in the selection of an instrument.

TOF GC/MS instruments use a *microchannel plate* for ion detection because the ion beam is much broader than that of the other three types of *m/z* analyzers. The microchannel plate has a series of hollow tubes that are coated with the same lead-doped glass used in the EM. Each tube acts as an EM. The electrons from all the channels are combined to produce the signal that is then digitized and recorded. Figure 4.23 is an illustration of a microchannel plate. It should be noted that the LECO TruTOF GC-MS does not use a microchannel plate but uses a discrete-dynode EM for ion detection.

4.7. *m/z* Scale Calibration

Unlike other spectrometric techniques used in analytical chemistry, the abscissa of the instrument's data output is not hard calibrated at the time of manufacture. The *m/z* scale must be calibrated on a regular basis; it must be checked for accuracy at least every 8 hours of operation. In the case of instruments used for accurate mass assignments, it may be necessary to calibrate at the time the data are acquired. The vast majority of GC/MS data are acquired at unit resolution although many of the data systems report the data to the 0.05 *m/z* unit. As was pointed out above, the best resolution that can be obtained with a QMF or QIT mass spectrometer is ~0.3 *m/z* units. The scale is calibrated at integer *m/z* values. Perfluorinated compounds are used for this because their ions have very little *mass defect*, the difference between the nominal mass and the monoisotopic mass. For example, the ion formed by fragmentation of the molecular ion of perfluorotributylamine (PFTBA)* (the compound used to calibrate the *m/z* scale of QMF and QIT GC/MS systems) that has a nominal mass of 502 Da has a monoisotopic mass of 501.971134 Da or a mass defect of 0.03 Da. A corresponding aliphatic ion $C_{36}H_{73}$ with a nominal mass of 505 Da would have a monoisotopic mass of 505.571225 Da or a mass defect of 0.6 Da. Most GC/MS instruments have a maximum *m/z* range to *m/z* 1,000. The maximum *m/z* value with an intensity observed in the mass spectrum of PFTBA is at *m/z* 614. Because of the linearity of the *m/z* for quadrupole-based instruments, it is possible to extrapolate from *m/z* 614 to *m/z* 1,000. PFTBA does not form a lot of fragments (*m/z* 50, 69, 100, 131, 164, 169, 219, 264, 414, 464, 502, and 604). Again, the linearity in the *m/z* scale interpolation between these points allows for a reliable calibration. In many cases, the peak at *m/z* 614 is not used because the abundance of this ion is very low. Figure 4.24 is an EI mass spectrum of PFTBA.

The double-focusing mass spectrometer does not have a linear *m/z* scale because it operates at a constant resolving power rather than a constant resolution. To calibrate the *m/z* scale of the instruments, perfluorokerosene** (PFK) is used. When ionized by EI, this material yields ions every

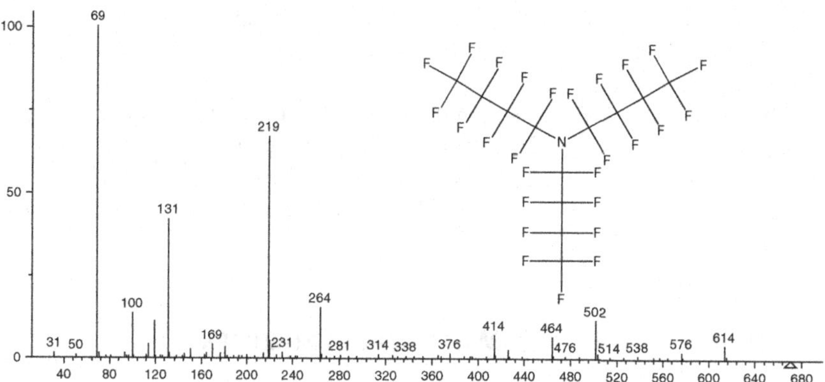

Figure 4.24 EI mass spectrum of PFTBA and its structure. This compound is used to calibrate the *m/z* scale in most QMF and QIT GC/MS systems. One of the reasons that it is used is that the mass defect* of fluorine relative to its mass is so small that the nominal mass and the monoisotopic mass of the various ions are the same.

---

*The commercial name for PFTBA is FC-43. This is often the name used for the tuning compound for the QMF and QIT GC/MS instruments.

**Unlike PFTBA, PFK is not a single compound. Just as kerosene is a mixture of alkanes, PFK is a mixture of alkanes in which the hydrogen atoms have been replaced with fluorine atoms.

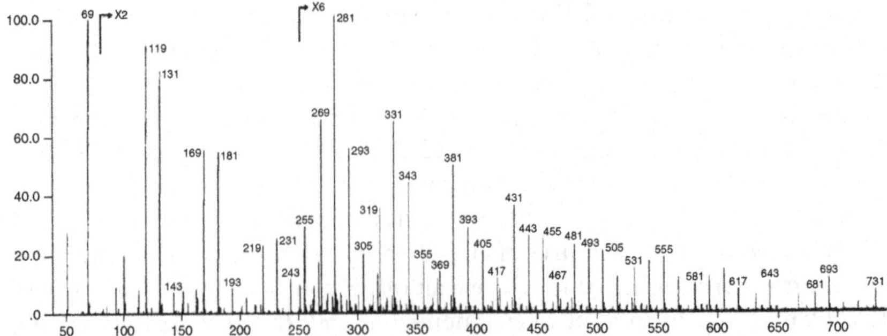

Figure 4.25 EI mass spectrum of PFK. PFK is not a pure compound like PFTBA but is a mixture of different fluorinated hydrocarbons. This is why the spectrum has as many peaks as it does.

few *m/z* units and allows for a calibration with far less interpolation than is associated with use of PFTBA. Another advantage of having PFK available is that its ions will have a negative mass defect; therefore, the peaks representing the ions of the calibrant can be separated from those of the analyte which, because of the high mass defect associated with hydrogen, will usually have a positive mass defect. This means that the PFK ions can be used for peak matching in accurate mass assignments. Figure 4.25 is an EI mass spectrum of PFK.

Theoretically, a TOF mass spectrometer can be calibrated using a single peak. This is something that is difficult for most users of GC/MS instrumentation to accept. Therefore, the *m/z* scale of the currently commercially available GC/MS instruments is calibrated using several ions in the mass spectrum of PFTBA.

4.8. Tuning the Mass Spectrometer

Before using the GC-MS, it must be determined that the instrument is performing properly. One of the best ways to make this determination is to acquire mass spectra of a known substance that provides peaks over the *m/z* range that will be used. An examination is made of the data presented graphically (a bar-graph spectrum) and in tabular format. The calibration compound is a good compound to use for this purpose. Most instruments have a reservoir where the liquid calibrant compound is stored. The vapor pressure above the liquid in the reservoir is sufficient to supply a constant partial pressure of the calibrant to the ion source during the spectral

---

* Mass defect is the difference in the integer mass and the exact mass of an isotope.

acquisition. Data are acquired at a rate of one spectrum per second for a minimum of 30 seconds and a maximum of 60 seconds. A continuous series of 5–10 spectra in the center of the acquisition interval are averaged and then examined. In the case of PFTBA, some of the examination points might be as follows:

- Does the bar-graph mass spectrum look like it usually does with respect to *m/z* peaks present and their relative intensities?
- Is the intensity of the base peak (a numerical value of signal strength) the same as it normally is?
- Are there any new peaks present that may be contributed by background in the ion source or that originate from higher than normal column bleed?
- After the use of the Auto Tune routine, does the spectrum look noisier than before? Is this level of noise acceptable?
- Are any of the significant PFTBA peaks missing?
- Does each of the major PFTBA peaks exhibit an isotope peak? To make this determination, it may be necessary to produce a graphic of an expanded display of the mass spectrum.
- From the tabular output, determine the percent intensity of the *m/z* 220 peak relative to the intensity of the *m/z* 219 peak. Do the same for the *m/z* 503 peak relative to the *m/z* 502 peak. Do these values match the theory for ions containing the 4 and 9 atoms of carbon, respectively? (see Chapter 5)
- From the tabular output, check the intensities of the *m/z* 100, 219, 264, 414, and 502 peaks relative to the base peak at *m/z* 69. Also see if the intensity ratio of the 502 peak relative to the 264 peak is consistent within acceptable limits.
- Assuming that data are acquired from approximately *m/z* 15 up, check the intensities of peaks at *m/z* 18, 28, and 32 relative to the base peak at *m/z* 69. These peaks represent the presence of water (*m/z* 18) and air (*m/z* 28 and 32) and are good indicators of an air leak.

The data systems supplied with all commercial GC/MS systems have what is called an *Auto Tune* routine. This is a computer program that does a lot of the itemized steps above and adjusts instrument parameters to correct any observed problems. The Auto Tune looks at the mass spectral peak shapes in the profile mode, adjusts voltages on various electrical components to bring the peak shapes to a condition that isotope peak ratios are correct, and makes adjustments to the voltage applied to the detector to obtain the *10^5 gain*[*] that provides consistent signal strength. Even with the Auto Tune making the adjustment, it is still a good idea to examine the spectra as

[*] A 10^5 gain means that one ion initially striking the detector will result in the subsequent emissions of 10^5 electrons.

suggested above. Some decisions like whether it is time to replace the EM must be made by the operator.

It is always best to make sure that there is a clear understanding of what a specific routine in a specific version of the software does. The importance of this is illustrated in what happened to users of the Hewlett–Packard's (Agilent Technologies as of 1999) GC/MS ChemStation in July, 1995. The HP ChemStation version G1034C was provided with the 5890/5972 GC/MS system. The Auto Tune program in this version was written to set the base peak of the PFTBA mass spectrum at *m/z* 69. In order to meet advertised detection limit specifications, another tuning program was provided that was called *Enhanced Sensitivity Tune*. This program set the base peak of the PFTBA mass spectrum at *m/z* 219, which gave increased observed intensity at *m/z* 282, the molecular ion peak of hexachlorobenzene, which was the compound used at that time to demonstrate the detection-limit specification. Customers refused to accept the demonstration of the detection-limit specification using the Enhanced Sensitivity Tune and wanted it demonstrated based on the use of the Auto Tune. In order to circumvent this situation, when a new version of ChemStation was introduced in July of 1995, HP renamed the Enhanced Sensitivity Tune as Auto Tune and named the old Auto Tune routine *Standard Spectrum Tune* without informing customers. The biggest impact occurred with those doing analysis of cocaine (Figure 4.26). The base peak in their mass spectra had always been at *m/z* 82 using the original version of the Auto Tune. With the new version of Auto Tune, the base peak was now *m/z* 182. Two 5972 mass spectrometers sitting side by side would give two different spectra for the same amount of cocaine analyzed under exactly the same conditions after they had been tuned using different versions of the manufacturer's

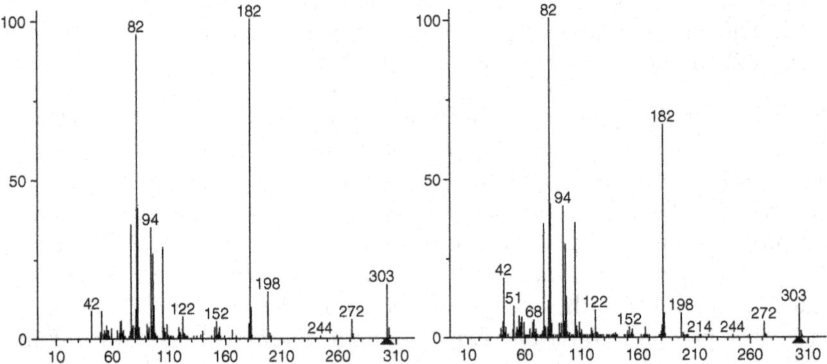

Figure 4.26 The EI mass spectrum on the left was obtained using the current *Auto Tune* file for the Agilent GC/MS ChemStation, and the spectrum on the right is obtained using the current *Standard Spectrum Tune* file (the old *Auto Tune* file).

Auto Tune routine. The only difference was that the instrument that produced the mass spectrum with *m/z* 82 as the base peak was using GC/MS ChemStation version G1034C (or earlier), and the instrument that produced the mass spectrum with *m/z* 182 as the base peak was using GC/MS ChemStation version G1701AA (or later).

Tuning with PFTBA or PFK is referred to as the *static tune procedure* because the partial pressure of the tuning substance remains constant during the acquisition of the spectrum. Instrument performance is also based on what could be referred to as a *dynamic tune*. These are instrument settings that will produce a spectrum with specific qualities when a specific compound is eluted from a GC column. These so-called dynamic tunes originated with the U.S. EPA and are specific to the analysis of groups of analytes called *volatiles* and *semivolatiles* in various matrices from drinking water to extracts from liquid pools and soil found at U.S. EPA Super Fund sites. These tunes are based on bromofluorobenzene (BFB) [24] and decafluorotriphenyl phosphine (DFTPP) [25]. Modern instruments have tune routines that use PFTBA or PFK to make adjustments so that when BFB or DFTPP is analyzed, the spectra meet the U.S. EPA tune requirements for specific methods. The tune requirements are included in the published methods.

4.9. Data Acquisition

In GC/MS, data acquisition is referred to as *continuous measurement of spectra* (a term coined by Bill Budde [26]) and SIM [27]. Budde selected the term *continuous measurement of spectra* as opposed to *scan mode* for the process where spectra are acquired one after another over a specified *m/z* range because nothing is scanned in the TOF instrument, and the QIT GC-MS is also a pulsed instrument. The term *SIM* refers to a process where ions of selected *m/z* values are monitored for a portion of the data acquisition cycle time. SIM is a technique limited to instruments that measure ion currents for specific *m/z* values at specific times during the acquisition mode; these instruments are the QMF and double-focusing GC-MS. By being able to spend more of the acquisition cycle time on ions of a few *m/z* values, more of these ions of each *m/z* value will be detected giving a stronger signal; therefore, a lower detection limit then would be possible if only short periods of the acquisition cycle were spent measuring the ions of these *m/z* values as would be the case where a portion of the acquisition time was spent measuring ion current of ion of all *m/z* values in a spectral acquisition range. SIM is used to obtain lower detection limits of target analytes in quantitative analyses. For the technique to be effective, ions of no more than eight different *m/z* values should be measured in a single cycle.

Instrument manufacturers have been increasing the number of ions of different *m/z* values that can be monitored in a single cycle to the point that some manufacturers claim the ability to monitor ions of as many as 100 different *m/z* values in the same acquisition cycle. This is self-defeating when considering that the purpose is to have as much time as possible measuring the ion current for a single *m/z* value. SIM can produce better chromatographic peak shapes resulting in better quantitative precision by increasing the number of data points that define the chromatographic peak. Depending on the data acquisition rate, SIM monitoring of the ion current of three different *m/z* values can reduce the limit of quantitation by three orders of magnitude or more.

A new feature that is being promoted by some QMF GC-MS manufacturers is the ability to acquire one cycle (spectrum) in a full-scan mode and the next cycle in an SIM mode on an alternating basis. The utility of such an acquisition must be carefully evaluated. If the amount of analyte is sufficient to allow for a full-spectrum acquisition, then SIM may not be necessary. If the amount of sample is so low that SIM will be required for the desired limit of quantitation, then spending time acquiring an unusable full-range acquisition is a waste. It may be possible to construct an analysis where SIM could be used for quantitation and full-spectrum acquisition could be used to detect presence of unknowns in the same, but this will have to be carefully developed.

A very important aspect of data acquisition regardless of using SIM or continuous measurement of spectra is the use of a delay between the time the sample is injected onto the GC column and when spectral acquisition begins. During this time, the ion-source filament and the detector are turned off. The first material to elute from the column is the solvent, which should pass through unretained. The amount of solvent usually far exceeds the amount of any analyte. By having the filament and detector turned off, ions that could be potentially formed at that time will not damage the filament and have a negative impact on the life of the detector.

4.9.1. Continuous Measurement of Spectra (Full-Spectrum Acquisition)

There are several considerations that must be made in setting up a full-spectrum acquisition mode. First, it should be noted that if a TOF instrument is being used, then the acquisition time is the same whether the *m/z* range is that of the instrument or a smaller subset. The only thing that changes when an acquisition range is set for a TOF instrument is the range of the stored spectra. The acquisition range will be a factor in the acquisition time for all of the other three types of instruments. With chromatographic peaks becoming narrower and a need for at least five data points to define these peaks, acquisition times are critical. In the QMF and the double-focusing GC-MS, the amount of analyte will always be changing during the data acquisition. This can lead to spectral skewing as illustrated in Figure 4.27. If the chromatographic peak represents a single compound, these spectra can be averaged to obtain an interpretable spectrum or a spectrum that can be matched against a spectrum of that compound in a mass spectral database. If the spectrum represents more than a single compound, then a software routine to bring about chromatographic peak deconvolution based on the mass spectra of each compound will be necessary.

Oftentimes, data acquisitions are started at *m/z* 35 or 40. This is done to avoid peaks in the mass spectrum at *m/z* 18, 28, and 32, which are due to air leaks. Air can do serious harm to the GC stationary phase, especially on a hot column. Acquisitions should be started as low as possible because peaks at *m/z* 30 and 31 are indicators that analytes are aliphatic amines and alcohols, respectively. It should be remembered that the QIT GC-MS has a minimum lower acquisition limit of *m/z* 20 when AGC is being used, and AGC should be used for all acquisitions of spectra of chromatographic elutions. The acquisition should begin at *m/z* 10 when using either a double-focusing or a QMF instrument, and the minimum stored *m/z*

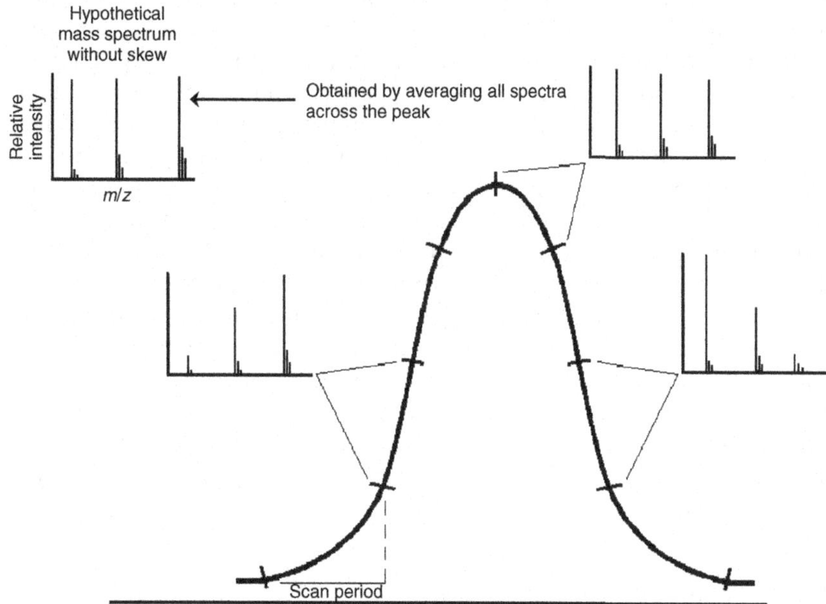

Figure 4.27 Chromatographic peak reconstructed from scanned mass spectral data illustrating the problem of spectral skewing.

value should be 10 when using a TOF instrument. This way, ion current can be observed for methyl ions, if formed. Data acquisitions with lower limits below *m/z* 5 should be avoided because the mass-to-charge ratio of a He ion is 4. There are a very large number of He ions formed and by sending them through to the EM can greatly shorten the life of this detector.

> One of the primary sources of air in the GC-MS is through overused septa. It is important to change the septum on the GC injector every 50–100 injections or, as explained in Section 2.2.1, use a septumless injection port cap.

Too high an acquisition rate in QMF or double-focusing instruments over too large of an *m/z* range can mean that too little time is spent measuring the ion current of ions of a single *m/z* value, which could result in very poor quality spectra. These spectra may not be suitable for interpretation or matching against mass spectral databases. Several QMF manufacturers have been allowing for higher spectral acquisition speeds; but it should always be kept in mind that even though these instruments have better electronic noise specifications, chemical noise (background) must be considered, and only so much ion current can be measured in any time interval. Most commercial GC/MS systems have maximum *m/z* ranges between 650 and 1,000. There are very few compounds that will make it through a GC column that have a nominal mass above 500. When compounds with nominal masses above 500 are likely to be encountered, most of the mass spectral information will be in a limited segment of the instrument's total range and the acquisition can be set accordingly. The acquisition range and the acquisition rate do not have to remain constant throughout the entire analysis. Chromatographic components eluting at the beginning of the analysis usually have lower nominal masses than those eluting later in the analysis. Even with temperature programming, the widths of chromatographic peaks representing the early eluting compounds tend to be less than those representing compound eluting later. This means that acquisition times over shorter *m/z* ranges can be used at the beginning of the analysis, and the acquisition range may be expanded later in the analysis.

4.9.2. Selected Ion Monitoring

SIM is a technique where the ion current for a few ions is monitored during a data acquisition cycle (spectrum acquisition) rather than monitoring the ion current for all *m/z* values within a given range. SIM allows more time to measure the ion current of selected ions during a data acquisition cycle allowing for a stronger signal, thus lowering the detection limit or the limit of quantitation. SIM only has meaning for QMF and double-focusing instruments because these instruments measure ion current for individual *m/z* values from a continuous beam that contains ions of all *m/z* values. In pulsed instruments, like the TOF and QIT GC-MS, all ions of all *m/z* values formed during a cycle are detected; therefore, there is no advantage to SIM. SIM also allows for shorter acquisition times producing better quality chromatographic peaks. The TOF GC-MS already has better acquisition rates than the QMF or double-focusing instrument (25–500 spectra per second compared to 10 or less spectra per second); therefore, the chromatographic peak quality in the TOF GC-MS should be sufficient for high precision in quantitation. The QIT GC-MS has a technique called *selected ion storage* (SIS) that will improve the detection limit and increase the linear dynamic range of a quantitation but does little with respect to chromatographic peak quality. In the QIT, manipulations of spectral ranges and acquisition segments can be applied that will assist with producing better quality chromatographic peaks.

SIM can also produce better specificity if a unique ion can be found to define the analyte. This increased specificity can also lower the detection limit. SIM is often used in conjunction with soft ionization techniques such as CI and ECNI. SIM plays a very important role in ECNI in that only analytes with high electro-negativities will be ionized, which offers the first degree of specificity. The use of SIM now offers a second degree, and this is why the detection limits using SIM and ECNI are as low as they are.

SIM in a GC-MS provides a retention time for the chromatographic peak, which is an indicator of an analyte's identity; and, in most cases, a nominal mass, which is a further indication of the analyte's identity. When using a double-focusing GC-MS, an accurate mass for the ion can be obtained, which is a far better confirmation of the analyte's identity. In some cases, SIM analyses are constructed so that three ions characteristic of the mass spectrum for the analyte are monitored; and when the ratios of the mass chromatographic peaks for each of these ions are compared with one another, they must fall within certain limits for the analyte to be confirmed. For quantitative purposes, deuterated, ^{13}C-labeled, or ^{15}N-labeled analogs of the analyte may be employed as an internal standard. This would mean that a series of four different *m/z* values would have to be monitored for each analyte. Care must be taken that analytes or their internal standards do not coelute with other analytes (deuterated analogs will elute slightly before the corresponding native isotopic analyte).

There can be multiple acquisition time segments for SIM just as there are for the continuous measurement of spectra. Each time, a segment can contain a different number of ions to be monitored. Each ion being monitored within a time segment can have a separate dwell time. A large percentage of GC/MS instruments in use today are used exclusively for SIM analyses. When talking about SIM, each letter should be individually pronounced. Avoid using the acronym *sim*. The chromatographic output from an SIM analysis is called a *selected-ion-monitoring chromatogram* or an *SIM chromatogram*. The term *SIM plot* should not be used. When using mass chromatograms from continuous measurement of spectral data, these presentations should be identified as *mass chromatograms* to clearly differentiate them from SIM chromatograms.

4.9.3. Alternate Full-Spectrum Acquisition

Due to improvements in the electronics of many of the modern transmission quadrupole instruments, it is now possible to acquire data alternately using a set of SIM ions for one data recording and acquiring a spectrum over a range of *m/z* for the next data recording. This usually results into separate data files for a single analysis. One of the advantages for this process is being able to screen for unknowns at the same time quantitation data is being acquired. SIM is usually employed to reduce the limits of detection; therefore, if the unknowns are at the same concentration level as the target compound, it may be difficult to see them. This process also reduces the number of points across the chromatographic peak, which can have an impact on quantitative precision. Caution must be taken in using this acquisition mode. This mode is only available on transmission quadrupole instruments.

4.10. Tandem Mass Spectrometry (MS/MS)

Tandem mass spectrometry, also known as MS/MS (mass spectrometry/mass spectrometry), is a process by which gas-phase ions of an analyte resulting from an initial ionization undergo separation according to their *m/z* values, using a first iteration of mass spectrometry. Ions of individual *m/z* values (*precursor ions*) are then subjected to collisional activation so that these precursor ions will fragment, producing *product ions* of different *m/z* values. These product ions then are separated and detected by a second iteration of mass spectrometry; thus the name *tandem mass spectrometry* or *MS/MS*. The process of fragmenting precursor ions is called collisionally activated dissociation (CAD) or collision-induced dissociation (CID). These terms are used interchangeably. CAD is a result of an inelastic collision between an ion of a specific *m/z* value (the precursor ion) and an inert atom or molecule such as He, Ar, Xe, or N_2.* Some of the kinetic energy of the ion is transformed into internal energy, and the activated ion will then dissociate.

MS/MS can be accomplished in a temporal or spatial domain. Spatial domain MS/MS is called *tandem-in-space* mass spectrometry. Temporal domain MS/MS is called *tandem-in-time* mass spectrometry. In the world of GC/MS, tandem-in-space mass spectrometry is carried out in a double-focusing instrument as described in the double-focusing GC-MS section of this chapter or in a *triple quadrupole* mass spectrometer, schematically illustrated in Figure 4.28. Other types of tandem-in-space MS/MS instruments exist, but they currently have no role in GC/MS. In this same world, tandem-in-time mass spectrometry is carried out in the QIT, illustrated in Figure 4.29. Tandem-in-time MS/MS can also be carried out in what is known as a Fourier transform ion cyclotron resonance mass spectrometer, but these instruments are really not very often, if at all, found in GC/MS.

The term *triple quadrupole* is somewhat misleading. The name implies that there are three QMFs in tandem; however, the instrument has two QMFs separated by a collision cell that was constructed from a quadrupole device used in the rf-only mode in the original design of the instrument. This instrument was developed at Michigan State University in the mid-1970s [28]. The development of the triple quadrupole had a significant impact on the field of mass spectrometry. Today, although still often referred to as the triple quadrupole, the collision cell can be any number of devices from an octupole to a hexapole to a traveling-wave device. Very few, if any, manufacturers use a quadrupole as the collision cell; therefore, a more appropriate description would be the *tandem quadrupole* mass spectrometer. What all of these devices have in common is that they have entrance and

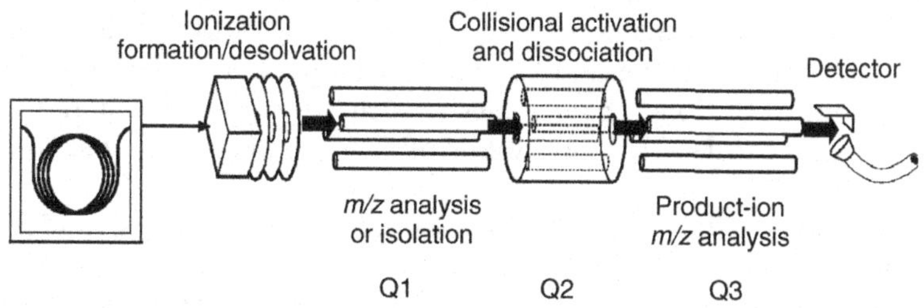

Figure 4.28 Schematic representation of tandem-in-space MS/MS.

*It should be noted that He is the only collision gas used for MS/MS in a double-focusing instrument. The same is true for the QIT GC-MS. The triple quadrupole is the only GC/MS/MS instrument that allows for use of different collision gases.

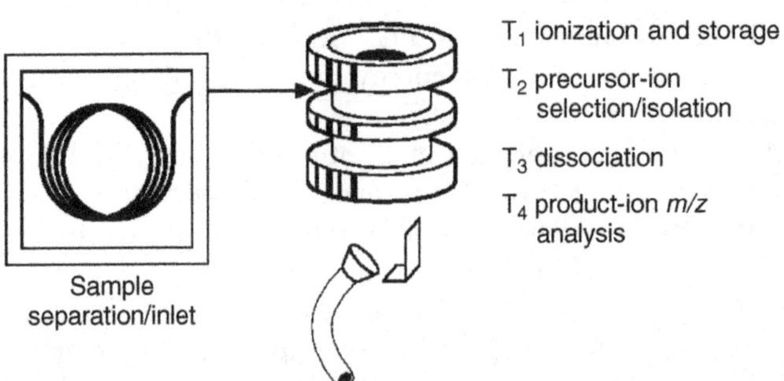

Figure 4.29 Schematic representation of tandem-in-time MS/MS.

exit lenses to speed up and slow down the ions entering and leaving the collision cell, respectively. The ion velocity is increased to increase its kinetic energy. The product ions will have the same velocity when they reach the exit of the collision cell as the precursor ions had upon entering. The product ions that are formed can also undergo CAD because they have kinetic energies as they are formed. The resulting spectrum can exhibit peaks representing product ions of the precursor ions and product ions of the product ions. The ions exiting the collision cell must be slowed so that the effect of the quadrupole field of the second QMF will be optimum for ion separation. The pressure in the collision cell is ~10^{-1} Pa, whereas the pressure in the two QMFs is ~10^{-3} Pa.

The operational pressure of QIT mass spectrometers is ~10^{-1} Pa; therefore, after isolation of the precursor ion of a specific *m/z* value in the ion trap, the energy of the ion is increased by applying waveforms to the end-cap electrodes to bring about collisions with the already present He atoms. These waveforms can be specific for a very narrow *m/z* range of ions; therefore, product ions that are formed and stored during the energizing of the precursor ion will not undergo collisional activation. However, it should be noted that because of the need to use rf voltages to both energize and store ions, the product ions that have an *m/z* value that is less than one-third of the *m/z* value of the precursor ions will not be sufficiently stored.

In the double-focusing GC-MS used for MS/MS, the collision cell is mounted between the ion source and the first field. This means that the ion will have the kinetic energy provided by the accelerating voltage, which is between 1 keV and 10 keV. This results in a single high-energy collision. In the QIT and triple quadrupole, CAD occurs for ions with <40 eV kinetic energy. This is called *low-energy CAD*. The difference is in the number of collisions. MS/MS in a triple quadrupole will result in <10 collisions; MS/MS in a QIT will result in a much larger number of collisions, several hundred to a few thousand.

4.10.1. Tandem-in-Space

In tandem-in-space mass spectrometry, MS/MS can take several forms. The first analyzer can be set to allow ions of only a specific *m/z* value to pass into the collision cell. The second analyzer is scanned to obtain a product-ion mass spectrum exhibiting ions of all *m/z* values resulting from CAD. This is called a *product-ion analysis*, and the results are a *product-ion mass spectrum*. The second *m/z* analyzer can be set to allow ions of only a single *m/z* value to pass to the detector. The first analyzer is scanned. This results in a signal for any ion of any *m/z* value that produces product ions of a specific *m/z* value. This is called a *precursor-ion analysis*, and the results are a *precursor-ion mass spectrum* for a specific product ion. The two *m/z* analyzers can be scanned simultaneously with the second *m/z* analyzer having a lower starting *m/z* than the first. This will result in a signal for ions of any *m/z* value passing the first analyzer that undergo a loss of a specific mass (a neutral loss) in the collision cell. This is called a *common-neutral-loss analysis*, and the results are called a *neutral-loss spectrum* of a specified offset.

The most widely used form of MS/MS is in a process called *selected reaction monitoring* (SRM).* SRM is a process where an ion of a lesser *m/z* value is allowed to pass the second *m/z* analyzer when ions of a specified higher *m/z* value pass the first analyzer. This process is analogous to SIM; however, because it involves the transition from an ion formed by an analyte to a fragment of that ion, it provides for a higher degree of selectivity. Manufacturers have said that SRM is a more *sensitive*** [*sic*] technique than SIM. Of course this is not possible for two reasons. The first is that there is a certain amount of ion loss going through each of two *m/z* analyzers associated with the instrument used for SRM, which will be twice the loss experienced when the ions pass through a single analyzer in SIM. The second is that all of the precursor ions may not be transformed into the product ion. The limit of detection or limit of quantitation can be much larger in SRM than in SIM because of the increased specificity associated with SRM.

In a single acquisition cycle of a tandem-in-space instrument, there can be multiple transition pairs. Just like SIM, too many pairs can defeat the increase in signal strength that will result from monitoring ions for longer

---

* In an effort to differentiate themselves from other manufacturers, Micromass (now Waters Corp.) decided to call their implementation of SRM multiple reaction monitoring (MRM). Today, most instrument manufacturers refer to the SRM process as MRM. It should be noted that SRM is the preferred term.

** Sensitivity, as used in analytical chemistry, is the slope (m) of a line defined by the equation $y = mx + b$; however, it is used by many instrument manufacturers to mean the limit of detection or a limit of quantitation.

periods. It is possible for transition pairs to have separate dwell times in an acquisition cycle. Different transition pairs can be implemented for different time segments in the analysis.

4.10.2. Tandem-in-Time

Tandem-in-time mass spectrometry only allows for product-ion analyses. A process can be carried out that will pass for SRM in the QIT.

The triple quadrupole GC-MS/MS has become somewhat obscure in recent years. Several years back, Waters and Thermo Scientific discontinued offering these instruments. This left only the Varian 1200. Both companies have now reintroduced their instruments as improved models; Agilent has introduced a GC/MS/MS instrument as of the 2008 Riva del Garda *32nd Symposium on Capillary Chromatography*. There is a company (CHROMSYS, LLC, the U.S. subsidiary of the German company CHROMTECH, GmbH) that has a third-party add-on for the Agilent GC/MS systems. This renewed interest is partially driven by interest in specificity for analytes in the area of food safety. The QIT GC/MS instruments from Varian and Thermo Scientific continue to be popular for MS/MS.

4.11. Conclusion

As was clearly illustrated over the last few pages, the mass spectrometer portion of the GC-MS is a complex device and must be treated with a great deal of care. Instrument performance is important and must be verified regularly. The quality of the data must be verified, and regular maintenance of the instrument is important. The GC-MS is not a black box to be used when needed. It must be thoroughly understood and watched over.

REFERENCES

1. Watson, J. T., Sparkman, O. D. (2007). *Introduction to Mass Spectrometry: Instrumentation, Applications, and Strategies for Data Interpretation*. 4th ed. Chichester, U.K.: Wiley; ISBN: 9780470516340.
2. de Hoffmann, E., Stroobant, V. (2007). *Mass Spectrometry: Principles and Applications*. 3rd ed. Chichester, U.K.: Wiley; ISBN; 978470033104 (hc) and 978470033111 (pb).
3. McEwen, C. N., McKay, R. G. (2005). A combination atmospheric pressure LC/MS: GC/MS ion source: Advantages of dual AP LC/MS:GC/MS instrumentation. *J. Am. Soc. Mass Spectrom.*, 16(11), 1730–38.
4. Kiser, R. W. (1965). *Introduction to Mass Spectrometry and Its Application*. Englewood Cliffs, NJ: Prentice-Hall.

5. Busch, K. L. (1995). Electron ionization, up close and personal. *Spectroscopy (Eugene, OR)*, 10, 39–42.
6. Munson, B., Field, F. H. (1966). Chemical ionization mass spectrometry. *J. Am. Chem. Soc.*, 88, 2621–30.
7. Munson, B. (1977). CI-MS: 10 years later. *Anal. Chem.*, 49, 772A–8A.
8. Hunt, D. F., Crow, F. W. (1978). ECNI. *Anal. Chem.*, 50, 1781–4.
9. Ong, V.S, Hites, R. A. (1994). Electron capture mass spectrometry of organic environmental contaminants. *Mass Spectrom. Rev.*, 13(3), 259–83.
10. Beckey H. D. (1963). Field ionization mass spectrometry. In *Advances in Mass Spectrometry* (R. M. Elliott, ed.), vol. 2. 1–24. Oxford, U. K.: Pergamon.
11. Horning, E. C., Carroll, D. I., Dzidic, I., Haegele, K. D., Horning M. G., Stillwell R. N. (1974). Atmospheric pressure ionization (API) mass spectrometry; solvent-mediated ionization of samples introduced in solution and a liquid chromatograph effluent stream. *J. Chromatogr. Sci.*, 12, 725–9.
12. Horning, E. C., Carroll, D. I., Dzidic, I., Haegele, K. D., Horning M. G., Stillwell, R. N. (1974). Liquid chromatography-mass spectrometry computer analytical systems; continuous-flow systems based on atmospheric pressure ionization mass spectrometry. *J. Chromatogr.*, 99, 13–21.
13. Carroll, D. I., Dzidic, I., Haegele, K. D., Stillwell, R. N., Horning E. C. (1975). Atmospheric pressure ionization mass spectrometry: corona-discharge ion source for use in liquid chromatography-mass spectrometry computer analytical system. *Anal. Chem.*, 47, 2369–73.
14. Horning, E. C., Carroll, D., Dzidic, I., et al., (1977). Development and use of bioanalytical systems based on mass spectrometry. *Clin. Chem.*, 13–21.
15. Horning, E. C., Horning, M. G., Carroll, D. I., Dzidic, I., Stillwell, R. N. (1973). New picogram detection system based on a mass spectrometer with an external ionization source at atmospheric pressure. *Anal. Chem.*, 45:936–43.
16. Carroll D. I., Dzidic, I., Stillwell, R. N., Horning, M. G., Horning, E. C. (1974). Subpicogram detection system for gas-phase analysis based upon atmospheric pressure ionization (API) mass spectrometry. *Anal. Chem.*, 46, 706–10.
17. Paul, W., Steinwedel, H. (1953), Ein neues Massenspektrometer ohne Magnetfeld. *Z. Naturforsch A*, 8, 448–450.
18. Paul, W., Raether, M. (1955). Das Elektrische Massenfilter. *Z. Physik.*, 140, 262–71.
19. Stafford, G. C. Jr, Kelley, P. E., Syka, J. E. P., Reynolds, W. E., Todd, J. F. J. (1984). Ion trap technology. *Int. J. Mass Spec. Ion Proc.*, 60, 85–98.
20. Cole, M. D. (2003). *The Analysis of Controlled Substances*. Chichester, U.K.: Wiley.
21. Wiley, W. C., McLaren, I. H. (1955). Time-of-flight mass spectrometer with improved resolution. *Rev. Sci. Instr.*, 26, 1150–7.
22. Thomson, J. J. (1921). *Rays of Positive Electricity and their Application to Chemical Analysis*. 2nd ed. London: Longmans Green (1st ed., 1913).
23. Aston, F. W. (1942). *Mass Spectrometry and Isotopes*. 2nd ed. London: Edward Arnold (1st ed., 1933)
24. Bartky, W., Dempster, A. J., (1929). Paths of charged particles in electric and magnetic fields. *Phys. Rev.*, 33, 1019.
25. Method 624, (1984). *Federal Register* 26 October, 49, 43234–43439; *Title 40 Code of Federal Regulations Part 136*.
26. Eichelberger, J. W., Harris, L. E., Budde, W. L. (1975). Reference compound to calibrate Ion abundance measurement in gas chromatography-mass spectrometry systems. *Anal. Chem.*, 47, 995.
27. Budde, W. L. (2001). *Analytical Mass Spectrometry: Strategies for Environmental and Related Applications*. Washington, DC: American Chemical Society and New York: Oxford.

28. Sparkman, O. D. (2006). *Mass Spectrometry Desk Reference*. 2nd ed. Pittsburgh, PA: Global View Publishing.
29. Yost, R. A, Enke, C. G. (1979). Triple quadrupole mass spectrometer for direct mixture analysis and structure elucidation. *Anal. Chem.,* 51, 1251A.

CHAPTER 5

Mass Spectral Data Interpretation

When the word "interpretation" is used in conjunction with "data acquired using the continuous measurement of spectra mode with an EI GC/MS system," it usually means the elucidation of the structure of an analyte based on the peaks that appear in the mass spectrum. The purpose of this structural determination is usually to identify an individual substance. Figure 5.1 is the EI mass spectrum of cocaine. The molecular ion ($M^{+\bullet}$)* peak can be used to identify the molecular weight of the compound and possibly its elemental composition. The peaks in the spectrum representing the fragment ions can reveal the structure of parts of the molecule. The differences in the *m/z* values of the $M^{+\bullet}$ peak and the fragment ion peaks (the *dark matter* of the mass spectrum) also hold information about the structure of the molecule.

In order for the exercise of structural determination to be successful, the mass spectrum should be of a pure analyte. GC/MS offers the advantage of being able to chromatographically separate a specific analyte from other analytes and from the sample matrix. Another major advantage that GC/MS has over mass spectrometry alone is that in addition to the mass spectral data, the retention index (RI) for the unknown substances can be determined. As was pointed out in Chapter 2, a database containing over 250,000 RI values of ~50 K compounds is now available from the National Institutes of Standards and Technology (NIST) along with the literature citations for each value and the GC method used in obtaining the values. This RI database is yet another tool to assist in the identification of an unknown.

---

* In mass spectrometry, the molecular ion is an ion that has the same atoms connected in the same way as the neutral molecule from which it was formed. The molecular ion is formed by the removal of the ($M^{+\bullet}$) or the addition of an electron ($M^{-\bullet}$) from or to (respectively) the original molecule. The symbol for the molecular ion includes a superscript + or − to indicate the fact that the species is an ion and to show the sign of the charge. This is followed by a dot, which shows that the ion has an odd number of electrons. All molecules have an even number of electrons; therefore, the addition or removal of an electron from the molecule results in a charged species that has an odd number of electrons. The nominal mass (see below for definition) of the molecular ion and the molecule are the same. For compounds that will be amenable to GC/MS, an unambiguous elemental composition of the molecular ion can be determined from its accurate mass without regard for the removed or added electron.

Gas Chromatography and Mass Spectrometry
DOI: 10.1016/B978-0-12-373628-4.00005-8

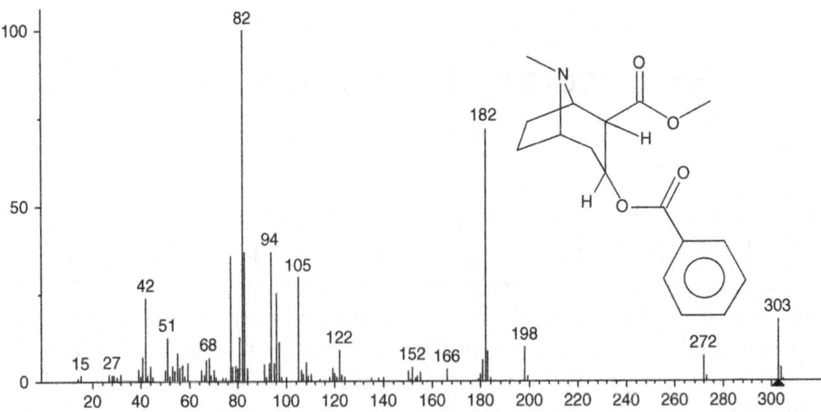

Figure 5.1 EI mass spectrum of cocaine. The $M^{+\bullet}$ peak at *m/z* 303 is easily identifiable because it is the highest *m/z* value peak in the spectrum that is not due to background or is not an isotope peak; and the peak at the next lower *m/z* value represents a loss of 31 Da, which could correspond to a $^{\bullet}CH_3$ radical. Because the numeric value of the $M^{+\bullet}$ peak is an odd number, the analyte is known to have an odd number of nitrogen atoms (*Nitrogen Rule*, which is explained later in this chapter).

There are many factors that are involved in the identification of an unknown substance. One of the most important of these factors is knowledge of the sample being analyzed. Where did it come from? How was it collected? What else was present where the sample was originally located? How has it been handled since it was collected? This knowledge of the sample prior to the GC/MS analysis is very important and may provide the key to an unknown substance's identity. Having said this, it must be pointed out that mass spectral interpretation is often clouded with making assumptions that can prejudice the result. Do not prejudice the data to reach a conclusion; let the data lead to the conclusion.

The first step to the identification of an unknown substance is to obtain quality data. Use acquisition parameters that will produce the best chromatographic separation possible in a reasonable time and allow for reasonable spectral quality. In a qualitative analysis using GC/MS, there should be at least two spectra on either side of the spectrum representing the apex of the reconstructed total ion current (RTIC) chromatographic peak which have a total ion current that exceeds the ion current of the background by at least 5% (Figure 5.2). To accomplish this may require adjustments to the chromatographic temperature program rate to broaden the peak to allow for the spectral acquisition time necessary. Try not to adjust the carrier gas flow rates because the mass spectrometer has an optimal linear velocity to produce the best data by maintaining a constant pressure in the ion source. Use mass spectral acquisition parameters that will produce high-quality

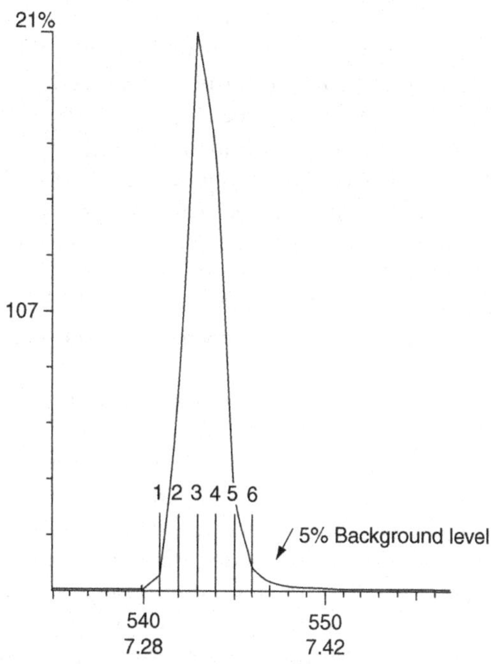

Figure 5.2 An illustration of multiple spectra constituting an RTIC chromatographic peak that provides data suitable for a mass spectral interpretation.

spectra; avoid scanning the *m/z* range too fast; and avoid using an excessively large *m/z* acquisition range. It is perfectly acceptable to have different spectral acquisition rates and mass spectral ranges for different time intervals when acquiring GC/MS data in the continuous measurement of spectra mode. The data requirements should dictate whether multiple time segments with different acquisition parameters are used.

The first step in the interpretation of the mass spectrum of an unknown substance is usually to perform a search of the spectrum against those in a mass spectral database, if available. There are many commercially available databases of EI spectra, as can be seen in Appendix I. The two most widely used are the *Wiley Registry of Mass Spectral Data* (now in its 9th edition) and the *NIST/EPA/NIH Mass Spectral Database*, which is updated every 3 years (last updated in 2008 and distributed as *NIST08*). Both of these collections are what are referred to as *general databases of mass spectra*. The other databases listed in Appendix I are what the authors refer to as *boutique databases*. These databases are used in conjunction with specific analysis areas such as flavor and fragrances or forensic toxicology. Oftentimes, these boutique databases contain spectra of compounds that are not contained in the general databases and can be very useful to the analyst in these specific areas.

Even with the use of a database search, it is best that the analyst be able to read the mass spectrum in a preliminary way, if for no other reason than to determine if the spectrum is suitable for being searched.

The RTIC chromatographic peak believed to represent the substance to be identified should be carefully examined. Does the peak shape indicate that it represents more than a single compound? Look at each spectrum that constitutes the chromatographic peak. Are the spectra consistent from spectrum to spectrum? If they are not, then there is a good possibility that the RTIC chromatographic peak represents more than a single compound; however, if a quadrupole mass filter (QMF) or double-focusing instrument is being used, do not be confused by possible spectral skewing (described earlier).

Another thing to look for in the spectra constituting the RTIC chromatographic peak is *noise spikes*. Noise spikes are signals represented as mass spectral peaks that have no isotope peak associated with them. These peaks are a result of signals produced by a neutral particle that was not removed by the vacuum system striking the detector at the time an ion of a specific *m/z* value should be reaching it. In some cases, these noise spikes may represent ions formed by the interaction of helium metastables and analyte molecules that occur in the area of the detector at a time when ions of specific *m/z* values should reach the detector. Noise spikes are easily distinguishable by the fact that they do not have associated isotope peaks and that they will not be present in the spectra just preceding or just following the spectrum where they appear. The intensity of the noise spike can have a broad range from several orders of magnitude greater than that of the base peak* of the analyte to as small as less than 1% of the intensity of the base peak.

Oftentimes, an individual spectrum or a group of spectra displayed as an average has so many peaks due to the chemical background that a database search will be meaningless, even if all the peaks representing the analyte are present. In other cases, two or more compounds may be represented by the RTIC chromatographic peak, and a search of the spectrum representing the apex of the chromatographic peak will result in erroneous results. As was pointed out earlier in this chapter, the use of mass chromatograms can help to determine whether or not mass spectral peaks actually represent ions produced by the still-unidentified analyte. Once extraneous mass spectral peaks have been identified, the analyst must figure out how to select a spectrum or a group of spectra to be subtracted as background from the spectrum (single or averaged) of interest. A help in such an endeavor is to use a computer program such as AMDIS *(Automated Mass spectral Deconvolution and Identification System)* developed by NIST [1]. AMDIS does a far

* The **base peak** in a mass spectrum is the most intense peak. The intensities of all the other peaks in the mass spectrum are reported relative to the base peak.

better job of background subtraction than can be accomplished by even the most experienced analyst.

Another potential problem in the interpretation of a mass spectrum is that important peaks at either end of the spectrum's *m/z* scale may be missing because of the data acquisition range. Care must be taken in performing the data acquisition to use a large-enough range to assure that all peaks are present in the recorded mass spectrum. However, when using instruments like the QMF or double-focusing mass spectrometer, too large of an *m/z* range can degrade the quality of the data due to too little time spent in the measurement of the ion current at each *m/z* value.

The analyst must also determine if the results of the database search are meaningful. The statement "The Library Search said this compound is such-and-such; therefore, it is such-n-such!" should never be made. This is the single biggest mistake an analyst can make with respect to EI mass spectral data obtained with a GC-MS. A close comparison of a graphic display of the sample spectrum and the database spectrum must be made. What peaks differ in the two spectra? Is the database compound an esoteric compound where only a single analysis has ever been performed on it, or is the database compound a substance that is commercially available and in wide use?

> When using the *NIST/EPA/NIH Mass Spectral Database* with the *NIST MS Search Program*, one way to answer the question about the commonality of the identified compound is its presence in other databases such as the Commercially Available Fine Chemical Index (Fine), Toxic Substances Control Act Inventory (TSCA), and European Index of Commercial Chemical Substances (EINECS). (See the *Help* file in the *NIST MS Search Program* for a complete listing of the nine Other Databases which can contain a listing for the compound.)

Another way of determining the commonality of a compound is to see if it has any trade or common names listed as synonyms. Both the *Wiley Registry* and the *NIST/EPA/NIH Database* have synonyms for many of the included compounds.

Even after being convinced that the database spectrum is that of the same compound as the unknown substance, before coming to a conclusion, an authentic sample of the identified compound must be obtained and analyzed using the same instrument that was used to obtain the spectrum of the unknown. It would be best to obtain the spectrum of the authentic sample under the same matrix conditions as the unknown. If the matrix without the unknown is not available, then the matrix containing the unknown can be "spiked" with the authentic sample of the compound identified by the database search to see if only the intensity of the mass spectral peaks represent the suspected analyte change; and, as stated above, another confirming factor is the RI.

If a satisfactory solution cannot be reached with the aid of a database search, then the analyte spectrum will have to be interpreted. Interpretation is sometimes very straightforward and easy; but, at other times, there are all sorts of traps that can trick an analyst into a false identification. Many books have been written on the interpretation of mass spectra over the last four decades. Two of the most recognized books dealing solely with the interpretation of EI mass spectra that are still in print are *Interpretation of Mass Spectra*, 4th ed., Fred W. McLafferty and František Tureček, University Science Books, Mill Valley, CA, 1993, ISBN: 0935702253 (the second edition by McLafferty alone, Benjamin/Cummings, London, 1973, ISBN: 080537047-1, is preferable, especially for beginners) and *Understanding Mass Spectra: A Basic Approach*, 2nd ed., R. Martin Smith, Wiley, Hoboken, NJ, 2004, ISBN: 047142949-X. Volume 4 of the Elsevier *Encyclopedia of Mass Spectrometry* [2] is dedicated to understanding EI mass spectra. Volume 6 [3] also has a great deal of information about the interpretation of EI mass spectra. One of the better treatments of interpretation of EI spectra is found in *Introduction to Mass Spectrometry: Instrumentation, Applications, and Strategies for Data Interpretation*, 4th ed. [4–7] Before embarking in earnest on the arduous task of interpreting EI mass spectra, a formal course* is highly recommended.

5.1. Using the Database Search

Different manufacturers use different mass spectral database search algorithms that report different values representing the quality of the match between the sample spectrum and the database spectrum. The algorithms that constitute the *NIST MS Search Program* have become the most widely used. Several manufacturers use the *NIST MS Search Program* as their database-matching routine. Others use the *NIST MS Search Program's* underlying search routines with their own proprietary user interface, therefore using the same reporting values as is used in the *NIST MS Search Program*. All commercially available GC/MS data systems have utilities for the output of spectra in a text format, which can then be imported into the *NIST MS Search Program*. All commercially available mass spectral databases (except for one obscure (Germany origin) database containing ~3,000

* O. David Sparkman teaches such a course in either a three-day format or over the Internet (http://www.LCResources.com). Other people and organizations also offer courses on the interpretation of EI mass spectra. Many times, such courses are a part of the graduate curriculum at a university.

spectra of flavor/fragrance compounds) are provided in the *NIST MS Search Program's* format or in a format that allows for their conversion to the NIST format. The conversion requires a utility, which is provided with the *NIST MS Search Program*, or it may be downloaded at no charge from the NIST Web site (http://chemdata.nist.gov). For these reasons, this section is based on the use of the *NIST MS Search Program*.

The *NIST MS Search Program* has several different methods of searching a sample spectrum against various databases. The most commonly used is called the *Identity Search*, which can be carried out by using a *Normal* presearch algorithm or a *Quick* presearch algorithm. The default settings for the *Program* include the Normal presearch and are usually never changed. The *Help* file and the *MS Search v.2.0 Manual* (which can be accessed in either a *Microsoft Word* or PDF format from the *NIST Mass Spectral Database* Program Group, Start Menu) contain information needed to understand all the *NIST MS Search Program's* algorithms and database search options for sample spectra. Only the *Normal Identity* search will be discussed in this section. It should be noted that in addition to retrieving spectra from a database as a result of a search of a sample spectrum, spectra can be retrieved from a database by a molecular weight (nominal mass*) formula (elemental composition), Chemical Abstracts Service registry number (CASrn), or identity number (number assigned to a spectrum based on its position in the database) search. A database can also be searched by entering *m/z* values and respective relative intensity values for specific peaks in a mass spectrum (*Any Peaks* search). Sample spectra searches and all other searches can be constrained as to compounds present in other databases, molecular weight (nominal mass) ranges, elements present, numbers of specific elements present, name fragments, and specific mass spectral peaks. The use of these constraints is found in the *Help* file and in the *MS Search v.2.0 Manual*. Figure 5.3 shows the selection of one of the ways of retrieving spectra from a mass spectral database using the *NIST MS Search Program*. Figure 5.4 shows a list of the various constraints that may be applied to one of the different types of searches shown in Figure 5.3 or that may be combined to create a custom search.

The *NIST/EPA/NIH Mass Spectral Database*, as provided for the *NIST MS Search Program*, is organized into two separate databases: the Main (Mainlib) Database, which contains one and only one spectrum for each and every compound (individual CASrn) in the database; and the Replicate (Replib) Database, which contains multiple additional spectra of some (more commonly encountered) compounds. The default configuration for

---

* Nominal mass of a molecule, ion, or radical is calculated from the nominal masses of the elements from which it is composed. The nominal mass of an element is the integer mass of the most abundant naturally occurring isotope of the element. The term *nominal mass* is discussed in more detail later in this chapter.

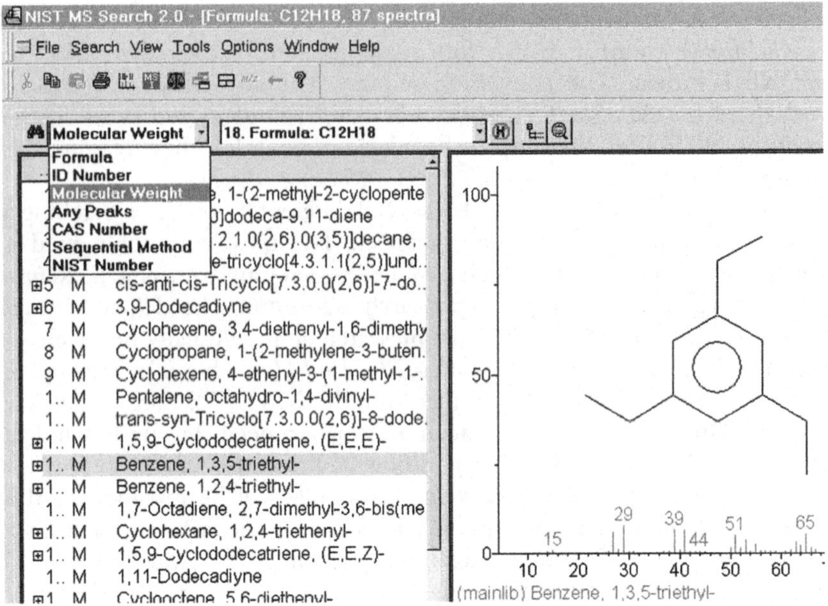

Figure 5.3 A search type other than that of a sample spectrum is selected in the Other Search view of the *NIST MS Search Program*. A dialog box appears for that search type allowing the selection of the databases to be searched and the application of desired constraints.

the *NIST MS Search Program* is to search both the Mainlib and Replib Databases. This means that multiple *hits* (matches of Database spectra to sample spectrum) of the same compound can be reported for a sample spectrum. These multiple hits can be an indication of a correct identification; however, care must be taken in coming to such a conclusion. Spectra can be imported from a text file or, in some cases, various GC/MS manufacturers' data systems can export a spectrum to a text file, call the *NIST MS Search Program*, after which the *Search Program* will automatically import the spectrum and search it. This is accomplished with a single command inside the manufacturer's proprietary data analysis software routine. This discussion will be limited to the search results obtained for sample spectra against the Mainlib and Replib Databases.

The use of a database search is very helpful in the case where no molecular ion peak is observed. Consider the RTIC chromatographic peak at a retention time of ~6.2 minutes in Figure 5.5. This chromatographic peak appears to be very symmetrical. It would be reasonable, based on an observation of the chromatogram, to select the spectrum representing the apex of the chromatographic peak for a database search.

Hitlist Search Options

Options | Constraints

Name Fragment: CYANO

Precursor m/z from 200 to 700

Peaks

Type	m/z	From	To
L	0	10	100

Mode: Absolute

Other databases

Fine
RTECS

Elements

Allowed number of atoms

O = 1
N > 0
C < 20

All of these

C
H
O
N

OK | Cancel | Help

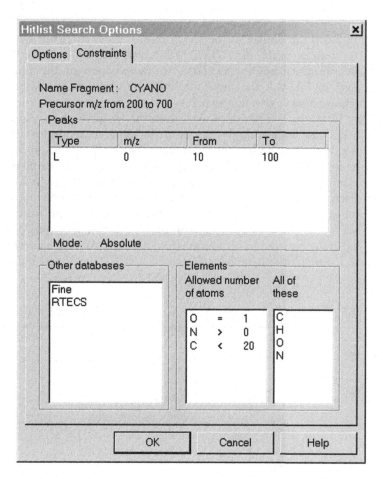

Figure 5.4 A displayed dialog box showing what constraints were used for a customized retrieval. Selecting the *Options* tab on this dialog box will display a list of the databases that were searched.

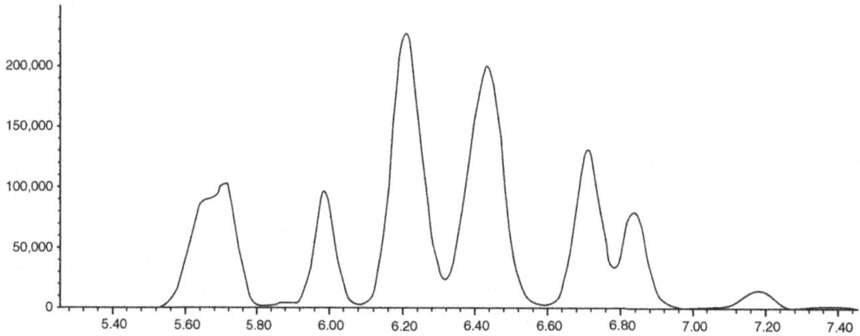

Figure 5.5 A reconstructed total ion current (RTIC) chromatogram of data acquired over the *m/z* range of 35–350 at a rate of one spectrum per second during a chromatographic analysis that used a temperature program of 50–200 °C at 10 °C min^{-1}.

Figure 5.6 is the display of the search results of the spectrum acquired at 6.216 minutes using the *NIST MS Search Program* and the Main and Replicate Databases in *NIST08*. An examination of the Hit List (lower left) reveals that the first four hits are for the same compound, 1,1,2-trichloroethane; the first two hits and the fourth hit are from the Replicate Database, and the third hit is from the Main Database. As stated above, a Hit List with multiple spectra of the same compound is a good indicator that the analyte has been identified. The first column on the left following the hit-ranking number in the Hit List provides the name of the database containing the spectrum. The next column to the right has the Match Factor. This is a numerical value between 0 and 999 that is an indication of how closely all the peaks in the sample spectrum match the peaks (m/z values and relative intensities) in the database spectrum. Values above 700 are considered to be confirmatory matches. All of these first four hits for 1,1,2-trichloroethane have Match Factors >700, another indicator

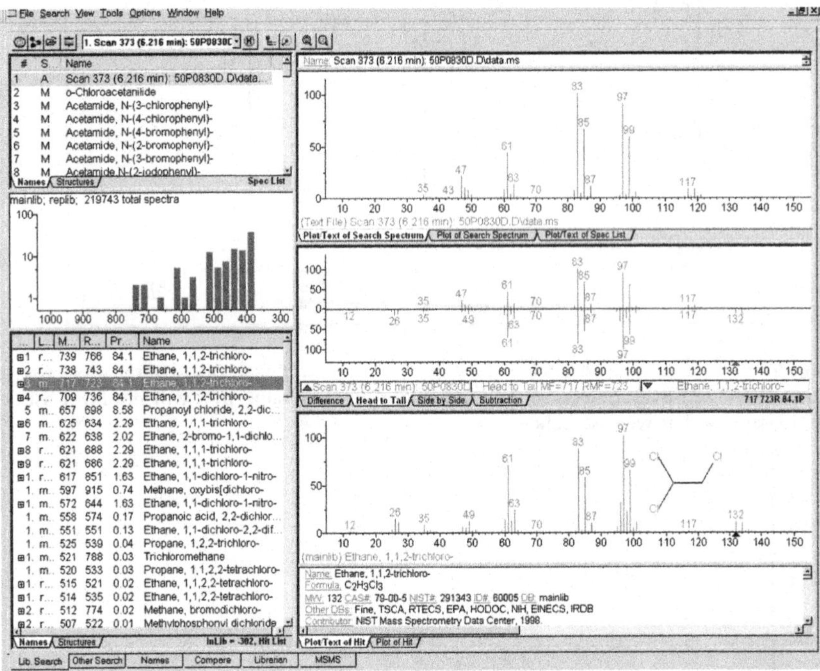

Figure 5.6 Results of a database search of the spectrum (upper right), representing the apex of the RTIC chromatographic peak at 6.216 minutes in Figure 5.5 of the Main and Replicate *NIST/EPA/NIH Databases* using the *NIST Mass Spectral Search Program*. The lower right spectrum is that of the third hit (which is the Mainlib spectrum); the panel on the lower left is the Hit List; and the display between the two spectra is a head-to-tail display of the sample (up) and Database (down) spectra.

along with multiple spectra of the same compound that the analyte has been identified.

If only the Hit List in this display was examined, it might be concluded that the analyte is 1,1,2-trichloroethane; however, examination of the sample spectrum, the spectrum of 1,1,2-trichloroethane that is in the Main Database, and the head-to-tail comparison (middle right side of Figure 5.5) of these two spectra would bring such a conclusion into question.

1. The sample spectrum has a cluster of peaks with nominal *m/z* 117 and an isotope peak pattern indicating an ion with three atoms of Cl. This cluster of peaks is missing the Main Database spectrum of 1,1,2-trichloroethane.
2. The Main Database spectrum of 1,1,2-trichloroethane has a cluster of peaks with nominal *m/z* 132 (the nominal mass of the compound) and an isotope peak pattern indicating an ion with three atoms of Cl, which is missing the spectrum of the sample.
3. The base peak in the sample spectrum is at *m/z* 83.
4. The base peak in the Main Database spectrum of 1,1,2-trichloroethane is at *m/z* 97.

These four differences are more than sufficient to question whether or not the sample spectrum is that of 1,1,2-trichloroethane. Returning to the data and displaying spectra at the beginning and end of the RTIC chromatographic peak strongly indicate that this RTIC chromatographic peak actually represents two different compounds—one whose mass spectrum has a base peak of *m/z* 83 and the other having a base peak of *m/z* 97 (Figure 5.7). Mass chromatograms for *m/z* 83 and 97 show that the two compounds actually have different retention times (Figure 5.8).

Performing a search of the mass spectrum obtained by subtracting the spectrum representing the apex of the mass chromatographic peak of *m/z* 83 from the spectrum representing the apex of the mass chromatographic peak of *m/z* 97 concludes with a different result. This result yields spectra of 1,1,1-trichloroethane as the top three hits (Figure 5.9). There are only two spectra for this compound in the Replicate Database. The Match Factor for this search ranges from a low of 773 to a high of 805 compared to a range of 709–739 for the previous search. More importantly, a visual inspection of the two spectra and the head-to-tail display show that there are no extraneous peaks present, a further indication of a much better probability of an identification. Reversing the subtraction process results in a spectrum, when searched, that gives the spectra of trichloromethane (a.k.a. chloroform) as the top three hits. Again, there are only two spectra for this compound in the Replicate Database.

The column to the right of the Match Factor column is the Reverse (R) Match Factor column. The R Match Factor is a match factor calculated disregarding any peaks in the sample spectrum that are not in the database spectrum. This value can be beneficial in determining the identity of two

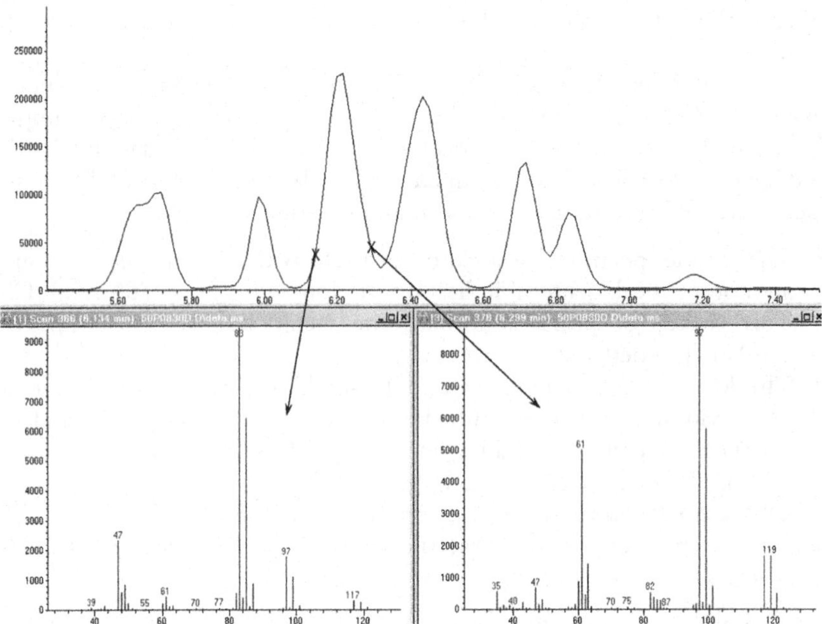

Figure 5.7 Spectra at the beginning (left side) and the end (right side) of the RTIC chromatographic peak indicate that two different compounds are represented. Note: This display from the Agilent ChemStation Data Analysis software results from a custom macro.

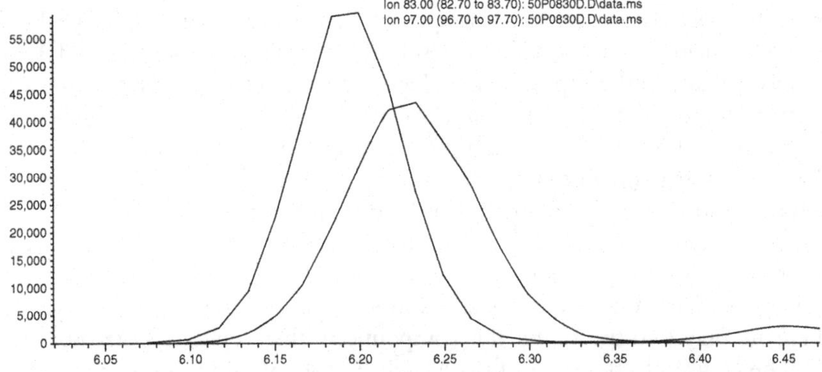

Figure 5.8 Mass chromatograms of m/z 83 (left) and m/z 97 (right), illustrating that the RTIC chromatographic peak with ~6.2 minutes retention time represents two different compounds.

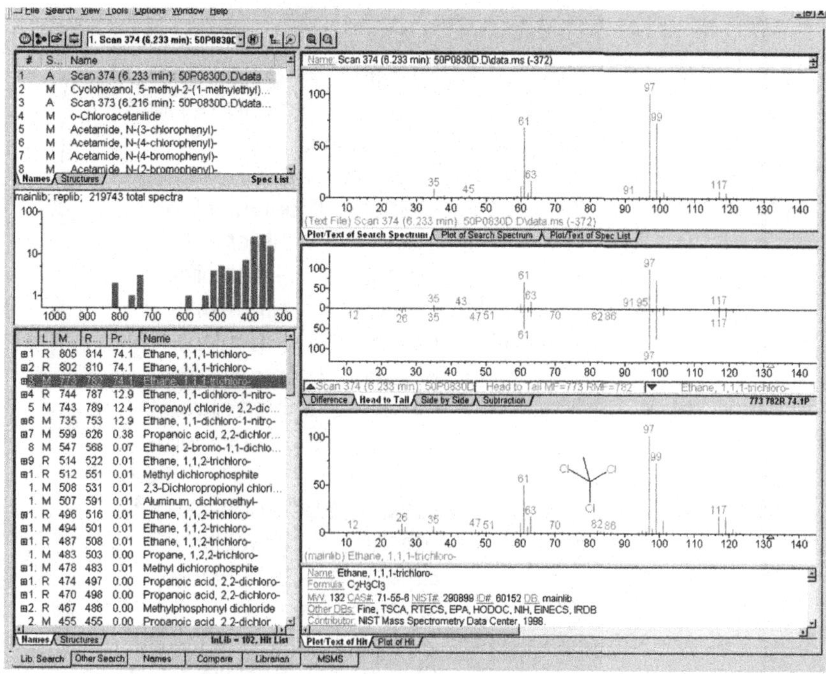

Figure 5.9 Result of a sample spectrum search of the spectrum obtained by the subtraction of the spectrum representing the apex of the mass chromatographic peak of *m/z* 83 from the spectrum representing the apex of the mass chromatographic peak of *m/z* 97.

compounds whose spectra are displayed as a single mass spectrum, as is the case in this example. In this example, one of the two components represented by the spectrum in Figure 5.6 is identified in the first three hits when the Hit List is ordered according to decreasing R Match Factor values (Figure 5.10). The comparisons of the sample spectrum with the first three hits of the Hit List ranked according to the R Match Factor illustrates how the value is determined (Figure 5.11). The spectrum of 1,1,1-trichloroethane was ranked as number 15 with an R Match Factor of 688 in the Hit List ordered according to the Match Factor values, whereas the spectrum of 1,1,2-trichloroethane, incorrectly identified by Match Factor order, had a number 6 ranking by the R Match Factor order with a value of 766. This shows that the R Match Factor has limited utility but should not be disregarded. When a Hit List has R Match Factor values that do not decrease with decreasing Match Factor values, the possibility of a mixed spectrum should be considered.

The other value in the Hit List (last column before the Names column) is a Probability value. This number represents the probability that if a spectrum is in the database that is of the same compound that generated the

...	L.	M...	▼ R	Pr...	Name
1	M	597	915	0.74	Methane, oxybis[dichloro-
⊞2	R	617	851	1.63	Ethane, 1,1-dichloro-1-nitro-
⊞3	M	521	788	0.03	Trichloromethane
⊞4	R	506	775	0.03	Trichloromethane
⊞5	R	512	774	0.02	Methane, bromodichloro-
⊞6	R	739	766	84.1	Ethane, 1,1,2-trichloro-
⊞7	R	738	743	84.1	Ethane, 1,1,2-trichloro-
⊞8	R	709	736	84.1	Ethane, 1,1,2-trichloro-
⊞9	M	717	723	84.1	Ethane, 1,1,2-trichloro-
⊞1.	R	488	722	0.02	Methane, bromodichloro-
1.	M	476	721	0.00	Methane, dichloronitro-
⊞1.	M	482	708	0.02	Methane, bromodichloro-
1.	M	657	698	8.58	Propanoyl chloride, 2,2-dic...
⊞1.	R	497	688	0.01	Ethane, 1,2,2-trichloro-1,1-...
⊞1.	R	621	688	2.29	Ethane, 1,1,1-trichloro-
⊞1.	R	621	686	2.29	Ethane, 1,1,1-trichloro-
⊞1.	R	505	677	0.03	Trichloromethane
1.	M	385	655	0.00	2-(1-Methylcyclopentyloxy)-...
⊞1.	M	572	644	1.63	Ethane, 1,1-dichloro-1-nitro-
2.	M	622	638	2.02	Ethane, 2-bromo-1,1-dichlo...
⊞2.	M	625	634	2.29	Ethane, 1,1,1-trichloro-

Names / Structures — InLib = -302, Hit List

Figure 5.10 Hit List of search results in Figure 5.6 sorted according to decreasing Reverse Match Factor.

sample spectrum, then an identification has occurred. The Probability value is based somewhat on the Match Factor but more on the uniqueness of the spectrum. As can be seen in Figure 5.12, the mass spectra of the three regioisomers* of xylene are essentially identical. When one of these spectra is searched against the *NIST08 Database*, hits with Match Factors >900 will result for the spectra of each of the three isomers and ethylbenzene in the Main and Replicate Databases. The Probability value may be low (<40%) because of the similarity of the spectra for these compounds. On the other hand, a search of a mass spectrum of *N,N*-dimethyl-*N'*-[3-(trifluoromethyl) phenyl]-urea acquired in an environmental matrix may result in a Match

* A term in organic chemistry used to describe isomers that have the same functional groups, but these groups are located at different positions; e.g., *n*-propyl alcohol and isopropyl alcohol or *m*-xylene and *p*-xylene.

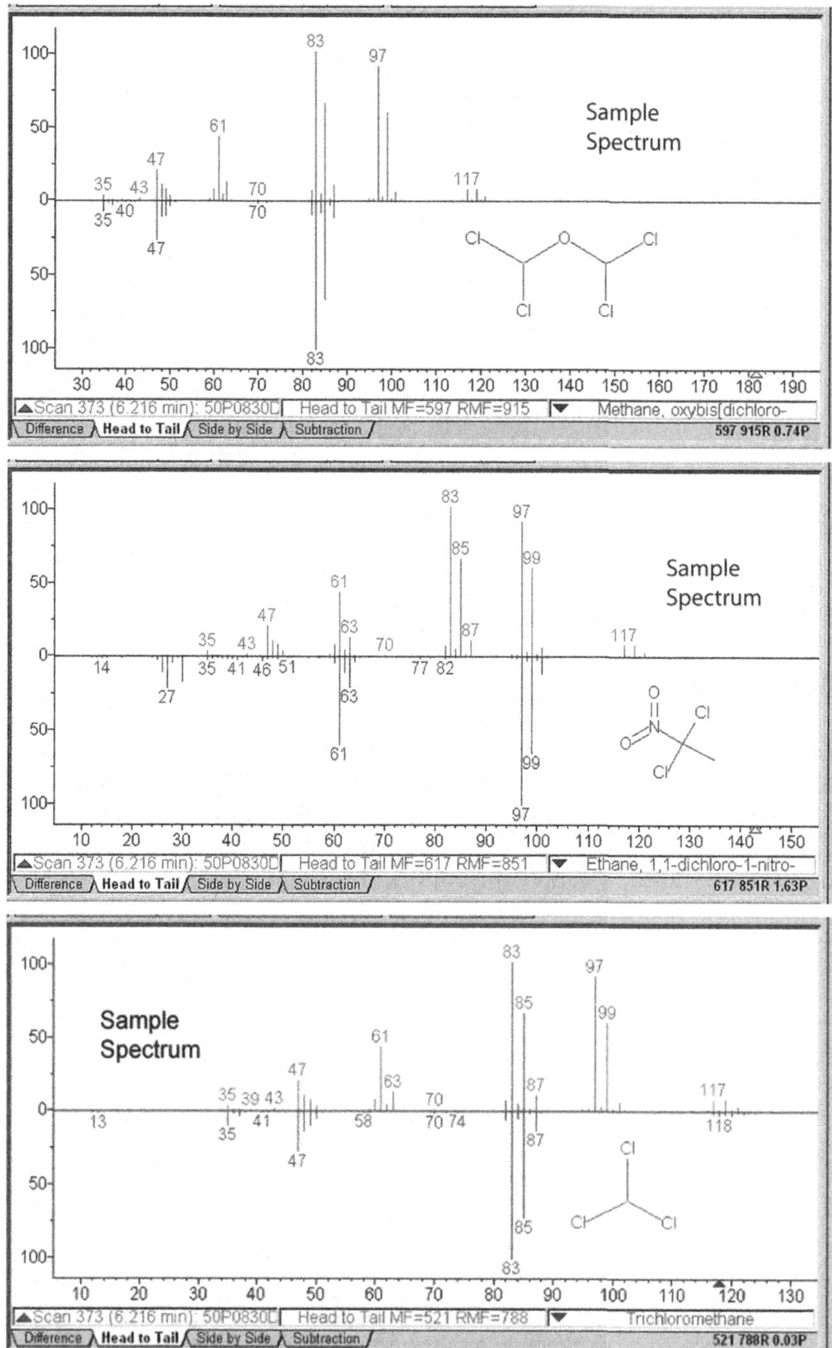

Figure 5.11 Comparison of the spectra from top three hits with the sample spectrum representing the apex of the RTIC chromatographic peak with a retention time of ~6.2 minutes.

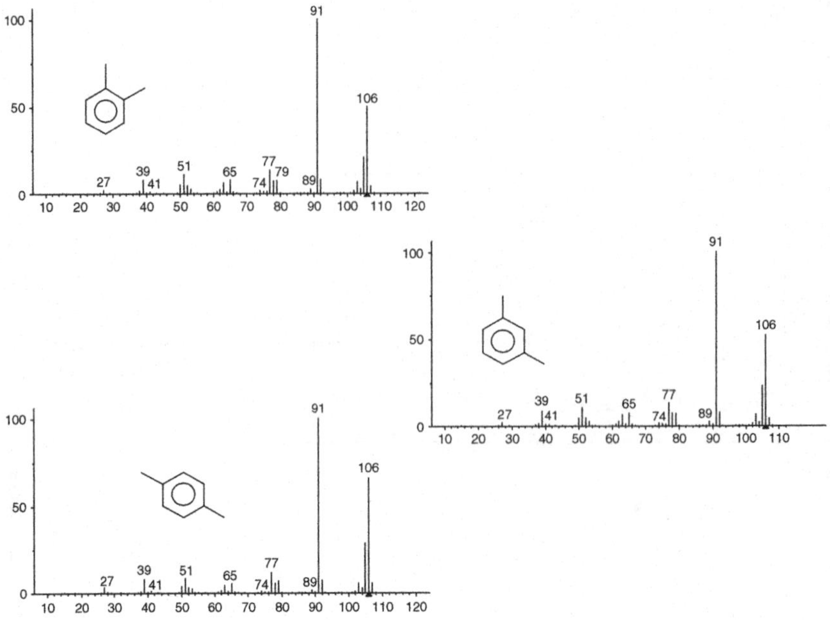

Figure 5.12 EI mass spectra of the three regioisomers of xylene.

Factor of <650 but a probability of 98% because of the uniqueness of the spectrum (Figure 5.13). The use of the term Probability can be confusing because this same term is used with the Probability Based Matching (PBM) algorithm (distributed by Agilent Technologies and Shimadzu) as an indication of how closely a sample spectrum and a database spectrum match.

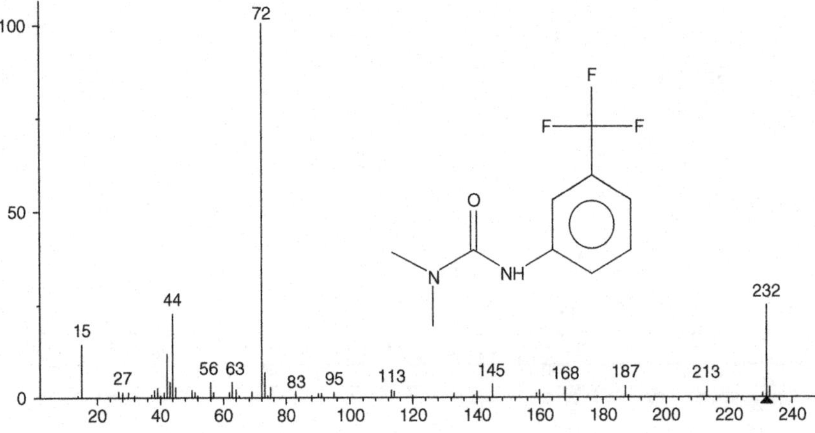

Figure 5.13 EI mass spectrum of *N,N*-dimethyl-*N'*-[3-(trifluoromethyl)phenyl]-urea.

Mass spectral database search routines are powerful tools. Their value is continually increasing due to the addition of spectra and the evaluation (and replacement or removal where necessary) of spectra that are carried out for the *NIST/EPA/NIH Mass Spectral Database* and many of the boutique databases. Analysts should be careful not to be abused by the database search. Many errors have resulted from database searches because the analyst did not take the time to look at the data. Use the mass spectral databases and the database searches, but do not let them do the job of the analyst, which is to identify the unknown.

If a satisfactory answer cannot be arrived at using a mass spectral database, it is necessary to try to match the spectrum to a structure based on the information in the mass spectrum. The most important piece of information is the compound's mass. If accurate mass is available, then an elemental composition may be possible. If only nominal mass is available, then the answer may lie in the spectrum's *dark matter* and the *m/z* values of the fragment ions. This leads to the first step in the evaluation of the mass spectrum: identification of the molecular ion ($M^{+\bullet}$) peak.

The $M^{+\bullet}$ peak can be used to provide information about the elemental composition of the analyte. It can also provide information as to the type of analyte that produced the mass spectrum. Intense $M^{+\bullet}$ peaks indicate that the analyte is an aromatic compound or, in some other way, resonantly stabilized (accompanied by few fragment ions many with low abundances) or represent an unsaturated ring (accompanied by a lot of fragment ions with high abundances). Low intensity or no $M^{+\bullet}$ peaks usually indicate straight-chained or branched aliphatic or allylic compounds. This latter case proves to be the most challenging in the interpretation of an EI mass spectrum.

5.2. Identification of a Molecular Ion Peak in an EI Mass Spectrum

The first step in determining the structure of an unknown analyte (an interpretation of an EI mass spectrum) is to try to identify the molecular ion ($M^{+\bullet}$) peak, if present. As already pointed out, about 20% of the compounds in the *NIST08 Mass Spectral Database* (*NIST08*) are represented by spectra that do not exhibit an $M^{+\bullet}$ peak. Always read the mass spectrum from the right to the left. The $M^{+\bullet}$ peak will be located on the right side of the mass spectrum. The peaks at the highest *m/z* values in the spectrum represent ions that contain the most information about the intact analyte. Table 5.1 provides a list of steps that should be taken in the identification of an $M^{+\bullet}$ peak in the mass spectrum obtained for an unknown analyte.

Table 5.1 Steps to be used in the determination of the *m/z* value for a molecular ion peak

1. The $M^{+\bullet}$ peak, if present, will be the peak at the highest *m/z* value in the mass spectrum that does not represent sample background or that is not an isotope peak.
   Corollary: The highest *m/z* value peak in the spectrum that is not due to background or that is not an isotope peak in a mass spectrum is not necessarily a $M^{+\bullet}$ peak.
2. The $M^{+\bullet}$ peak must be followed (right to left) by nonisotope peaks that represent losses of logical groups of atoms based on nominal mass and valence rules (see Table 5.2 for a list of some of the more common logical losses). Other logical losses can be found in Section 5.5.2. These logical losses are called the *dark matter* of the mass spectrum because the values are not of the detected ions but of the difference between the detected ions.
3. In an EI mass spectrum, the $M^{+\bullet}$ peak represents an odd-electron ion because all neutral molecules have an even number of electrons, and the $M^{+\bullet}$ is formed by the loss of a single electron from the neutral molecule to form a particle that has a net-positive charge. These odd-electron ions can fragment by the loss of a molecule, having a mass less than that of the analyte to produce an odd-electron fragment ion ($OE^{+\bullet}$) or they can fragment by the loss of a radical (a neutral species that has an odd number of electrons) to produce an even-electron fragment ion (EE^{+}). Molecular ions more frequently fragment through the loss of a radical. For all organic compounds containing only atoms of O, Si, S, P, or halogens in addition to C and H, the nominal mass will be an even number, and therefore the $M^{+\bullet}$ peak will have an even *m/z* value; and fragment ions formed by the loss of a radical will have an odd *m/z* value.
4. The *Nitrogen Rule*: All compounds that contain an odd number of N atoms in addition to C, H, O, Si, S, P, and halogens will have an *odd nominal mass*; therefore, the $M^{+\bullet}$ peak for these compounds will have an odd *m/z* value. Fragment ions retaining the odd number of nitrogen atoms will have even *m/z* values if they are formed by the loss of a radical. Fragment ions that do not retain the nitrogen atom when formed by the loss of a radical will have an odd *m/z* value.
5. Any compound containing an even number of nitrogen atoms will have an even *m/z* value $M^{+\bullet}$ peak. When molecular ions of these compounds fragment through the loss of a radical, the fragment ions will have an odd *m/z* value if they retain an even number of nitrogen atoms. They will have an even *m/z* value if they retain an odd number of nitrogen atoms.
6. No fragment ion can contain a larger number of atoms of any element than are contained by the molecular ion.

Table 5.2 Common neutral losses observed in EI mass spectrometry

M − 1	Loss of hydrogen radical	M − $^{\bullet}$H
M − 15	Loss of methyl radical	M − $^{\bullet}CH_3$
M − 17	Loss of hydroxyl radical	M − OH
M − 27	Loss of vinyl radical	M − $H^{\bullet}C{=}CH_2$
M − 29	Loss of ethyl radical	M − $^{\bullet}CH_2CH_3$ or M − $H^{\bullet}C{=}O$
M − 31	Loss of methoxyl radical or methoxy radical	M − $^{\bullet}OCH_3$ M − $^{\bullet}CH_2OH$
M − 43	Loss of propyl	M − $^{\bullet}CH_2CH_2CH_3$ or M − $CH_3^{\bullet}C{=}O$
M − 45	Loss of ethoxyl	M − $^{\bullet}OCH_2CH_3$
M − 57	Loss of butyl radical	M − $^{\bullet}CH_2CH_2CH_2CH_3$ or M − $C_2H_5^{\bullet}C{=}O$
M − 2	Loss of a hydrogen molecule	M − H_2
R − 16	Loss of a molecule of ammonia	R − CH_4
M − 17	Loss of a molecule of methane	M − NH_3
M − 18	Loss of a water molecule	M − H_2O
M − 28	loss of CO or ethylene or loss of formaldehyde	M − CO or M − C_2H_4 M − $H_2C{=}O$
M − 32	Loss of methanol	M − CH_3OH
M − 44	Loss of CO_2 or acetaldehyde	M − CO_2 or $CH_3(C{=}O)H$
M − 46	Loss of formic acid or ethanol	M − H(CO)OH or M − C_2H_5OH
M − 60	Loss of acetic acid	M − CH_3CO_2H
M − 90	Loss of silanol: HO − $Si(CH_3)_3$	M − O − Si − $(CH_3)_3$

The losses can be from the $M^{+\bullet}$ or from fragment ions through secondary fragmentation.

Once a peak has been identified as being a good candidate for the $M^{+\bullet}$ peak, the next step is to determine as much about this ion's elemental composition as possible. First apply the *Nitrogen Rule*. If the integer *m/z* value of the $M^{+\bullet}$ peak is an odd number, then the analyte molecule contains an odd number of nitrogen atoms. It is important to keep in mind that, for ions that have a nominal mass of greater than ~500 Da, the hydrogen mass defect (the difference between the nominal mass and the monoisotopic mass of hydrogen) in instruments that report *m/z* values to the nearest integer, the integer *m/z* value can represent an ion that has a nominal mass of 1 less than the observed *m/z* value, i.e., an $M^{+\bullet}$ with a reported *m/z* value of 721 probably has a nominal mass of 720 Da, *not* 721 Da, due to the hydrogen mass defect.*

The *Nitrogen Rule* is based on the nominal mass of the molecule. If the *m/z* value of the $M^{+\bullet}$ peak is even, the analyte does not contain an odd number of nitrogen atoms. One should be careful that this statement does not mean that the analyte molecule does not contain nitrogen. The analyte molecule may well contain nitrogen; however, if any atoms of nitrogen are present, there will be an even number of them.

5.2.1. Elemental Composition Based on the Relative Intensity of Isotope Peaks

An ion of a given elemental composition is not represented by a single peak in the mass spectrum. It is represented by a cluster of peaks. The lowest *m/z* value of these peaks is the monoisotopic mass/nominal mass peak. The peaks at successively 1 *m/z* unit intervals higher are *isotope peaks*. The pattern produced by this series of peaks and the intensity of these peaks relative to the monoisotopic peak can be informative as to the analyte's elemental composition, especially if there are X+2 elements (defined below) present.

The common series of elements encountered in compounds analyzed by GC/MS (C, H, N, O, Si, P, S, F, Cl, Br, and I) are categorized into three groups, X, X+1, and X+2**:

---

*The nominal mass of a H atom is 1 Da. Its monoisotopic mass is 1.007825 Da. This difference of 7.825 millimass units means that the nominal mass of H is 0.7825% less than the monoisotopic mass. This is the largest percent mass defect of any of the nuclides normally encountered in mass spectrometry. A saturated hydrocarbon with a nominal mass of ~500 Da will have about 35 atoms of carbon and about 72 atoms of hydrogen. The exact *m/z* value of the $M^{+\bullet}$ of this molecule will be 493.5634. If the mass spectrometer reports the *m/z* value to the nearest integer, then the reported value for this ion will be *m/z* 493 due to the hydrogen mass defect. Some manufacturers' data systems allow for mass defects to be applied to the integer *m/z* values of ions before the value is stored. This can be both positive and negative, depending on the mass defect correction value used (which is operator settable).

**As opposed to the X, X+1, and X+2 symbolism used in this book, many authors use the symbols A, A+1, and A+2. A is the IUPAC symbol for the mass number of an element (the number of neutrons and protons in the nucleus of the most abundant naturally occurring stable isotope of an element). Because this nomenclature is applied to the peaks in a mass spectrum that can have any *m/z* value (a general characterization of a number is the use of the symbol X) and not just to the individual elements, the use of the X, X+1, X+2 is more appropriate.

1. X elements: elements that have one and only one naturally occurring stable isotope (F, P, and I; H is also considered to be an X element even though it has two naturally occurring stable isotopes because the abundance of deuterium relative to protonium is so low that there is not a significant X+1 contribution in ions with a mass of <1,000 Da).
2. X+1 elements: elements that have two and only two naturally occurring stable isotopes, and these two isotopes differ in mass by 1 Da (C and N).
3. X+2 elements (O, Si, S, Cl, Br): elements that have two naturally occurring stable isotopes that differ in mass by 2 Da. Some of the X+2 elements have three naturally occurring stable isotopes, each having 1 Da higher mass. However, two of these isotopes are 2 Da apart; therefore, the elements are considered to be X+2 elements (O, Si, and S). Both Cl and Br have only two naturally occurring isotopes and they differ in mass by 2 Da.

All of the naturally occurring stable isotopes commonly encountered in mass spectrometry, along with their abundances and factors used to estimate the number of atoms present, are found in Table 5.3.

An examination of the abundances of the various isotopes relative to the nominal mass isotope of this limited list of elements shows that the elements with the most abundant X+2 isotopes are Br, Cl, S, and Si; it should also be noted that Si has the most abundant X+1 isotope of all of the elements. The high X+2 values result in very characteristic isotope peak patterns for ions containing atoms of these elements, with ions containing Cl and Br being the most characteristic. Examples of the X+2 isotope peak patterns for Cl/Br-containing ions can be seen in Figure 5.14; the peak-pattern appearances for ions containing various numbers of atoms of Cl and/or Br are very unique. When such an X+2 pattern is observed in the series of peaks representing the $M^{+\bullet}$, a safe conclusion is that the analyte molecule contains atoms of Cl and/or Br.

In some cases, such as analytes containing only a single atom of Br, the determination of the element and number of atoms of that element is easy; however, in the case of an ion containing six atoms of Cl, the determination is not so easy. Examination of the various patterns in Figure 5.14 reveals several possible combinations (Cl_6, Cl_2Br_2, and Cl_3Br). This can only be sorted out through the use of the tabular data (Appendix E). This is one of the reasons that, in addition to a graphic display of the spectrum, a tabular display is important when trying to perform an interpretation.

It is important to make sure that the spectrum is correctly evaluated. An initial examination of the two EI mass spectra in Figure 5.15 might lead to the conclusion that both analytes contain a single atom of bromine; however, a more careful examination reveals that the two peaks of almost equal intensity at the high end of the *m/z* scale in the spectrum on the top are only one *m/z* apart, whereas the two peaks similarly located in the spectrum

Table 5.3 List of elements and their masses, types, abundances, and their X factors

Type	Element	Symbol	Integer mass	Exact mass	Abundance	X+1 factor	X+2 factor
X	Hydrogen	H	1	1.0078	99.99		
		D or ^{2}H	2	2.0141	0.01		
X+1	Carbon	^{12}C	12	12.0000	98.91		
		^{13}C	13	13.0034	1.1	$1.1n_C$	$0.0060n_C^2$
X+1	Nitrogen	^{14}N	14	14.0031	99.6		
		^{15}N	15	15.0001	0.4	$0.37n_N$	
X+2	Oxygen	^{16}O	16	15.9949	99.76		
		^{17}O	17	16.9991	0.04	$0.04n_O$	
		^{18}O	18	17.9992	0.20		$0.20n_O$
X	Fluorine	F	19	18.9984	100		
X + 2	Silicon	^{28}Si	28	27.9769	92.2		
		^{29}Si	29	28.9765	4.7	$5.1n_{Si}$	
		^{30}Si	30	29.9738	3.1		$3.4n_{Si}$
X	Phosphorus	P	31	30.9738	100		
X+2	Sulfur	^{32}S	32	31.9721	95.02		
		^{33}S	33	32.9715	0.76	$0.8n_S$	
		^{34}S	34	33.9679	4.22		$4.4n_S$
X+2	Chlorine	^{35}Cl	35	34.9689	75.77		
		^{37}Cl	37	36.9659	24.23		$32.5n_{Cl}$
X+2	Bromine	^{79}Br	79	78.9183	50.5		
		^{81}Br	81	80.9163	49.5		$98.0n_{Br}$
X	Iodine	I	127	126.9045	100		

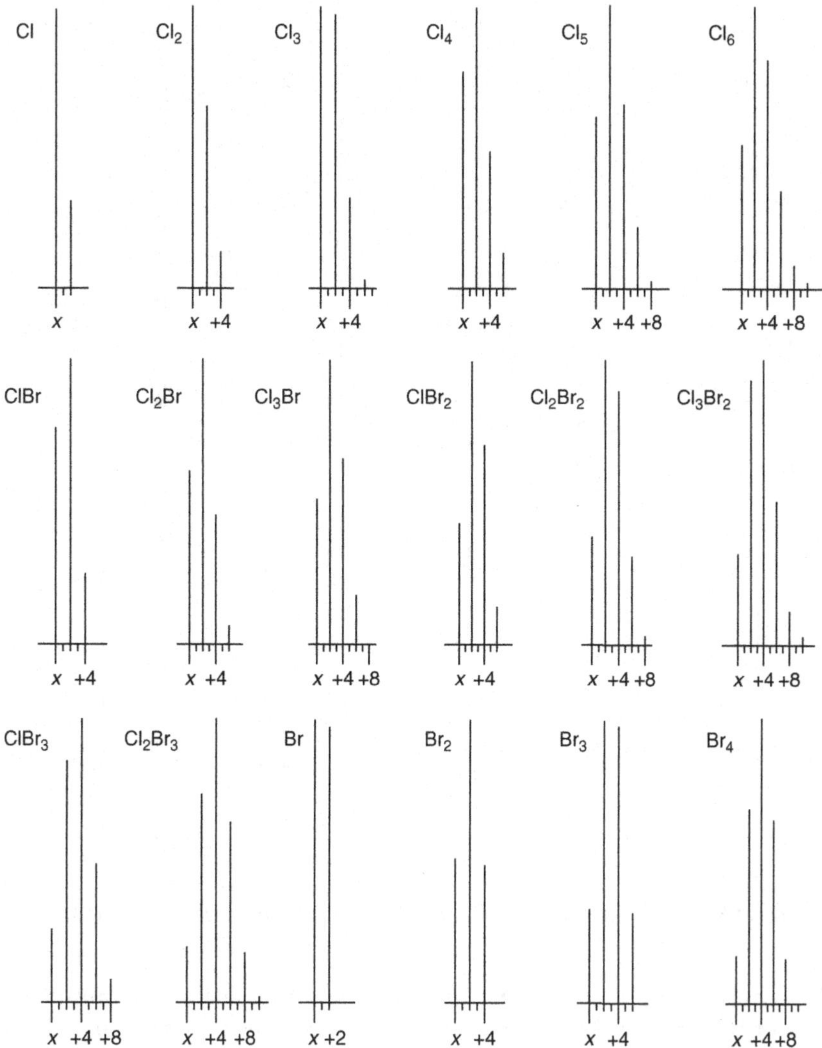

Figure 5.14 Graphic presentation of X+2 isotope peak patterns for ions containing atoms of Cl and/or Br.

on the bottom are 2 *m/z* units apart. This means that the spectrum on the bottom does represent an analyte that contains an atom of Br, and the spectrum on the top represents an analyte that does not contain any atoms of Br. What the two analytes do have in common is that neither contains an odd number of N atoms because the *m/z* value of the nominal mass $M^{+\bullet}$ peak is an even number.

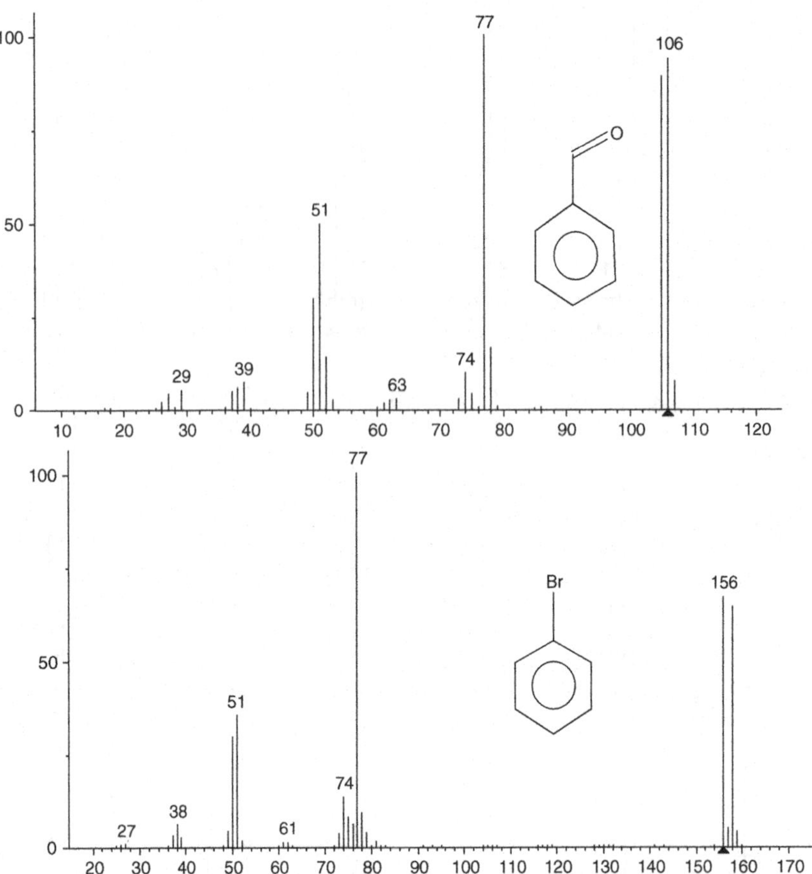

Figure 5.15 The EI mass spectra of benzaldehye (top) and bromobenzene (bottom).

When talking about the X+2 patterns for ions containing atoms of Cl and/or Br, take care not to call them "halogen isotope peak patterns." Two of the halogens, F and I, are *X* elements and make no contribution at X+2. These patterns are "X+2 isotope peak patterns" or "Cl/Br isotope peak patterns" *not* "halogen isotope patterns."

The presence of S is not as obvious from the X+2 pattern of an $M^{+\bullet}$ as is the presence of Cl and/or Br. It may be necessary to make this determination from the tabular display of the spectrum. If it is obvious that there are no atoms of Cl or Br present, look first at the relative intensity of the X+1 peak as determined from the tabular display of the spectrum. This value can be used to estimate the number of atoms of C that might be present (see Appendix D). Look at the value of the relative intensity of the X+2 peak (Table 5.3). If that value is much higher than what can be attributed to the number of carbon atoms* indicated by the X+1 peak, then there is something else contributing at X+2; that something else must be another X+2 element (S, Si, or O). After some practice, this will become more intuitive.

Try to use these isotope peak patterns associated with a $M^{+\bullet}$ peak to develop a possible elemental composition. This same technique can be used with peak patterns representing fragment ions. There are several important factors that must always be considered with respect to determining an elemental composition.

1. The precision of the measurement of the relative intensity of an isotope ion in most GC/MS instruments is about $\pm 10\%$ at best; therefore, once the observed isotope peak's relative intensity has been reconciled to within $\pm 10\%$ of the observed value, consider the reconciliation complete. This does not mean that it will not be necessary to reconsider the reconciliation; but, as a first attempt, consider it complete.
2. As the abundance of the monoisotopic ion decreases, the accuracy of the measurement of the relative intensity of the isotope peak becomes less.
3. Contribution at X+1 due to the presence of three atoms of N atoms equals the X+1 contribution of one C atom.
4. The contribution from C should be determined at X+2 before trying to assign the number of atoms of O. The contribution at X+2 for an ion containing 10 C atoms is 0.6% relative to the value for X. If the contribution due to the 10 atoms of carbon is not considered for an X+2 peak that has a 1.0% relative intensity to the monoisotopic peak before assigning the number of atoms of O, five atoms of O rather than two may be assigned.
5. When ions of two different elemental compositions are separated by 1 *m/z* unit (as is the case for the ions with *m/z* 91 and 92 in the mass spectrum of toluene (Figure 5.16)), the X+1 contribution of the lower *m/z* value ion must be subtracted from that of the ion one unit higher before determining the X+1 relative intensity of the higher *m/z* value ion.

A more detailed treatment of the determination of elemental compositions of ions based on isotope peak patterns is found in Chapter 5 of *Introduction to Mass Spectrometry: Instrumentation, Applications, and Strategies for Data Interpretation*, 4th ed.; and in the first article "Rules for Mass Spectral Interpretation: The Standard Interpretation Procedure" in Chapter 7 "Structure Determination of Organic Compounds" of the Elsevier *Encyclopedia of Mass Spectrometry*, Vol. 4 *Fundamentals of and Applications to Organic (and Organometallic) Compounds*, Nico M.M. Nibbering, Editor. The steps to

* Although carbon has no X+2 isotope peak, there is an abundance at X+2 attributable to the probability that the ion will have two atoms of ^{13}C present. As seen in Table 5–3, the X+2 factor for the probability of two atoms of ^{13}C being present is 0.006 times the number of C atoms squared.

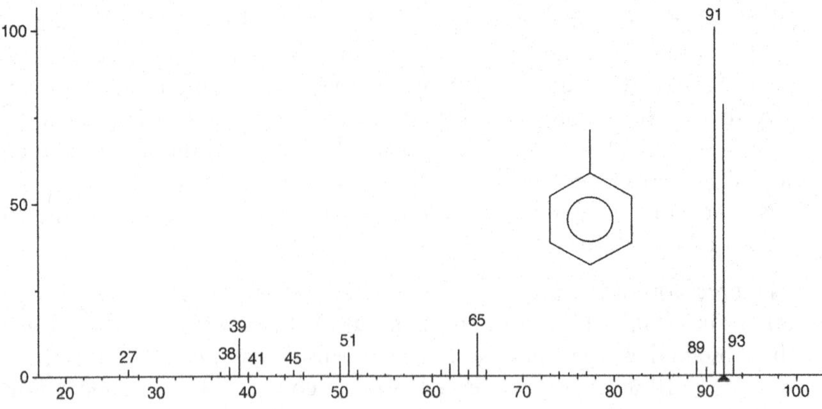

Figure 5.16 EI mass spectrum of toluene.

use in the determination of an elemental composition are detailed in Appendix F. These steps should be followed in the exact order that they are listed when using isotope peak intensities to determine an elemental composition. Examples of these determinations are also found in Appendix F.

5.2.2. Elemental Composition Based on the Accurate Mass Assignment of an Ion

An accurate mass* assignment of an ion to the nearest 0.1 millimass units (0.0001) is required for the assignment of an unambiguous elemental composition of ions encountered in the EI spectra obtained by GC/MS. Even though the mass of an electron is 0.0005 Da (0.5 millimass units), which means that the difference in the mass of the molecule and the mass of a molecular ion is 0.0005 Da, for purposes of determining the elemental composition of an ion, this difference is disregarded. In a 1994 editorial in *The Journal of the American Society of Mass Spectrometry*, Michael Gross [8] reports the mass accuracy ranges for unambiguous elemental compositions of ions of various masses up to about 600 Da. This is sufficient considering 600 Da is the upper limit of mass for most compounds analyzed by GC/MS (polyfluorinated derivatives and polybrominated biphenyls are an

*Two terms are often erroneously used interchangeably; *accurate mass* and *exact mass*. An *accurate mass* is a measured value determined based on a measured *m/z* value and the charge state of an ion. The *exact mass* is a value calculated from the published masses of the elements and number of atoms of these elements in an elemental composition. Both are usually used in terms of a monoisotopic mass.

exception; however, these are usually target analytes and not encountered as unknowns).

Accurate mass assignments have usually been associated with instruments that have high resolving power. The two mass spectrometers used in GC/MS that have a high resolving power are the TOF-MS and the double-focusing instrument. As the TOF mass spectrometer has gained popularity in GC/MS due to its ease of operation, accurate mass data are being reported more often. The elemental composition is arrived at using a computer program, such as that provided with the data systems from some manufacturers of high resolving power instruments or that are found in a suite of mass spectrometry utilities such as *MS Tools* (available from ChemSW [http://www.ChemSW.com]). The accurate mass is entered, along with a list of elements and the range of the number of atoms of each element that might be present. The software returns a series of possible elemental compositions along with the exact mass for each composition and whether the ion represented by the elemental composition is an odd-electron or even-electron ion.

Through the use of software developed by Cerno Bioscience, *MassWorks*, accurate mass assignments can be obtained for data acquired with transmission quadrupole mass spectrometers where the data have been acquired and stored in the *profile mode* (10–20 measurements per each integer *m/z* value in the spectrum). *MassWorks* uses an average of several spectra of perfluorotributylamine (PFTBA) acquired in the profile mode to calibrate the *m/z* scale of the instrument by defining the mass spectral peak shape at multiple *m/z* values. These PFTBA spectra can be in a separate data file, or they can be in the same data file that contains spectra of the analyte which has peaks representing ions that are to have accurate *m/z* values assigned. The best results, using the Agilent 5975 MSD, have been observed when the PFTBA calibration spectra are in the same data file as the analyte spectra.

An example of the benefit of accurate mass assignment can be seen in the consideration of an ion with *m/z* 99. This ion can have an elemental composition of C_7H_{15} or $C_6H_{11}O$, both of which have a nominal mass of 99 Da. The two ions have respective exact masses of 99.1174 Da and 99.0810 Da; a difference of 36.4 millimass units is easily distinguishable from one another.

If *MassWorks* is going to be used with the data, the acquisition must be set to acquire data in the profile mode. The data systems on most commercially available GC/MS instruments based on the QMF have an option for the storage of data in the profile mode. The Agilent Technologies ChemStation data acquisition software's **MS SIM/Scan Parameters** dialog box, **Aq. Mode** setting selection box, in the *MS Instrument Parameters* area of this dialog box, has three options: **Scan**, **SIM**, and **Raw Scan**. The **Raw Scan** option stores the data in the profile mode (Figure 5.17). Care must also be taken in the value used for the acquisition threshold (accessible by selecting the **Edit Scan Parameters** button in the **MS SIM/Scan Parameters** dialog box to display the **Edit Scan Parameters** dialog box and then selecting the

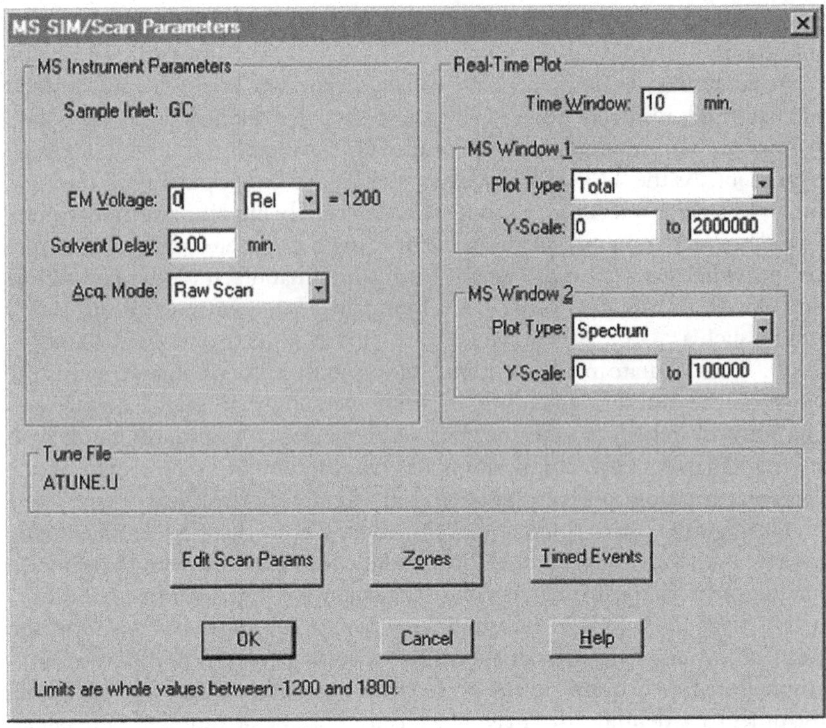

Figure 5.17 Agilent MS SIM/Scan Parameters dialog box with settings for acquiring and saving data in the profile mode.

Threshold and Sampling Rate tab. The default acquisition *Threshold* (the single intensity below which data are disregarded) is set to 150 counts. This should be changed to zero (0). Figures 5.18 and 5.19 show the difference in the appearance of spectra acquired and stored in the profile mode and those stored in the centroid mode. One significant disadvantage of storing data in the profile mode is the size of the data file. Two files were acquired for the same sample under the same conditions with the same amount of sample being injected. One was saved as centroid data and the other as profile data. The size of the data file with centroid data was 962 KB; the size of the profile data file was 43,068 KB, ~40 times larger than the centroid data file.

5.2.3. After the Molecular Ion's Elemental Composition Has Been Determined

If it is possible to deduce an elemental composition for the molecular ion, the next step in the interpretation is to determine if that composition makes

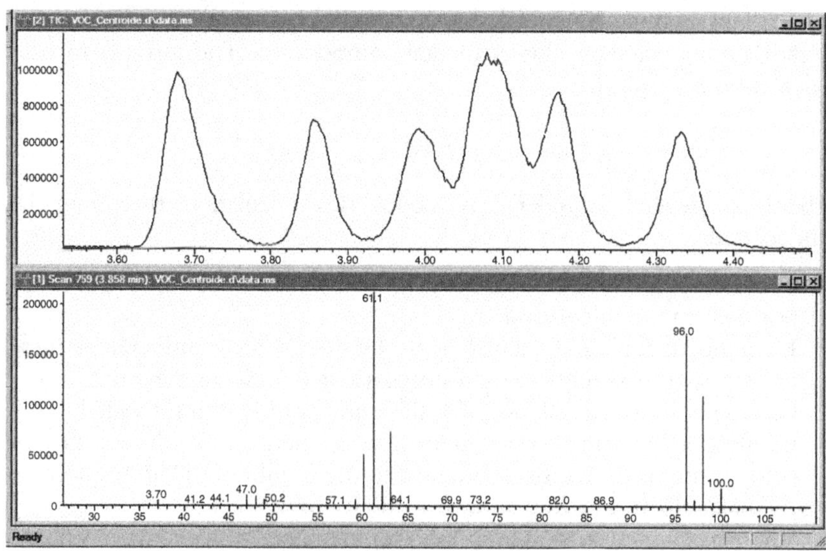

Figure 5.18 An RTIC chromatogram (top) produced from centroid mass spectra and a centroid mass spectrum (bottom).

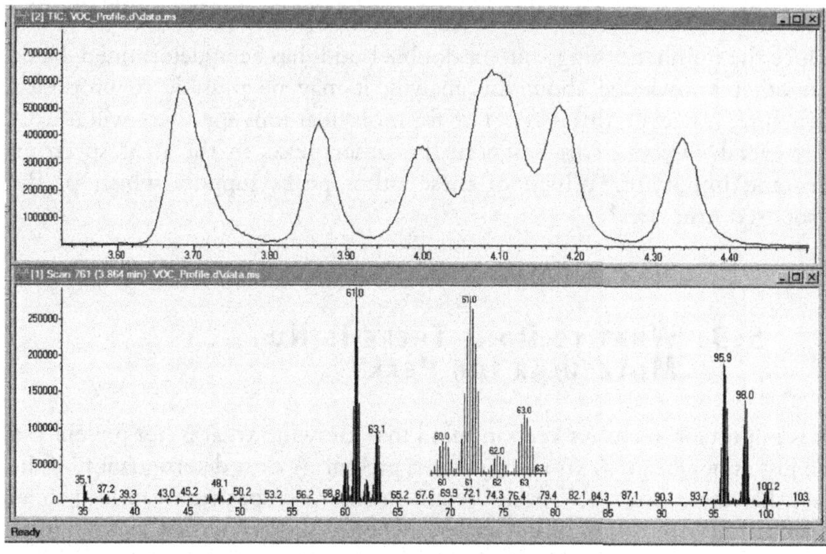

Figure 5.19 An RTIC chromatogram (top) produced from profile mass spectra and a profile mass spectrum (bottom). The insert is to illustrate the actual appearance of the mass spectral peaks in the profile mode.

sense. This is done by calculating the number of *rings plus double bonds* (R+dB) represented by this elemental composition. The formula for this calculation is as follows:

$$R{+}dB = C_{\#} - \tfrac{1}{2}H_{\#} + \tfrac{1}{2}N_{\#} + 1$$

There are several considerations about this calculation that must be understood:

1. The elements O and S, which are considered to have a valence of 2, are not used in the calculation.
2. Elements of the same valence as C, H, and N are considered as these elements; i.e., the number of Si atoms is added to the number of C atoms because both have a valence of 4; the number of P atoms is added to the number of N atoms because both have a valence of 3; the number of halogen atoms (F, Cl, Br, I) is added to the number of H atoms because the halogens and hydrogen have a valence of 1.
3. The R+dB calculation is only valid for elements in their lowest valence state. When S has a valence of 4 or 6 or P has a valence of 5, any double bonds associated with these elements will not be determined using the R+dB calculation.
4. The R+dB calculation will always end as a whole number for molecular ions and any odd-electron fragment ions.
5. The R+dB calculation will end in ½ for all even-electron ions.
6. Triple bonds are equivalent to and considered to be two double bonds.

Once the number of rings and/or double bonds has been determined, based on other knowledge about the analyte, it may be possible to propose a structure; however, this is very rare for molecular ions and there will usually be several choices. This is where the other peaks in the mass spectrum become important. Which of these other peaks support which of the proposed structures?

5.3. What to Do If There Is No Molecular Ion Peak

It is important to always keep in mind that knowing what is not present can be just as significant as knowing what is present. A clear determination of the fact that there is no $M^{+\bullet}$ peak present in the mass spectrum can lead to an eventual determination of the analyte, whereas suspecting that a peak represents a molecular ion and it does not can lead to frustration. Sometimes when it is obvious that there is no $M^{+\bullet}$ peak, peaks representing ions formed by the losses from the molecular ion can be used to deduce the nominal mass of the

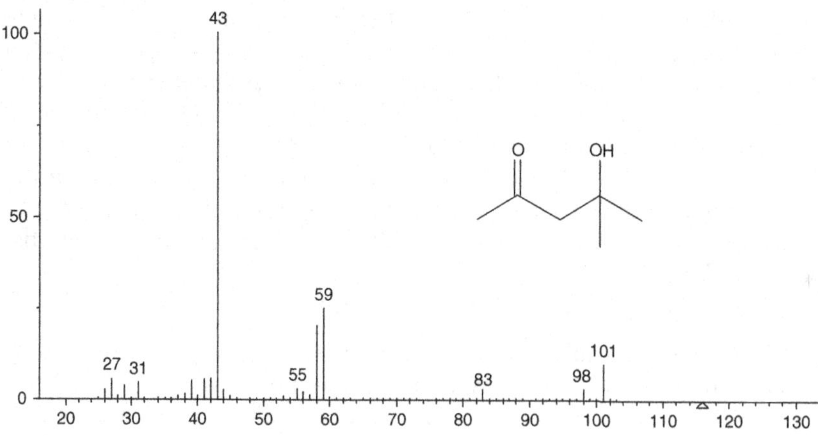

Figure 5.20 EI mass spectrum of 4-methyl-4-hydroxy-2-pentanone.

analyte. A good example of this is seen in the mass spectrum of 4-methyl-4-hydroxy-2-pentanone (Figure 5.20). Reading the spectrum from the right to the left, the first peak that is not an isotope peak or that does not represent an ion due to sample background is at *m/z* 101. This becomes a candidate for the $M^{+\bullet}$ peak. If this is a $M^{+\bullet}$ peak, then the analyte molecule contains an odd number of N atoms. The peak at the next lowest *m/z* value is at *m/z* 98. If the peak at *m/z* 101 is a $M^{+\bullet}$ peak, then this loss of 3 *m/z* units does not make sense. The next peak is at *m/z* 83, which is 18 *m/z* units less than the suspected $M^{+\bullet}$ peak at *m/z* 101. This could represent the loss of a molecule of water, and the resulting odd-electron fragment ion has retained the odd number of nitrogen atoms. The next peak is at *m/z* 59, which is 42 *m/z* units less than that of the suspected $M^{+\bullet}$ peak. The ion with *m/z* 59 has to be an odd-electron fragment ion that has retained the nitrogen atom or an even-electron ion that does not contain an odd number of nitrogen atoms. There is also a peak at *m/z* 58 that could be a substituted immonium ion (an even-electron ion with a single atom of N) or an odd-electron fragment ion with no atoms of N, which is often characteristic of an aliphatic ketone. The next peak is the base peak at *m/z* 43, which could represent a methyl acylium ion often found in the mass spectrum of methyl ketones, another indication that the analyte may have a ketone moiety. There is a very noticeable peak at *m/z* 31 even though its intensity is low. Peaks at *m/z* 31 are considered to be diagnostic for the mass spectra of aliphatic alcohols. The low intensity of the peak at *m/z* 31 indicates that this could be the mass spectrum of a secondary or tertiary aliphatic alcohol. It is also a known fact that the mass spectra of aliphatic alcohols do not exhibit a $M^{+\bullet}$ peak.

At this point, it is fairly safe to say that the peak at *m/z* 101 is not a $M^{+\bullet}$ peak. It is possible that the peak at *m/z* 101 does not belong to the mass spectrum being interpreted; however, before discarding it, consider some other possibilities. The EI mass spectra of aliphatic alcohols can exhibit an $[M - H_2O]^+$ peak. This peak would represent an odd-electron ion and therefore have an even *m/z* value if it did not contain any atoms of nitrogen. This means that the peak at *m/z* 98 is a good candidate for representing the $[M - H_2O]^+$ ion. If this assumption is correct, then the nominal mass of the analyte is 116 Da (98 + 18). Now the peak at *m/z* 101 makes sense as representing an $[M - CH_3]^+$ ion, 15 *m/z* units less than that of the $M^{+\bullet}$ peak.

Take inventory of what the mass spectrum now indicates.

1. The analyte probably contains a secondary or tertiary alcohol moiety. These moieties will result in substituted oxonium ions ($H_2C{=}O^+H$) that have an *m/z* value of 31. The peak at *m/z* 59 could represent a dimethyl- or ethyl-substituted oxonium ion.
2. The peak at *m/z* 58 indicates that the analyte could be an aliphatic ketone with a C atom located in the gamma position relative to the carbonyl. Ions of *m/z* 58 are formed by a β cleavage induced by a γ-hydrogen shift (sometimes called a McLafferty rearrangement). The presence of this ketone moiety is supported by the peak at *m/z* 43, which could represent a methyl acylium ion ($H_3C{\equiv}O^+$).

Fred McLafferty makes a point in the introduction for the computer-based training program produced by the then Hewlett–Packard, Inc. (manufacturer of GC and GC/MS instruments; now known as Agilent Technologies) that "Mass spectrometry is about the masses of the molecules and the masses of the pieces of the molecules." With this statement in mind, the mass of the substituted oxonium ion (59 Da) and the methyl ketone moiety (43 Da) adds to 102 Da. If the nominal mass of the analyte is suspected to be 116 Da, the difference is 14 Da. This means that the analyte must contain a CH_2 moiety separating the methyl keto group and the secondary or tertiary aliphatic alcohol group. The presence of the CH_2 group is necessary for a H atom to be present on a C atom that is gamma with respect to the carbonyl group, which results in the ion with *m/z* 58.

The only remaining question for the interpretation of this mass spectrum is whether the alcohol moiety is a secondary alcohol with a methyl substitution on the carbinol carbon or a tertiary alcohol with two methyl groups on the carbinol carbon. This question is easily answered because the mass spectrum exhibits an $[M - 15]^+$ peak and not an $[M - 29]^+$ peak, meaning that the substitution on the carbinol carbon must be two methyl groups and not an ethyl group. Schemes 5.1 and 5.2 explain the formations of the ions represented in the mass spectrum with the charge on the alcohol O atom or on the O atom of the ketone.

The above interpretation required the knowledge of a number of somewhat esoteric facts. This knowledge only comes with the continued practice

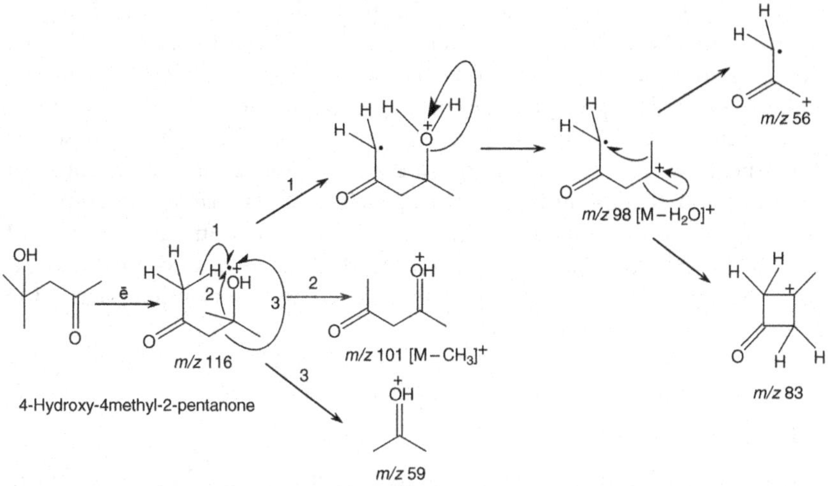

Scheme 5.1 Fragmentation initiated, based on the charge and radical sites being associated with the hydroxyl oxygen.

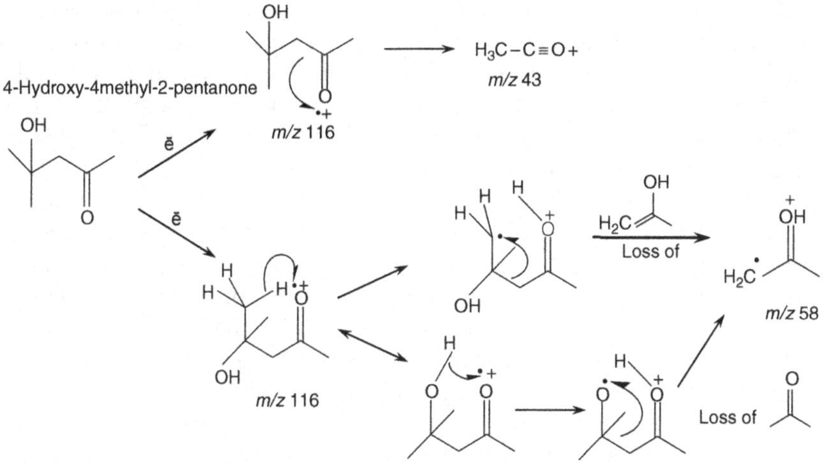

Scheme 5.2 Fragmentation initiated, based on the charge and radical sites being associated with the carbonyl oxygen.

of interpreting EI mass spectra. Interpretation of mass spectra is not a skill that can be pulled out and used occasionally. It is something that requires continued use until it becomes a gestalt with the analyst and, even then, must be exercised often. This is why mass spectral database searches are so important.

Another way to guess the nominal mass of an analyte is to look for information about the analyte's elemental composition based on what elements may be present in ions that have *m/z* values lower than the highest *m/z* value peak in the mass spectrum. For example, if peaks representing a fragment ion at a lower *m/z* value than the highest *m/z* value peak in the spectrum indicate the presence of three atoms of chlorine, and the highest *m/z* value peak indicates that it represents an ion with two atoms of chlorine, try adding 35 to the nominal *m/z* value of the highest *m/z* value peak to get an idea of the analyte's nominal mass. If none of these tactics work to guess the analyte's nominal mass, use derivatization or soft ionization.

5.3.1. Soft Ionization

One way of determining the nominal mass of an analyte when no $M^{+\bullet}$ peak is present is to use one of the soft ionization techniques. The more popular of the two commercially available techniques is chemical ionization (CI). If instrumentation is available that has field ionization (FI) capability, refer to that equipment's manuals and application notes.

As was pointed out earlier in Chapter 4, one of the main advantages of the quadrupole ion trap (QIT) GC-MS is the ability to switch between EI and CI without any hardware changes or worrying about ion-source fouling. This is because the QIT uses a time domain to accomplish CI and therefore needs only a reagent gas partial pressure of ~10^{-3} Pa. Some commercially available GC/MS instruments based on the QMF and double-focusing geometry have ion-source designs that allow for switching between EI and CI through the use of a probe-like device that either replaces the ion volume of the ion source or changes its geometry to a closed configuration (required for EI) or to an open configuration (required for EI). However, this arrangement does not do anything for the potential problem of source contamination due to the large amount of reagent gas required in such systems (a partial pressure of 10–100 Pa).

Many of the analytes that do not produce $M^{+\bullet}$ peaks are polar and therefore very amenable to CI because of their high proton affinity. Behavior of different analytes under CI conditions is sometimes a source of confusion. As an example, CI is often used in the analysis of aliphatic alcohols to establish their nominal mass. These compounds do not exhibit $M^{+\bullet}$ peaks when subjected to EI under GC/MS conditions (small amounts of time to measure ion current for specific *m/z* values). Sometimes the EI mass spectrum for aliphatic alcohols exhibits an $[M - H_2O]^+$ peak, and other times the highest *m/z* value peak in the mass spectrum represents an $[(M - H_2O) - C_2H_4]^+$ ion (an M – 46 peak). The nominal mass of 2-ethyl-1-hexanol is 130 Da. Under CI conditions, using methane as the reagent gas, the highest *m/z* value peak in the mass spectrum represents what appears to be an $[M - H]^+$ ion at *m/z* 129. Herein lies the source of confusion. The 129 value is an odd number; this means that the ion represented by this peak is an even-electron ion (provided it is understood that the analyte has no atoms of N); therefore, it might be assumed that the nominal mass of the analyte is 128 Da because it is assumed that a protonated molecule (MH^+) is formed under CI conditions. However, the next lowest *m/z* value peak appears at what would represent an $[M - 17]^+$ ion (Figure 5.21). The peak at *m/z* 129 actually represents an $[(M + H) - H_2]^+$ ion, whereas the peak at *m/z* 113 represents an $[(M + H) - H_2O]^+$ ion and the peak at *m/z* 111 probably represents the loss of water (H_2O) from the $[(M + H) - H_2]^+$ ion (*m/z* 129). This example illustrates the importance of carefully evaluating the mass spectrum and understanding what is going on in the ion source before reaching a conclusion.

Even though the mass spectrum obtained by CI usually contains the information needed to deduce the nominal mass of the analyte, as seen

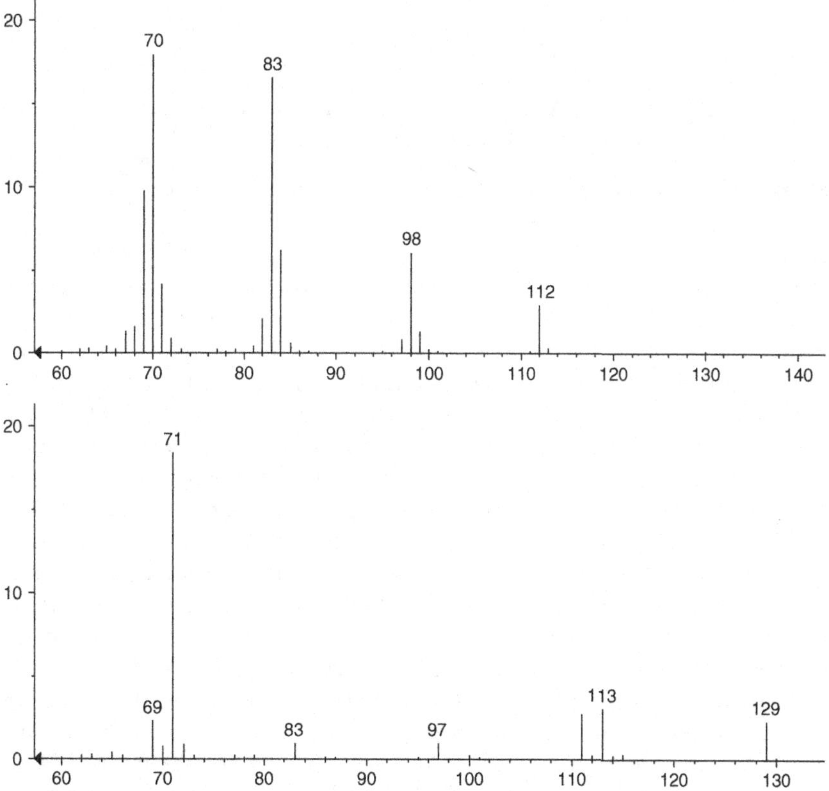

Figure 5.21 EI (top) and methane CI (bottom) of 2-ethyl-1-hexanol.

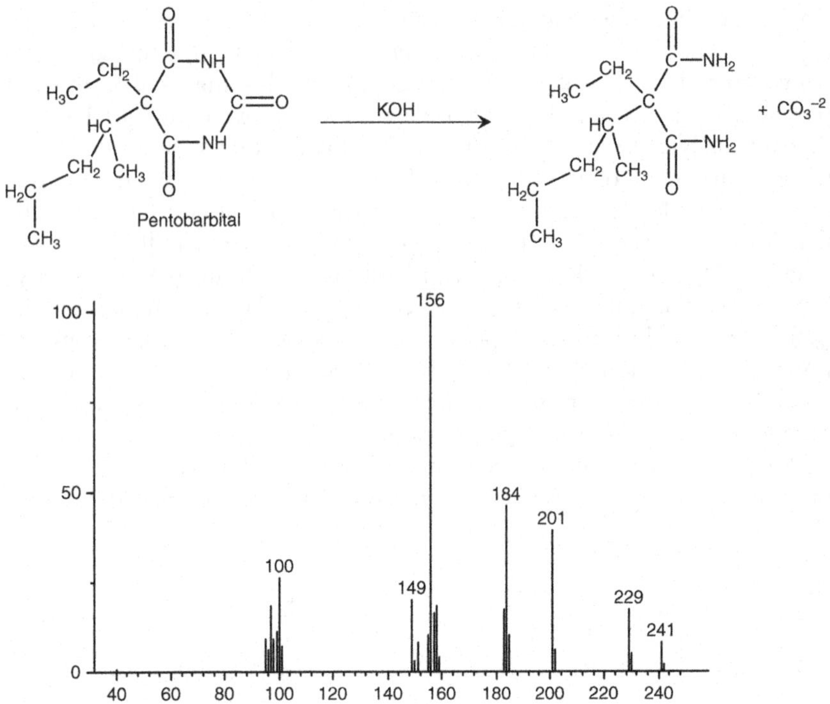

Figure 5.22 Methane CI mass spectrum of a malonamide of pentobarbital.

from the above example, the answer may not be straightforward. Another good example of the need to interpret the CI mass spectrum is seen in Figure 5.22—the methane CI mass spectrum of a malonamide of pentobarbital, nominal mass 200 Da. The peak at *m/z* 201 represents the protonated molecule. The peaks at *m/z* 229 and 241 represent the $[M+C_2H_5]^+$ and $[M+C_3H_5]^+$ adduct ions, respectively. The peaks at *m/z* value below 201 represent fragments of the protonated molecule that result from the excess energy coming from the difference in the proton affinities of the methane and the analyte when the protonated molecule is formed. If this was a mass spectrum of an unknown, each peak in relation to each other peak would have to be evaluated individually. The spectrum would be read from right to left. The difference between 241 and 229 is 12. This has no immediate meaning. The difference between 241 and 201 is 40. This is one less than 41, which is the mass added by the formation of the $[M+C_3H_5]^+$ adduct ion, which can be formed in methane CI. The difference between 229 and 201 is 28, which is one less than 29, the mass added when the $[M+C_2H_5]^+$ adduct ion is formed in methane CI. The peak at *m/z* 184 is 17 *m/z* units less than the peak at *m/z* 201, which could represent the loss of a molecule of ammonia from the protonated molecule. All of this indicates that the nominal mass of the unknown analyte is 200 Da. Knowledge of the sample being analyzed would be very helpful in this case.

It should be pointed out that another advantage of CI in the internal ionization QIT is the ability to use the vapor of various liquid-phase reagents at room temperature. This means that some rather esoteric substances such as acetonitrile can be used as a reagent gas. The use of acetonitrile CI has proved to be very beneficial in confirming the molecular weight of straight-chain hydrocarbons, which often do not exhibit a $M^{+\bullet}$ peak under the rapid data acquisition techniques used with QMF instruments or the long storage times before detection associated with ions in the QIT [9]. The QIT has an advantage when ammonia is to be used as the reagent gas in CI. Rather than having to pump in NH_3 gas or 5–10% NH_3 in methane, the vapor pressure above a sealed vial of NH_4OH can be used as a source for the 10^{-3} Pa reagent gas pressure. This means a much more pleasant working atmosphere.

5.3.2. Derivatization

As was pointed out in Chapter 2, in order to make some analytes volatile enough to pass through the GC column, they are derivatized. Compounds that are usually too involatile to pass through the gas chromatograph are also often very polar and even if they can be separated using GC, their mass spectra do not exhibit $M^{+\bullet}$ peaks. Derivatization of these analytes cannot only facilitate the GC process but can allow for the determination of the analyte's nominal mass. One of the more universal types of derivatization reagents is one that results in alkylsilyl adducts. The most widely formed derivatives is the trimethylsilyl (TMS) adduct; however, adducts of *tert*-butyldimethylsilyl (TBDMS) and pentamethyldisilyl moieties are also used. Figure 5.23 shows the structure of some of the more widely TMS reagents. Figure 5.24 shows the products resulting from the reaction of various functionalities and lists the order of reactivity.

The mass spectra of TMS derivatives are very characteristic in that they usually do not exhibit a $M^{+\bullet}$ peak. There is usually a distinctive $[M-15]^+$ peak in the spectra of TMS derivatives. In the case of aliphatic alcohols, phenols, and carboxylic acids, this is the very stable siloxonium ion ($R-O^+{=}Si(CH_3)_2$), which is characterized by fairly intense X+1 and X+2 peaks due to the presence of Si in the ion. In the case of amines and amides, this $[M-15]^+$ peak represents a silimmonium ion. Provided atoms of nitrogen are not an issue, the $[M-15]^+$ peak (which is the highest *m/z* value peak in the mass spectrum that is neither a background nor an isotope peak) will be at an odd *m/z* value, and the nominal mass of the derivative can be calculated by adding 15 to this *m/z* value to obtain an even number.

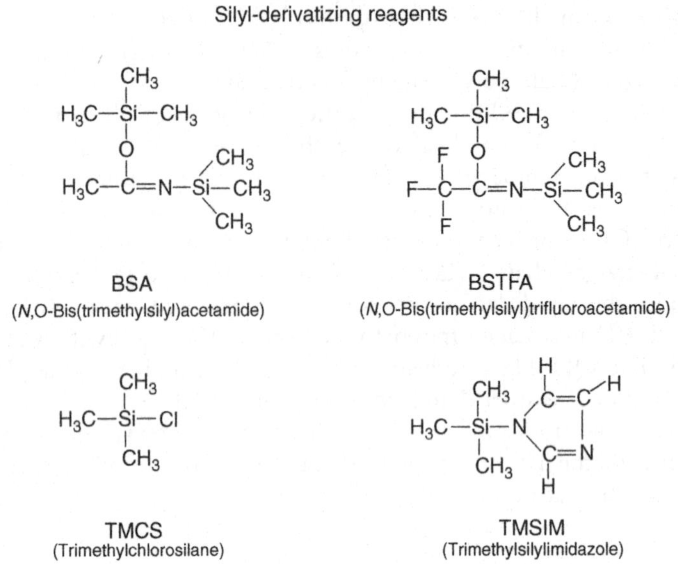

Figure 5.23 Commonly used silyl derivatizing reagents.

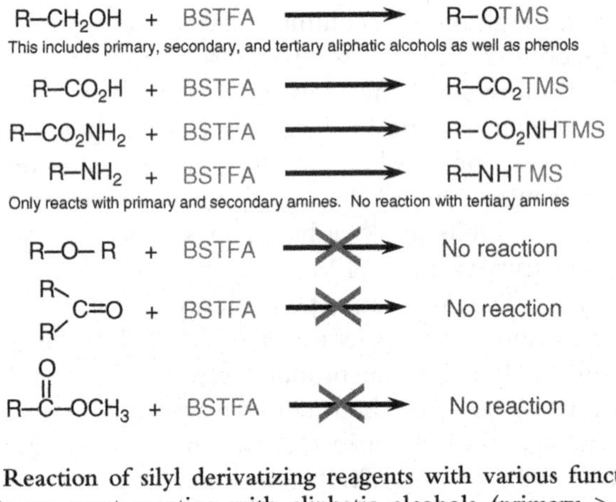

Figure 5.24 Reaction of silyl derivatizing reagents with various functional groups. These reagents are most reactive with aliphatic alcohols (primary > secondary > tertiary) followed by phenols, then carboxylic acids, amines (primary more reactive than secondary) followed by amides.

The TMS derivative adds 72 to the mass of the analyte—73 from the TMS group and the loss of a H atom from the analyte in the formation of the derivative. Therefore, care must be taken in using this technique not to add too much to the mass of the analyte resulting in a compound that has a mass that exceeds the upper limit of the *m/z* analyzer when there are multiple derivatizable sites. Regardless of how many sites are derivatized, the mass spectrum will still exhibit just a single $[M - 15]^+$ peak. When there are multiple derivatizable sites, there may be multiple derivatives formed with successively increasing numbers of TMS groups added. Therefore, assuming the nominal mass of an analyte to be a number that is 72 less than 15 more than the *m/z* value of the $[M-15]^+$ peak can lead to an erroneous conclusion. All the data must be carefully evaluated.

When more than a single group is derivatized using a TMS agent, peaks at *m/z* 147 and 149 representing ions formed by fragment ions with *m/z* 177 and 165, respectively, are present in the mass spectrum [10]. A requirement for the formation of these ions is that the $(CH_3)_3SiO$ moieties be on adjacent carbon atoms. But remembering that the ions in a mass spectrometer are free from any interactions with other matter that may be present allows for the understanding that the gas-phase geometry of the ion can be such that the two groups are actually adjacent to one another as would be the case for the mass spectrum shown in Figure 5.25. The only thing that can be deduced from the presence of peaks at *m/z* 147 and 149 is that more than one silyl group has been added. A conclusion that only two groups were derivatized could be false.

If it is suspected that more than a single moiety has been derivatized with a TMS reagent, a good way to determine how many groups have been derivatized is to repeat the derivatization using a deuterated analog of the derivatizing reagent. The difference in mass achieved by using a $(CD_3)_3Si$ reagent instead of a $(CH_3)_3Si$ reagent is 9 Da per silyl group added. If one silyl group is added, then the nominal mass of the derivative formed with the deuterated reagent will be 9 Da greater than the derivative formed with

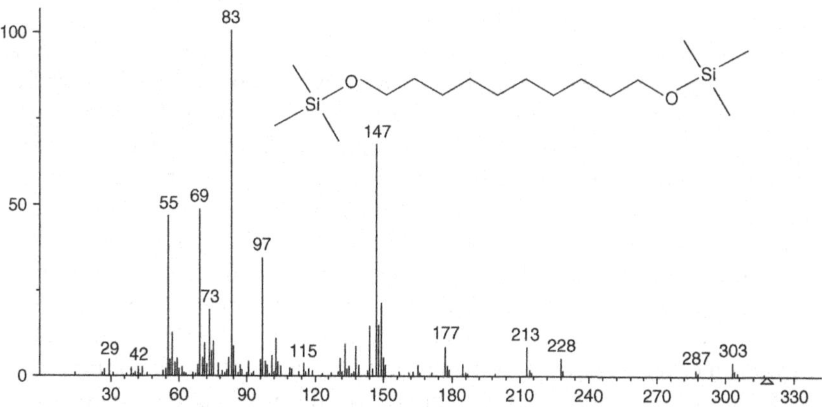

Figure 5.25 EI mass spectrum 1,10-decanediol bis(trimethylsilyl) ether exhibiting the peaks at *m/z* 147 and 149 characteristic of two or more TMS additives to moieties on adjacent carbon atoms.

the nondeuterated reagent; if two groups are derivatized, the difference in the two derivatives is 18 Da; three groups, 27 Da; four groups, 36 Da; and so on.

In the mass spectra of all TMS derivatives, there will be a peak at *m/z* 73, which represents the $(H_3C)_3Si^+$ ion. In the mass spectra of TMS derivatives of aliphatic alcohols (1°, 2°, and 3°), there is also a peak at *m/z* 75 $[H^+O{=}Si(CH_3)_2]$. In the mass spectra of the TMS derivative of primary aliphatic alcohols, there will also be peaks at *m/z* 89 and 103. Peaks representing homologs of these ions are found in the mass spectra of 2° and 3° aliphatic alcohols. Care must be taken in reaching conclusions based on the presence of peaks at *m/z* 89 and 103 because the peaks can represent ions other than those associated with a 1° aliphatic alcohol. As an example, a methyl-substituted oxonium ion will have a mass of 103 Da. The mass spectrum of a TMS of a methyl-substituted 2° aliphatic alcohol will have a peak at *m/z* 103, but there will be no peak at *m/z* 89 because the peak at *m/z* 103 is a homolog of the peak at *m/z* 89 in the mass spectrum of a TMS derivative of a 1° alcohol. The homolog of the peak representing the ion at *m/z* 103 in the mass spectrum of this 2° alcohol with a methyl moiety on the carbinol carbon atom will be at *m/z* 117.

The mass spectrum of TMS ether formed by the fragmentation of a derivatized secondary alcohol will exhibit two peaks that represent two very stable TMSO-substituted secondary carbenium ions, which can be used to determine the site of the hydroxyl group. This is illustrated in the mass spectra of the TMS derivatives of 3-decanol and 4-decanol in Figure 5.26. One secondary carbenium ion is formed by the loss of an aliphatic radical on one side of the carbon bearing the hydroxyl group in the original alcohol and the other secondary carbenium ion is formed by the loss of an aliphatic radical from the other side of the erstwhile hydroxyl-bearing carbon atom.

When dealing with an unknown analyte, after obtaining GC/MS data, the next step could be to see if the analyte(s) react(s) with a TMS derivatizing reagent. Inject the sample that has been treated with the derivatizing reagent to see if new RTIC chromatographic peaks appear. Do the RTIC chromatographic peaks have the same retention time as the peaks in the original data? If the unknown forms a derivative (indicated by the presence of a peak at *m/z* 73 in its mass spectrum), look to see if any of the various peaks indicating an aliphatic alcohol are present. If not, look to see what other possible moieties may have been derivatized. Apply the *Nitrogen Rule* to the mass spectrum of the derivative. Look to see if the unknown analyte reacts with other types of derivatizing reagents such as those that form esters.

Perfluoro-, propanoic-, and acetic-anhydride are often used to derivatize amines. These reagents react with amines more readily than TMS reagents. The perfluoronated derivatives are highly electrophilic compounds and are very suitable for electron capture negative ionization (ECNI) techniques.

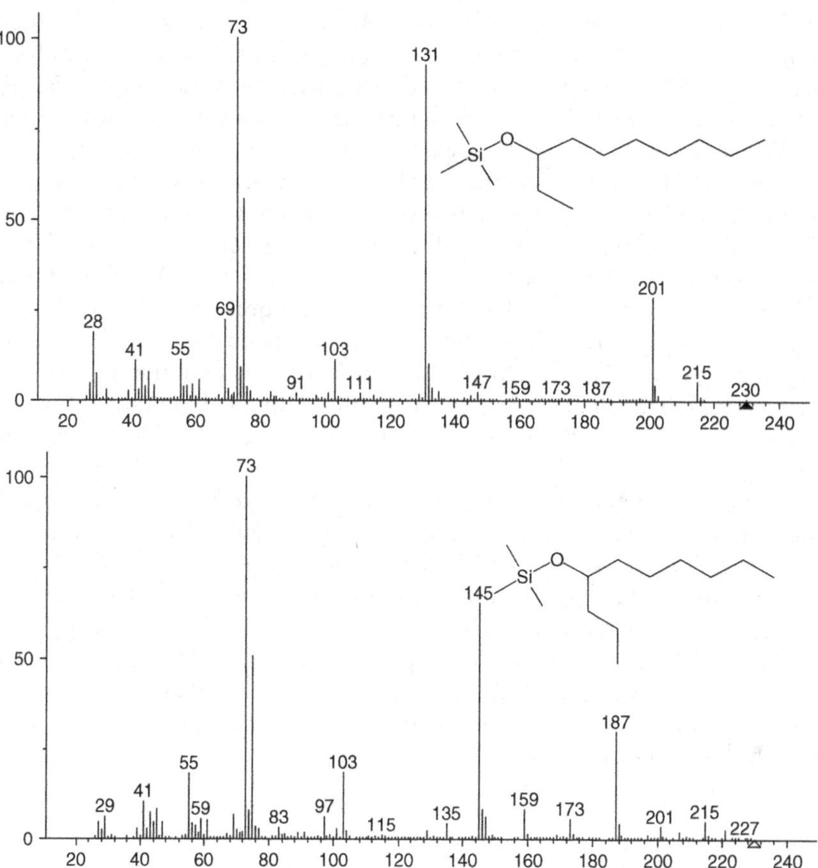

Figure 5.26 EI mass spectra of [(1-ethyloctyl)oxy]trimethylsilane (top) and [(1-propylheptyl)oxy]trimethylsilane (bottom). The peaks at *m/z* 201 and *m/z* 131 in the top spectrum and at *m/z* 187 and *m/z* 145 in the bottom spectrum can be used to deduce the location of the hydroxyl group.

These derivatives are especially useful for quantitation of analytes in dirty matrices because of the high specificity of ECNI. Esters can easily be formed by reacting diazomethane with carboxylic acids. When an analyte has more than a single functionality, mixed derivatizing reagents can be used to obtain the necessary information.

A good example of using mixed derivatizing reagents would be the analysis of a mixture of fatty acids. The first step might be to react the mixture with diazomethane* to form methyl esters. The resulting GC/MS

* Before using diazomethane, read all the cautionary statements found in the literature such as *Handbook of Derivatives for Chromatography*, 2nd ed., Karl Blau and John Halket, Eds., Wiley, Chichester, U.K., 1993, 047192699-X.

analysis yielded four RTIC chromatographic peaks. The mass spectra representing the apex of three of the RTIC chromatographic peaks were characteristic of the methyl esters of a straight-chain fatty acid (Figure 5.27); however, the mass spectrum representing the apex of the last of the four peaks had a very different mass spectrum (Figure 5.28 top). The mixture resulting from the derivatization with diazomethane was treated with a TMS derivatizing reagent and the sample analyzed a second time. The fourth chromatographic peak had an increased retention time, and the mass spectrum was very different. An intense peak at *m/z* 73 indicated that this was the spectrum of a TMS derivative. The peak at *m/z* 75 indicated that it was the spectrum of a derivative of an aliphatic alcohol. The lack of the pair of peaks at *m/z* 89 and 103 indicated that it was not the spectrum of a primary aliphatic alcohol. As an exercise, based on the data in the two mass spectra in Figure 5.28, propose a structure for this double-derivatized analyte.

If the mass spectrum of an analyte does not exhibit a $M^{+\bullet}$ peak and the database search is not successful, the best option is derivatization. Derivatization should be used before resorting to CI. Derivatization is nothing more than an organic synthesis. Everyone who has taken a college-level organic chemistry course that included a laboratory has prepared derivatives to identify pure unknowns. Derivatization in GC/MS is the same, with the only difference being that there may be multiple compounds to be identified and a matrix containing these analytes has to be dealt with.

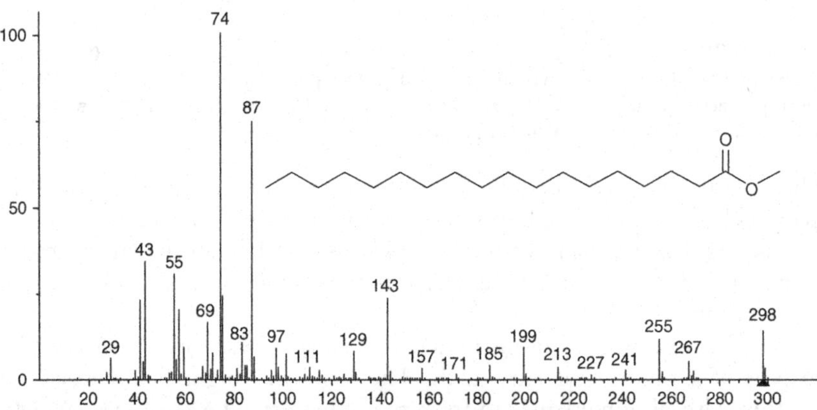

Figure 5.27 EI mass spectrum of methyl stearate (methyl ester of octadecanoic acid). This spectrum exhibits the typical pattern for spectra of methyl esters straight-chain fatty acids, i.e., base peak at *m/z* 74 followed by prominent peaks every 56 *m/z* units (*m/z* 87, 143, 199, 255), a $M^{+\bullet}$ peak, and an $[M-31]^+$ peak.

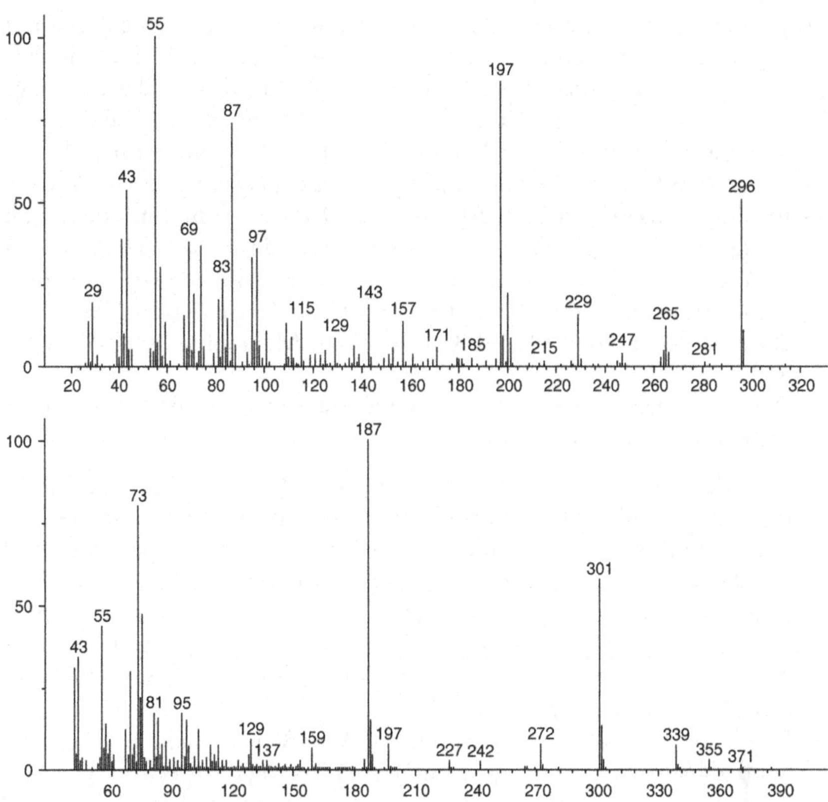

Figure 5.28 EI mass spectrum representing the apex of the RTIC chromatographic peak after derivatization with diazomethane (top) and after the diazomethane derivative mixture was treated with a TMS derivatizing reagent (bottom).

5.4. Selecting the Spectrum to Be Interpreted

A number of factors have already been discussed with respect to selecting a mass spectrum from a data set (RTIC chromatogram). It will soon become obvious that the challenge of a single spectrum selected by someone from a data set will only lead to frustrations when attempting an interpretation. At a minimum, both a graphic display and a tabular display of the spectrum should be used when an interpretation is attempted. The graphic display presents the picture of the analyte. The tabular display contains the details of the analyte such as information about the elemental composition of various ions.

It is best to have the data file containing the spectrum and a way to manipulate and interrogate this data file, if for no other reason than to be able to get a background-subtracted spectrum into a database search program. If the software from the data system used to acquire the data is not available, there are alternatives that can be used to read almost any instrument manufacturer's data file format or that can read data files that have been exported to the netCDF format. AMDIS (mentioned above and discussed in more detail below) is one such alternative and is a free download from NIST (http://chemdata.nist.gov/mass-spc/amdis). AMDIS is also a part of the *NIST/EPA/NIH Mass Spectral Database* when distributed with the *NIST MS Search Program.* Another free download alternative is *Wsearch32 2005* (http://www.wsearch.com.au) developed and maintained by Frank Antolasic in the Chemistry Department of the Royal Melbourne Institute of Technology (RMIT University) at Melbourne 3001, Victoria, Australia.

Points to keep in mind when selecting a spectrum for interpretation or submission to a database search:

1. Does the spectrum appear to be interpretable? Coming to the conclusion that a spectrum is not interpretable will take some experience in viewing a larger number of spectra (both interpretable and uninterpretable) over an extended period of time; practice, practice, practice.
2. Is spectral skewing an issue? If so, determine if the RTIC chromatographic peak represents a single compound and use an average of one or two spectra on either side and the spectrum that represents the apex of the peak along with the apex spectrum.
3. Eliminate peaks that represent noise spikes. A noise-spike peak can be verified by examining the spectrum just preceding the spectrum believed to contain a noise-spike peak and the spectrum just following it.
4. Are there peaks in the selected spectrum or the averaged spectrum that could well be attributed to background, i.e., *m/z* 149 representing a fragment ion of phthalate plasticizers, *m/z* 207, 281, 355, 421, etc. that are often due to bleed of the column's stationary phase or spectrum bleed, material that may be common to the sample matrix, etc.? If these peaks are present, consider a background subtraction, either a single background spectrum or an average of spectra that will be considered to be part of the background.
5. Is it possible that the spectrum represents more than one compound? Pick the *m/z* values of mass spectral peaks that do not look like they belong to the mass spectrum of the same compound and display mass chromatograms of them. If the mass chromatographic peaks do not rise and fall together, then there is a high likelihood that there is more than one compound represented by the spectrum.
6. Watch for peaks that may represent the compound used in most GC/MS instruments to calibrate the *m/z* scale (perfluorotributylamine, PFTBA, or FC-43) at *m/z* 69, 131, 219, 264, 502, etc. It is possible that the calibration gas valve has been accidently left open or is malfunctioning.

Once the appropriate spectrum has been selected and a tabular and a graphic display have been prepared, the interpretation should be fairly straightforward via a database search or the old fashion way of rationalizing all the peaks with respect to what ions they represent. Do not be intimidated by the mass spectrum. It is a language of its own and can be learned, read, and spoken in the context of the chemistry that it represents.

5.4.1. Use of Mass Chromatograms

The mass chromatogram [11], as already described, is a plot of the ion current at a specific *m/z* value, a range of *m/z* values, or the sum of noncontiguous *m/z* values contained in a conservative series of mass spectra acquired in the full-spectrum mode (measurement of ion current at each *m/z* value in a range of *m/z* values specified in the acquisition) versus spectrum number, which is a function of time. The mass chromatogram (sometimes called an extracted ion chromatogram (EIC) or an extracted ion current chromatogram (EICC)*) is one of the most significant aspects of GC/MS data when it comes to the identification of unknowns. An example of mass chromatograms being used to show that an apparently very symmetrical RTIC chromatographic peak actually represents two compounds is seen in Figures 5.7 and 5.8. Another good example of the use of the mass chromatogram is to use it to verify that an ion of a specific *m/z* value is inherent to the background. Plot a mass chromatogram for the ion with *m/z* 149 and a mass chromatogram believed to actually represent the analyte. The plot of the ion believed to represent the analyte will rise and fall to form a chromatographic peak. If *m/z* 149 is part of the background, its mass chromatographic plot will be a wavy line just above the zero line for the data plots. If *m/z* 149 is a part of the mass spectrum characterized by the other ion, the *m/z* 149 ion's mass chromatographic peak will rise and fall in parallel with the plot of the other ion.

Some samples may have such a strong background that there are no obvious RTIC chromatographic peaks when the data are viewed. A mass chromatogram for an ion characteristic of a specific analyte can reveal the presence or absence of the analyte. Mass chromatograms are also very useful for quantitation. Quantitative results are usually calculated based on mass chromatographic peak areas rather than RTIC chromatographic peak areas. This will avoid possible interferences for variable compositions of matrices.

---

* When spectra are acquired over a continuous range of *m/z* values using a device such as the Chromatoprobe or a direct insertion probe, this display is called an *extracted ion current profile*, an EIC profile.

5.4.2. Background Subtraction

The chromatographic process can only do so much with respect to separating the sample from its matrix. The GC column will become contaminated with less volatile materials as one analysis after another is performed. GC column stationary phase that bleeds from the column will adhere to the walls of the ion source and be a constant source of background (Figure 5.29). The background associated with an individual mass spectrum may be the result of only a few ions that have a relative low abundance, or it can be from a near-coeluting substance that is present in a much higher concentration than the analyte. The data system used to view GC/MS will allow one spectrum to be subtracted from another. This can be an individual spectrum subtracted from another individual spectrum or from an average of several spectra. The background can be an average of several spectra from both sides of the RTIC chromatographic peak representing the analyte.

In a GC/MS analysis, like a GC analysis using a conventional GC detector, the baseline (the signal recorded when no sample is eluting from the column) will rise. This rise is usually due to the increase of the column temperature during the analysis, which results in increased stationary-phase bleed and elution of low-volatile materials that have accumulated on the column from multiple analyses, one after the other. The common contaminant peaks due to column bleed are observed at *m/z* 207, 281, and 355. These peaks are characterized by intense X+1 and X+2 peaks due to the presence of multiple atoms of Si in the ions that they represent.

If the RTIC chromatographic peak represents a single analyte, it is best to choose a background spectrum following the selection of the analyte. If there is a solvent tail associated with the elution of the analyte, a spectrum before the analyte's RTIC chromatographic peak may be the best place to

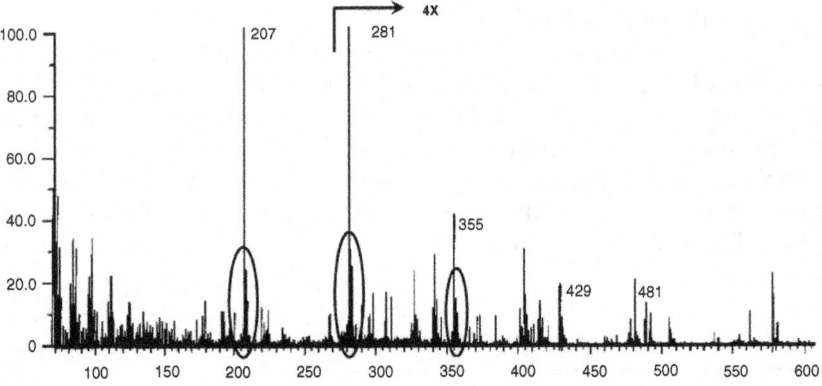

Figure 5.29 EI mass spectrum of GC column bleed. The intensity of the peaks above *m/z* 275 has been expanded by 4×.

select the background so that interference from any ions formed by the solvent is removed. When the background is subtracted, the intensity of each individual *m/z* value in the background spectrum is subtracted from the intensity of the corresponding *m/z* value in the sample spectrum. This subtraction will never result in a negative intensity in the sample spectrum (an *m/z* value <0). A peak at *m/z* 149 is another potential contaminant ion peak observed in a nonbackground-subtracted mass spectrum. This peak represents the most abundant ion in the mass spectrum of alkoxy esters of phthalic acid. Phthalates are commonly used as plasticizer and can not only contaminate individual samples, but can also accumulate in the GC-MS and be represented by the *m/z* 149 peak in every spectrum in the data set.

In the case of near coelution, the background spectrum should be the spectrum that represents the valley between two chromatographic peaks. Figure 5.30 is a good example of where the appropriate background spectrum should be located. Rather than taking the background spectrum after the elution of the analyte, the background spectrum is selected at the lowest point in the valley formed by the elution of a previous component and the analyte.

When preparing a background-subtracted spectrum for a database search, it may be wise to try several different options. Selecting the background is very subjective. If the background spectrum contains peaks that are in common with peaks in the analyte spectrum and the intensity of these peaks is greater in the background spectrum than in the analyte spectrum,

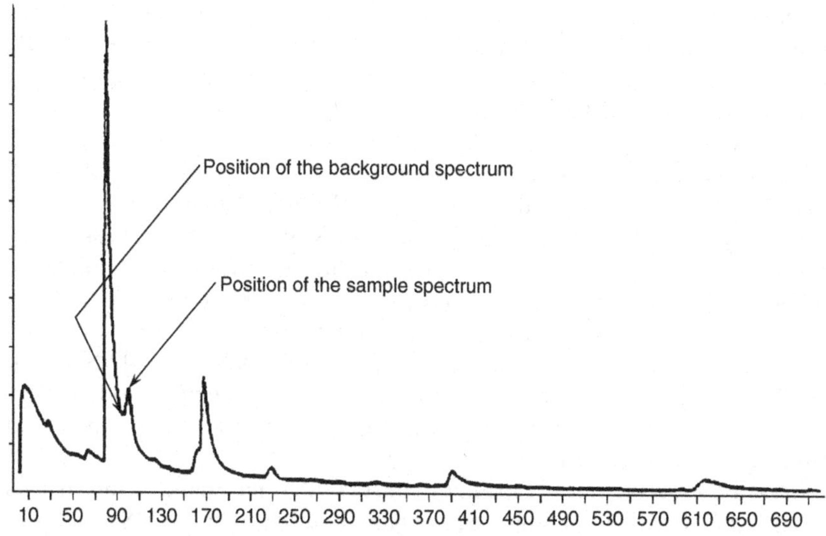

Figure 5.30 An RTIC chromatogram with the location of a sample spectrum and the position for the appropriate background spectrum.

then peaks in the sample spectrum can be lost, thus possibly compromising a database search or an interpretation.

Some instrument manufacturers' data systems have automated background-subtracting routines, but they can produce the same problems that an operator background subtraction can cause. Background subtraction is tricky at best and can be disastrous at its worst. If a spectrum of the analyte is believed to be in the database being searched using the *NIST MS Search Program*, it may be best to search a nonbackground-subtracted spectrum and look to the R Match Factor value. Another very reliable method of producing pure mass spectra (spectra that represent only single compounds) is through the use of AMDIS.

5.4.3. AMDIS

AMDIS is an acronym for *Automated Mass spectral Deconvolution and Identification System*, a software routine developed by NIST for use by the *Organization for the Prohibition of Chemical Weapons* (OPCW), an international agency, located in The Hague, The Netherlands, in its efforts to verify compliance by member nations with the *Convention on the Prohibition of the Development, Production, Stockpiling and Use of Chemical Weapons and on their Destruction* (CWC treaty). AMDIS analyzes data acquired by GC/MS by preparing mass chromatograms for each integer *m/z* value in the data's acquisition range. The mass chromatographic peak shapes are compared with one another to allow for an assignment of ions to a spectrum representing a single compound. This process is very effective in the deconvolution of multiple components represented by a single RTIC chromatographic peak and in removing peaks from a mass spectrum that can be attributed to background. AMDIS does a far better job of background subtraction than can be done by an analyst or the automated background substraction systems provided by the various instrument manufacturers. The algorithms used by AMDIS take into account the possibility of coeluting substances having ions in common and assign the appropriate amount of the ion current to the spectrum of each substance when spectra of two different compounds are represented by a single spectrum. AMDIS can be used to search for the presence of low levels of target analytes in very complex matrices. In such an analysis, AMDIS uses an AMDIS database that can be constructed from mass spectra in other databases or from spectra that are deconvoluted (and/or background subtracted) from data files of standards.

A good example of the capabilities of AMDIS is illustrated in the analysis of an RTIC chromatographic peak obtained during the GC/MS analysis of an environmental water sample. The spectrum in Figure 5.31 represents the apex of the RTIC chromatographic peak. Based on an experienced

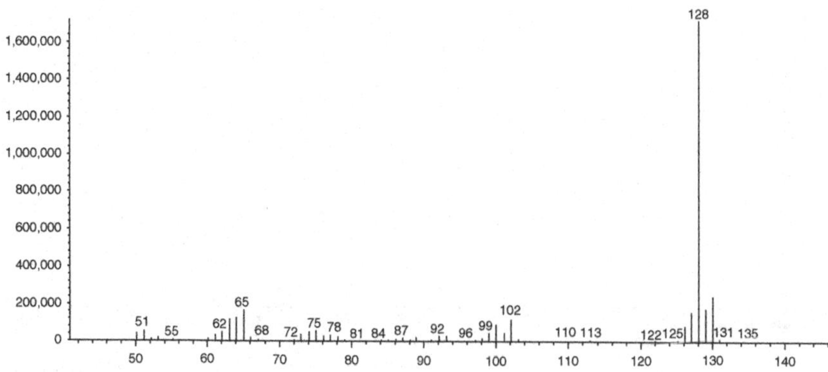

Figure 5.31 Mass spectrum representing the apex of an RTIC chromatographic peak.

understanding of EI mass spectra, it appears that the base peak in the spectrum represents a molecular ion. This means that the base peak and the $M^{+\bullet}$ peak are the same. Intense $M^{+\bullet}$ peaks are strong indications that the analyte is an aromatic compound. A database search of the analyte using the *NIST MS Search Program* and the *NIST08 Mass Spectral Database* (mainlib and replib) resulted in the first three hits being for naphthalene with Match Factors of 820 or higher. The next four hits were all for *p*-chlorophenol with Match Factors between 806 and 817, which is essentially the same as the Match Factors obtained for naphthalene. Examination of the sample spectrum revealed that the X+2 peak was far too intense to belong to the mass spectrum of naphthalene, and it was far too low to belong to the mass spectrum of *p*-chlorophenol.

An examination of spectra at the beginning of the RTIC chromatographic peak and at the end (Figure 5.32) indicated that this chromatographic peak might represent two compounds; however, if it did represent naphthalene and *p*-chlorophenol, then there was no way for the operator to manipulate the data to show this because, with the exception of the differences in the intensities of the X+2 peak, the two spectra were too similar.

This situation allows for the demonstration of one of the more spectacular performances by AMDIS by the separation of the two compounds that constituted this particular RTIC chromatographic peak (Figure 5.33). An analysis of the RTIC chromatographic peak was carried out using an AMDIS target analyte search on the data acquired with an Agilent Technologies 5975 MSD using a 7890 GC. The RTIC chromatographic peak for both compounds has its maximum at spectrum (scan) 1,064, retention time of 10.160 minutes. The AMDIS deconvoluted retention time for *p*-chlorophenol is 10.159 minutes. The deconvoluted retention time for naphthalene is 10.161 minutes. The difference in the deconvoluted retention times for the two components is 0.12 seconds.

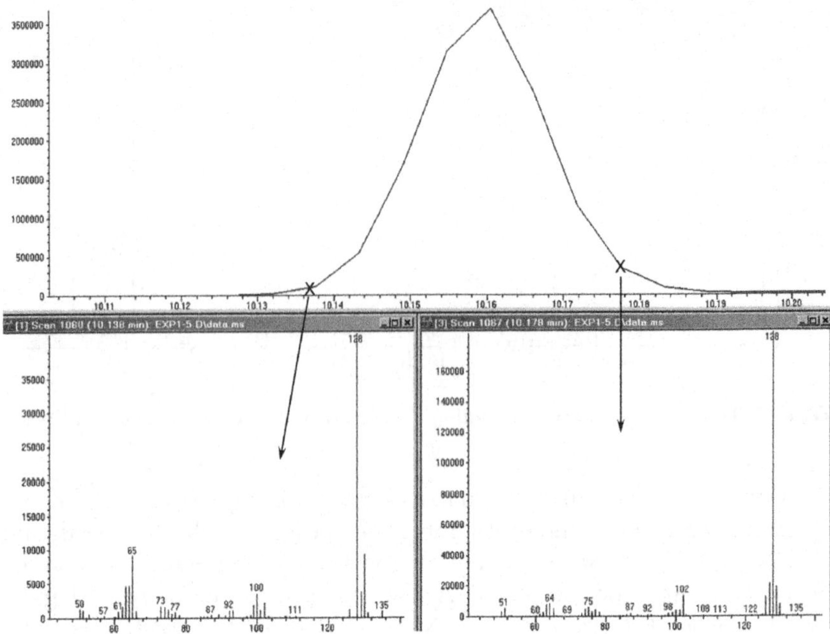

Figure 5.32 Spectra at the beginning (left side) and the end (right side) of the RTIC chromatographic peak indicate two different compounds are represented. Note: This display from the Agilent ChemStation Data Analysis software results from a custom macro.

AMDIS is a very powerful program and has a lot of features and a lot of settings that have to be understood. Agilent Technologies has based their *Deconvolution Reporting Software* on AMDIS and supply application-specific AMDIS databases with the program. They have also optimized the settings in AMDIS for specific analyses using specific GC columns and GC and mass spectral data acquisition settings, making the use of AMDIS more user-friendly (simpler). Before trying to use AMDIS, the *Tutorial* in its *Help* file should be reviewed in detail.

5.5. Reading an EI Mass Spectrum

As described in Section 5.2, the mass spectrum is read from right to left. If present, the $M^{+\bullet}$ peak will be the highest nonisotope, nonbackground peak in the mass spectrum. The molecular ion produced by EI is a positive-charge odd-electron ion because it is formed by the loss of a single electron from the gas-phase analyte molecule in a vacuum, and it has the same nominal mass as the analyte. All molecules are neutral and have an even

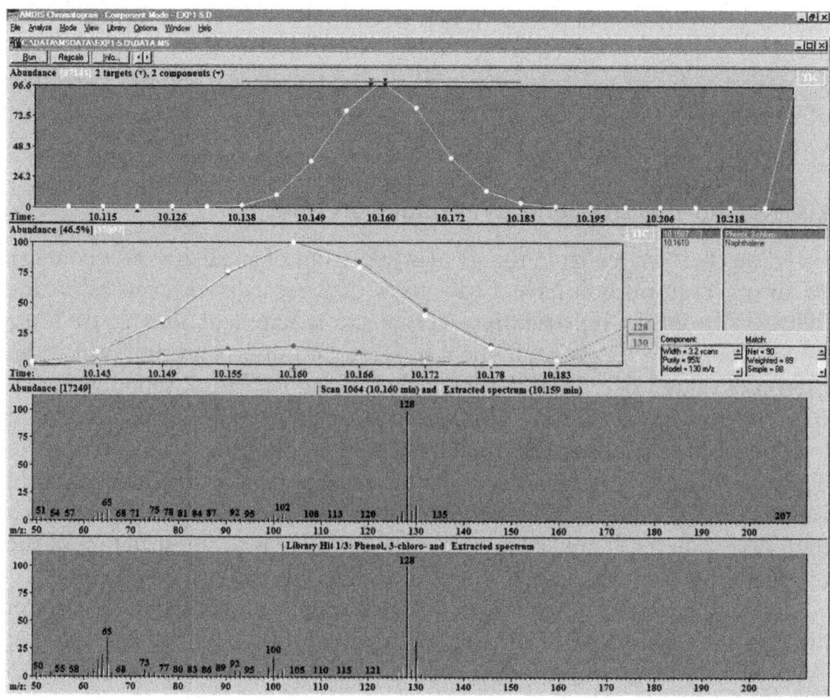

Figure 5.33 AMDIS Target Analysis results for two coeluting analytes.

number of electrons. All molecular ions formed by EI have a positive charge and an odd number of electrons. The molecular ions formed from these molecules can fragment through the loss of a radical (a neutral species with an odd number of electrons) to form an even-electron fragment ion or through the loss of a molecule (a neutral species with an even number of electrons), smaller than the analyte molecule, to form an odd-electron fragment ion. Most often, the molecular ion fragments to form an even-electron ion with the loss of a radical.* Special geometric considerations, usually involving the presence of heteroatoms (but not always) bearing the charge and radical site in the molecular ion, are required for the formation

* The unimolecular reaction that occurs in the mass spectrometer involves a single reactant and two products, another ion of smaller mass and a neutral, which is either a radical (a neutral species with an odd number of electrons) or a molecule. The resulting ion will be an even-electron ion, if formed by the loss of a radical from the odd-electron molecular ion. The loss of a species with an odd number of electrons from a species with an odd number of electrons results in a species with an even number of electrons. If the odd-electron ion fragments through the loss of a molecule (an even-electron species), the result is an odd-electron fragment ion.

of odd-electron fragment ions. When an even-electron ion is formed, it is a result of a single-bond cleavage. When an odd-electron ion fragment is formed, it is the result of cleavage of more than one bond and the formation of new bonds.

5.5.1. Odd-Electron and Even-Electron Ions

Based on the *Nitrogen Rule*, the EI mass spectrum of an analyte that contains no nitrogen atoms will have a $M^{+\bullet}$ peak (if present) at an even *m/z* value followed by peaks representing even-electron fragment ions at odd *m/z* value. There may occasionally be a nonisotope nonbackground peak at an even *m/z* value lower than the *m/z* value of the $M^{+\bullet}$ peak representing an odd-electron fragment ion. The presence of these odd-electron fragment ions peaks is very important; therefore, after reading the mass spectrum to determine if a $M^{+\bullet}$ peak is present, the next step in reading the mass spectrum is to determine if any odd-electron fragment ion peaks are present and what they may indicate. These peaks can often be clues as to the structure of the original analyte molecule and whether or not heteroatoms are present.

Again, based on the *Nitrogen Rule*, if the analyte contains a single atom of nitrogen, the *m/z* value of the molecular ion will be odd, the *m/z* value of even-electron fragment ions retaining the nitrogen atom will be even, and the *m/z* value of odd-electron fragment ions retaining the nitrogen atom will be odd. If an even-electron fragment ion is formed from a $M^{+\bullet}$ that has a single atom of nitrogen and the nitrogen atom is retained with the radical neutral loss, then that fragment ion will have an odd *m/z* value. If an odd-electron fragment ion is formed from a $M^{+\bullet}$ that has a single atom of nitrogen and the nitrogen atom remains with the molecular neutral loss, then that fragment ion will have an even *m/z* value.

5.5.2. Logical Losses

The term *logical loss* relates to what is represented by the numeric value of the loss from the molecular ion in terms of possible elemental compositions and valence rules. Most fragmentation results from single-bond cleavage; rearrangements that produce odd-electron fragment ions usually result from the breaking of two chemical bonds and forming new bonds. It would be possible to rationalize an $[M - 3]^+$ peak as representing an ion formed by first the loss of a molecule of H_2 from the molecular ion (a rearrangement fragmentation) followed by the loss of a $^{\bullet}H$ (a single-bond cleavage); however, the loss of 7 from the molecular ion is not something that can be rationalized as being logical. The logic of the loss is determined by the mass of the atoms associated with loss and how they have to be connected after a single-bond cleavage to form a radical or after they are expelled as a molecule. These logical losses are the *dark matter* of the mass spectrum. Values of some of the more common neutral losses are found in Table 5.2. Other values are found in Appendix Q.

The next step in reading the mass spectrum, after assigning the $M^{+\bullet}$ peak and determining if there are any $OE^{+\bullet}$ peaks present, is to look at its *dark matter* to see if there is any evidence of obvious structural moieties. The *dark matter* of a mass spectrum is seen as the difference between the *m/z* value of the $M^{+\bullet}$ peak and the *m/z* value of a fragment ion peak. A peak that is 15 *m/z* units lower than the $M^{+\bullet}$ peak represents an ion that was formed by the loss of a methyl radical. Methyl radicals are not very stable; therefore, an $[M - 15]^+$ ion peak usually represents the presence of a *special methyl* such as that associated with a methyl ketone or that can result in the formation of a secondary or tertiary carbenium ion. As the size of the *dark matter* increases, the ambiguity of what it represents also increases; i.e., a peak 29 *m/z* units lower than the $M^{+\bullet}$ peak could represent the loss of an ethyl radical ($^\bullet CH_2CH_3$) or the loss of a $H^\bullet C{=}O$ radical. The meaning of a number of *dark matter* values and specific *m/z* values are found in Appendix Q.

5.6. Final Remarks

One very important statement from the first edition of this book must be repeated:

> ***Compare the Predicted Mass Spectra of the Postulated Structures with the Unknown Mass Spectrum:*** *After the possible structures are obtained, predict their mass spectra by examining the mass spectra of similar structures. Also, the GC retention time may eliminate certain structures or isomers. Discuss these results with the originator of the sample to determine the most probable structure. With experience, it is usually possible to determine which fragment peaks are reasonable for a given type of structure.*

Since publication of the first edition of this book, a number of features have been added to the *NIST MS Search Program* to facilitate the task outlined in this statement. The *NIST MS Search Program* has an algorithm designed specifically to generate a Hit List of compounds that have produced spectra similar to sample spectrum when it is believed that a spectrum of the compound that generated the sample spectrum is not present in the Database. A structure (in MOL file format*) can be drawn and associated

* The MOL file format is the most commonly used system to encode chemical structures, substructures, and conformations as text-based connection tables. It was developed by *MDL Information Systems Inc.* for their MACCS or ISIS programs.

with the sample spectrum. When put into a user database and retrieved, a calculated RI is reported. The spectrum, along with the proposed structure, can be sent to a *NIST MS Search Program* utility, *MS Interpreter*, which will predict fragments of the proposed structure and match them to peaks in the sample spectrum. None of these utilities take away from the requirements for the analyst to think, ask questions, and make judgments; they just make some of these tasks easier.

The MS Interpreter Program can be used with spectra to assist in an interpretation of an unknown spectrum even before a structure has been assigned. The spectrum is sent to *MS Interpreter* from the NIST Spec List. It is helpful to edit the spectrum and assign a $M^{+\bullet}$ peak if possible and enter the *m/z* value as part of the spectra record before sending the spectrum to *MS Interpreter. MS Interpreter* has a utility to calculate the *dark matter* (the neutral losses) in the spectrum. *MS Interpreter* also has a *Formula Calculator* and an *Isotope-Peak Intensity Calculator/Comparator* as shown in Figure 5.34.

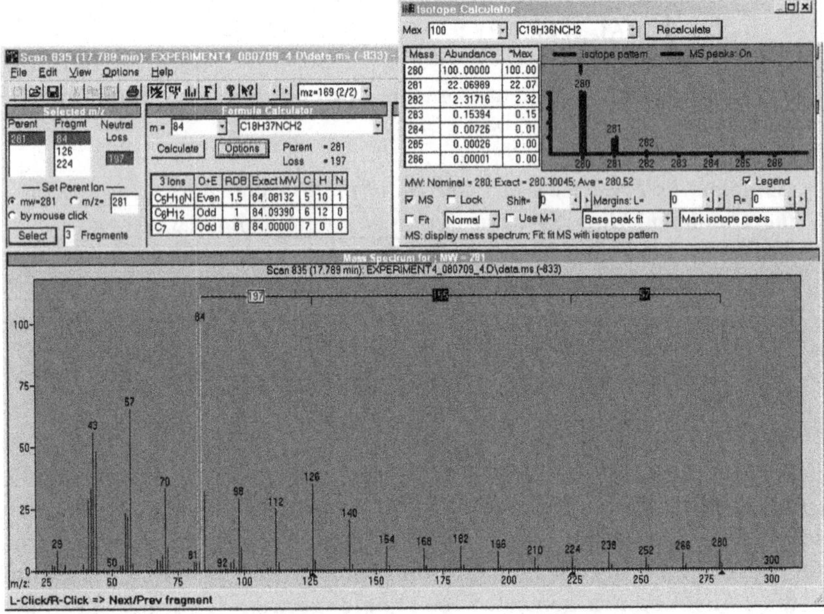

Figure 5.34 Screen views of MS Interpreter. The numbers shown at the top of the bottom panel are the difference between the *m/z* values of the designated peaks and the assumed $M^{+\bullet}$ peak. The top left panel is a formula calculator to calculate the formula of ions formed by the suspected molecular ion. The right panel is an isotope calculator showing the theoretical intensities (left) alongside the intensities for peaks with those *m/z* values in the mass spectrum.

Once an elemental composition and probable structure have been developed, it is a good idea to search that elemental composition against a mass spectral database to see if that or a similar compound is present. This should be done, even if no hit was obtained when the spectrum of the unknown was searched against the database. An example of the importance of this exercise is shown in Figure 5.35. The top spectrum was obtained for an unknown analyte resulting from a dichloromethane solution containing stearyl amine and some 2° and 3° aliphatic amine. The structure shown on the bottom spectrum in Figure 5.35 was determined to be a good candidate. Even though the spectrum of the unknown had been searched against the *NIST08 Mass Spectral Database*, the formula (elemental composition) was searched against the same *Database.* The result was 24 hits, only one of which matched the more probable of the two possible structures determined by the mass spectrometrist. As can be seen by a comparison of the two spectra, it is obvious the mass spectrometrist needs to start over. It turns out that the compound was

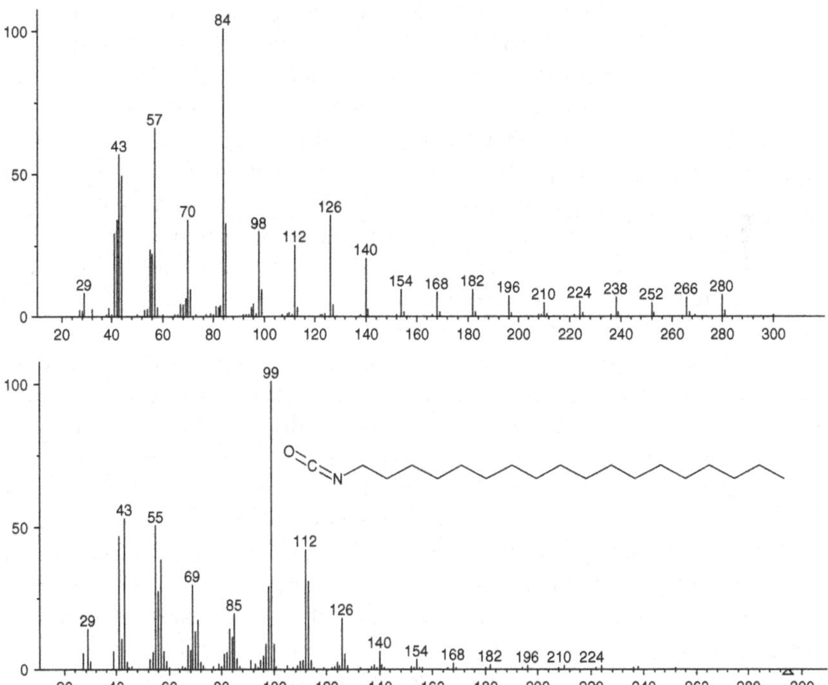

Figure 5.35 EI mass spectrum of an unknown component in a mixture obtained by GC/MS (top); EI mass spectrum and structure based on the interpretation of the unknown mass spectrum by a mass spectrometrist (bottom). The spectrum of the unknown compound was believed to not be in the *NIST08 Mass Spectral Database* based on an Identity Search of the unknown spectrum.

later identified as having the formula n-$C_{17}H_{35}$–CH_2–N=CH_2. The peak at *m/z* 280 represents an $[M - H]^+$ ion and the peak at *m/z* 281 is both a $M^{+\bullet}$ peak and an X+1 isotope peak relative to the peak at *m/z* 280.

Mass spectrometrists (which most of the people using this book will not be) like to think that all analytical problems can be solved through the use of mass spectrometry and, for the types of analytes covered in this book, specifically GC/MS alone. The mass spectrometrist will concede the importance of prior knowledge about the sample, but has a tendency to disregard other analytical techniques such as nuclear magnetic resonance (NMR) and infrared (IR) spectroscopy. It is true that these techniques require far more sample than is needed by the GC-MS; but when data are available from these techniques, the data should not be ignored. Another important parallel technique involves the splitting of the eluate from the GC column between the mass spectrometer and a GC nitrogen–phosphorous detector to quickly determine if the analyte contains these elements. This process does not require an excessive amount of additional sample. Also, derivatization, other than that used to make the analyte volatile and/or to determine molecular weight, can be used to indentify specific functionalities provided that a pure sample is available, which can sometimes be collected by preparative GC. Perform the analysis in the old-fashioned way—the way it was done in the undergraduate qualitative organic analysis lab.

Always keep in mind that the final confirming test is to obtain an authentic sample of the proposed analyte and obtain an EI mass spectrum on the same instrument used to generate the data for the unknown. If either derivatization or CI was involved in coming to the original conclusions, then also repeat these tests. The answer may not be for a known commercially available substance and, in such a case, the authentic sample will have to be the result of a synthesis.

This treatise on the interpretation of mass spectra is far from exhaustive. It is meant to provide enough information to get started with the data that is generated by the GC-MS. There has been no discussion of fragmentation mechanisms, no explanation of a γ-hydrogen shift-induced β cleavage, no explanation of homolytic or heterolytic cleavages as imitated by a radical and charge site, no mention of charge retention in sigma-bond cleavage, or secondary fragmentation of fragment ions. There is some detail contained in the pages for various compound types in Section II. For those interested in a more in-depth understanding of EI fragmentation, there are many good references [2,4,12–16]. There are also several courses on the subject offered by LC Resources and by the American Society for Mass Spectrometry at their annual meeting held each year at the end of May/beginning of June in various locations around North America. There is usually a course offered at the triennial International Mass Spectrometry Conference held in various European countries (the 2012 meeting will be held in Osaka, Japan).

Just always keep in mind that the data should lead to the conclusion; always follow the *KIS* (*keep it simple*, the simplest answer is usually the best answer) *philosophy* and practice, practice, practice!!!

REFERENCES

1. Stein, S. E. (1999). An Integrated method for spectrum extraction and compound identification from gas chromatography/mass spectrometry. *J. Am. Soc. Mass Spectrom.*, 10, 770–81.
2. Nibbering, N. M. M., ed. (2005). Fundamentals of and applications to organic (and organometallic) compounds. In: *The Encyclopedia of Mass Spectrometry*. Vol. 4. (Gross, M. L., Caprioli, R, editors-in-Chief), Oxford, UK: Elsevier; ISBN:0080438466.
3. Caprioli, R. M., Gross, M. L., eds. (2006). Ionization methods. In: *The Encyclopedia of Mass Spectrometry*. Vol.3. (Gross, M. L., Caprioli, R. M., editors-in-Chief), Oxford, UK: Elsevier; ISBN:9780080438016.
4. Watson, J. T., Sparkman, O. D. *Introduction to Mass Spectrometry: Instrumentation, Applications and Strategies for Data Interpretation*. 4th ed. Chichester, UK: Wiley: ISBN:9780470516348.
5. Watson, J. T., Sparkman, O. D. (1976). *Introduction to Mass Spectrometry: Biomedical, Environmental, and Forensic Applications*. 1st ed. New York: Raven; ISBN:0890040567.
6. Watson J. T., Sparkman O. D. (1985). *Introduction to Mass Spectrometry*. 2nd ed. New York: Raven; ISBN:0881670812.
7. Watson, J. T., Sparkman, O. D. (1997). *Introduction to Mass Spectrometry*. 3rd ed. Philadelphia/New York: Lippincott-Raven; ISBN:0397516886.
8. Gross, M. L. (1994). Editorial: accurate masses for structure confirmation. *J. Am. Soc. Mass Spectrom.*, 5, 57.
9. Moneti, G., Pieraccini, G., Dani, F. R., Catinella, S., Traldi, P. (1996). Acetonitrile as an effective reactant species for positive-ion chemical ionization of hydrocarbons by ion-trap mass spectrometry. *RCM*, 10(2), 167–70.
10. McCloskey, J. A., Stillwell, R. N., Lawson, A. M. (1968). Use of deuterium-labeled trimethylsilyl derivatives in mass spectrometry. *Anal. Chem.*, 40(10), 233–6.
11. Hites, R. A., Biemann, K. (1970). Computer evaluation of continuously scanned mass spectra of GC effluents. *Anal. Chem.*, 42, 855–60.
12. Smith, R. M. (2004). *Understanding Mass Spectra: A Basic Approach*. 2nd ed. Hoboken, NJ: Wiley; ISBN:047142949X (reviewed JASMS 16:792).
13. Smith, R. M., Busch KL, eds. (1999). *Understanding Mass Spectra: A Basic Approach*. New York: Wiley; ISBN:0471297046 (reviewed JASMS 11:664).
14. McLafferty, F. W, Tureček, F. (1993). *Interpretation of Mass Spectra*. 4th ed. Mill Valley, CA: University Science; ISBN:0935702253 (reviewed JASMS 5:949).
15. McLafferty, F. W. (1973). *Interpretation of Mass Spectra*. 2nd ed. Reading, MA: Benjamin.
16. Budzikiewicz, H, Djerassi, C, Williams, D. H. (1967). *Mass Spectrometry of Organic Compounds*, San Francisco, CA: Holden-Day.

REFERENCES

CHAPTER 6

Quantitation with GC/MS

6.1. Introduction

In GC/MS, the quantity of an analyte can be determined (quantitation) using data acquired with any ionization source on any type of *m/z* analyzer. The data can be acquired using the continuous monitoring of full-spectrum mode, selected ion monitoring (SIM), or selected reaction monitoring (SRM) with MS/MS. When quantitation is done using data acquired in the continuous monitoring of full-spectrum mode, a *quantitation ion* or, more simply, a *quant ion* is selected. The quant ion is an ion that is uniquely characteristic of the analyte and has as great an abundance. The quant ion is often represented by the base peak in the mass spectrum provided this ion does not have the *same m/z* value as a coeluting compound. A mass chromatogram of the quant ion results in a chromatographic peak area that is used in quantitation.* Selection of the quant ion is discussed in Section 6.2.

There are several advantages that GC/MS has over GC when it comes to quantitative analyses. In quantitative GC, it is crucial to obtain a good separation of the components of interest. This is not critical when a mass spectrometer is used because unique ions for identification and quantitation can be mass selected; nevertheless, chromatographic separation is a practice many continue to follow. When GC/MS is used, trace components that coelute with other analytes and matrix can still be quantitated as long as the analyte has at least one exclusive characterizing ion (that is not represented by an isotope peak).

Another advantage GC/MS has over GC for quantitation is the confirmation of the analyte. In GC, the only confirming factor is the analyte's retention time. In GC/MS, in addition to the retention time of the analyte, there is the *m/z* value of the ion used for the quantitation calculation and the intensity of other mass spectral peaks that are unique to the analyte when data are acquired in the continuous monitoring of full-spectrum

* Sometimes, the *quant ion* is the sum of several *m/z* values or a range of *m/z* values. An example would be cases where the quant ion contains multiple atoms of chlorine. In such a case, the use of the nominal mass ion and several of the X+2 isotope ions might be desirable.

Gas Chromatography and Mass Spectrometry
DOI: 10.1016/B978-0-12-373628-4.00006-X

mode. When full-spectrum data are used in a quantitative analysis, a database search of the spectrum representing the target analyte can be used to confirm the identity of the analyte.

The quant ion may also be monitored in SIM to produce an SIM chromatographic peak. Additional ions characteristic of the analyte may also be monitored and the area ratios of their SIM chromatographic peaks compared with that of the quant ion's chromatographic peak to confirm the presence of the analyte. This method of confirmation is frequently used in the analysis of drugs of abuse where three ions are monitored, and the ratio of the three ions in samples must match the ratio of the ions in a standard. This same process of matching ion ratios to a standard can be done with full-spectrum data. James Sphon showed that most organic compounds could be characterized by the peak intensity ratios of as few as three different ions acquired by SIM using electron ionization (EI) and that full-spectrum data was not necessary for an unambiguous identification [1]. More recently, this proposition has been questioned by Stein and Heller [2].

In an SRM analysis, chromatographic peaks generated by transition pairs (a precursor ion of a specific *m/z* value that fragments in the collision cell (process) to a product ion of a different *m/z* value) are used in the same manner for quantitation and calibration. The transition of one *m/z* value to another, which is what is monitored in an SRM analysis, at a specific retention time produces a higher degree of analyte confirmation than does the presence of an ion of a specific *m/z* value at a specific retention time. Just as monitoring multiple ions (usually not more than three or four) and comparing their abundances for added analyte confirmation in SIM, multiple transition pairs can be used in the same way for an SRM analysis to produce an even higher confirmation.

In an SIM analysis using a scanning mass spectrometer like the transmission quadrupole or the double-focusing instrument, more time can be spent measuring the ion current of the ions being used for quantitation and confirmation of the analyte, rather than dividing the measurement time over a large range of *m/z* values containing many meaningless *m/z* values. SIM will result in lower detection limits for these types of instruments. The TOF and QIT mass spectrometers will provide the same detection limits with full-spectrum acquisition as can be achieved by SIM with transmission quadrupole and double-focusing instruments. Monitoring too many ions in a single time interval (>10) in an SIM analysis can greatly degrade the limit of detection for an analyte.

Due to the high specificity of SRM, oftentimes lower limits of quantitation in complex matrices are achieved than with SIM. This does not mean that the detection limit, per se, for the analyte is lower with SRM than SIM. As a matter of fact, more ions representing the analyte will reach the detector when using SIM. However, due to the possibility of the matrix producing ions of the same *m/z* value as the analyte, the SRM process offers better signal to background, producing a lower limit of quantitation. The same limit for the number of transition pairs used in an SRM analysis that applies to the maximum number of ions monitored in an SIM analysis should be enforced.

The analytes in a quantitative analysis are often referred to as *target analytes* because their identities are known; it is just their amounts that are being determined. In GC/MS, the term *analyte* is usually reserved for a substance whose identity is to be determined in the analysis.

If the GC eluate is split between the mass spectrometer and a selective detector such as a nitrogen–phosphorous detector or a pulsed-flame photometric detector, the signal from either the detector or the mass spectrometer can be used for quantitation. Of course, if the GC detector signal is to be used for quantitation, analyte separation is essential as stated above.

6.2. Selection of the Quantitation Ion

Regardless of the data acquisition technique (SIM, SRM, or full-spectrum acquisition), a characteristic ion for the target analyte must be selected as the quant ion. If EI is being used, a spectrum of the analyte can usually be found in the *NIST08 Mass Spectral Database* (or later version). This spectrum can be used to select the quant ion. If SRM is going to be used with EI, a precursor ion or ions will have to be selected from the EI mass spectrum; then a product-ion spectrum will have to be obtained for each precursor ion to establish the transition pairs.

Chemical ionization (CI) results in a mass spectrum with little or no fragmentation; the spectrum consists of peaks representing the intact molecule (MH^+, $[M - H]^+$) or peaks representing ions characteristic of the intact molecule and simple loss from the ion representing the intact molecule. Therefore, interferences from the matrix are minimal, resulting in better limits of quantitation. CI is used in quantitative work, but the technique of positive-ion CI using a regular CI source can yield data that is not very precise. This is due to variations in the ion-source pressure caused by the needed high pressure of reagent gas. Electron capture/negative ionization (ECNI) in these same instruments under the same conditions is much more stable.

Positive-ion CI in the internal ionization QIT mass spectrometer is much more reliable for quantitation because of the extreme low pressure required for the reagent gas, which means that the ionization conditions are less variable than in a conventional CI source.

The quant ion should be unique to the *target analyte* at its elution time; i.e., the matrix should not produce any ions (even those represented by isotope peaks) that are *isobaric** with the quant ion. The quant ion should constitute as much of the ion current of the mass spectrum of the analyte as possible and should have as high an *m/z* value as possible. Interferences from matrix ions are more often encountered at low *m/z* values. In the case of compounds containing bromine or chlorine, an X+2 ion (an ion containing an atom of ^{37}Cl or ^{81}Br) oftentimes is the best choice because it is more likely to be unique than the monoisotopic ion. Aromatic compounds make for good *target analytes* because their $M^{+\bullet}$ peaks are very intense, often the base peak. That is the good news; the bad news is that aromatic compounds often do not produce very abundant fragment ions, which means that confirming signals may not available.

Limits of detection can be lowered by increasing the resolution (having a single *m/z* value represent several integer *m/z* values). An example of this is seen in the EI mass spectrum of chlorpyrifos. If the resolution is degraded to the point that a single peak represents 4 or 5 integer values, the peak at *m/z* 199 could provide for a lower limit of detection (Figure 6.1). Even though the peak with nominal *m/z* 197 represents an ion with three atoms of chlorine and has a greater intensity than the peak with *m/z* 314, which represents an ion with two atoms of chlorine, the unresolved cluster characterized by the peak at *m/z* 316 would be the better choice for quantitation because of the decreased probability of interference from ions

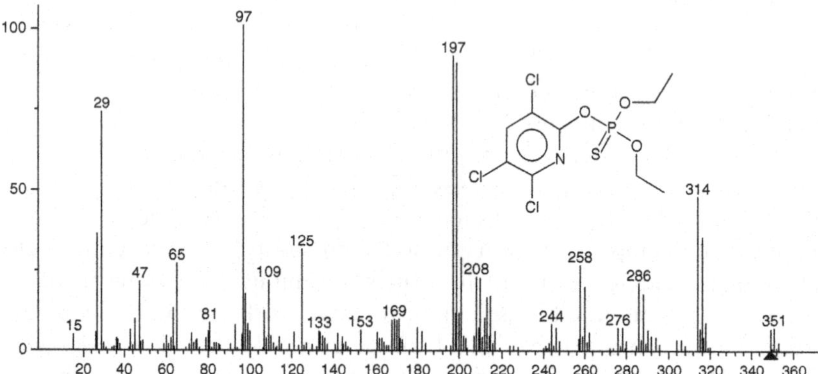

Figure 6.1 EI mass spectrum of chlorpyrifos (diethyl 3,5,6-trichloro-2-pyridyl ester of phosphoric acid). *From NIST08 (National Institute of Standards and Technology) with permission.*

* *Isobaric* means two or more ions, molecules, or radicals that are assigned the same mass. Species that have the same nominal mass are *isobaric*; however, these same two species have different exact masses; i.e., CO and N_2 have the same nominal mass, 28 Da, and are isobaric; however, the exact mass of CO is 27.9949 Da and that of N_2 is 28.0062 Da; therefore, these two species are not *isobaric*. Whether two species are *isobaric* is relative.

formed by the matrix. This would be especially true in this particular example because pesticides are often associated with agricultural commodities, which will produce significant ion current at a lot of *m/z* values in the range below *m/z* 250.

The mass spectrum of methyl stearate (a highly fragmented analyte) is a good example of how difficult it can be to select appropriate quant ions (Figure 6.2). The best ions would be *m/z* 298, 255, and 199, and possibly *m/z* 143. Ions below *m/z* 200 will have possible interference from matrix components in biological samples. The percent of the total ion current represented by the ions at the higher *m/z* values is low, which means that the quantitative response factor (RF) for this compound will be low.

The base peak in the ECNI mass spectra of electrophilic compounds is usually the $M^{-\bullet}$ peak or an $[M - \text{halogen}]^-$ peak. For this reason, perfluorinated derivatives are often prepared to increase specificity and lower limits of detection. The $M^{-\bullet}$ peak of perfluorinated derivatives, such as perfluoropropyl (PFP) esters, will appear at high *m/z* values (because of the added mass due to the presence of multiple atoms of fluorine, PFP adds 168 Da to the mass of a compound that can be converted to an ester). There will be less likelihood of interferences because the matrix is usually not very electrophilic.

The $[M-15]^+$ ion is often a good choice for the quant ion when analyzing trimethyl silyl derivatives using EI. The analyst must be careful when working with analytes that have multiple sites that can be derivatized. When selecting a derivatization procedure, verify that the analyte has been totally converted to the derivative and that only one compound has resulted from the derivatization procedure. One class of compounds that can yield multiple derivatives is mono- and disaccharides. Consult the references cited in Appendix G for further information.

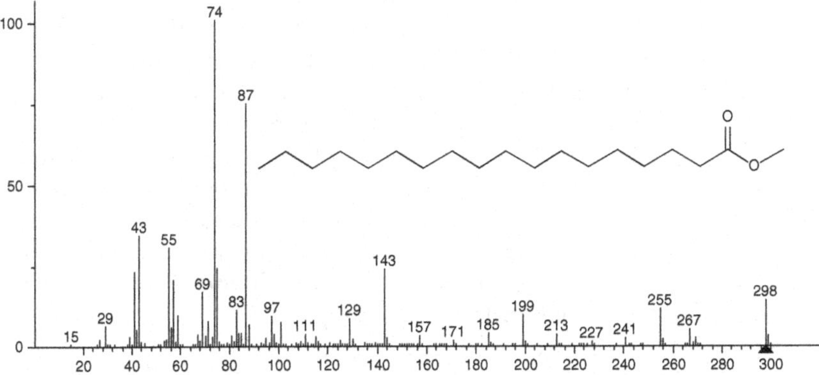

Figure 6.2 EI mass spectrum of methyl stearate. *From NIST08 (National Institute of Standards and Technology) with permission.*

6.3. Quantitation Methods

Once the chromatography and mass spectrometry procedures have been established, the analyst must decide on a quantitation method and prepare analytical standards. In all cases, a blank should be prepared and processed in the same manner as the samples and standards. The blank should be analyzed at the beginning of the series of analyses and after the highest concentration calibration standard in order to verify that the system is free of contamination and there is no carryover of the analyte.

6.4. Making Standard Solutions

The following procedure may be useful in making standard solutions.

- Accurately weigh 10–20 mg of the standard using an analytical balance.*
- Quantitatively transfer this amount to a clean dry 100-mL volumetric flask. Fill the flask about 5 cm below the mark with solvent.
- Shake the volumetric flask vigorously to allow the material to dissolve and assure homogeneity.
- Carefully fill the flask to the mark (using a pipette dropwise for the final addition) and invert the stoppered flask 10-20 times to assure a homogenous solution.
- Immediately transfer the contents of the volumetric flask to a clean dry storage bottle and label it with the name of the compound(s), concentration, the solvent, and the date.

The resulting solution has an accurately known concentration that is close to 100 ppm = $100\,\mu g\,mL^{-1} = 100\,ng\,\mu L^{-1}$. Label the storage bottle using the concentration determined from the actual weights, cover with aluminum foil if the compound is light-sensitive, and store in a refrigerator. This is known as the "stock solution." The analyst should determine by analysis over a period of weeks, how long the stock solution may be used.

To dilute to lower concentrations and prepare a series of standards of different concentrations, allow the storage bottle containing the stock solution to come to room temperature; pour a small amount of the solution from the storage bottle (never put a pipette into the storage bottle); and, using a micropipette, transfer appropriate quantities of the standards to a volumetric flask containing solvent for further dilution or to volumetric flasks containing the blank matrix of the sample.

* In many cases, the analyte is not available as a pure compound for making standards and must be purchased from a chromatography supply company in a sealed vial, containing the analyte dissolved in an appropriate solvent (usually methanol).

For diluting stock solutions containing volatile analytes into a water matrix, do not use volumetric flasks. A very significant portion of the analyte will be lost in the headspace in the volumetric flask. To prepare standards of volatiles in water, use Erlenmeyer flasks with ground-glass stoppers. Determine the total volume in the flask by weighing the flask empty and when it is completely filled with water. Then fill the calibrated flask with water almost to the top, add the volatile analyte (usually in a methanol solution) with a micropipette, fill with water to just where the ground-glass stopper will enter the flask. Seal the flask with Parafilm®, shake, and store in a refrigerator. These diluted standards should be transferred quickly to a vial for headspace, purge and trap, or solid-phase microextraction (SPME) analysis and used only once.

The following methods are used for quantitative GC/MS.

6.5. External Standard Method

With the external standard method, the peak areas of the analytes in the samples are compared to the peak areas of the same analytes in a standard solution. An external calibration analysis is conducted as follows:

1. Prepare a series of calibration standards containing the compound(s) of interest at concentrations in the expected range of the unknown samples to be analyzed. Normally, a calibration curve would consist of about five or six concentration levels (duplicates at each level), and the standards would be in the same matrix as the unknown.
2. Analyze the standards including blanks as mentioned above.
3. Then calculate the RF for each analyte:

$$\mathrm{RF} = \frac{\mathrm{Area}_{\mathrm{std}}}{C_{\mathrm{std}}}$$

   where $\mathrm{Area}_{\mathrm{std}}$ is the peak area of the analyte standard and C_{std} is the concentration of the standard.
4. Calculate the mean of the RFs and use the mean RF to calculate the quantitative value for each of the analytes in the samples as follows:

$$C_x = \frac{\mathrm{Area}_x}{\mathrm{RF}_{\mathrm{mean}}}$$

   where C_x is the concentration of the analyte in the sample, Area_x is the peak area of the analyte, and $\mathrm{RF}_{\mathrm{mean}}$ is the mean response factor.

With most modern software, the user should be able to specify external standard calculation, and the data system will prompt for the entry of a value corresponding to the concentration of each of the calibration standards. After the analysis, the data system should plot a calibration curve of the concentration of each analyte versus the peak area and do a linear regression analysis, showing the correlation coefficient to a straight line. The data system should also calculate the mean and standard deviation of the RFs and use the mean RF for calculating a value for each analyte in the samples.

An example of calibration results from a modern data system will be shown in Section 6.6 of this chapter. Although the data system is very good at handling the tedious calculations, it is up to the analyst to examine the results and verify how valuable the data are. For example, is the curve linear over the expected range of the analysis and are replicate analyses in reasonable agreement with one another? The analyst must decide on the criteria for acceptable data. The external standard method depends on two criteria to give good quantitative results.

1. Injection volumes must be the same for each analysis; therefore, automatic injection is strongly recommended.
2. There must not be a complicated sample preparation procedure. If the sample preparation procedure involves steps such as solvent extraction, derivatization, and evaporation and reconstitution of solvents, then the external standard will not give reliable results and internal standard calibration should be used.

A good example of a sample that would be suited for quantitation by the external standard method would be drinking water containing trace solvents to be analyzed by automated headspace or purge and trap.

6.6. Internal Standard Method

For the most accurate quantitation in GC/MS, the internal standard method is advisable. The internal standard corrects for losses during subsequent separation and concentration steps as well as variation in the amount of sample injected into the GC. This method requires the addition of a known quantity of a compound known as an internal standard to an accurately measured aliquot of the sample being analyzed. The best internal standard is one that is chemically similar to the compound to be measured but that elutes in an empty space in the chromatogram. With mass spectrometry, it is possible to work with isotopically labeled analogs of the analyte that coelute with the analyte but are distinguished from one another by the differences in their *m/z* values.

Note: When using deuterated internal standards, the deuterated variant will elute from most GC columns slightly before the nondeuterated analog. This same retention time difference is not observed when ^{13}C-labeled analogs of the analyte are used as the internal standard.

The analytical procedure is as follows:

A series of calibration standards is prepared by preparing solutions containing the compound(s) of interest at concentrations in the expected range of the unknown samples to be analyzed. Normally, a calibration curve would consist of about five or six concentration levels (duplicates at each level) and the standards would be in the same matrix as the unknown. Next, a known weight of the internal standard is dissolved in a solvent; and equal volumes of the solution containing the internal standard* are added to carefully measured and equal volumes of each of the calibration standards and each of the unknown samples. The final concentration of the internal standard should be approximately in the mid-range of the expected concentration of the compound being measured. Note that the internal standard solution can contain a separate internal standard compound for each analyte of interest. A blank should be prepared by adding the internal standard to the solvent used for the sample.

Once the internal standard has been added to the calibration standards and the unknown samples and is thoroughly mixed, the solution can be concentrated, if necessary. Often, samples and reference standards are evaporated to dryness and then redissolved in a carefully measured quantity of solvent. At this point, chemical derivatization can be performed. As stated above, the use of an internal standard corrects for significant sample losses during the sample processing. After the sample processing is completed, the calibration method should be validated by demonstrating a linear calibration curve.

The GC/MS analysis should begin with a blank, followed by the calibration samples, and then another blank. For each individual standard at each calibration level, a relative response factor (RRF) is calculated (normally by the data system).

$$\mathrm{RRF} = \frac{\mathrm{Area}_{\mathrm{std}} \times C_{\mathrm{is}}}{C_{\mathrm{std}} \times \mathrm{Area}_{\mathrm{is}}}$$

where $\mathrm{Area}_{\mathrm{std}}$ is the area of the peak of the analyte in the calibration standard, C_{std} is the concentration of the calibration standard, $\mathrm{Area}_{\mathrm{is}}$ is the area of the peak representing the internal standard in the calibration standard, and C_{is} is the concentration of the internal standard in the calibration standard.

As with the external standard method, a calibration curve and linear regression data are calculated by the data system, and a mean RRF is

*The volume of the internal standard is usually much smaller than the volumes of the standards and samples; for example, 20 μL of internal standard solution is added to 2 mL of standard or sample. However, if there are slight differences in the matrices of the various samples, the internal standard may be added to a matrix-modifying solution such as water saturated with sodium sulfate; and relatively large volumes of the combined internal standard-matrix modifier may be added to the standards and samples to minimize the matrix variations. This method is especially important in headspace analysis.

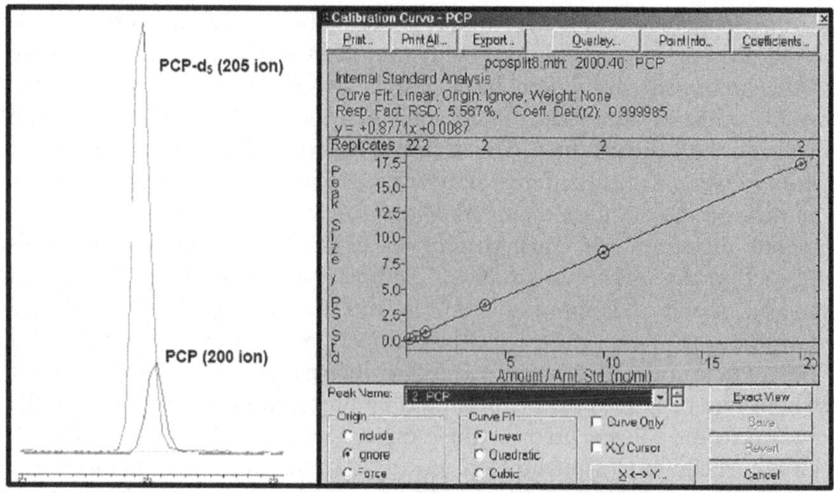

Figure 6.3 An internal standard calibration curve for phencyclidine. Note that the deuterated internal standard elutes before the standard. The data system has calculated the slope and intercept of the curve, the correlation to a straight line, and the relative standard deviation of the RRFs. *From the Varian Toxicology Manual, printed with permission.*

computed for all of the calibration standards (Figure 6.3). This mean RRF is used to calculate the concentration of the analytes as follows:

$$C_x = \frac{\text{Area}_x \times C_{\text{is(smp)}}}{\text{RRF}_{\text{mean}} \times \text{Area}_{\text{is(smp)}}}$$

where C_x is concentration of the analyte in the unknown sample, Area_x is the peak area of the analyte in the unknown sample, $C_{\text{is(smp)}}$ is the concentration of the internal standard in the unknown sample, RRF_{mean} is the mean relative response factor, and $\text{Area}_{\text{is(smp)}}$ is the peak area of the internal standard in the unknown sample.

6.7. Standard Additions

The method of standard additions is a variation of the external standard method. In some analytical procedures where there is virtually no sample preparation such as static headspace and SPME, the accuracy of the analysis depends on the sample matrix being the same for all of the samples and standards. For these samples, a blank matrix is necessary in order to prepare the calibration standards. However, many samples are very complex and a blank matrix (i.e., one that contains everything but the analyte of interest) is not available. Examples of such samples are paint mixtures, polymers containing volatiles that must be monitored, food products containing volatiles, and so forth.

The method of standard additions works as follows:

1. A series of standards containing known concentrations of the analyte (including a blank) are prepared in a solvent that is compatible with the sample.
2. The same precise amount of the sample containing the analyte is added to a series of sample vials.
3. Each sample vial is spiked with one of the standards prepared in step 1. The volume of each standard should be the same for each sample. One of the samples should be spiked with the blank. The concentration of the analyte in the most-concentrated spiking standard should be high enough so that when the sample is spiked with this standard, the analyte is at least twice the expected concentration in the sample containing the unknown amount.

The spiked samples are then analyzed. From an increase in the area of the chromatographic peak, the amount of the analyte present in the original sample can be deduced. The standard additions method requires a linear analyte response; i.e., a 10% increase in the amount of analyte present results in a 10% increase in the peak area. In order to calculate the concentration of the analyte in the original unspiked sample, the peak area of the analyte for each of the analyses is plotted on the y axis versus the total quantity of the standard added to each of the spiked samples on the x axis (Figure 6.4). To

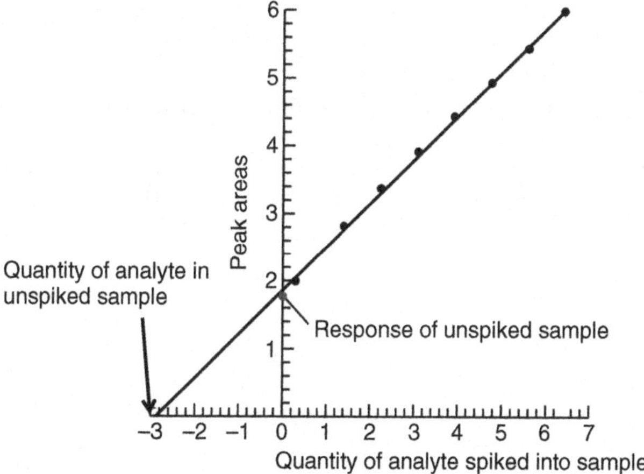

Figure 6.4 A calibration curve for the method of standard additions. The quantity of the analyte in the sample is obtained by extrapolating to the x axis. To determine concentration, divide the quantity by the volume of the sample.

determine the quantity of analyte in the original sample, extrapolate the curve to the x axis. Divide this value by the volume of sample that was added to the vial to determine the concentration of the analyte in the unspiked sample.

With standard additions, there are no internal standards; and injection volumes must be the same. This is not a problem with automated headspace or SPME where the technique would normally be used.

6.8. Concluding Remarks

Quantitation in the GC/MS laboratory requires a great deal of care to assure that the answers being generated are correct and that results are consistent from day to day. Most commercially available GC/MS data systems have routines to provide quantitative data (including elementary statistics but usually not control charts) from the GC/MS analyses. These data systems are also capable of generating reports that meet individual laboratory needs. Before starting on a quantitative project, a review of Chapters 3 (Experimental Error), 4 (Statistics), and 5 (Quality Assurance and Calibration Methods) in *Quantitative Chemical Analysis*, 7th ed. (Harris, DC; Freeman, WH; New York, 2007, ISBN: 978-07167-7041), or a later edition should be made.

Additional information regarding quantitation in GC/MS is found in *Trace Quantitative Analysis by Mass Spectrometry* (Boyd, RK; Basic, C; Bethem, RA; Wiley: Chichester, U.K., 2008, ISBN: 978-470-05771-1).

REFERENCES

1. Sphon, J. (1978). Use of mass spectrometry for confirmations of animal drug residues. *J. Assoc. Anal. Chem.*, 61, 1247–1252.
2. Stein, S. E., Heller, D. N. (2006). On the risk of false positive identification using multiple ion monitoring in qualitative mass spectrometry: large-scale intercomparisons with a comprehensive mass spectral library. *J. Am. Soc. Mass Spectrom.*, 17(6), 823–835.

SECTION 2

GC Conditions, Derivatization, and Mass Spectral Interpretation of Specific Compound Types

CHAPTER 7

ACIDS

7.1. GC SEPARATIONS OF UNDERIVATIZED CARBOXYLIC ACIDS

The chromatography of free acids can be challenging because these compounds are very polar. As such, they will often produce chromatographic peaks with a high degree of tailing. Most of the time, acids are derivatized prior to separation by GC.

A. Aliphatic acids
   1. Capillary columns
      a. C_1–C_5: 30-m DB-FFAP or HP-FFAP column at 135 °C. C_2–C_{10}: 30-m DB-FFAP column (or equivalent), 50–240 °C at 10 °C min^{-1}, run for 1 hour (~100 ppm can be detected).
      b. C_2–C_7: 25–30-m OV-351 column at 145 °C.
      c. C_1–C_7: 30-m DB-WAX column, 80–230 °C at 10 °C min^{-1}.

B. Simple mixtures (which include free acids)
   1. Formic acid, acetic acid, and propionic acid 2-m Porapak QS at 170 °C.
   2. Acetaldehyde, ethyl formate, ethyl acetate, acetic anhydride, and acetic acid 25-m CP-WAX 52CB column, 50–200 °C at 5 °C min^{-1}.

C. Aromatic carboxylic acids (see the following derivatization procedures)

7.2. GENERAL DERIVATIZATION PROCEDURE FOR C_8–C_{24} CARBOXYLIC ACIDS

A. Aliphatic acids–TMS derivatives

   For low-molecular-weight aliphatic acids, try TMSDEA reagent. Otherwise, use MSTFA, BSTFA, or Tri-Sil/BSA (Formula P). For analysis of the keto acids, methoxime (MO) derivatives should be prepared first, followed by the preparation of the tetramethylsilane (TMS) derivatives using BSTFA reagent. This results in the MO–TMS derivatives.

Gas Chromatography and Mass Spectrometry
DOI: 10.1016/B978-0-12-373628-4.00007-1

7.3. GC Separation of Derivatized Carboxylic Acids

A. Krebs cycle acids
   1. Derivatives: Krebs cycle acids have been analyzed using only the TMS derivatives, even though some are keto acids.
   2. GC conditions: 30-m DB-1 column, 60–250 °C at 5 °C min^{-1}.

Acid	Nominal mass of TMS derivatives (Da)
Malic	350
Fumaric	260
Succinic	262
2-Ketoglutaric	290
Oxalsuccinic	406
Isocitric	480
cis-Aconitic	390
Citric	480

B. α-Keto acids–MO–TMS derivatives
   1. Derivatives: Add 0.25 mL of MO hydrochloride in pyridine and let stand at room temperature for 2 hours. Evaporate to dryness with clean dry nitrogen. Add 0.25 mL of BSTFA, MSTFA, or BSA reagent and let stand for 2 hours at room temperature.
   2. GC conditions: 30-m DB-1 column, 60 (2 minutes)–200 °C at 10 °C min^{-1}, 200–250 °C at 15 °C min^{-1}.
   3. The following components are in the order of elution using the GC conditions given previously.
      a. Pyruvic acid–MO–TMS
      $CH_3C(NOCH_3)C(O)OTMS$.
      Major peaks: *m/z* 174.0586, *m/z* 73 base peak, $M^{+\bullet}$ peak at *m/z* 189.0821 (<2%).
      b. α-Ketobutyric acid–MO–TMS
      $CH_3CH_2C(NOCH_3)C(O)OTMS$.
      Major peaks: *m/z* 73 (base peak), 89 (~20%), 188 (~40%).
      For selected ion monitoring (SIM), *m/z* 188.0742 ($[M - CH_3]^+$).
      c. α-Ketoisovaleric acid–MO–TMS
      $(CH_3)_2(NOCH_3)C(O)OTMS$.
      Major peaks: *m/z* 73, 89, and 100.

For SIM, monitor 186.0948 ($[M - CH_3]^+$).
Highest *m/z* value peak observed: *m/z* 202 or 217.

d. α-Keto-β-methylvaleric acid–MO–TMS
$CH_3CH_2CH(CH_3)C(NOCH_3)C(O)OTMS$.
Major peaks: *m/z* 73 and 89.
For SIM, monitor *m/z* 200.1107 ($[M - CH_3]^+$).
m/z value peak observed: *m/z* 216.
m/z 203 distinguishes this isomer from the α-ketoisocaproic acid–MO–TMS. Both isomers have peaks at *m/z* 189, 200, and 216.

e. α-Ketoisocaproic acid–MO–TMS
$(CH_3)_3\ CHCH_2C(NOCH_3)C(O)OTMS$.
Major peaks: *m/z* 189, 200, and 216.
For SIM, monitor *m/z* 216.1056.

f. 2,3-Dihydroxyisovaleric acid–TMS

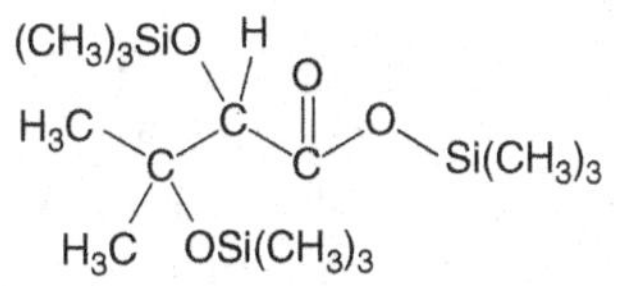

Base peak: *m/z* 131.
For SIM, monitor *m/z* 292.1346.

g. α-Isopropylmaleic acid–TMS

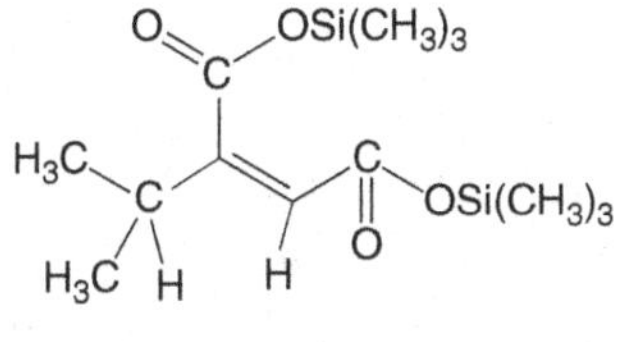

For SIM, monitor *m/z* 287.1135 ($[M - CH_3]^+$).

h. α,β-Dihydroxy-β-methylvaleric acid–TMS

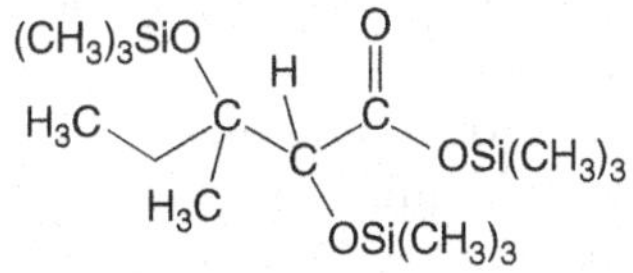

Base peak: *m/z* 145.
For SIM, monitor *m/z* 292.1346.

i. α-Isopropylmalic acid–TMS

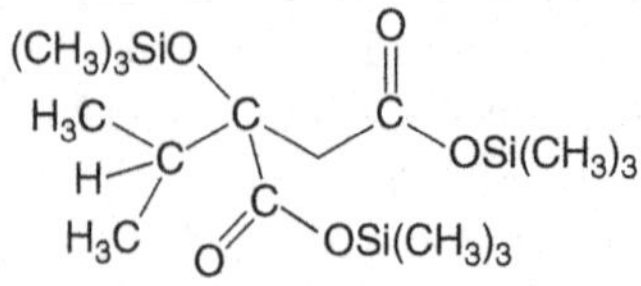

Major peaks: *m/z* 275, 261, and 349.
For SIM, monitor *m/z* 275.1499 ([M – C(O)OTMS]$^+$).

j. β-Isopropylmalic acid–TMS

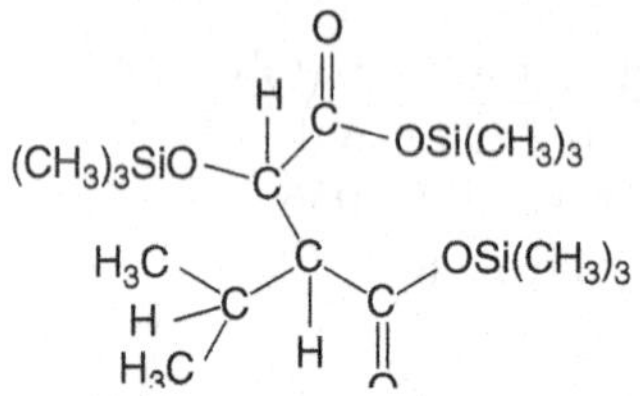

Note: The α isomer elutes slightly ahead of the β isomer.
Major peaks: *m/z* 275, 191, 231, and 305.
For SIM, monitor, *m/z* 275.1499 ([M – C(O)OTMS]$^+$).

C. Itaconic acid, citraconic acid, and mesaconic acid
 1. Derivatives: Add 0.25 mL MTBSTFA reagent to less than 1 mg of sample and heat at 60 °C for 30 minutes.
 2. GC conditions: 30-m DB-210 column, 60–220 °C at 10 °C min^{-1}.

D. Higher boiling acids such as benzoic and phenylacetic acids
 1. Derivatives: Add 0.25 mL of MSTFA or Tri-Sil/BSA (Formula P) to the dried extract and heat at 60 °C for 30 minutes.
 2. GC conditions: 25-m CPSIL-5 column, 100–210 °C at 4 °C min^{-1}.

E. Organic acids in urine
 1. Derivatives: 1 mL of urine adjusted to pH 8 with $NaHCO_3$ solution. Add MO hydrochloride or ethoxime hydrochloride. Dissolve and mix thoroughly, and then saturate the solution with NaCl. Adjust the solution to pH 1 with 6 N HCl.
 Extract with three 1-mL volumes of diethyl ether (top layer) followed by three 1-mL volumes of ethyl acetate. Combine the extractions and evaporate to dryness with clean dry nitrogen. Add 10 µL of pyridine and 20 µL of BSTFA reagent. Cap the vial and heat at 60 °C for 7 minutes.
 2. GC conditions: 30-m DB-1 column, 120 (4 minutes)–290 °C at 8 °C min^{-1} and hold for 25 minutes.
 3. Acids commonly found in urine: Some of the acids found in urine are given in the proceeding text. We have found as many as 100 GC peaks in urine samples, which include urea and other nonacids. The following components are in order of elution.

Component	Nominal mass of TMS derivatives (Da)
Phenol	166
Lactic acid	234
Glycolic acid	220
Oxalic acid	234
Hydroxybutyric acid	262
Benzoic acid	194
Urea	204
Phosphoric acid	314
Phenylacetic acid	208
Succinic acid	262
Glyceric acid	322
Fumaric acid	260
Glutaric acid	276
Capric acid	244
Malic acid	350
Hydroxyphenylacetic acid	296
Pimelic acid	304
Tartaric acid	438
Suberic acid	318
Aconitic acid	390
Hippuric acid	323
Citric acid	480
Isoascorbic acid	464
Indoleacetic acid	319
Gluconic acid	628
Palmitic acid	328
Uric acid	456

F. Bile acids

1. Derivatives: Acetylated methyl esters are the most suitable derivatives (e.g., deoxycholic acid and cholic acid). Evaporate the sample to dryness with clean dry nitrogen. Add 250 μL of methanol and 50 μL of concentrated sulfuric acid. Heat at 60 °C for 45 minutes. Add 250 μL of distilled water and allow cooling. Then add 50 μL of chloroform or methylene chloride. Shake the mixture for 2 minutes. Remove the bottom layer with a syringe. Evaporate to dryness with clean dry nitrogen. Acetylate with 50 μL of three parts acetic anhydride and two parts pyridine for 30 minutes at 60 °C. Evaporate to dryness with clean dry nitrogen. Dissolve the residue in 25 μL of ethyl acetate.
2. GC conditions: 25-m DB-1 column 200–290 °C at 4 °C min^{-1}.

G. C_6–C_{24} monocarboxylic acids and dicarboxylic acids as methyl esters
  1. Derivatives
    a. BF_3/methanol: Add 1 mL of BF_3/methanol reagent to less than 1 mg of the dry extract. Let the reaction mixture stand overnight or heat at 60 °C for 20 minutes. Cool in an ice-water bath and add 2 mL of water. Within 5 minutes, extract twice with 2 mL of methylene chloride. Evaporate the total methylene chloride extracts (if necessary).
    b. Methanol/acid (preferred method for trace analysis): Using a 1- or 2-mL reaction vial, add less than 1 mg of the sample. Add 250 μL of methanol and 50 μL of concentrated sulfuric acid. Cap the vial, shake, and heat at 60 °C for 45 minutes. Cool and add 250 μL of distilled water using a syringe. Add 500 μL of chloroform or methylene chloride and shake the mixture for 2 minutes. Inject a portion of the chloroform layer into the GC.
    c. Diazomethane: To less than 1 mg of the dry extract, add 200 μL of an ethanol-free solution of diazomethane in diethyl ether. (Caution: Do not use ground glass fittings when running this reaction.) This solution is stable for several months if stored in small vials (1–10 mL) in the freezer at −10 °C. Evaporate the methylation mixture and dissolve the residue in methanol.
    d. Methyl-8® reagent: Add 0.25 mL of Methyl-8® reagent to less than 1 mg of the dry extract. Cap the vial and heat at 60 °C for 15 minutes.
  2. GC conditions
    a. C_{14}–C_{22} unsaturated dibasic acids: 30-m DB-23 or CPSIL-88 column, 75–220 °C at 4 °C min^{-1}.
    b. 30-m DB-WAX column, 60–200 °C at 4 °C min^{-1}.

H. Bacterial fatty acids
  1. Derivatives (Appendix G)
  2. GC conditions
    a. C_8–C_{20} methyl esters of bacterial acids: 30-m DB-1 column, 150 (4 minutes)–250 °C at 6 °C min^{-1}.
  3. Types of bacterial fatty acids
    a. Saturated, straight chain: CH_3–$(CH_2)_n$–COOH.
    b. Unsaturated, straight chain: CH_3–$(CH_2)_n$–CH=CH–$(CH_2)_n$–COOH.
    c. Branched chain:
      1. Iso: $(CH_3)_2CH(CH_2)_nCOOH$.
      2. Anteiso: $C_2H_5CH(CH_3)(CH_2)_nCOOH$.

d. Cyclic:

$$H_3C-(CH_2)_n-CH(-CH_2-)CH-(CH_2)_n-COOH$$

e. Hydroxy:

$$H_3C-(CH_2)_n-CH(OH)-COOH$$

1. α (2-OH):

$$H_3C-(CH_2)_n-CH(OH)-CH_2-COOH$$

2. β (3-OH):

I. Cyanoacids: $N{\equiv}C(CH_2)_nCOOH$

1. GC conditions: 30-m DB-1 column, 100–275 °C at 10 °C min^{-1}.

7.4. Mass Spectral Interpretation

A. Underivatized carboxylic acids

Although carboxylic acids are more often analyzed as methyl esters, there are occasions when they are more easily analyzed as free acids, such as in water at the ppm level.

Intense peaks are observed in the mass spectra of straight-chain carboxylic acids at *m/z* 60 and 73 from *n*-butanoic to *n*-octadecanoic acid. The formation of an abundant rearrangement ion with *m/z* 60 requires a hydrogen in position four of the carbon chain. Most mass spectra of acids are easy to identify with the exception of 2-methylpropanoic acid, which does not have a hydrogen at the C-4 position and cannot undergo the γ-hydrogen-shift rearrangement, which induces a β cleavage (cleavage of the bond between C_2 and C_3). If C_2 is substituted, instead of *m/z* 60, the rearrangement ion will have *m/z* 74, 88, and so forth, depending on the number of CH_2 units in the substitution. Even though methyl esters have a characteristic peak at *m/z* 74, the mass spectrum of an acid can be distinguished from that of an ester by examining the losses of OH, H_2O, and COOH from the molecular ion of acids in contrast to the loss of $^{\bullet}OCH_3$ radical in the case of methyl esters. Also, in the higher-molecular-weight aliphatic acids, the intensities of the molecular ions increase from *n*-butanoic to *n*-octadecanoic acid.

B. Derivatized carboxylic acids (Appendix G)

C. Mass spectra of underivatized cyano acids

The molecular ion peak is usually not observed. Intense ions are observed with *m/z* 41 and 55. Other characteristic peaks are observed representing the ions *m/z* 60, $[M - CH_3]^+$, $[M - 40]^+$, $[M - 46]^+$, and $[M - 59]^+$.

D. Sample mass spectrum

Examination of the mass spectrum of *n*-decanoic acid (Figure 7.1) shows peaks with major intensities at *m/z* 60 and 73. A peak at *m/z* 60 (Appendix Q) suggests the mass spectrum may represent an aliphatic carboxylic acid. This peak in combination with a peak at *m/z* 73 (Appendix Q) is a strong indication of a carboxylic acid. Small peaks at *m/z* 31 and 45 also suggest the presence of oxygen. The molecular ion appears to be at *m/z* 172. Subtracting 32 for the two atoms of oxygen of the carboxylic acid group leaves 140 Da, which is $C_{10}H_{20}$. The compound is decanoic acid.

As already stated, acids are usually derivatized prior to chromatographic analyses. This results in far-less tailing and much better chromatographic peak shapes. Figure 7.2 is the EI mass spectrum of the TMS derivative of decanoic acid, and Figure 7.3 is a mass spectrum of the methyl ester of decanoic acid. The two most often used derivatives are organosiloxanes, such as trimethyl silyl derivatives and methyl esters.

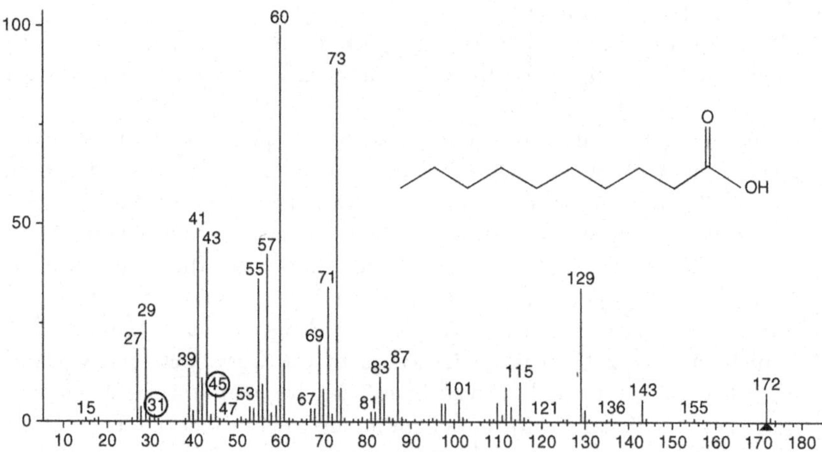

Figure 7.1 EI mass spectrum of decanoic acid.

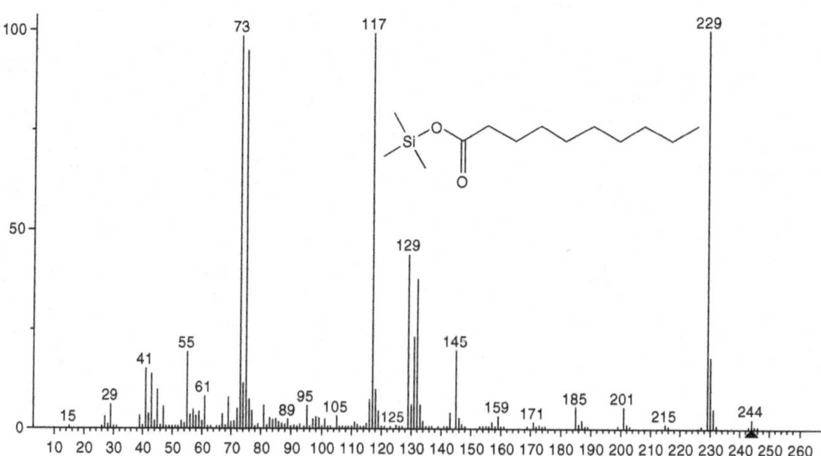

Figure 7.2 Mass spectrum of the TMS derivative of decanoic acid.

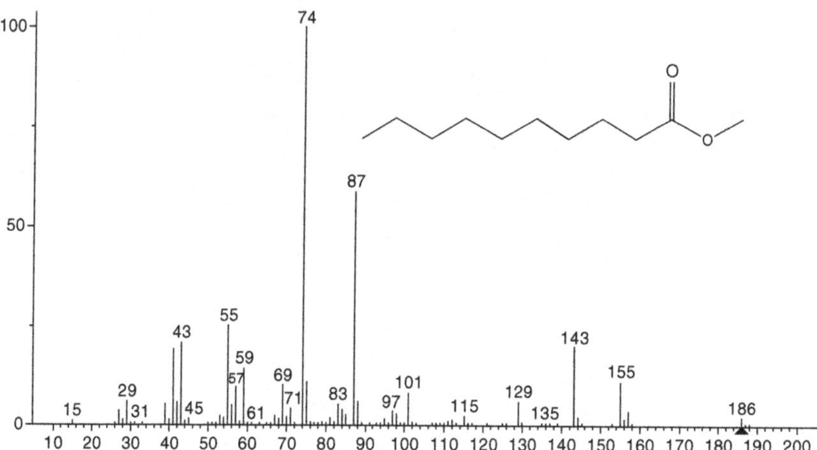

Figure 7.3 Mass spectrum of the methyl ester of decanoic acid.

CHAPTER 8

ALCOHOLS

Like aliphatic acids, aliphatic alcohols are very polar. Those with more than eight atoms of carbon have a tendency to produce chromatographic peaks that exhibit a great deal of tailing; therefore, many of these compounds are derivatized before chromatographic separations are carried out.

8.1. GC CONDITIONS FOR UNDERIVATIZED ALCOHOLS

A. General GC separations

1. C_1–C_5 alcohols: 2-m Carbowax 1500 on a Carbopak C column, 60–175 °C at 5 °C min^{-1}, or isothermal at 135 °C.
2. C_4–C_8 alcohols: 30-m CP-WAX 52CB column, 50–200 °C at 10 °C min^{-1}.
3. C_8–C_{18} alcohols: 30-m DB-5 column, 50–140 °C at 10 °C min^{-1}, then 140–250 °C at 4 °C min^{-1}.

B. Separation examples

1. Ethanol, 1-propanol, 2-methyl-1-propanol, 2-pentanol, isoamyl alcohol: 50-m CP-WAX 52CB column, 60–70 °C at 2 °C min^{-1}, then 70–200 °C at 10 °C min^{-1}.
2. Methanol, ethanol, isopropyl alcohol, *n*-propyl alcohol, *sec*-butyl alcohol, *n*-butyl alcohol: 30-m GS-Q column, 60–200 °C at 6 °C min^{-1}.
3. 3-methyl-1-butanol, 2-methyl-1-butanol: 30-m CP-WAX 51 (or CP-WAX 57CB) column, from 60–175 °C at 5 °C min^{-1}.
4. Methanol, ethanol, isopropylalcohol, *n*-propylalcohol, *tert*-butyl alcohol, 2-butanol, 2-methyl-1-propanol, 1-butanol, 2-pentanol, 2-methyl-1-butanol, 1-pentanol: 30-m Poraplot Q column, 135–200 °C at 2 °C min^{-1}.
5. 2-butanol, 1-butanol, 1,3-butanediol, 2,3-butanediol, 1,4-butanediol: 3-m 3% Carbowax 1500 column on 80–200 mesh Carbopack B, 80–225 °C at 8 °C min^{-1}.
6. Acetaldehyde, methanol, acetone, ethanol, isopropyl alcohol, *n*-propyl alcohol: 2-m Carbowax 20M column on Carbopack B at 75 °C.

Gas Chromatography and Mass Spectrometry
DOI: 10.1016/B978-0-12-373628-4.00008-3

8.2. TMS Derivative of >C_{10} Alcohols

A. Preparation of trimethylsilyl (TMS) derivative using less than 1 mg of alcohol, add 250 μL MSTFA reagent. Heat at 60 °C for 5–15 minutes.
B. GC separation of TMS derivative 30-m DB-5 column, 60–250 °C at 10 °C min^{-1}.

8.3. Mass Spectral Interpretation

A. Primary aliphatic alcohols
   1. General formula: ROH.
   2. Molecular ion peak: The intensity of the molecular ion peak in both straight-chain and branched alcohols decreases with increasing molecular weight. Beyond C_5, in the case of branched primary alcohols, and C_6, in the case of straight-chain primary alcohols, the molecular ion peak is usually insignificant and not really observed in the mass spectrum.

      The peak representing the loss of water from the molecular ion can easily be mistaken for the molecular ion peak. The spectrum is similar to that of an olefin below the $[M - H_2O]^+$ peak except that the peaks at *m/z* 31, 45, and 59 (along with other alkyl side chains) indicate an oxygen-containing compound.
   3. Fragmentation: A primary alcohol is indicated when the *m/z* 31 peak is intense and will be the base peak for C_1–C_4 straight-chain primary alcohols. The mass spectra of C_4 alcohols, and higher straight-chain primary alcohols, will exhibit peaks that represent the loss of 18, 33 (loss of H_2O followed by the loss of a $^{\bullet}CH_3$ radical), and 46 (loss of water followed by the loss of $H_2C{=}CH_2$) *m/z* units from the molecular ion. The intensity of the peaks representing the $[M-H_2O]^+$ and the $[(M-H_2O)-CH_3]^+$ ions can be very low and difficult to distinguish. Branched aliphatic alcohols do not appear to lose 46 *m/z* units from the molecular ion. Branching at the end of the chain, especially when an isopropyl or *tert*-butyl group is present, results in intense peaks at masses 43 and 57. In addition, the mass spectra of these compounds can exhibit an $[M-15]^+$ peak.
   4. Characteristic fragment ions: *m/z* 19 and 31, *m/z* 41, 55, etc., similar to 1-olefins.
   5. Characteristic losses from the molecular ion: $[M-18]^+$, $[M-33]^+$, $[M-46]^+$
B. Secondary and tertiary alcohols
   1. General formula: R_2CHOH and R_3COH.

2. Molecular ion peak: The molecular ion peak is slightly more intense in the mass spectra of secondary alcohols than in those of tertiary alcohols; but even in secondary alcohols, the $M^{+\bullet}$ peak is very small and may be impossible to distinguish.
3. Fragmentation: Many low-molecular-weight ($<C_8$) secondary and tertiary alcohols exhibit no $[M-18]^+$ peaks. C_8 and higher secondary alcohols exhibit $[M-18]^+$ peaks. The $[M-46]^+$ peak is usually missing in the mass spectra of secondary and tertiary alcohols. In secondary and tertiary alcohols, the loss of the largest alkyl group results in intense fragment ion peaks.
4. Characteristic fragment ion peaks: $[M-18]^+$ $>C_8$, $[M-33]^+$, no $[M-46]^+$.

   Peak at *m/z* 45 in spectra 2° alcohols with a methyl on the carbinol carbon (in aliphatic alcohols, the carbon atom with the hydroxyl attached is the carbinol carbon; carbon atoms attached to the carbinol carbon are α-carbons).

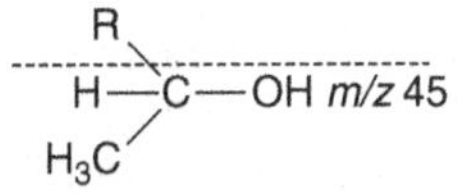

   R
   H—C—OH *m/z* 45
   H_3C

   Peak at *m/z* 45 in spectra 3° alcohols with two methyls on the carbinol carbon.

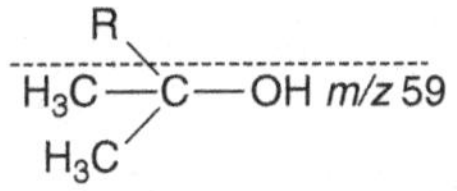

   R
   H_3C—C—OH *m/z* 59
   H_3C

C. Cyclic alcohols

1. General formula:

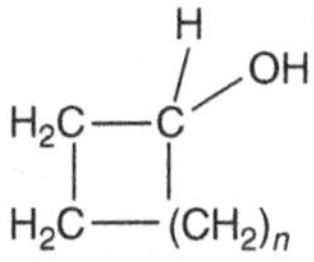

   H, OH
   H_2C—C
   H_2C—$(CH_2)_n$

2. Molecular ion: The intensity of the $M^{+\bullet}$ peak is generally <2.5%.
3. Fragmentation: The intensity of the peak at *m/z* 31 is sufficient to suggest the presence of oxygen. Peaks at *m/z* 44 and 57 are usually present, and an $[M-18]^+$ peak is also detectable. A peak at *m/z* 44 usually suggests an aldehyde, which is unbranched on the α-carbon; but this peak is also prominent in the mass spectra of cyclobutanol, cyclopentanol, cyclohexanol, and so forth. A peak at *m/z* 57

(C_3H_5O) is also fairly intense in the spectra of C_5 and larger cyclic alcohols. If the spectrum is of an aldehyde, peaks representing $[M-1]^+$, $[M-18]^+$, and $[M-28]^+$ are observed.

4. Characteristic fragment ions: $[M-18]^+$. Peaks at *m/z* 44 and 57 are fairly intense. No peaks representing $[M-1]^+$ or $[M-28]^+$ ions.

D. Mass spectra of TMS derivatives of aliphatic alcohols

The mass spectrum of the TMS derivative is used to determine the molecular weight of unknown alcohols, even though the molecular ion peak of the derivative may not be observed. If two peaks at high *m/z* values are observed and are 15 *m/z* units apart, then the highest *m/z* value peak (excluding isotope peaks) is the molecular ion peak of the TMS derivative. However, if there is only one peak at the high end of the *m/z* scale of the mass spectrum, add 15 to the *m/z* value of this peak to deduce the molecular mass of the analyte. There is a very high probability that the *m/z* value of this peak could be an odd number because the peak represents an EE^+ formed by the loss of a $^{\bullet}CH_3$ radical from the TMS moiety. The molecular weight of the alcohol is then determined by subtracting 72 (C_3H_8Si) from the molecular weight of the TMS derivative. Peaks representing ions with *m/z* 73, 89, and 103 are also usually present in the mass spectra of the TMS derivatives of aliphatic alcohols.

E. Sample mass spectra

1. In the mass spectrum of 1-octanol (Figure 8.1), the peaks at *m/z* 31 and 45 show that the compound contains oxygen. The presence of an intense *m/z* 31 peak further suggests that the analyte is a primary aliphatic alcohol, ether, or possibly a ketone. However, the lack of a peak at *m/z* 58 (explained later) rules out a ketone as a possibility.

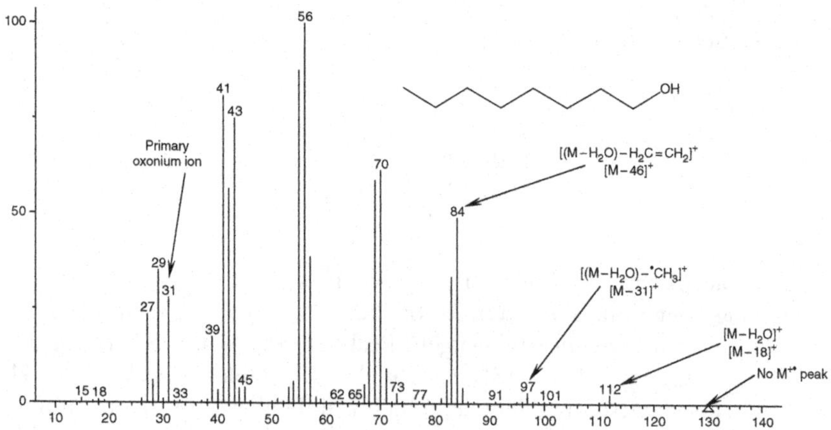

Figure 8.1 EI mass spectrum of 1-octanol.

By adding 18 to the highest *m/z* value peak observed (*m/z* 112), the deduced molecular weight would be 130 Da. Now check to see if peaks representing $[M-33]^+$ (loss of H_2O followed by the loss of a $^{\bullet}CH_3$ radical) and $[M-46]^+$ (loss of H_2O followed by the loss of $H_2C{=}CH_2$) are present at *m/z* 97 and *m/z* 84. This mass spectrum suggests a primary aliphatic alcohol with a molecular weight of 130, which is 1-octanol (caprylic alcohol), $C_6H_{18}O$.

2. For the mass spectrum of the TMS derivative of 1-octanol (Figure 8.2), note the large *m/z* 187 ion and the low-intensity peak 15 *m/z* units higher. The molecular ion of the TMS derivative is at *m/z* 202. Subtract 72 from 202 to obtain the molecular weight of the alcohol (202 − 72 = 130).
3. The peaks at *m/z* 77, 65, 51, and 39 in Figure 8.3 suggest a phenyl group. The ion at *m/z* 91 suggests a benzyl group, and the molecular ion peak 17 *m/z* units higher suggests benzyl alcohol. Aromatic alcohols do lose water; and, unlike aliphatic alcohols, their mass spectra exhibit very intense molecular ion peaks. The peak at *m/z* 31 representing a primary oxonium ion is almost nonexistent in this mass spectrum. Also see Chapter 29 on Phenols for more on aromatic alcohols.

If the EI mass spectra of the homologues series (C_4, C_5, C_6, C_7, and C_8) of ω-phenyl-*n*-alkanols are examined, it will be seen that all the spectra have peaks at *m/z* 91, 92, 104, and 117, $[M-H_2O]^+$, and that representing the molecular ion ($M^{+\bullet}$). Unlike the spectra of the corresponding *n*-alkanols, the $[M-H_2O]^+$ and $M^{+\bullet}$ are of sufficient intensities to be easily recognized. In all cases, the base peak is at *m/z* 91, which represents the tropylium ion

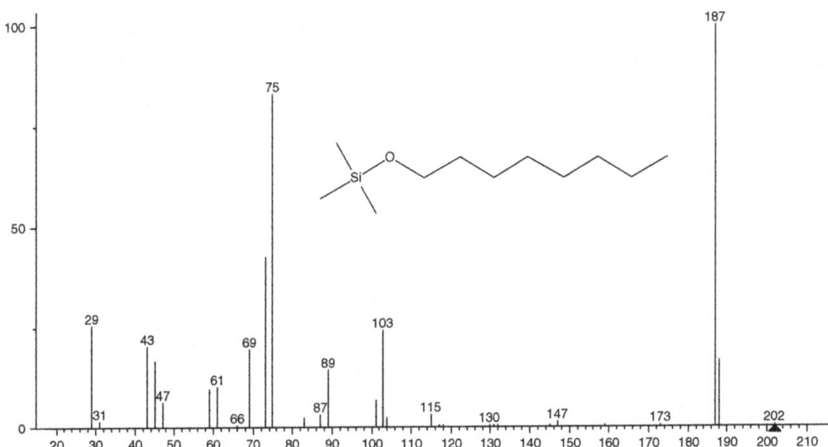

Figure 8.2 EI mass spectrum of the TMS derivative of 1-octanol.

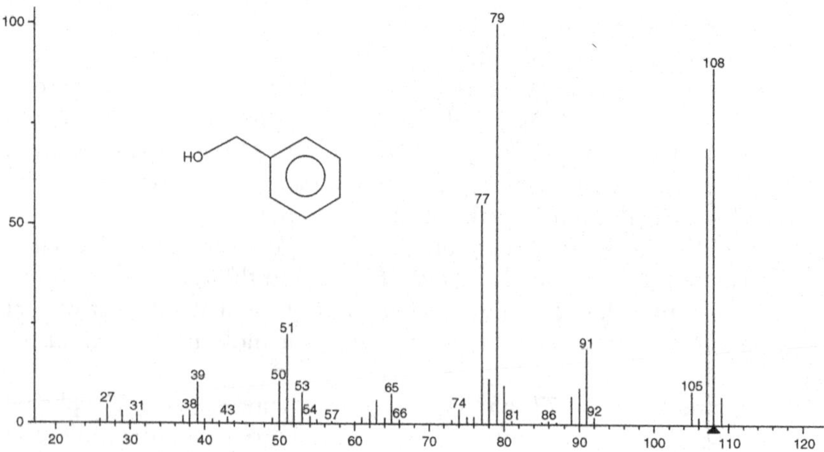

Figure 8.3 EI mass spectrum of benzyl alcohol.

(see Chapter 21 on Hydrocarbons), and is dominant in the mass spectra of all compounds that have a benzyl moiety (C_6H_5–CH_2–). The peak at *m/z* 92 is present in the mass spectra of all compounds that have a benzyl moiety and has hydrogen atoms on an atom that is three positions away from the ring (γ-hydrogens). The peak at *m/z* 104 in the mass spectrum of 4-pheny-*n*-butanol (nominal mass 150) represents the ion formed by the loss of $H_2C{=}CH_2$ from the ion that was formed by the loss of H_2O (the $[M - 46]^+$ peak). The peak at *m/z* 104 in the mass spectra of all the higher mass phenyl-*n*-alkanols represents ions formed by the successively larger 1-alkenes losses from the ion formed by the loss of water. In the mass

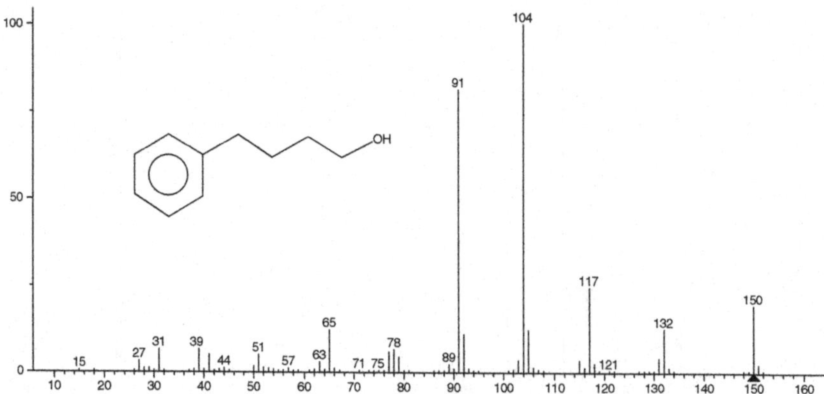

Figure 8.4 EI mass spectrum of 4-phenyl-*n*-butanol.

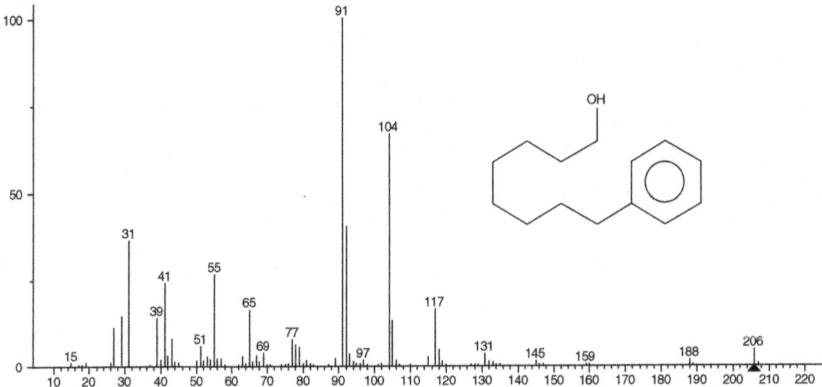

Figure 8.5 EI mass spectrum of 8-phenyl-*n*-octanol.

spectrum of 4-pheny-*n*-butanol, the peak at *m/z* 117 represents the ion formed by the loss of a $^{\bullet}CH_3$ radical from the ion formed by the loss of H_2O (the $[M - 33]^+$ peak). In all the other mass spectra, the *m/z* 117 peak represents ions that are formed by the loss of successively larger alkyl radicals from the $[M - H_2O]^+$ ion (Figures 8.4 and 8.5).

8.4. Aminoalcohols

See Chapter 11, Amines.

CHAPTER 9

Aldehydes

9.1. GC Separation of Underivatized Aldehydes

A. Capillary columns

1. Acetic acid, isobutyraldehyde, methylethyl ketone, isobutyl alcohol, *n*-propyl acetate, and isobutyric acid: 30-m Poraplot Q column, 100–200 °C at 10 °C min^{-1}.
2. Acetaldehyde, acetone, tetrahydrofuran, ethyl acetate, isopropyl alcohol, ethyl alcohol, 4-methyl-1,3-dioxolane, *n*-propyl acetate, methyl isobutyl ketone, *n*-propyl alcohol, toluene, *n*-butyl alcohol, 2-ethoxyethanol, and cyclohexane: 30-m DB-WAX column, 75 °C (16 minutes)–150 °C at 6 °C min^{-1}. Although the DB-FFAB column is similar to the DB-WAX column, it should not be used to separate aldehydes because it may remove them from the chromatogram.
3. Acetaldehyde, acetone, isopropyl alcohol, ethyl acetate, methyl isobutyl ketone, toluene, butyl acetate, isobutyl alcohol, and acetic acid: 30-m FFAP-DB column, 50–200 °C at 6 °C min^{-1}.
4. Aromatic aldehydes
   a. Benzyl alcohol, 1-octanol, benzaldehyde, octanoic acid, benzophenone, benzoic acid, and benzhydrol: 30-m DB-WAX column, 60 °C (1 minute)–230 °C at 10 °C min^{-1}.
   b. Tolualdehydes: *Ortho*- and *meta*-isomers do not separate very well. *Para*-isomers elute last. 50-m DB-WAX column, 60–180 °C at 6 °C min^{-1}.

9.2. Derivatization of Formaldehyde

A. Formaldehyde is derivatized for trace analyses. React 2-hydroxyethylpiperidine with formaldehyde to form 2-oxaindolizidine ($C_7H_{13}NO$).

   Selected ion monitoring of *m/z* 127 is used to determine the concentration of formaldehyde.

Gas Chromatography and Mass Spectrometry
DOI: 10.1016/B978-0-12-373628-4.00009-5

9.3. Mass Spectra of Aldehydes

A. Aliphatic aldehydes

1. General formula: RCHO.
2. Molecular ion: Both straight-chain and branched aliphatic aldehydes show molecular ion peaks up to a minimum of C_{14} aldehydes.
3. Fragmentation: Above C4, aliphatic aldehydes undergo the McLafferty rearrangement, resulting in a peak at *m/z* 44, provided the α- or β-carbon is not substituted. Substitution on the α- or β-carbon results in a peak at a higher *m/z* value (see the preceding text).

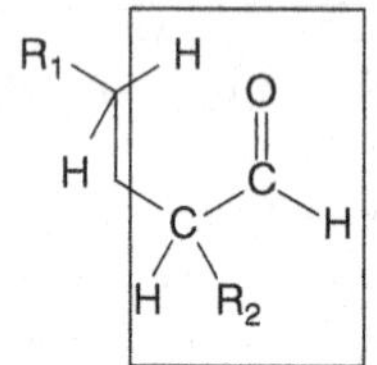

Note: Subtract 43 from the *m/z* value of the rearrangement ion to determine the mass of R_2

When $R_2 = H$, observe *m/z* 44.
When $R_2 = CH_3$, observe *m/z* 58.
When $R_2 = C_2H_5$, observe *m/z* 72, etc.

Small peaks at *m/z* 31, 45, and 59 indicate the presence of oxygen in the compound. Also, the mass spectra of aldehydes exhibit peaks at *m/z* values representing the losses of 28 and 44 from the molecular ion.

4. Characteristic fragment ions

The mass spectra of aliphatic aldehydes show *m/z* 29 (CHO) for C_1–C_3 aldehydes and *m/z* 44 for C_4 and longer chain aldehydes.

Characteristic losses from the molecular ion: $[M - 1]^+$ (loss of H), $[M - 18]^+$ (loss of H_2O), $[M - 28]^+$ (loss of CO), and $[M - 44]^+$ (loss of CH_3CHO).

Aldehydes are distinguished from alcohols by the losses of 28 and 44 from the molecular ion. The $[M - 44]^+$ ion results from the McLafferty rearrangement with the charge remaining on the carbonyl oxygen.

B. Aromatic aldehydes

1. General formula: ArCHO.
2. Molecular ion: Aromatic aldehydes exhibit a very intense molecular ion peak.
3. Fragmentation: The $[M - 1]^+$ peak due to the loss of the aldehyde hydrogen through the special case of homolytic cleavage (α cleavage) is usually very intense. An $[M - 29]^+$ peak is characteristic of an aldehyde group attached to an aromatic moiety. Peaks at *m/z* 39, 50,

51, 63, and 65, and the abundance of the molecular ion peak show that the compound is aromatic. Accurate mass measurement data indicate the presence of an oxygen atom.
4. Characteristic losses from the molecular ion: $[M - 1]^+$ (loss of H) and $[M - 29]^+$ (loss of CHO).

C. Sample mass spectra
1. An intense peak at *m/z* 44 in the mass spectrum of hexanal suggests an aliphatic aldehyde. $[M - 18]^+$, $[M - 28]^+$, and $[M - 44]^+$ peaks (at *m/z* 56, 72, and 82, respectively) suggest an aliphatic aldehyde unbranched at the α-carbon (see Appendix Q: Ions for Determining Unknown Structures). The molecular ion peak at *m/z* 100 confirms that this is the spectrum of hexanal (see Figure 9.1).
2. The mass spectrum of 2-methylbenzaldehyde suggests an aromatic compound because of the intensity of the molecular ion peak, the peak at *m/z* 91, and the peaks at *m/z* 39, 51, and 65 (see Figure 9.2). The peaks representing the loss of a hydrogen atom and loss of 29 (•CH=O radical) from the molecular ion indicate that this is an aromatic aldehyde. The peak at *m/z* 91 suggests the following structure:

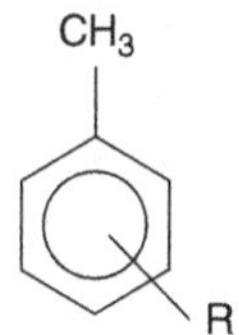

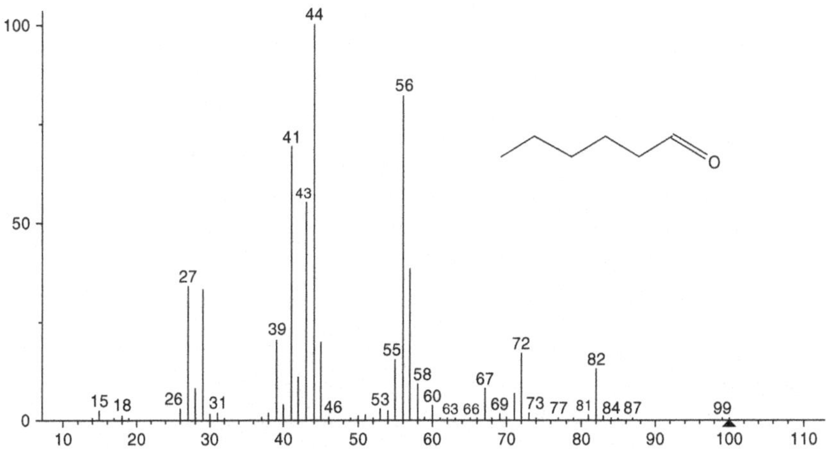

Figure 9.1 EI mass spectrum of hexanal.

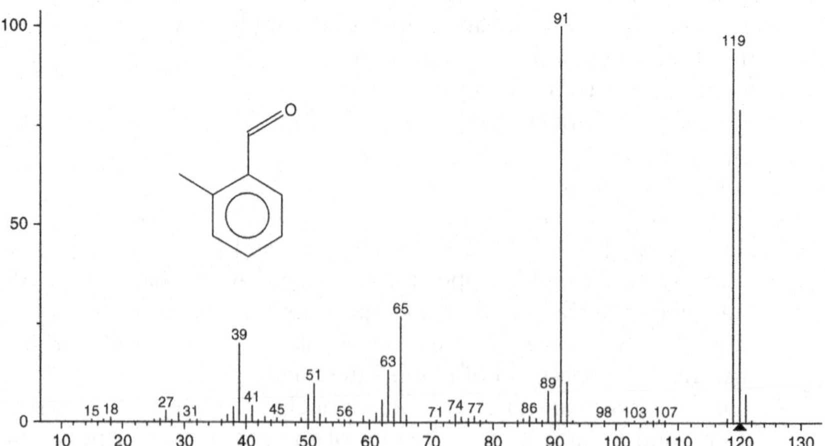

Figure 9.2 EI mass spectrum of 2-methylbenzaldehyde (2-tolualdehyde).

From the *m/z* value of the molecular ion (*m/z* 120), the structure for the aromatic aldehyde is

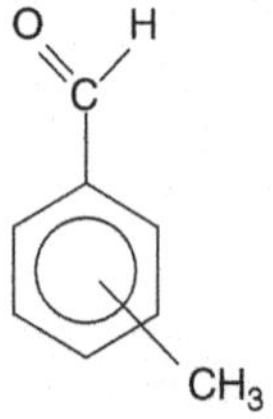

CHAPTER 10

Amides

10.1. GC Separation of Underivatized Amides

A. Capillary columns
   1. General conditions for separation of amides: 30-m FFAP-DB column, 80–220 °C at 12 °C min^{-1}.
   2. Hexamethylphosphoramide, pentamethylphosphoramide, tetramethylphosphoramide, trimethylphosphoramide, and $[(CH_3)_2N]_2$ P(O) NHCHO: 30-m DB-WAX column, 60–220 °C at 10 °C min^{-1}.
   3. *N,N*-dimethylacetamide (DMAC) impurities: *N,N*-dimethylacetonitrile, DMF, DMAC, *N*-methylacetamide, and acetamide—60-m DB-WAX column, 60–200 °C at 7 °C min^{-1}.

10.2. Derivatization of Amides

A. Primary amides

   Derivatized primary amides are more volatile. Typically, TMS of *N*-dimethylaminomethylene derivatives are prepared.
   1. TMS derivatives of amides: Add 250 µL of TRI-Sil/BSA (Formula P) reagent to less than 1 mg of the sample. Heat at 60 °C for 15–20 minutes.
   2. *N*-dimethylaminomethylene derivatives of primary amides: Add 250 µL of Methyl-8® reagent to less than 1 mg of sample. Heat at 60 °C for 20–30 minutes.

$$R{-}C(=O){-}NH_2 \xrightarrow{\text{Methyl-8}^{®}} R{-}C(=O){-}N{=}CH{-}N(CH_3)_2$$

99

Gas Chromatography and Mass Spectrometry
DOI: 10.1016/B978-0-12-373628-4.00010-1

This derivative also works well with diamines or amino amides (e.g., 6-aminocaproamide).

$$H_2N{-}(CH_3)_5{-}C(=O){-}NH_2 \xrightarrow{\text{Methyl-8®}} (CH_3)_2N{-}CH{=}N{-}(CH_2)_5{-}C(=O){-}N{=}CH{-}N(CH_3)_2$$

B. Aromatic amides (ArC(O)NHR or RC(O)NHAr where R can represent H, an alkyl group, or an aromatic group)
   1. Derivatization of aromatic amides: Except for simple aromatic amides such as benzamide and acetanilide, derivatization is recommended.[*] The most common derivatives used in this laboratory are TMS, acetate, and *N*-dimethylaminomethylene (for primary amides).
      a. Preparation of TMS derivatives of primary and secondary aromatic amides. Add 250 μL of TRI-Sil/BSA (Formula D) reagent to less than 1 mg of sample. Heat at 60 °C for 15–30 minutes.
      b. Preparation of acetate derivatives of primary and secondary aromatic amides. Add 150 μL of acetic anhydride and 100 μL of pyridine to less than 1 mg of sample. Heat at 60 °C for 30 minutes. Evaporate to dryness with clean dry nitrogen. Dissolve residue in 25 μL of DMF or other suitable solvent.
      c. *N*-dimethylaminomethylene derivative of primary aromatic amides. Add 250 μL of Methyl-8® reagent to less than 1 mg of sample dissolved in DMF.

10.3. GC Separation of Derivatized Amides (TMS or Methyl-8®)

A. 30-m DB-5 column, 60–275 °C at 8 °C min^{-1}
B. Aromatic amides: 30-m DB-5 column, 150–300 °C at 10 °C min^{-1}

10.4. Mass Spectra of Amides

A. Primary amides
   1. General formula: $RC(O)NH_2$.

[*] Even when they are underivatized, benzamide and benzanilide have very similar mass spectra.

2. Molecular ion: The mass spectra of underivatized amides generally show molecular ion peaks.
3. Fragmentation: For straight-chain amides (>C_3) having a γ-hydrogen, the base peak is *m/z* 59 (C_2H_5NO), which is a result of a McLafferty rearrangement.

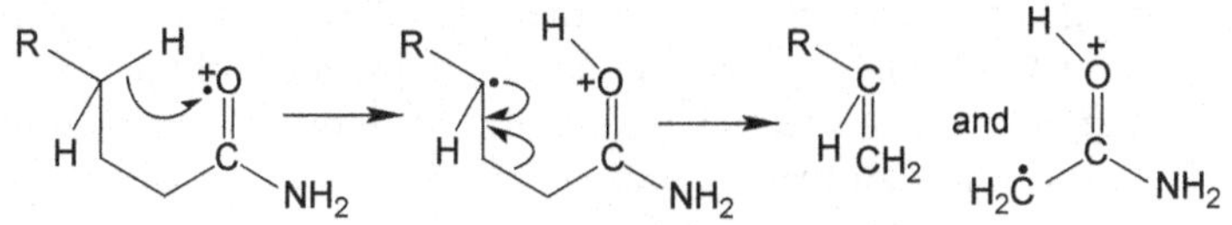

Peaks are also observed at *m/z* 44, 58, 72, and so forth with the peaks at *m/z* 59 and 72 being the most intense.

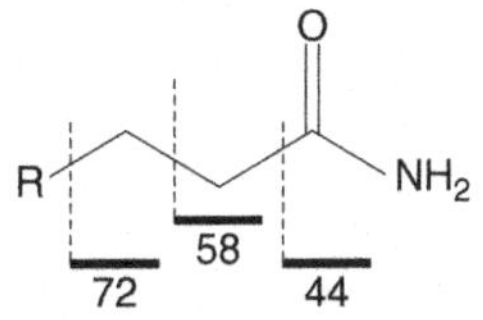

In summary, if the unknown mass spectrum has an intense peak at *m/z* 59, and *m/z* 72 with an odd *m/z* value molecular ion peak, this suggests a primary amide.

B. Secondary amides
1. General formula: $R_1C(O)NHR_2$.
2. Molecular ion: As would be expected, the molecular ion decreases in intensity as R_1 or R_2 increases in length.
3. Fragmentation: The mass spectra of secondary amides have an intense peak at *m/z* 30 representing a rearrangement ion. The fragmentation of the R_2 chain can occur to yield $R_1C(O)NH_2^+$ or $R_1C(O)NH_3^+$. Secondary amides also undergo the McLafferty rearrangement:

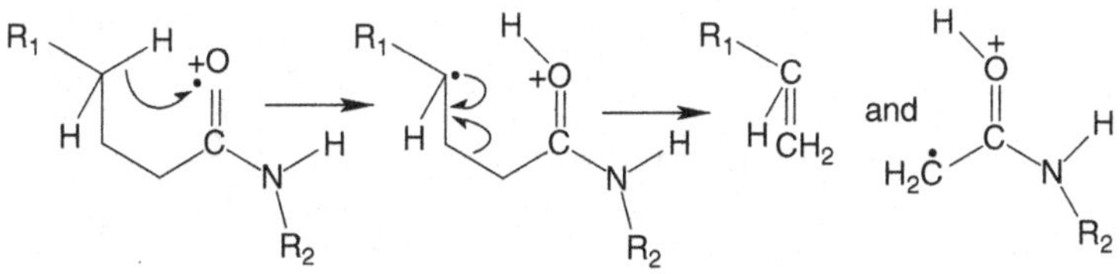

Subtract 58 from the *m/z* value of this rearrangement ion to find the mass of R_2.

C. Tertiary amide
   1. General formula: $R_1C(O)NR_2R_3$.
   2. Molecular ion: A molecular ion peak is observed when R_1, R_2, and R_3 are less than or equal to C_4.
   3. Fragmentation: Providing the R_1 group has a γ-hydrogen, $[CH_3C(O)NR_2R_3]^+$ is a common fragment. Subtract 57 from the *m/z* value of this rearrangement ion to find the mass of $R_2 + R_3$.

D. Mass spectra of aromatic amides
   In simple aromatic amides, fragmentation occurs on both sides of the carbonyl group. If a hydrogen is available in N-substituted aromatic amides, it tends to migrate and form an aromatic amine and the loss of a ketene ($H_2C{=}C{=}O$). Some simple aromatic amides include: benzamide, dibenzamide, *N*-phenyldibenzamide, nicotinamide, *N,N*-diethylnicotinamide, acetanilide, and benzanilide.

Benzamide (MW = 121 Da)

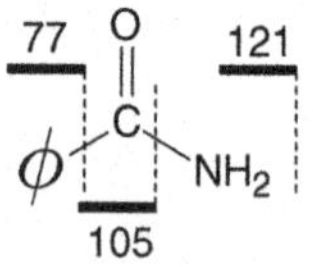

77 O 121
ϕ C NH2
105

m/z 105, 77, 121

Dibenzamide (MW = 225 Da)

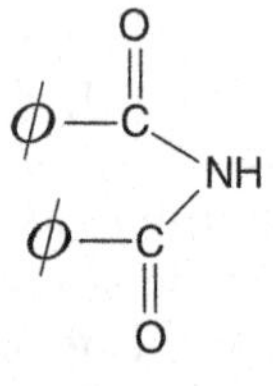

ϕ—C(=O)—NH—C(=O)—ϕ

m/z 107, 77, 225

N-Phenyldibenzamide (MW = 301 Da)

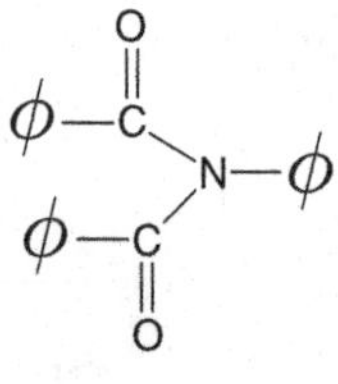

ϕ—C(=O)—N(—ϕ)—C(=O)—ϕ

m/z 105, 77, 197

Nicotinamide (MW = 122 Da)

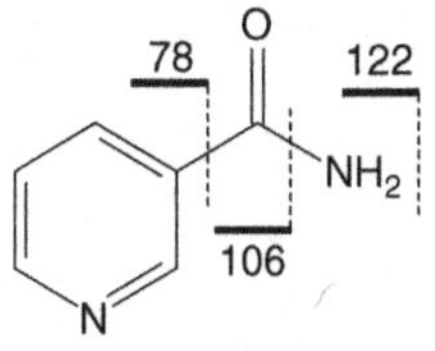

m/z 122

N,N-Diethylnicotinamide (MW = 178 Da)

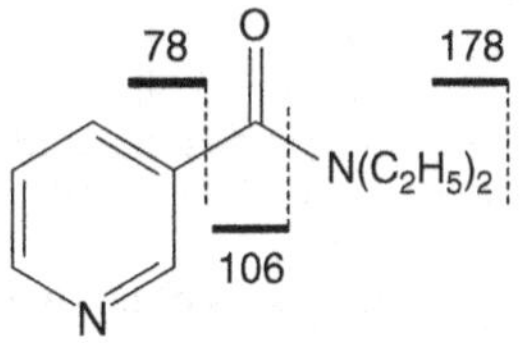

m/z 106, 78, 177, 178

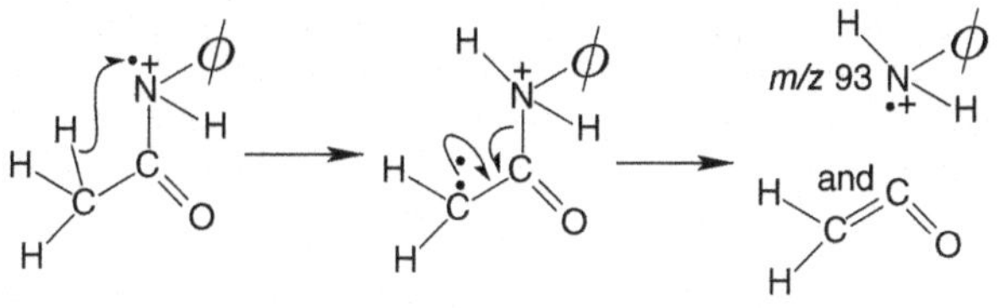

Benzanilide (MW = 197 Da)

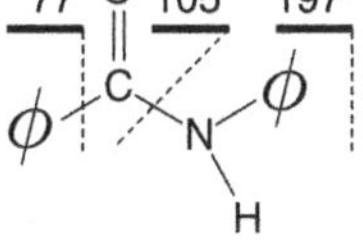

m/z 105, 197, 77

E. Sample mass spectrum
Notice the ions at *m/z* 59, 44, and 72 in the mass spectrum of hexanamide (Figure 10.1). Looking them up in the tables in Appendix Q suggests an aliphatic primary amide. Looking at the mass spectrum very closely, the highest mass ion occurs at *m/z* 115. These data suggest a hexanamide.

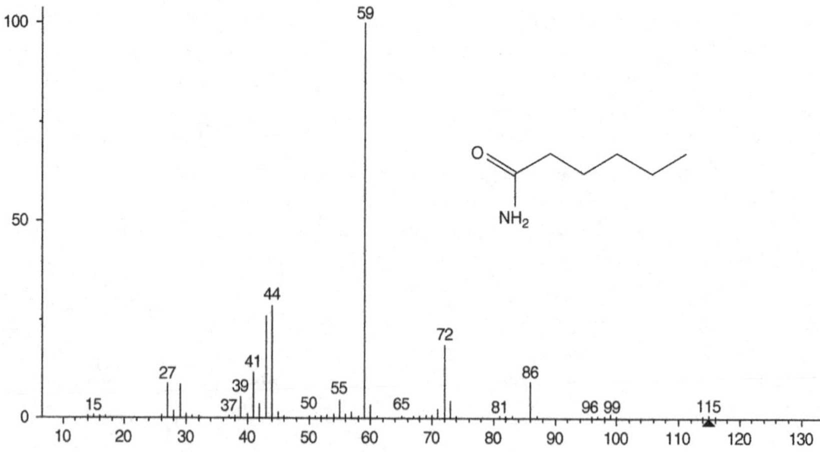

Figure 10.1 EI mass spectrum of hexanamide.

10.5. Mass Spectra of Derivatized Amide

Benzamide–TMS (MW = 193 Da) *m/z* 73, 178, 135, 193, and 192.
Acetanilide–acetate (MW = 177 Da) *m/z* 93, 135, 43, and 77.

CHAPTER 11

AMINES

11.1. GC SEPARATIONS OF UNDERIVATIZED AMINES

Authors' Note: *Due to the current nonuse of packed columns in mass spectrometry laboratories, references to them have been removed from the first edition of this book; however, the original authors had a very strong opinion about their belief that packed columns were able to bring about superior chromatographs of amines. Based on this opinion of the original authors, the comments and information regarding packed columns have been retained for this chapter only.*

Although capillary columns are generally preferred, there are many examples where separation is better by using packed columns, especially for low-boiling amines.

A. Low-boiling aliphatic amines
   1. Amines from C_1 (methylamine) to C_6 (cyclohexylamine): 2-m 4% Carbowax 20M column + 0.8% KOH on 60–80 mesh Carbopack B, 75–150 °C at 6 °C min^{-1}.
   2. Isopropylamine, *n*-propylamine, diisopropylamine, di-*n*-propylamine: 2-m Chromosorb 103 column, 50–150 °C at 8 °C min^{-1}.
   3. Methylamine, ethyleneimine, dimethylamine, trimethylamine: 2-m Chromosorb 103 column, 60–180 °C at 6 °C min^{-1}.
   4. 1-Aminooctane, 2-aminooctane: 2-m Chromosorb 103 column at 135 °C.

B. Higher boiling aliphatic amines and diamines
   1. Diaminoethane, diaminopropane, diaminobutane, diaminopentane, diaminohexane, diaminooctane: 25-m CP-WAX column for amines and diamines (Chrompack cat. no. 7424), 75–200 °C at 6 °C min^{-1}.
   2. Diaminoethane, diaminopropane, 1-amino-2-propanol, diaminobutane, diaminopentane, *n*-decylamine: 25-m CP-WAX column for amines and diamines at 135 °C.

C. Aromatic amines and diamines
   1. Aniline, 2,3,4-picolines: 2-m Carbowax 20M column or Carbopack B column, 75–150 °C at 3 °C min^{-1}.

Gas Chromatography and Mass Spectrometry
DOI: 10.1016/B978-0-12-373628-4.00011-3

2. Dimethylanilines and trimethylanilines: 25-m DB-1701 column, 60–270 °C at 5 °C min^{-1}.
3. Toluidine, nitrotoluene isomers, diaminotoluene, and dinitrotoluene isomers: 30-m DB-17 column, 100–250 °C at 8 °C min^{-1}.
4. Phenylenediamines
   a. Lower boiling impurities (sample dissolved in acetonitrile): Aniline (MW = 93 Da), quinoxaline (MW = 130 Da), dimethylquinoxaline (MW = 158 Da) from the phenylenediamines (MW = 108 Da). 25-m CP-WAX column for Amines (Chrompack) at 200 °C.
   b. Higher boiling impurities: Some impurities may be found under these GC conditions: quinoxaline, phenazine, tetrahydrophenazine, nitroanilines, hydroxyanilines, chloronitrobenzenes, hydroquinone, diaminophenazine, aminohydroxyphenazine. 30-m DB-17 column, 100–275 °C at 6 °C min^{-1}.
5. Diisopropylamine, diisobutylamine, dibutylamine, pyridine, dicyclohexylamine, aniline, 2,6-dimethylaniline. 25-m CP-WAX-51 column, 70–210 °C at 5 °C min^{-1}.

11.2. Derivatization of Amines and Diamines

MTBSTFA is the recommended reagent for silylation of the amine functionality because this reagent forms a more stable derivative than MSTFA, BSTFA, or BSA. The solvent used is important because amines can be difficult to silylate.

A. TBDMS and TMS derivatives
   Add 0.1 mg of the sample in 50 μL of acetonitrile (or THF) to 50 μL reagent. Let the solution stand at room temperature for 10–20 minutes.
   Reagents: MTBSTFA (recommended)
   MSTFA (recommended for amine hydrochlorides)
   BSTFA
   BSA
B. Preparing Methyl-8® derivatives: Add less than 0.1 mg of sample to 50 μL of acetonitrile and then add 50 μL of Methyl-8® reagent $[(CH_3)_2NCH(OCH_3)_2]$. Heat at 60 °C for 30 minutes or at 100 °C for 20 minutes.

11.3. GC Separation of Derivatized Amines

A. Diamines
   1. TBDMS or TMS derivatives of diamines: 30-m DB-225 column, 75–225 °C at 8 °C min^{-1}.

2. Methyl-8® derivatives of diaminohexanes and diaminooctanes: 30-m DB-5 column, 80–270 °C at 8 °C min^{-1}.

11.4. Mass Spectral Interpretation of Amines

A. Underivatized

Organic compounds with an odd number of nitrogen atoms will have an odd nominal mass; therefore, the $M^{+\bullet}$ peak in the mass spectrum of compounds having an odd number of nitrogen atoms will be at an odd *m/z* value. There will be fragment ion peaks at even *m/z* values as long as the ion is an EE^{+} ion peak and retains the odd number of nitrogen atoms. When a fragment ion is formed from a molecular ion containing an odd number of nitrogen atoms and a nitrogen atom is lost, this ion will have an even *m/z* value now that it contains an odd number of nitrogen atoms. For aliphatic amines, the most important fragmentation is cleavage of the bond that is alpha to the carboamino carbon atom (the carbon atom attached to the nitrogen) with the charge remaining on the nitrogen-containing fragment. In electron ionization, when the molecular ion of an aliphatic amine is formed, the most likely site of the charge and radical will be on the nitrogen atom due to the loss of one of the two nonbonding electrons associated with that atom. The first reaction involving fragmentation of the molecular ion is a cleavage of the bond between C_1 and C_2. This is called α cleavage. Aliphatic amines will also undergo β cleavage, which is the breaking of the bond between C_2 and C_3 (Scheme 11.1).

Examination of the EI mass spectrum of *n*-octyl amine in Figure 11.1 shows that there are peaks representing these two ions (*m/z* 30 and 44).

The mass spectrum for this same compound in the first edition of this book showed a relative intensity of about 25% for the peak at *m/z* 44. In this mass spectrum, the intensity of the peak at *m/z* 44 is ~5% of base peak. The same is true for all the spectra of this compound in the NIST08 Database. Beta cleavage does play a significant role in the mass spectra of 2° and 3° aliphatic amines; however, for straight-chain

Scheme 11.1

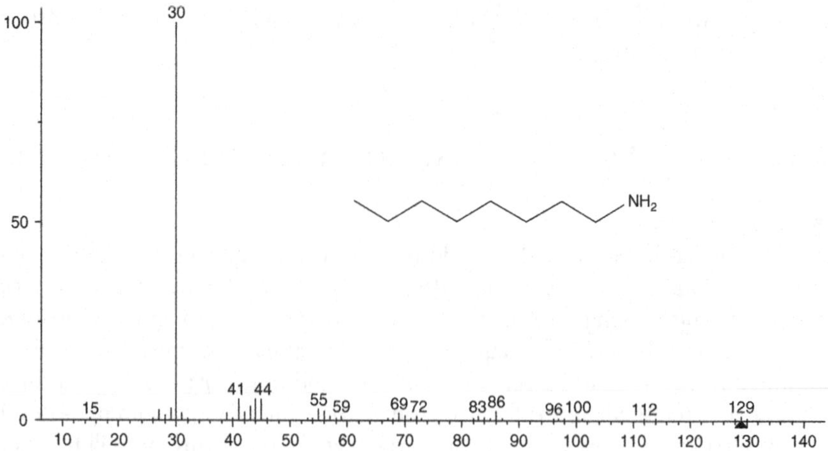

Figure 11.1 EI mass spectrum of *n*-octylamine.

compounds with six or more carbon atoms, the ion with *m/z* 44 is primarily the result of a complex series of rearrangements. The ion has the structure rather than the structure shown in the above example of β cleavage. The tendency for this rearrangement fragment ion to form is highly dependent on analyte ion-source pressure. The spectrum in the first edition of this book was from the authors' laboratory and was probably acquired using a direct insertion probe. This could well have meant that the partial pressure of the analyte was very high, which would favor the formation of the rearrangement product [1].

$$H(H_3C)C{=}\overset{+}{N}H_2$$

Although EI mass spectrometry is considered to be the most reproducible of all the mass spectrometry techniques, this example is a good illustration of variabilities that can exist and have to be watched for in all GC/MS analyses.

Underivatized aliphatic diamines are difficult to identify by their mass spectra alone because of the low abundance of the molecular ion (<3%). However, in some cases, an $[M - 17]^+$ (loss of NH_3) peak representing a common fragment ion formed by a hydrogen shift from the α-carbon followed by a heterolytic cleavage as shown in Scheme 11.2.

1. Primary amines

   The mass spectra of primary aliphatic amines show characteristic peaks at *m/z* 18 ($[NH_4]^+$ difficult to distinguish from the $[H_2O]^+$) and *m/z* 30 $[H_2C{=}NH_2]^+$. If the carboamine carbon is alkyl substituted, then intense peaks are observed at *m/z* 44, 58, or 72,

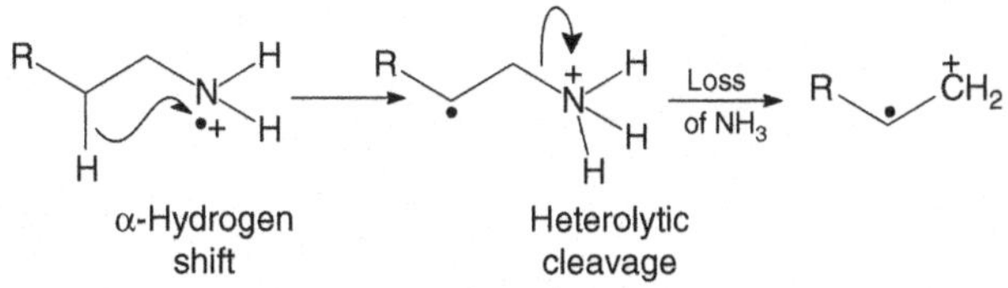

Scheme 11.2

and so on. If the unknown amine reacts with acetone or Methyl-8®, then it is a primary amine (Scheme 11.3).

2. Secondary amines
   The molecular ion of a secondary amine will undergo α cleavage to form a secondary immonium ion, which in turn will undergo a hydride-shift rearrangement fragmentation to lose an olefin and result in a primary immonium ion, provided both chains have at least two carbon atoms (Scheme 11.4 and Figure 11.2).
3. Tertiary amines
   Tertiary amines will undergo α cleavage, preferentially in the loss of the largest hydrocarbon chain. If there are two or more carbon atoms present in both of the other two chains, then two hydride-shift rearrangements occur to result in the primary immonium ion (Scheme 11.5). The peak representing the primary immonium ion is not as intense as in the spectra of primary and secondary amines because all of the intermediate ions are not converted as can be seen by the intensity of the peaks that represent each of them (Figure 11.3).

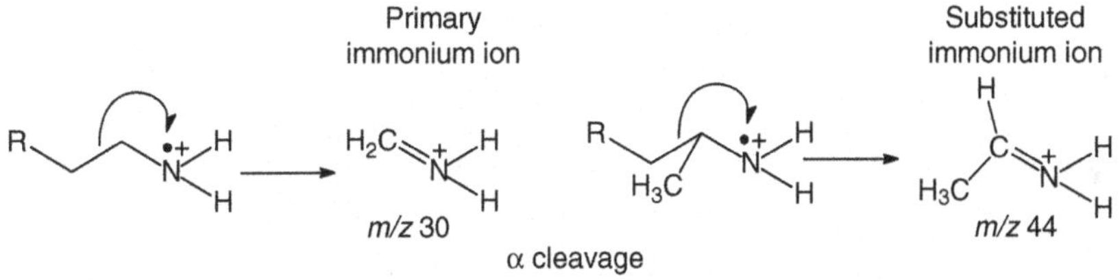

Scheme 11.3

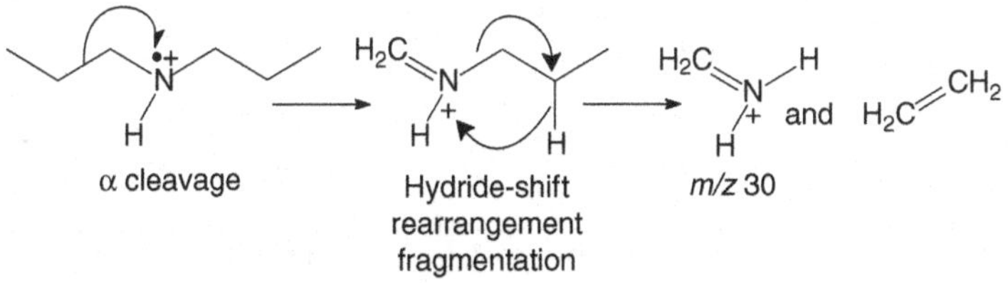

Scheme 11.4

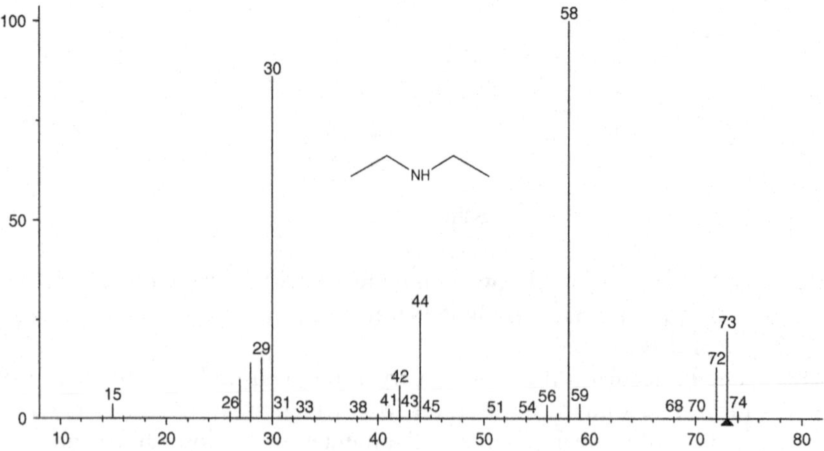

Figure 11.2 EI mass spectrum of *N,N*-diethylamine exhibiting peaks at *m/z* 58 (α cleavage) and *m/z* 30 (a primary immonium ion); the result of a hydride-shift rearrangement fragmentation of the ion with *m/z* 58.

Loss of $\bullet C_3H_7$
α cleavage
Loss of $R_1HC{=}CH_2$
Hydride-shift rearrangement fragmentation
Loss of $R_1HC{=}CH_2$
Hydride-shift rearrangement fragmentation
m/z 30

Scheme 11.5

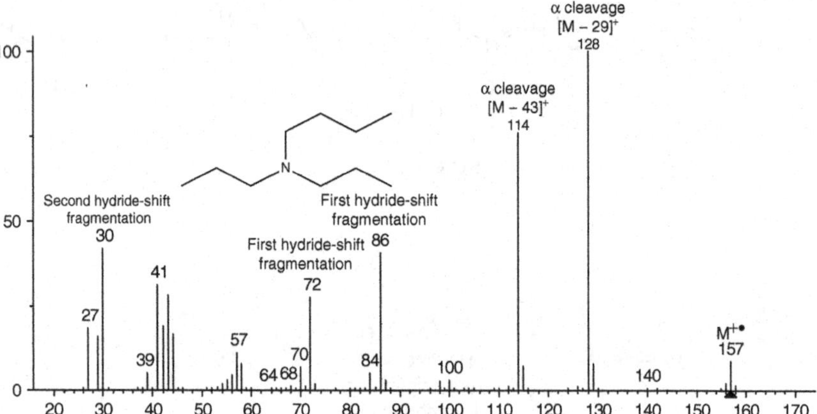

Figure 11.3 EI mass spectrum *N,N*-dipropyl-1-butanamine.

Summary: If the molecular weight is odd, then the compound contains an odd number of nitrogen atoms. Fragment ions observed at even-mass values suggest the presence of nitrogen. The loss of ammonia is fairly common in nitrogen compounds and may not indicate exclusively that an amine is present. Chemical derivatization will easily determine if the unknown is a tertiary amine.

4. Mass spectra of cyclic amines
   $M^{+\bullet}$ peaks as well as peaks representing $[M - H]^+$ ions are observed in the mass spectra of cyclic amines.

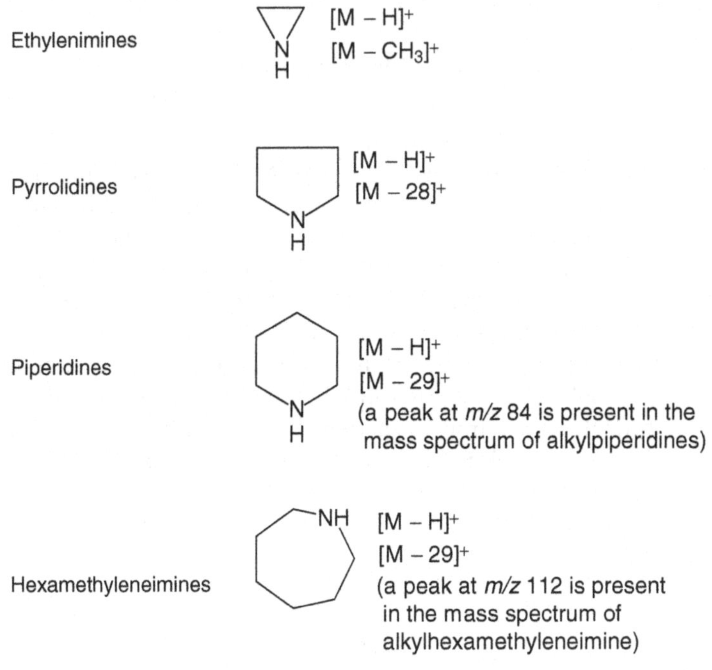

The peak at *m/z* 28 in the mass spectra of cyclic amines represents the $H_2C{=}N^+$ ion. The peak at *m/z* 30 is fairly intense in the mass spectra of nonsubstituted cyclic amines.

5. Mass spectra of cycloakylamines
   $M^{+\bullet}$ peaks are easily detected in the mass spectra of most cycloalkylamines; the spectra of these compounds also have a characteristic peak at *m/z* 30 representing a primary immonium ion. In the mass spectrum of methylcyclopentylamine, *m/z* 30 is the base peak; whereas in the mass spectrum of cyclohexylamine, a peak representing an ion formed by a rearrangement fragmentation is the base peak (Scheme 11.6).

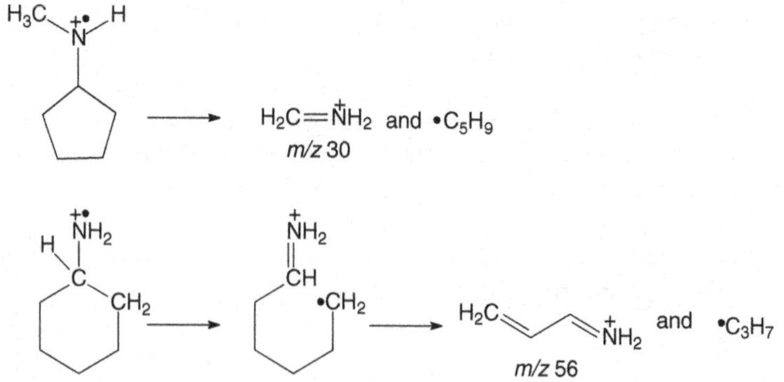

$H_2C{=}\overset{+}{N}H_2$ and $\bullet C_5H_9$
m/z 30

H₂C=CH–CH=N⁺H₂ and •C₃H₇

m/z 56

Scheme 11.6

6. Mass spectra of aromatic amines
   Aromatic amines show intense $M^{+\bullet}$ peaks (characteristic of all aromatic compounds). When alkyl side chains are present, the $M^{+\bullet}$ peak intensities decrease with increasing alkyl chain length, but the $M^{+\bullet}$ peaks are still fairly intense. The mass spectra of aromatic amines (including the naphthylamines) exhibit peaks representing ions formed by the loss of 1, 27, and 28 from the molecular ion, but the peaks represented by these losses also decrease in intensity as the alkyl side chain increases in size. From the mass spectrum alone, it is difficult to determine whether the alkyl group is on the ring or on the nitrogen. A peak at *m/z* 106 represents an abundant ion when one alkyl group is on the nitrogen:

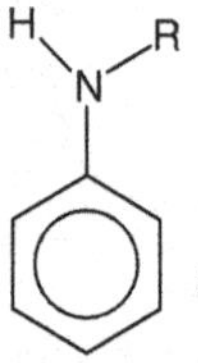

m/z 106

An important member of this class of compounds is drugs related to amphetamine. These compounds have both a benzyl ($C_6H_5CH_2$–) and an aliphatic amine moiety (R–CH_2–N(R_1)R_2) where R_1 and R_2 can be a H or another alkyl group. If the benzylic carbon and the carboamino carbon are the same, the fragmentation is dominated by benzylic cleavage (cleavage following the benzylic carbon resulting in the formation of a tropylium ion; see Chapter 21). If the benzylic carbon and the carbinamino carbon are separated by one or more other alkyl carbon atoms, the fragmentation is dominated by α cleavage initiated by the charge and radical sites being on the nitrogen atom of the molecular ion. This is clearly shown in Figures 11.4 and 11.5.

A third class of compounds that must be considered in this category is the halo anilines.

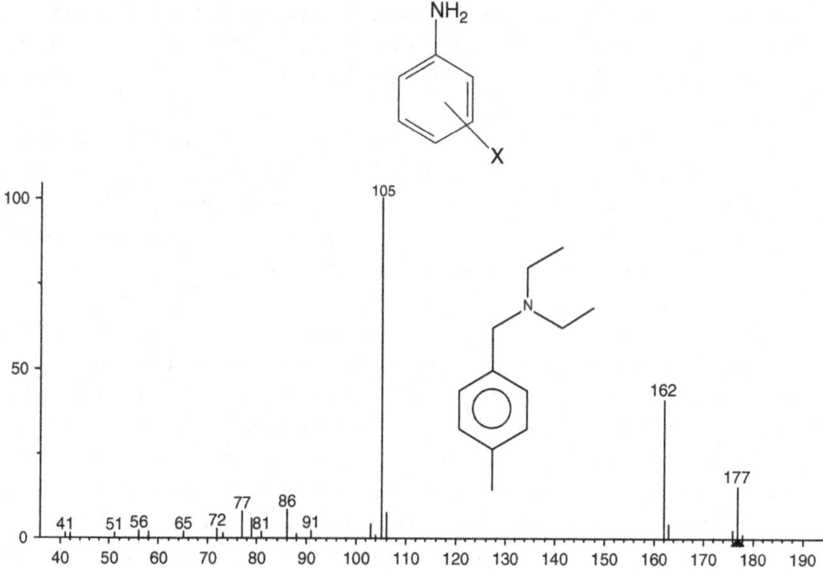

Figure 11.4 EI mass spectrum of diethyl(4-methylbenzyl)amine. The benzyl and carbinamino carbon is the same atom. The spectrum has all the characteristics of one produced by an aromatic compound (intense $M^{+\bullet}$ peak, very little fragmentation, the base peak represents a methyl-substituted tropylium ion).

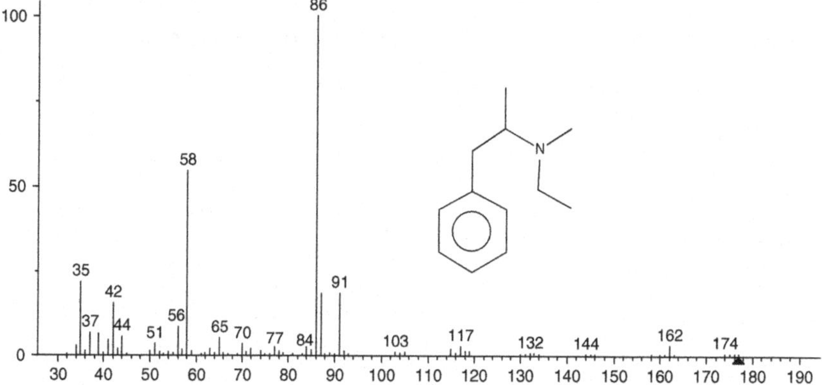

Figure 11.5 EI mass spectrum of *N*-ethyl-*N*,α-dimethyl-benzeneethanamine. The benzyl and carbinamino carbon atoms are not the same in this molecule. The spectrum is dominated by fragmentation characteristic of an aliphatic amine with the charge and radical sites on the nitrogen atom.

The mass spectra of bromo- and chloroaniline are distinguishable from one another by the unique X+2 patterns exhibited by their $M^{+\bullet}$ peaks. The iodo- and fluoroanilines have unique molecular weights, and the $M^{+\bullet}$ peak for all four compounds is the base peak. For any one given compound, there is no real difference in the mass spectra of the three regioisomers (*o*–, *m*–, and *p*–); however, these isomers can be distinguished from one another by the differences in their retention indices.

These compounds can be reacted with organic acids to form halophenyl N-substituted amides [2]. The mass spectra of the *ortho*-isomers of the chloro-, bromo-, and iodo-isomers of the acetyl, formyl, and benzoyl derivatives all exhibit a peak with a significant intensity representing the loss of the halo radical. The mass spectra of various compounds with *ortho*-substitution will exhibit an ortho effect which manifests itself in the loss of a molecule and the formation of odd-electron fragments, whereas the mass spectra of the *meta*- and *para*-isomers do not exhibit such behavior. This conventional ortho effect is used to distinguish the *ortho*-isomer from the other two regioisomers of many multiple-substituted aromatic compounds. Apparently this variation on the ortho effect that occurs under electron ionization of these derivatized haloanilines can also be used in a similar manner (Figure 11.6 and Figure 11.7). The loss of a halo radical is not observed when the halogen is a fluorine.

7. Mass spectral fragmentation

Amine	Characteristic Fragments	Rearrangement Ions	Characteristic losses From the Molecular Ion
RNH_2	*m/z* 30	*m/z* 18 (NH_4)	$[M - NH_3]^+$ (especially diamines)
R_2NH	α cleavage longest chain	*m/z* 30 ($H_2C{=}NH_2^+$)	
R_3N	α cleavage longest chain	$(NHRCH_2)^+$	

B. Derivatized

Preparing derivatives of amines can make the identifications much easier.

Functional Group	Derivative	Increase in MW
$—NH_2$	$—NH–Si(CH_3)_3$ (TMS)	72
$—(CH_2)_3NH_2$	$–N[Si(CH_3)_3]_2$ (TMS)	144
$—NH_2$	$NH–Si(CH_3)_2C(CH_2)_3$ (TBDMS)	114
$—NH_2$	$NHCOCF_3$ (trifluoroacetyl)	96
$—NH_2$	$N=CHN(CH_3)_2$ (Methyl-8®)	55

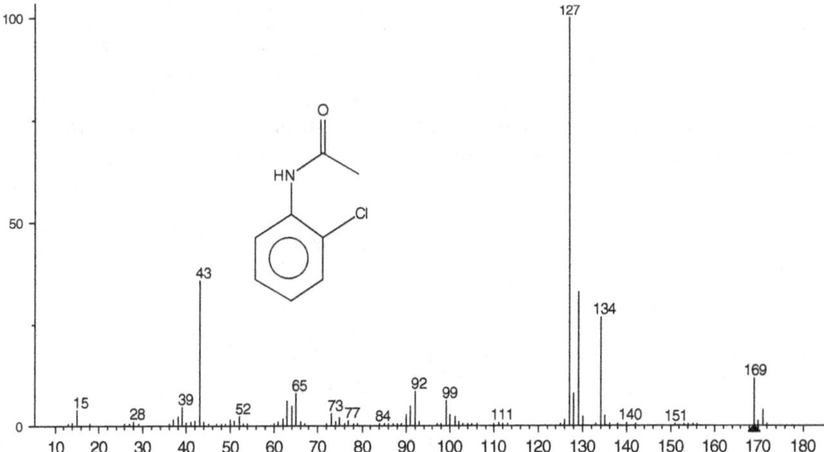

Figure 11.6 EI mass spectrum of the acetyl derivative of *o*-chloroaniline. The peak at *m/z* 134 represents the loss of a chlorine radical from the molecular ion, an ortho effect.

If at least three CH_2 groups are present between the amino group and another functional group, it is possible to add two TMS groups to the amine functional group. The presence of an intense peak representing an ion with *m/z* 174 confirms the addition of two TMS groups on the same nitrogen. By adding 15 to the *m/z* value of the peak with a significant intensity that represents the highest mass ion, the molecular weight of the TMS derivative is determined. The molecular weight of the amine is then determined by subtracting 114 from this value. The Methyl-8® derivative is excellent for use in analyzing diaminohexanes, diaminooctanes, and so forth. Only one Methyl-8® derivative adds per nitrogen so that multiple derivatives are not obtained as is the case with TMS. The $M^{+\bullet}$ peaks of these derivatives are relatively intense.

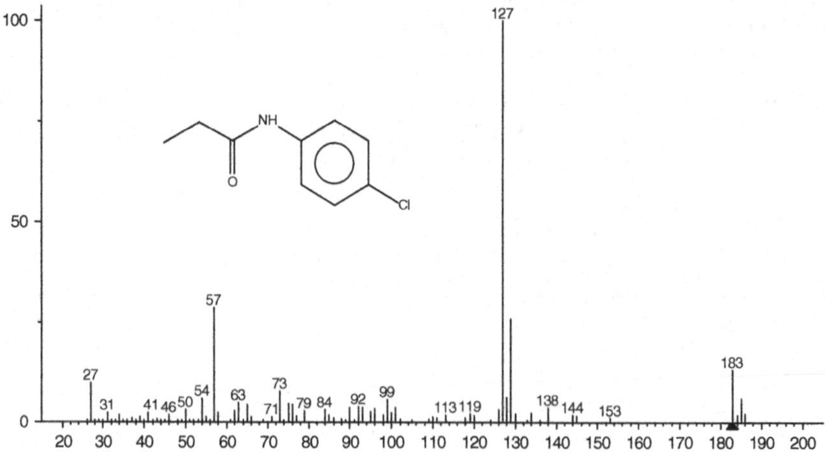

Figure 11.7 EI mass spectrum of the acetyl derivative of *p*-chloroaniline. The peak at *m/z* 134, seen in Figure 11.6, is not present in this spectrum or the spectrum of the *meta*-isomer. This supports this unusual variation on the ortho effect.

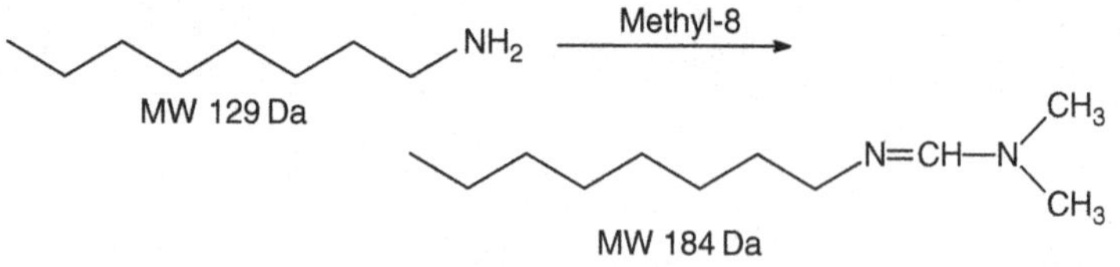

NH_2 —Methyl-8→

MW 129 Da

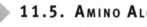

$N{=}CH{-}N(CH_3)_2$

MW 184 Da

Scheme 11.7

C. Sample mass spectrum
   The most prominent peak in the mass spectrum of 1-octylamine is at *m/z* 30 (Figure 11.1). From Appendix Q, a very intense peak at *m/z* 30 suggests a primary or secondary amine, or a nonsubstituted cyclic amine. A very small $M^{+\bullet}$ peak is barely visible at *m/z* 129. If the unknown reacts with Methyl-8® reagent, it is a primary amine (not a secondary amine) or a cyclic amine (Scheme 11.7).

11.5. Amino Alcohols (Aliphatic)

A. GC separation of underivatized amino alcohols
   1. Monoethanolamine, diethanolamine, triethanolamine, and impurities: 30-m, 1.0-µm Rtx-35 (Restek) column, 40 (2 minutes) 0–250 °C at 6 °C min^{-1} (hold 15 minutes).

2. 1-Amino-2-propanol, 3-amino-1-propanol, 2-amino-2-methyl-1-propanol, and similar compounds: 25-m CP-WAX-51 column or CP-WAX column for amines, 50–210 °C at 5 °C min^{-1}.

B. GC separation of derivatized amino alcohols

1. 30-m DB-1701 column, 45 (10 minutes) 250 °C at 10 °C min^{-1}.

C. Mass spectra of amino alcohols

In the mass spectra of aliphatic amino alcohols, the peak at *m/z* 30 is intense, whereas the peak at *m/z* 31 has a low intensity. The amino group dominates the fragmentation, making it difficult to recognize the alcohol group. If only an *m/z* 30 peak is found in the mass spectrum of an unknown, it does not mean that no alcohol group is present. Sometimes the peak at *m/z* 31 due to the presence of an alcohol moiety can be mistaken for the ^{13}C isotope peak relative to the peak at *m/z* 30. The presence of a distinguishable peak at *m/z* 31 suggests the presence of a terminal alcohol group. The mass spectra of these compounds usually do not exhibit an $M^{+\bullet}$ peak of a reasonable or even discernible intensity.

To identify the presence of an amino alcohol, use the following procedure. Prepare a TMS derivative of the unknown using MSTFA reagent and obtain a mass spectrum of the resulting TMS derivative. Prepare a second TMS derivative of the unknown using TRI-Sil Z reagent. When the molecular weight of the unknown increases by 144 mass units using MSTFA reagent and the TMS derivative using TRI-Sil Z reagent only increases the molecular weight by 72 mass units, this suggests the presence of both an amino group and a hydroxyl group in the unknown—TRI-Sil Z reagent silylates alcohols and carboxy hydroxyl groups but not amino groups.

A method to determine the number of amino groups present in the molecule requires the formation of a TMS derivative with MSTFA which silylates both hydroxyl and amino groups [3]. First, obtain a mass spectrum of the unknown using GC/MS. Next, add MBTFA reagent to the previously prepared TMS reaction mixture and let stand approximately 30 minutes. Obtain a mass spectrum of the resulting reaction product. A trifluoroacetyl group will replace each TMS group on primary and secondary amino groups because the amino-TMS group is less stable. From the mass differences obtained before and after the reaction with MBTFA (24 for each amino group), the number of amino groups present can be determined.

MBTFA	$CF_3CON(CH_3)COCF_3$ *N*-methyl-bis (trifluoroacetamide)
MSTFA	$CF_3CON(CH_3)$-TMS *N*-methyl-*n*-trimethylsilyl trifluoroacetamide

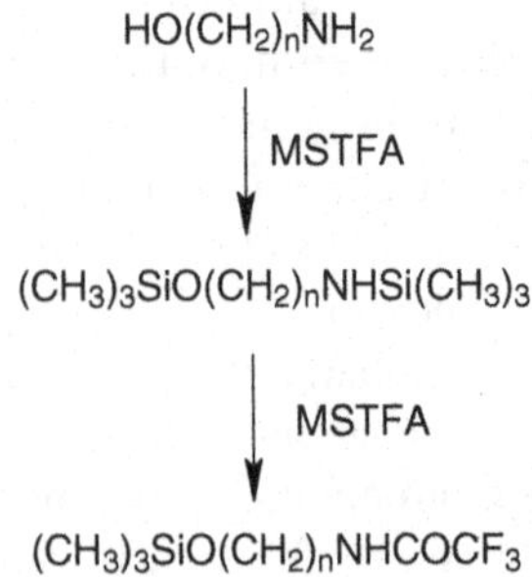

11.6. Aminophenols

A. GC conditions

1. Aminophenols (underivatized): *o*-aminophenol, *p*-aminophenol, and acetanilide: 30-m DB-1701 column, 45–250 °C at 10 °C min^{-1}.
2. Aminophenols (as acetates; see Chapter 35.1.C for acetate derivatization procedure) acetanilide (C_8H_9NO), *o*- and *p*-aminophenol ($C_{10}H_{11}O_3N$) 30-m DB-1 column, 100–250 °C at 10 °C min^{-1}.

B. Mass spectra of underivatized aminophenols

$M^{+\bullet}$ peaks of unsubstituted aminophenols are intense. The peaks representing the $[M - 28]^+$ and $[M - 29]^+$ ions are observed. Peaks of prominent intensities are seen at *m/z* 80 and 109 ($M^{+\bullet}$ in the mass spectrum of all three regioisomers of aminophenol). There is no apparent ortho effect exhibited by the mass spectra of any of the three isomers of aminophenol; however, the mass spectrum of 2-aminoresorcinol exhibits an $[M - H_2O]^+$ at *m/z* 107, which has an intensity that is ~40% of that of the base peak. This peak is not present in the mass spectrum of 4-aminoresorcinol, the only other amino resorcinol spectrum in NIST08. This is a clear example of the ortho effect.

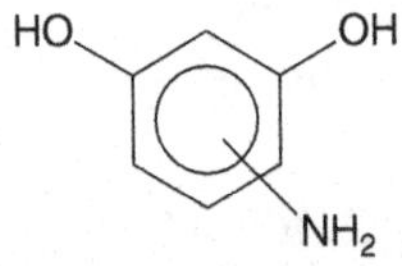

C. Mass spectra of aminophenols as acetates

The $M^{+\bullet}$ peaks of the unsubstituted phenols are observed in the mass spectra of these compounds (acetate adds to amino group), but they have a smaller intensity than the $M^{+\bullet}$ peaks in the spectra of the underivatized aminophenols. The $[M - 42]^+$ peak is characteristic of acetates. Peaks with prominent intensities are observed at *m/z* values of 109, 151, and 193. The mass spectrum of the *ortho*-isomer of the derivative does exhibit an $[M - H_2O]^+$ peak, which is missing from the mass spectra of the acetate derivatives of the *para*- and *meta*-isomers of aminophenol. Again, this is a good example of the ortho effect.

11.7. Solvent Consideration

Amines tend to be very reactive compounds. Long-chain aliphatic primary amines like stearylamine ($C_{18}H_{37}NH_2$) will form $C_{17}H_{35}CH_2{-}N{=}CH_2$, which has a molecular weight that is 12 Da higher than the molecular weight of stearylamine. The compound is difficult to identify because there is no $M^{+\bullet}$ peak, just a very discernable $[M - 1]^+$ peak. When 3,4-(methylenedioxy)-phenethylamine hydrochloride was dissolved in methanol at a level of about 10–100 ng$\,\mu L^{-1}$ and injected at an injection-temperature of 290 °C, there was almost a 100% conversion to the $R{-}H_2C{-}N{=}CH_2$ analog. The first time this was encountered, it was unexpected and resulted in quite a surprise. When ethanol was substituted, the corresponding $R{-}H_2C{-}N{=}CH_2CH_3$ analog was formed. Similar results have been reported [4]. According to this citation, secondary amines appear to undergo methylation under these conditions.

In the GC/MS analysis of amines, care must be taken to avoid these potential reactions with solvent.

REFERENCES

1. Hammerum, S., Christensen, J., Egsgaard, H., Larsen, E., Derrick, P., Donchi, K. (1983), Slow alkyl, alkene, and alkenyl loss from primary alkylamines: ionization of the low-energy molecular ions Prior To fragmentation in the Mu-sec timeframe. *Int. J. Mass Spectrom. Ion Phys.*, 47, 351–354.
2. Jariwala, F. B., Figus, M, Attygalle, A. B. (2008). Ortho effect in electron ionization mass spectrometry of N-acylanilines bearing a proximal halo substituent. *J. Am. Soc. Mass Spectrom.*, 19, 1114–1118.
3. Sullivan, J. E., Schewe, L. R. (1977), Preparation and gas chromatography of highly volatile trifluoroacetylated carbohydrates using N-methyl bis(trifluoro acetamide). *J. Chromatogr. Sci.*, 15(6), 196, 197.
4. Clark, R. E., DeRuiter, J., Noggle, F. T. (1992). GC–MS identification of amine–solvent condensation products formed during analysis of drugs of abuse. *J. Chromatogr. Sci.*, 30, 399–404.

CHAPTER 12

Amino Acids

12.1. GC Separation

A. GC separation of *t*-butyldimethylsilyl (TBDMS)-derivatized amino acids
   30-m DB-1 (Methylsilicone) column, 150–250 °C at 3 °C min^{-1}, then 250–275 °C at 10 °C min^{-1}. Run time is 1 hour.
   Retention times and suggested ions for selected ion monitoring (SIM) of TBDMS derivatives of amino acids (AAs) are given in Table 12.1 and the GC separation shown in Figure 12.1. AAs should be derivatized prior to separation. The TBDMS derivatives are preferred and stable for at least 1 week.
B. GC separation of PTH–amino acids
   1. The only phenylthiohydantion (PTH)–AAs that can be analyzed easily by GC/MS without derivatization are alanine, glycine, leucine, isoleucine, methionine, phenylalanine, proline, and valine. 30-m DB-1 column, 75–275 °C at 8 °C min^{-1}. Dissolve the PTH–AA in the minimum amount of ethyl acetate.
   2. Order of elution and molecular weight (MW): alanine (206 Da), glycine (142 Da), valine (234 Da), proline (232 Da), leucine (248 Da), isoleucine (248 Da), methionine (266 Da), and phenylalanine (282 Da).
C. GC conditions for TBDMS-derivatized PTH–amino acids*
   15-m DB-5 column (or equivalent), 80–300 °C at 15 °C min^{-1}.
D. GC conditions for *N*-PFP isopropyl esters of D- and L-amino acids
   A 25- to 50-m CHIRASIL-L-VAL column (Chrompack cat. no. 7495), 80 (3 minutes)–190 °C at 4 °C min^{-1}. Separates D- and L-isomers of AAs (see Figure 12.2).

* If the reaction time for the TBDMS derivatives is not long enough, a mixture of mono- and di-TBDMS derivatives is observed, resulting in more than one GC peak and thus reduced sensitivity.

Gas Chromatography and Mass Spectrometry
DOI: 10.1016/B978-0-12-373628-4.00012-5

Table 12.1 Retention times and accurate masses for TBDMS-derivatized AAs (separation conditions)

Approximate retention time (minutes)	Amino acid	*m/z* values for selected ion monitoring
5:55	Alanine	158.1365, 232.1553, 260.1502
6:17	Glycine	218.1396, 246.1346
8:28	Valine	186.1678, 288.1815, 260.1866
9:24	Leucine	200.1834, 302.1972, 274.2022
10:12	Isoleucine	200.1834, 302.1972, 274.2022
10:51	Proline	184.1522, 286.1659, 258.1709
16:00	Methionine	292.1587, 320.1536
17:10	Serine	390.2316
18:06	Threonine	376.2523, 404.2473
19:29	Phenylalanine	234.1678, 336.1815
21:40	Aspartic acid	418.2265
22:40	Hydroxyproline	314.2335, 416.2473
22:58	Cysteine	406.2088
25:00	Glutamic acid	432.2422
25:35	Asparagine	417.2425
28:11	Lysine	300.1815, 431.2945
28:55	Glutamine (peak 1)	431.2581
30:52	Arginine	442.2741
33:03	Histidine	338.2448, 440.2585
34:17	Glutamine (peak 2)	413.2476
34:34	Tyrosine	466.2629
38:51	Tryptophan	489.2789
45:57	Cystine	348.1849

12.2. Derivatization of Amino Acids and PTH–Amino Acids

A. *t*-Butyldimethylsilyl derivative [1]
Reagents: *N*-methyl-*N*-(*tert*-butyldimethylsilyl), trifluoroacetamide, and MTBSTFA: Add 0.25 mL of *N,N*-dimethylformamide (DMF) to the dried hydrolyzate. Add 0.25 mL of MTBSTFA reagent and cap tightly. Heat at 60 °C for 60 minutes or overnight at room temperature (longer reaction times prevent mixtures of derivatives). Sample will need to be concentrated prior to injection. For trace analyses, it is important to use the minimum amount of solvent.

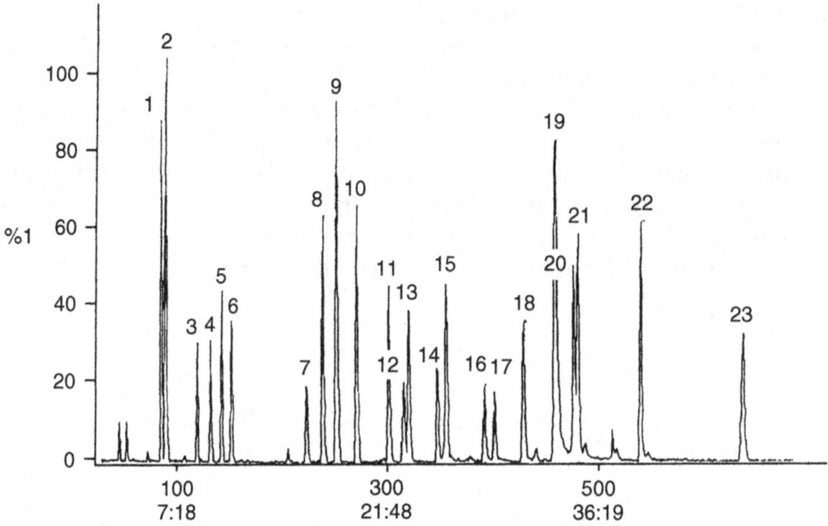

Figure 12.1 GC separation of TBDMS-derivatized AAs (see Table 12.1). 1 = alanine, 2 = glycine, 3 = valine, 4 = leucine, 5 = isoleucine, 6 = proline, 7 = methionine, 8 = serine, 9 = threonine, 10 = phenylalanine, 11 = aspartic acid, 12 = hydroxyproline, 13 = cysteine, 14 = glutamic acid, 15 = asparagine, 16 = lysine, 17 = glutamine (peak 1), 18 = arginine, 19 = histidine, 20 = glutamine (peak 2), 21 = tyrosine, 22 = tryptophan, and 23 = cystine.

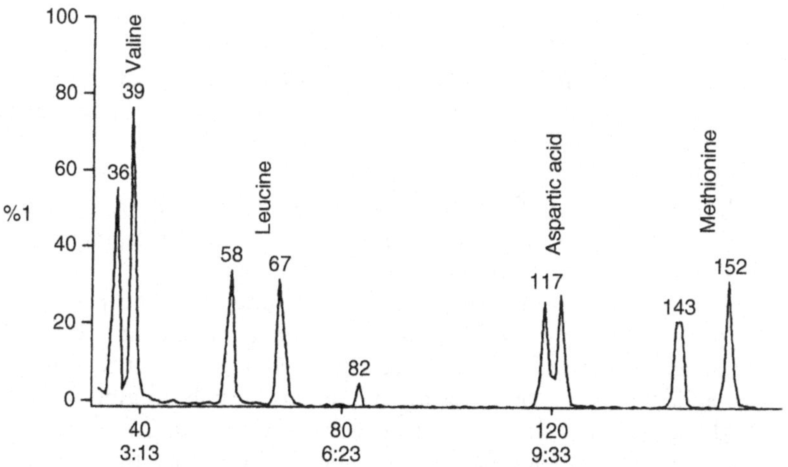

Figure 12.2 GC separation of *N*-PFP isopropylesters of D- and L-AAs.

B. *N*-PFP isopropyl ester derivative

Evaporate the sample to dryness with clean dry nitrogen. Add 0.5 mL of 2 M HCl in isopropyl alcohol. Heat at 100 °C for 1 hour. If tryptophan and/or cystine are suspected of being present, add 1 mL of ethyl mercaptan to prevent oxidation. Evaporate the reaction mixture to dryness with clean dry nitrogen. Add 0.5 mL ethyl acetate and 50 μL of pentafluoropropionic anhydride (PFPA). Heat at 100 °C for 30 minutes. Evaporate again with clean dry nitrogen. Dissolve residue in the minimum amount of methylene chloride.

H_2N–CH(R)–C(=O)–OH $\xrightarrow[\text{2. PFPA}]{\text{1. }(CH_3)_2CHOH}$ C_2F_5–C(=O)–NH–CH(R)–C(=O)–O–CH$(CH_3)_2$

MW = 262 + R

This procedure works well with alanine, valine, threonine, isoleucine, glycine, leucine, proline, serine, aspartic acid, cystine, methionine, phenylalanine, tyrosine, ornithine, and lysine.

12.3. Mass Spectral Interpretation

A. Mass spectra of TBDMS derivatives of amino acids

Typical fragment ions are $[M-15]^+$ (loss of a $^{\bullet}CH_3$ radical), $[M-57]^+$ (loss of a $^{\bullet}C_4H_9$ radical), $[M-85]^+$ (loss of a $^{\bullet}C_4H_9$ followed by the loss of a molecule of CO), and $[M-159]^+$ (loss of $^{\bullet}$C(O)–O–TBDMS radical).

In general, if two fragment ions are observed that are 28 *m/z* units apart, then 57 (C_4H_9) is added to the highest *m/z* value fragment ion to deduce the molecular weight of the TBDMS derivative.

Identification of unknowns, as well as confirming the presence of known AAs, is more reliable if accurate mass measurement data are also available. Identification is easily accomplished if mass spectra of the AAs are added to the computer-assisted library search routine (also see Table 12.2). SIM of characteristic ions using previously determined retention time windows is generally used for trace analyses (see Table 12.1 and Figure 12.1). Accurate mass SIM reduces chemical

Table 12.2 Characteristic ions of TBDMS-derivatized AAs

Nominal mass of derivative	*m/z* values of other ions	Amino acid
303	218, 246	Glycine
317	158, 232, 260	Alanine
343	184, 258, 286	Proline
345	186, 260, 288	Valine
359	200, 274>302	Leucine (elutes first)
	200, 274<302	Isoleucine
377	218, 292, 320	Methionine
393	234, 302, 336	Phenylalanine
447	362, 390	Serine
461	303, 376, 404	Threonine
463	378, 406	Cysteine
473	314, 388, 416	Hydroxyproline
474	302, 417	Asparagine
475	316, 390, 418	Aspartic acid
488	300, 329, 431	Lysine (elutes first)
	299, 329, 357, 431	Glutamine (first of two peaks)
489	272, 330, 432	Glutamic acid
497	196, 338, 440, 459	Histidine
499	199, 340, 442	Arginine (first of two peaks)
523	302, 364, 438, 466	Tyrosine
546	244, 302, 489	Tryptophan
696	348, 537, 589, 639	Cystine

noise at the expense of transmitted ion current in a double-focusing instrument but not in a TOF analyzer.

B. Mass spectra of underivatized PTH–amino acids
The PTH derivative plus the AA backbone has a mass of 191 Da. Therefore, subtracting 191 from the *m/z* value of the molecular ion of the PTH–AA derivative gives the mass of the AA side chain. A peak representing a characteristic fragment ion is observed at *m/z* 135. Leucine and isoleucine can be differentiated using their mass spectra by the absence of a peak at *m/z* 205 in the spectrum of isoleucine.

C. Mass spectra of TBDMS derivatives of PTH–amino acids
Multiple TBDMS derivatives may form depending on the R group of the AA and the reaction time. Some of the PTH–AAs form TBDMS derivatives in less than 1 hour, whereas others require overnight reaction times.

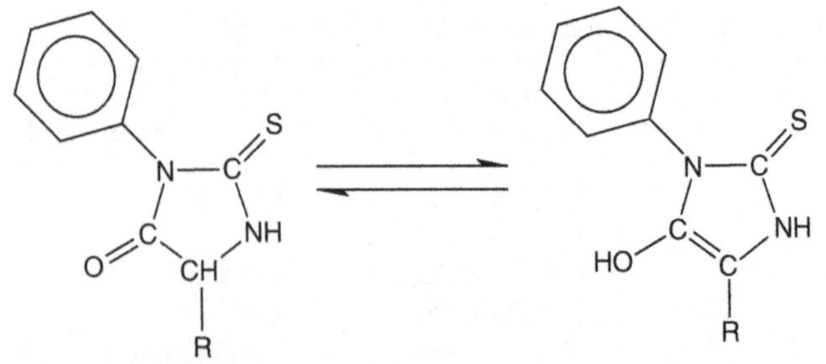

The MW of the TBDMS derivatives of PTH–AAs can be calculated using the following formula: MW = R + 191 + n(114) where n is the number of TBDMS groups. Arginine loses NH_3 during, or prior to, the derivatization reaction. Therefore, the characteristic loss of 57 m/z units ($^{\bullet}C_4H_9$) from the TBDMS derivative occurs, resulting in a peak representing an $[M-74]^+$ ion (57 + 17).

Table 12.3 SIM of some PTH–AAs as the TBDMS derivatives

Component	Suggested m/z values to monitor
PTH–alanine–TBDMS	377, 734
PTH–glycine–TBDMS	363, 420
PTH–valine–TBDMS	291, 405, 462
PTH–leucine–TBDMS	419, 433, 476
PTH–isoleucine–TBDMS	419, 447, 476
PTH–proline–TBDMS	232, 346
PTH–methionine–TBDMS	437, 494
PTH–serine–TBDMS	507, 564
PTH–threonine–TBDMS	521, 578
PTH–phenylalanine–TBDMS	453, 510
PTH–aspartic acid–TBDMS	507, 535, 592
PTH–cysteine–TBDMS	—
PTH–glutamic acid–TBDMS	549, 606
PTH–asparagine–TBDMS	534, 591
PTH–lysine–TBDMS	548, 605
PTH–glutamine–TBDMS	416, 446, 473, 662 (MW = 719 Da)
PTH–arginine–TBDMS	559, 616
PTH–histidine–TBDMS	557, 614
PTH–tyrosine–TBDMS	563, 640
PTH–tryptophan–TBDMS	606, 663
PTH–cystine–TBDMS	–
PTH–*S*-carboxymethylcysteine–TBDMS	353, 410

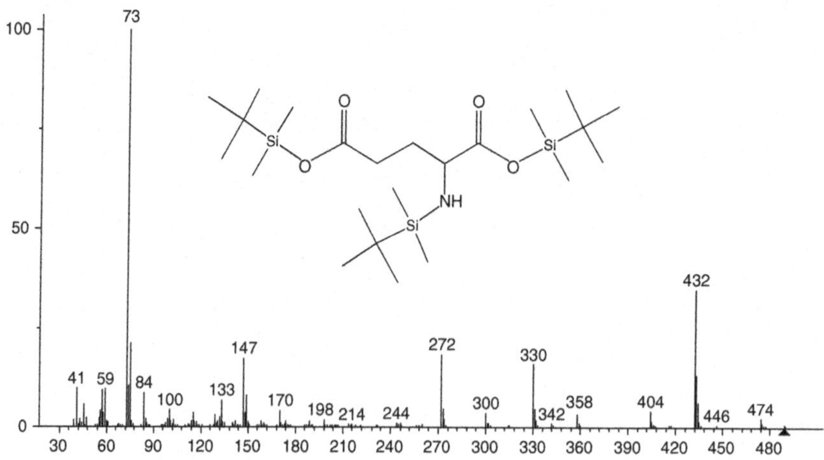

Figure 12.3 Mass spectrum of TBDMS-derivatized glutamic acid.

If the derivatization does not go to completion, it is a good idea to plot the values in Table 12.3 and the *m/z* values of peaks that are 114 *m/z* units less to determine the presence of a particular PTH–AA.

D. Mass spectra of *N*-PFP isopropyl esters of D- and L-amino acids
The molecular ion peak is usually not observed but can be deduced by adding 87 ($COOC_3H_7$) to the *m/z* value of the most intense, high-*m/z* value peak. Peaks at *m/z* 69 ($^+CF_3$) and *m/z* 119 ($^+C_2F_5$) may also be observed [2].

E. Sample mass spectrum TBDMS-derivatized amino acids
Examining the mass spectrum of glutamic acid–TBDMS shows two high-*m/z* value peaks that are 28 units apart (see Figure 12.3). By adding 57 (C_4H_9) to 432, the MW of the TBDMS derivative is deduced to be 489 Da. Characteristic fragment ions of the TBDMS derivative include the following:

M – 15 (loss of CH_3) *m/z* 474
M – 57 (loss of C_4H_9) *m/z* 432
M – 85 (loss of C_4H_9 + CO) *m/z* 404
M – 159 (C(O) – TBDMS) *m/z* 330

REFERENCES

1. Kitson, F. G., Larsen, B. S. (1990). In: Capillary GC/MS for the Analysis of Biological Materials McEwen, C. N., Larsen, B. S., eds. *Mass Spectrometry of Biological Materials*. 437–468. New York: Marcel Dekker.
2. Gelpi, E., Koenig, W. A., Gilbert, J, Oro, J. (1969). Combined GC-MS of amino acid derivatives. *J. Chromatogr.*, 7, 604.

CHAPTER 13

Common Contaminants

13.1. Contaminants Occasionally Observed after Derivatization with TMS Reagents

m/z = 73, 75, 201, 117 (MW = 216 Da): octanoic acid–TMS
m/z = 73, 75, 313, 328 (MW = 328 Da): palmitic acid–TMS
m/z = 73, 75, 341, 356 (MW = 356 Da): stearic acid–TMS
m/z = 73, 99, 241, 147, 256 (MW = 256 Da): uracil–TMS
m/z = 73, 75, 111, 147, 275 (MW = 290 Da): adipic acid–TMS
m/z = 73, 117, 147 (MW = 234 Da): lactic acid–TMS
m/z = 73, 130, 45, 59 (MW = 247 Da): aminobutyric acid–DiTMS
m/z = 73, 147, 205 (MW = 308 Da): glycerol–TMS
m/z = 73, 147, 233, 245 (MW = 350 Da): malic acid–TMS
m/z = 73, 178, 135, 193, 192 (MW = 193 Da): benzamide–TMS
m/z = 73, 273, 147 (MW = 480 Da): citric acid–TMS
m/z = 73, 332, 147 (MW = 464 Da): ascorbic acid–TMS
m/z = 75, 73, 67, 55 (MW = 352 Da): linoleic acid–TMS
m/z = 75, 73, 131, 45, 146 (MW = 146 Da): propionic acid–TMS
m/z = 75, 73, 145, 45 (MW = 160 Da): *n*-butyric acid–TMS
m/z = 75, 73, 159 (MW = 174 Da): valeric acid–TMS
m/z = 75, 117, 45, 43, 73 (MW = 132 Da): acetic acid–TMS
m/z = 91, 165, 135 (MW = 180 Da): benzyl alcohol–TMS
m/z = 105, 206, 73, 308 (MW = 323 Da): benzaminoacetic acid–TMS
m/z = 147, 189, 73 (MW = 204 Da): urea–TMS
m/z = 147, 227, 73, 93 (MW = 242 Da): sulfuric acid–TMS
m/z = 151, 166 (MW = 166 Da): phenol–TMS
m/z = 165, 91, 180, 135 (MW = 180 Da): *o*-cresol–TMS
m/z = 165, 180, 91 (MW = 180 Da): *m*-cresol–TMS
m/z = 174, 59, 75, 147 (MW = 319 Da): aminobutyric acid–TriTMS
m/z = 179, 105, 77, 135 (MW = 194 Da): benzoic acid–TMS
m/z = 187, 75, 73, 69 (MW = 202 Da): octanol–TMS
m/z = 189, 174 (MW = 189 Da): indole–TMS
m/z = 195, 120, 210 (MW = 252 Da): acetylsalicylic acid–TMS
m/z = 203, 188 (MW = 203 Da): skatole–TMS
m/z = 255, 73, 113, 270, 147 (MW = 270 Da): thymine–TMS
m/z = 266, 281, 192 (MW = 281 Da): aminobenzoic acid–TMS

Gas Chromatography and Mass Spectrometry
DOI: 10.1016/B978-0-12-373628-4.00013-7

$m/z = 267, 73, 193, 223$ (MW = 282 Da): hydroxybenzoic acid–TMS
$m/z = 285, 117, 132, 145$ (MW = 300 Da): tetradecanoic acid–TMS
$m/z = 299, 314, 73$ (MW = 314 Da): phosphoric acid–TMS

13.2. Contaminants Occasionally Observed in Underivatized Samples

Peaks observed at	Compound
$m/z = 84, 133, 42, 162, 161$	Nicotine
$m/z = 98, 112, 30, 129$	BHMT
$m/z = 99, 155, 211$	Tributyl phosphate
$m/z = 122, 105, 77$	Benzoic acid
$m/z = 149, 167, 279$	Dioctyl phthalate
$m/z = 194, 109, 55, 67, 82$	Caffeine
$m/z = 205, 220, 57$	Di-*tert*-butyl cresol
$m/z = 221, 57, 236, 41, 91$	Ionol 100
$m/z = 225, 93, 66, 65, 39$	Tinuvin-P
$m/z = 530, 57, 43, 515, 219$	Irganox 1076

13.3. Column Bleed

GC column bleed is a frequently encountered contaminant of mass spectra when high column temperatures are employed. Modern data systems offer the best way to eliminate this type of contamination by subtracting a spectrum showing column bleed from all other spectra in the GC/MS run.

The two peaks most often due to GC column are observed at m/z 207 and 281. Often the m/z 207 is 4× more intense than the m/z 281 peak. Both peaks represent ions with multiple atoms of silicon; therefore, the intensity of the X + 1 and X + 2 isotope peaks is very high. Figure 13.1 is the EI mass spectrum of column bleed.

Another peak that is often due to contamination of the sample or the GC/MS system is observed at m/z 149. This is the base peak in the mass spectra of dialkyl phthalates (plasticizers).

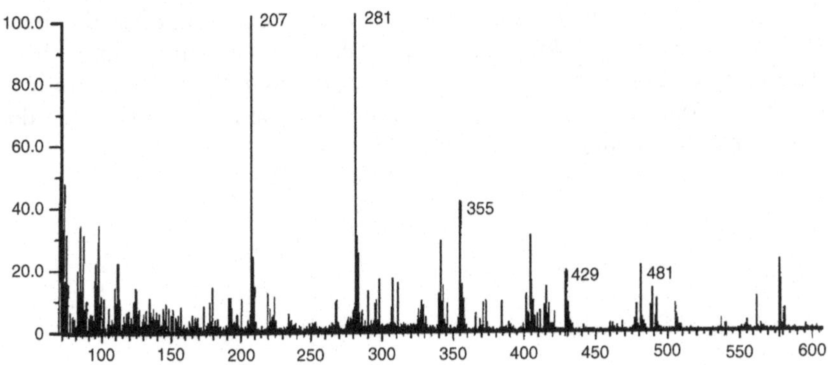

Figure 13.1 EI mass spectrum of column bleed. Values above *m/z* 250 have their intensities increased by 4×.

Peaks at *m/z* 69, 131, 219, 264, 502, and 606 may not belong to the mass spectrum being interpreted. These peaks are common in the compound used to assign *m/z* values to the instrument's *m/z* scale and represent ions used in the computerized tuning of the mass spectrometer. These peaks represent ions formed by the fragmentation of perfluorotributylamine (PFTBA a.k.a. FC-43). Figure 13.2 is the mass spectrum of PFTBA.

If a peak is suspected to represent a system contaminant, generate a mass chromatogram for that *m/z* value. If the chromatogram is at a fairly consistent level over the entire data file, the ion represented by that *m/z* value in

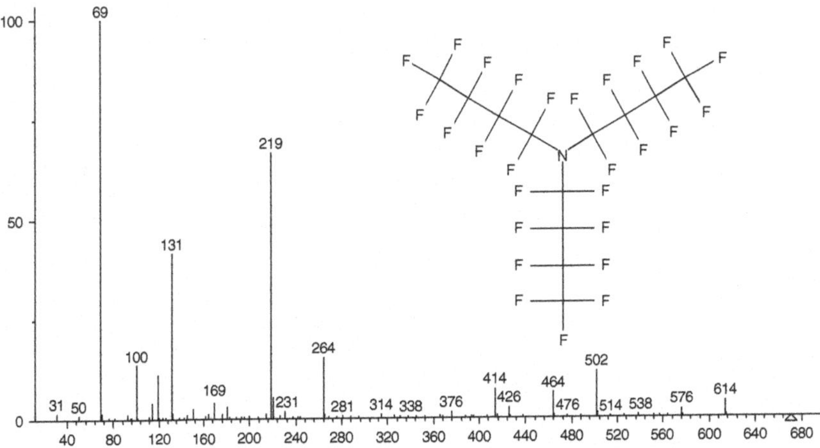

Figure 13.2 EI mass spectrum of perfluorotributylamine. No molecular ion peak is observed for this compound.

the mass spectrum is probably due to system background. If the mass chromatographic peak does not rise and fall together with the mass chromatographic peak of the other *m/z* values observed in the mass spectrum, it means that the suspected peak may belong to a compound other than the one being identified.

CHAPTER 14

Drugs and Their Metabolites

Liquid chromatography/mass spectrometry (LC/MS) rather than GC/MS may be a more desirable approach to the analysis of drugs and metabolites because molecules of these compounds are typically polar and thermally labile. However, if GC/MS is used, more structural information may be obtained, particularly using accurate mass measurement electron ionization (EI) and chemical ionization (CI) combined with derivatization. Using wide-bore (0.53 mm) GC columns in place of narrow-bore (0.25 mm) GC columns results in greater sample capacity and less adsorption. The disadvantages are lower GC resolution and the possible need to use a jet separator or open-split interface [1].

14.1. GC Separations*

A. Underivatized
1. Basic drug screen
   30-m DB-1 column, 100–290 °C at 6 °C min^{-1} or 15-m DB-1301 column, 150–250 °C at 15 °C min^{-1} (injection port at 280 °C).
2. Nicotine, pentobarbital, secobarbital, caffeine, oxazepam, and diazepam redissolved in methanol
   15-m DB-5 or HP-5 column, 150–300 °C at 10 °C min^{-1}.
3. Cocaine (MW = 303 Da), codeine (MW = 299 Da), and morphine (MW = 285 Da)
   25-m DB-1 column, 100–280 °C at 15 °C min^{-1}.
4. Cocaine metabolites (e.g., ecogonine, benzoylecogonine)
   30-m DB-5 column, 200–280 °C at 10 °C min^{-1}. Preparation of methyl ester and TMS derivatives is recommended.
5. Naloxone (MW = 327 Da) and nalbuphine (MW = 357 Da)
   30-m DB-1 column, 150–280 °C at 15 °C min^{-1}.
6. Amphetamines
   30-m DB-1 column at 150 °C.
7. Fentanyl
   30-m DB-1701 column at 270 °C.

* For metabolite work, first test the separation on the precursor drug or chemical.

Gas Chromatography and Mass Spectrometry
DOI: 10.1016/B978-0-12-373628-4.00014-9

8. Hexamethylphosphoramide and its metabolites
30-m FFAP-DB or DB-WAX column, 60–220 °C at 10 °C min^{-1}.
9. *o*-Phenylenediamine (OPD) metabolites
30-m DB-225 column, 75–225 °C at 10 °C min^{-1}.
Urine extracts from rats exposed to OPD were examined without derivatization. The major metabolites were identified as methylbenzimidazole, methylquinoxaline, and dimethylquinoxaline.

B. TMS derivatives
1. Naloxone–TMS (MW = 471 Da), nalbuphine–TMS (MW = 573 Da)
30-m DB-1 column, 60–270 °C at 10 °C min^{-1}.
2. Methadone metabolites, methadone, cocaine (underivatized), morphine, and heroin
30-m DB-1 column, 100–250 °C at 10 °C min^{-1}.
3. Daidzein and its metabolites (as TMS derivatives) 30-m DB-1 column, 100–280 °C at 10 °C min^{-1}

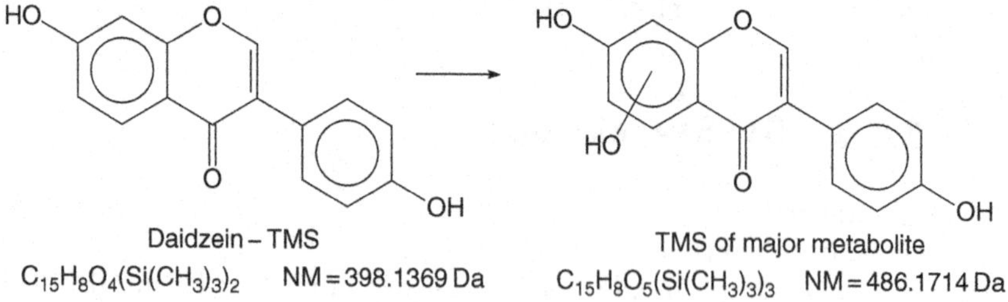

Daidzein – TMS
$C_{15}H_8O_4(Si(CH_3)_3)_2$ NM = 398.1369 Da

TMS of major metabolite
$C_{15}H_8O_5(Si(CH_3)_3)_3$ NM = 486.1714 Da

C. Acetate derivatives from acid hydrolysis
Injection port: 280 °C, 30-m DB-1 column, 100–300 °C at 10 °C min^{-1}.

D. Methylated barbituates and sedatives
15- or 30-m DB-17 column, 100–250 °C at 10 °C min^{-1}.

14.2. Sample Preparation

A. Extraction with solvents
Drugs and metabolites can be extracted from cultures and urine by adding two drops of concentrated HCl to 1 mL of urine for a pH 1–2 [2–4]. Extract with three 1-mL volumes of diethyl ether (top layer) or methylene chloride (bottom layer). Combine extractions and evaporate with clean dry nitrogen. Adjust to a pH of 8–10 by adding 250 µL of 60% KOH to 1 mL of urine. Extract with three 1-mL volumes of diethyl ether (top layer) or three 1-mL volumes of methylene chloride (bottom layer). Combine extractions and evaporate to dryness with clean dry nitrogen (See also reference 2).

B. Solid-phase extraction
   1. Prepare the solid-phase extraction (SPE) tube (1-mL LC-18 SPE tube) by conditioning with 1 mL of methanol followed by 1 mL of water.
   2. Extract the drugs and metabolites by diluting 1 mL of serum with 1 mL of 0.1 M sodium carbonate buffer (pH 9). Force the mixture dropwise through the SPE tube previously prepared.
      Wash the SPE tube packing with three 200-μL aliquots of water, dry it with nitrogen for 5 minutes, and elute the drugs with three 100-μL aliquots of 90 parts ethanol and 10 parts diethyl ether. Concentrate the recovered drugs by evaporating some or all of the solvent before analysis by GC/MS (See Supelco Bulletin 810B).
   3. Clean or change injection port liners frequently because nonvolatile materials in extracts from body fluids can accumulate in the injection port and/or head of the GC column and cause separation problems.

14.3. Derivatization of Drugs and Metabolites

A. TMS derivatives of drugs and their metabolites
   Add 100 μL of MSTFA reagent to less than 1 mg of dry extract [5]. Heat at 60 °C for 15–20 minutes. If necessary, add 250 μL of acetonitrile or other suitable solvent. For additional structural information, prepare the methoxime–TMS derivative to determine if one or more carbonyl groups are present.

B. MO–TMS derivative of drugs
   Add 250 μL of methoxime hydrochloride in pyridine (MOX) reagent to less than 1 mg of dry extract. Let this solution stand at room temperature for 2 hours. Evaporate to dryness with clean dry nitrogen. Add 250 μL of MSTFA reagent and let stand for 2 hours at room temperature.

C. Acetyl derivatives of drugs
   Add 60 μL of acetic anhydride and 40 μL of pyridine to less than 1 mg of dry extract. Heat for 1 hour at 60 °C. Add excess methanol and evaporate to dryness with clean dry nitrogen. Dissolve the residue in the minimum amount of butyl acetate or ethyl acetate.

D. Methylation of barbituates and sedatives
   Heating MethElute (Pierce 49300X) with drug-containing extracts from body fluids gives quantitative methylation of barbiturates, sedatives, and so on. Follow the procedure provided by Pierce Chemical Company using the MethElute reagent.
   15- or 30-m DB-17 column, 100–250 °C at 10 °C min^{-1}.

14.4. Mass Spectral Interpretation

A. Metabolites

The mass spectra of metabolites will usually follow similar fragmentation pathways to those prevalent in the mass spectra of the precursor molecule [6–8]. Therefore, knowledge of the possible biotransformations that can lead to metabolites is important. La Du et al. [3] list oxidation, reduction, and hydrolysis reactions that are common to living organisms. Some of the more common biotransformations are listed in the following text with the exception of conjugated metabolites, which have insufficient volatility to be observed by GC/MS and are not considered. If conjugation is expected, it will be necessary to cleave the conjugate by hydrolysis before GC/MS analysis of the metabolite [4].

1. Side-chain oxidation and hydroxylation of toluene

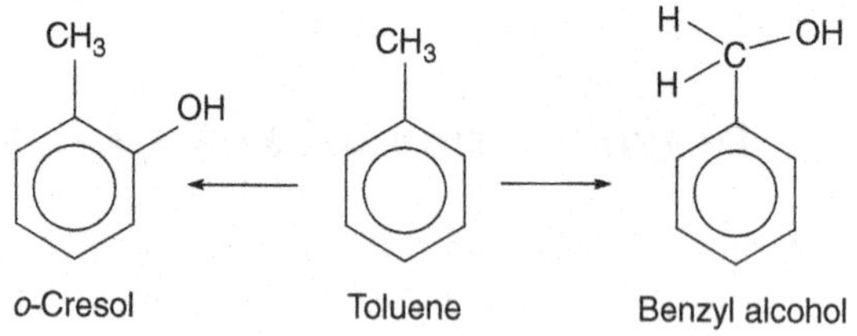

2. Epoxide formation and hydroxylation of benzene

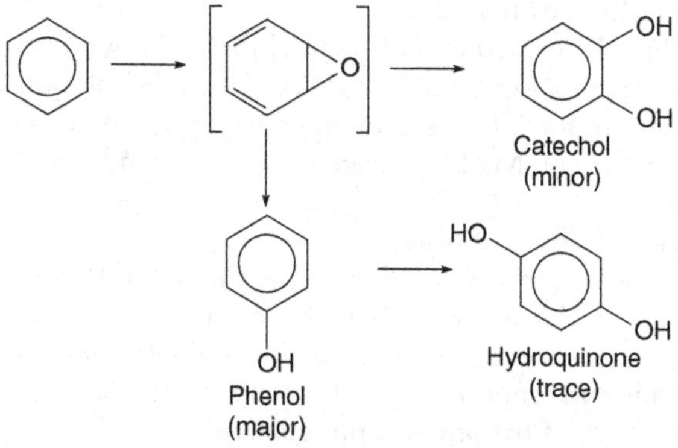

GC conditions should be used, which separate phenol, hydroquinone, resorcinol, and catechol.

3. *o*-Dealkylation of 7-ethoxycoumarin

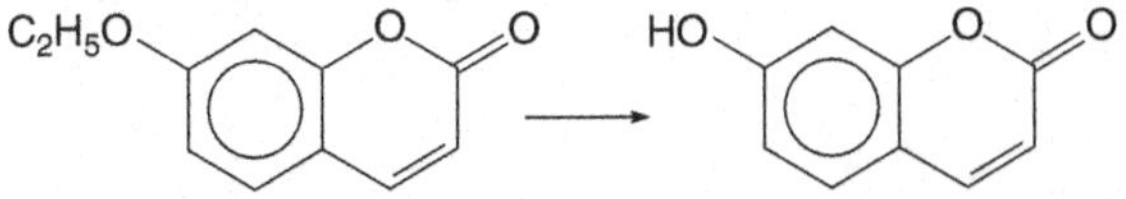

C_2H_5O O O HO O O

7-Ethoxycoumarin 7-Hydroxycoumarin

4. Hydroxylation and ketone formation of cyclohexane

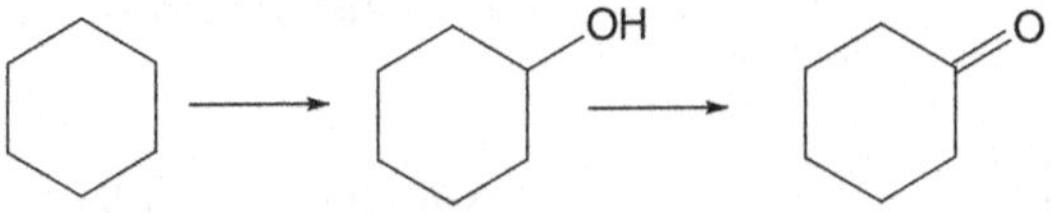

OH O

Cyclohexane Cyclohexanol Cyclohexanone (trace)

5. N-Dealkylation

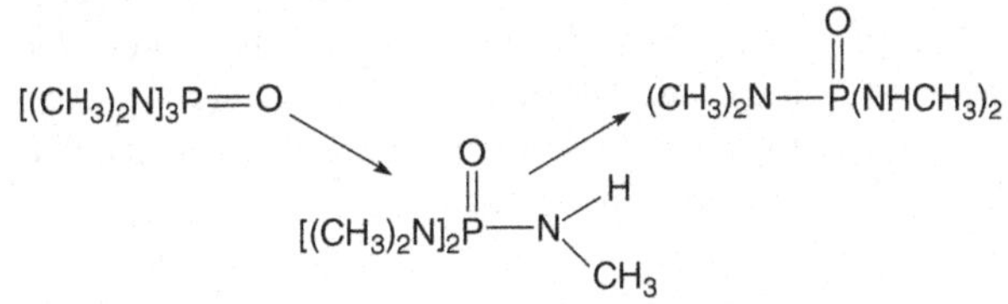

$[(CH_3)_2N]_3P{=}O$ → $[(CH_3)_2N]_2P(=O){-}N(H)CH_3$ → $(CH_3)_2N{-}P(=O)(NHCH_3)_2$

Be careful when examining the fragmentation pattern of the metabolite and comparing the mass spectra of the precursor molecule; it is often possible to determine not only the nature of the biotransformation but also its position in the molecule. In the following example, accurate mass measurement was used to determine that a hydroxyl group had been added to the benzene ring containing the fluorine substituent.

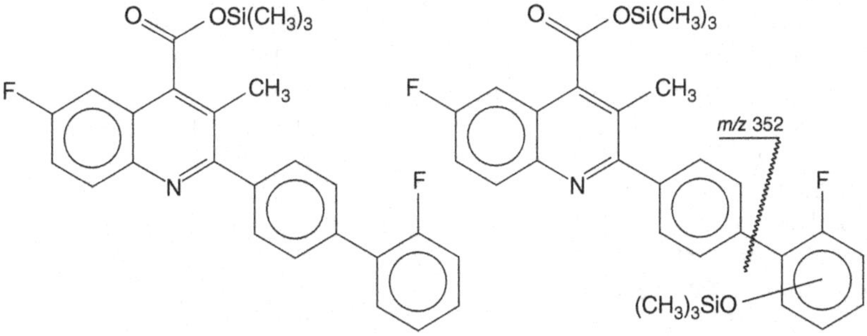

O $OSi(CH_3)_3$ F CH_3 N F

O $OSi(CH_3)_3$ F CH_3 *m/z* 352 N F $(CH_3)_3SiO$

TMS of precursor
$C_{26}H_{23}O_2NF_2Si$
NM = 447.1466 Da

TMS–major metabolite
$C_{29}H_{31}O_3NF_2SI_2$
NM = 535.1810 Da

Molecular ion (NM = 535.1810 Da)	$C_{29}H_{31}O_3NF_2Si_2$
Intense ion (NM = 352.1169 Da)	$C_{20}H_{19}O_2NFSi$
Loss from the molecular ion	$C_9H_{12}OFSi$

The loss of 183.0641 Da from the molecular ion showed that the OH group was on the benzene ring containing the fluorine:

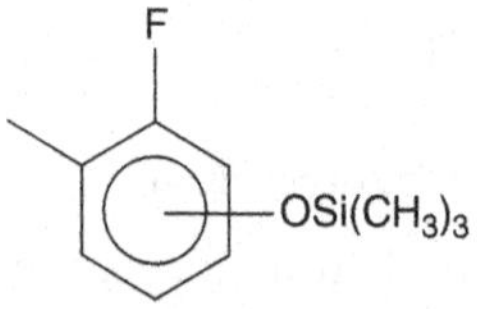

F

$OSi(CH_3)_3$

$C_9H_{12}OFSi$ (NM = 1830641)

B. Drugs

The mass spectra of drugs are as varied as the molecules from which they are formed (see Table 14.1). Major sources that are available for identifying drugs are computer library search routines using the *NIST08 Mass Spectral Database* and/or the *Wiley Registry of Mass Spectra Data 9th ed.; Mass Spectral and GC Data of Drugs, Poisons and Their Metabolites* [2] and *Mass Spectra of Designer Drugs* [6]. Check Appendix I for a list of commercially available mass spectral databases that can be used with library search programs.

Table 14.1 Fragmentation and elution order of underivatized drugs

Drug	*m/z* value of typical fragment ions
Amphetamine (MW = 135 Da)	44, 91, 120
N-Methylamphetamine (MW = 149 Da)	58, 91, 134
Nicotine (MW = 162 Da)	84, 133, 162
Ephedrine (MW = 165 Da)	58, 77, 146
Barbital[a] (MW = 184 Da)	156, 141
Aprobarbital (MW = 210 Da)	162, 124, 195
"Tylenol" (MW = 151 Da)	109, 151, 80
Phenacetin (MW = 179 Da)	108, 109, 179
Mescaline (MW = 211 Da)	181, 182, 211
Amobarbital[a] (MW = 226 Da)	156, 141
Pentobarbital[a] (MW = 226 Da)	156, 141
Meprobamate (MW = 218 Da)	55, 83, 96, 114, 144
Secobarbital (MW = 238 Da)	168, 167, 195
Caffeine (MW = 194 Da)	194, 109, 55
Glutethimide (MW = 217 Da)	189, 117, 132
Hexobarbital (MW = 236 Da)	221, 181, 157, 236
Lidocaine (MW = 234 Da)	86, 58, 72, 234
Phencyclidine (MW = 242 Da)	200, 91, 84
Doxylamine (MW = 270 Da)	58, 71, 183, 182, 200
Theophylline (MW = 180 Da)	180, 95, 68

(*Continued*)

Table 14.1 (*Continued*)

Drug	*m/z* value of typical fragment ions
Phenobarbital (MW = 232 Da)	204, 117, 232
Cyclobarbital (MW = 236 Da)	207, 141
Procaine (MW = 236 Da)	86, 99, 120
Methaqualone (MW = 280 Da)	235, 250, 91
Methadone (MW = 309 Da)	72, 294, 309
Cocaine (MW = 303 Da)	82, 182, 303
Imipramine (MW = 280 Da)	58, 235, 280
Desipramine (MW = 266 Da)	44, 195, 235, 266
Scopolamine (MW = 303 Da)	94, 138, 154, 303
Codeine (MW = 299 Da)	299, 162, 229, 124
Morphine (MW = 285 Da)	285, 162
Chlordiazepoxide (MW = 299 Da)	282, 283, 284
Heroin (MW = 369 Da)	327, 369, 268
Flurazepam (MW = 387 Da)	86, 387, 315
Papaverine (MW = 339 Da)	339, 338, 324
Hydroxyzine (MW = 374 Da)	201, 299, 374
Thioridazine (MW = 370 Da)	98, 70, 370

GC conditions given in Part **I**.A.1.

[a] These drugs can be differentiated by retention time and *m/z* 156 and 157 abundance ratios.

C. Sample mass spectrum

The mass spectrum in Figure 14.1 shows a dominant molecular ion peak at *m/z* 151. The odd nominal *m/z* value suggests the presence of an odd number of nitrogen atoms. The observed loss of 42 *m/z* units from the

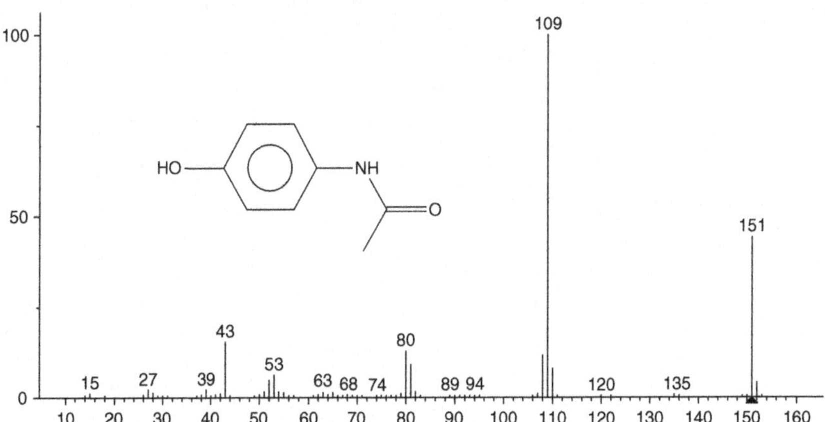

Figure 14.1 EI mass spectrum of acetaminophen.

molecular ion suggests an N-acetylated compound, and the peak with a significant intensity at *m/z* 109 (see Appendix Q) suggests that the ion with *m/z* 109 is an aminophenol. By acetylating with acetic anhydride and pyridine, the presence of an OH group will be confirmed with this assumption. The molecular ion for the acetylated material will now be represented by a peak at *m/z* 193 with corresponding fragments at *m/z* 151 and 109.

Another way to derivatize the OH group is by silylation, using the Tri-Sil Z reagent that will silylate the hydroxyl group but not silylate the secondary amino group.

REFERENCES

1. Smith, F. P., Siegel, J. A. eds. (2005). *Handbook of Forensic Drug Analysis*. San Diego, CA: Elsevier Academic Press; ISBN:0126506418.
2. Maurer, H., Pfleger, K., Weber, A. (2007). *Mass Spectral and GC Data of Drugs, Poisons and Their Metabolites*. Weinheim, Germany: Wiley-VCH; (Data are available in electronic format).
3. LaDu, B. N., Mandel, A. G., Way, E. L. (1972). *Fundamentals of Drug Metabolism and Drug Disposition*. Baltimore, MD: Williams & Wilkins Company.
4. Sunshine, I., Caplis, M., eds. (1981). *CRC Handbook of Mass Spectra. of Drugs*. Boca Raton, FL: CRC Press, Inc.
5. Ahuja S. (1976). Derivatization in gas chromatography. *J. Pharm. Sci.*, 65, 163–182 (See also Hewlett–Packard Ion Notes, 6(2), 1991).
6. Rösner, P., Jung, T., Westphal, F., Fritschi, G. (2007). *Mass Spectra. of Designer Drugs*. Weinheim, Germany: Wiley-VCH; (subsequent data releases in electronic format in 2008 and 2009).
7. Smith, R. M. (2004). *Understanding Mass Spectra: A Basic Approach*, 2nd ed.; Hoboken, NJ: Wiley; ISBN:047142949X (reviewed JASMS 16:792).
8. Smith, R. M. (1999). *Understanding Mass Spectra: A Basic Approach*, Busch KL, Tech. ed. New York: Wiley; ISBN:0471297046 (reviewed JASMS 11:664).

CHAPTER 15

Esters

15.1. GC Separation of Esters of Carboxylic Acids

A. Capillary columns
1. Methyl esters (general)
   30-m FFAP column, 60–200 °C at 4 °C min^{-1}.
2. C_{14}–C_{24} methyl esters
   30-m DB-23 column, 150–210 °C at 4 °C min^{-1}.
3. Methyl esters (C_8–C_{20} polyunsaturated)
   30-m DB-WAX column, 80–225 °C at 8 °C min^{-1} or Omegawax 320 column, 100–200 °C at 10 °C min^{-1}. Run for 50 minutes.
4. Butyl esters
   30-m DB-1 column, 70–250 °C at 6 °C min^{-1}.
5. Diesters: dimethyl malonate, dimethyl succinate, dimethyl glutarate, and dimethyl adipate
   30-m CPSIL-88 column, 150–220 °C at 4 °C min^{-1} or 30-m DB-WAX column, 50–100 °C at 8 °C min^{-1}, 100–200 °C at 10 °C min^{-1}.
6. Dimethyl terephthalate (DMT) impurities
   30-m DB-1 column, 100–250 °C at 10 °C min^{-1}.

Compound	Retention Time (minutes)	*m/z*
Acetone	0.9	58
Benzene	1.3	78
Toluene	2.2	92
Xylene	2.7	106
Methyl benzoate	4.8	150
$O=C(H)-C_6H_4-C(=O)OCH_3$	5.9	164

(*Continued*)

Gas Chromatography and Mass Spectrometry
DOI: 10.1016/B978-0-12-373628-4.00015-0

(*Continued*)

Compound	Retention Time (minutes)	*m/z*
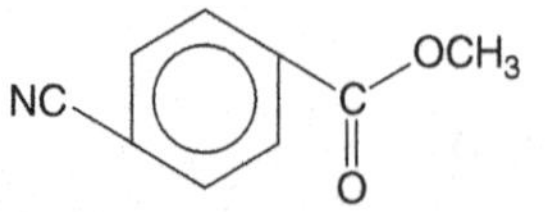	6.0	161
Methyl toluate	6.8	164
$CH_3C(O)OC_6H_4C(O)OCH_3$ (DMT)	10.7	194
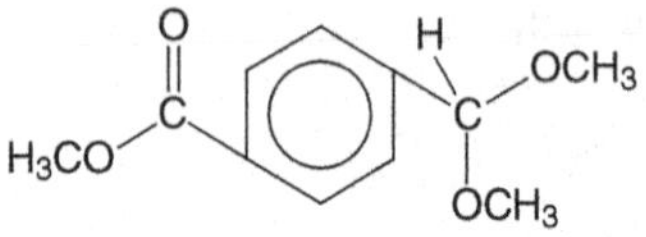	12.3	210
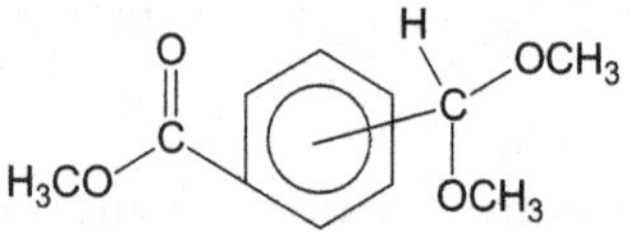	13.6	210
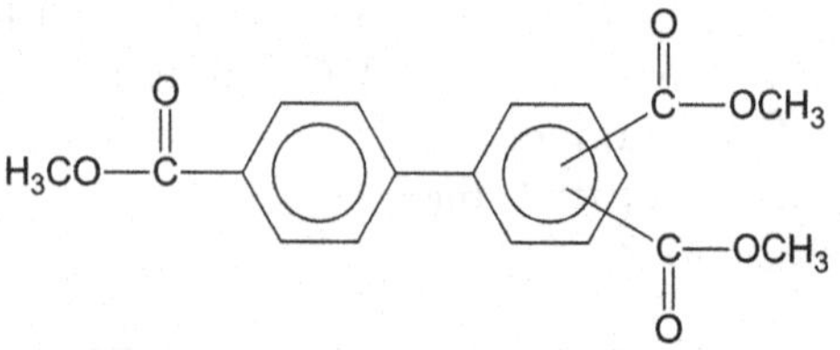	14.5	328

7. Glycol methacrylates (and retention times): ethylene glycol dimethacrylate (RT = 13 minutes), diethyleneglycol dimethacrylate (RT = 18 minutes), triethylene glycol dimethacrylate (RT = 22 minutes), and tetraethyene glycol dimethacrylate (RT = 30 minutes) 30-m DB-17 column, 60–230 °C at 8 °C min^{-1}.
8. Methyl esters of cyano acids
   30-m DB-17 column, 75–275 °C at 10 °C min^{-1}.

15.2. Mass Spectra of Esters

A. Methyl esters of aliphatic acids
   1. General formula
      RCO_2Me

2. Molecular ion
   If a molecular ion peak is not observed, it can be deduced by adding 31 *m/z* units to the highest peak observed (neglecting isotopes).
3. Fragmentation
   Cleavage of bonds that are adjacent to the carbonyl group gives rise to R^+, $[RCr \equiv O]^+$, and $[OCH_3]^+$. An intense peak in the mass spectra of C_6–C_{26} methyl esters results from the γ-hydrogen-shift β-cleavage (McLafferty) rearrangement fragmentation:

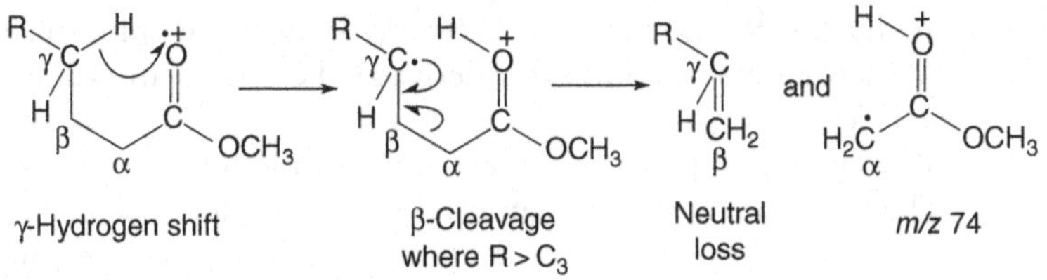

γ-Hydrogen shift → β-Cleavage where R > C_3 → Neutral loss and *m/z* 74

The McLafferty rearrangement results from a β-cleavage caused by the transfer of a γ-hydrogen to the carbonyl oxygen. A peak at *m/z* 87 is also characteristic of the mass spectra of methyl esters. The combination of the peaks at *m/z* 74 and 87 suggests a methyl ester. Another notable peak in the mass spectrum of a methyl ester is the $[M - OCH_3]$ peak. The intensity of this peak is low compared to the intensity of the *m/z* 74 and 87 peaks, but it serves a very diagnostic purpose. Branching at the α-carbon may give rise to peaks at *m/z* 88, 102, 116, and so forth, rather than the peak at *m/z* 74. These ions are also observed with ethyl and higher esters (see the proceeding text). If even small peaks are observed at *m/z* 31, 45, 59, and so forth, then it is likely that oxygen is present. Accurate mass measurement data would establish the presence of oxygen, especially in the ions represented by the more intense peaks. The following list summarizes the mass spectral characteristics for methyl esters:

In summary:

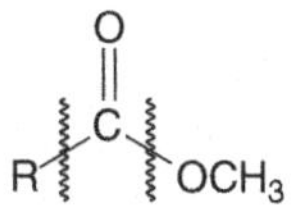

- $[M - 31]^+$ ($^{\bullet}OCH_3$)
- $[M - 43]^+$ ($^{\bullet}CH_2CH_2CH_3$)
- $[M - 59]^+$ ($^{\bullet}C(O)OCH_3$)
- R^+ is intense in methyl acetate, methyl propionate, methyl butyrate, and methyl pentanoate.
- *m/z* 74 and 87 are characteristic of straight-chain C_4–C_{26} methyl esters. Unsaturation in an ester is normally apparent from the molecular weight and the more abundant molecular ion.

B. Determination of the double-bond position in C_{10}–C_{24} monounsaturated methyl or ethyl esters

1. Derivatization

   Dissolve less than 1 mg of methyl or ethyl ester in 1 mL of freshly distilled pyrrolidine and 0.1 mL of acetic acid. Heat the mixture in a sealed or capped tube (able to withstand high temperatures) at 100 °C for 30 minutes. Cool the reaction mixture to room temperature and add 2 mL of methylene chloride. Wash the methylene chloride extract with dilute HCl, followed by ion exchange water. Dry the methylene chloride extract with anhydrous magnesium sulfate [1].
2. GC separation conditions for derivatized esters 30-m DB-1 column, at 175 °C.
3. Mass spectra of pyrrolidine derivatives of methyl or ethyl esters:

   The molecular ion peaks are intense with characteristic fragment ions represented by peaks at *m/z* 70, 98, and 113. The position of the double bond can be determined by locating two peaks that differ by 12 *m/z* units instead of the ubiquitous periodicity of every 14 *m/z* units. A relatively intense peak should be observed 26 *m/z* units higher than the lower value peak of the pair separated by 12 *m/z* units. The double bond lies between the two peaks that are separated by 26 *m/z* units.

   In Figure 15.1, peaks are observed at *m/z* 126, 140, 154, 168, 182, 196, 210, 224, 236, 250, 264, 278, 292, 306, 320, and 335 (MW).

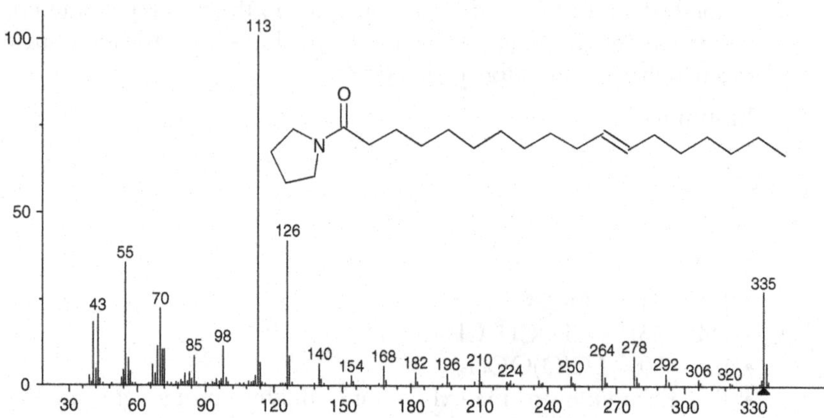

Figure 15.1 Pyrrolidine derivative of monounsaturated methyl or ethyl ester for determining the double-bond position.

Based on the presence of a molecular ion peak at *m/z* 335, the number of carbons in the fatty acid is determined by subtracting the mass of the pyrrolidine moiety (70 Da), the mass of a terminal hydrogen on the end of the molecule (1 Da), and the mass of the carbonyl moiety (28 Da) from 335 Da. This results in a value of 236 Da. The mass of the two hydrogen atoms that are missing due to the presence of the double bond is added to give a mass of 238 Da. This value is divided by the mass of a $-CH_2-$ unit (14 Da) resulting in 17 carbon atoms. Add back the carbon atom associated with carbonyl, and it is clear that the fatty acid contains 18 atoms of carbon.

Subtracting the *m/z* value of the higher value of the two peaks separated by 26 *m/z* units (*m/z* 250 in this example) from the *m/z* value of the molecular ion minus 1 gives a value of 84. This number divided by 14 results in the number 6, which is the number of atoms between the end of the analyte (a $-CH_3$ group) and the double bond, which will be between the number 11 and 12 carbon atoms, as seen on the structure shown above.

This procedure also applies to the case when two double bonds are present with only a CH_2 group separating them. (Not all cases have been examined.)

C. Esters ($R' > C_1$)

The molecular ion peak can be very small or nonexistent. Esters where R′ is greater than methyl form a protonated acid that aids in the interpretation (e.g., *m/z* 47, formates; *m/z* 61, acetates; *m/z* 75, propionates; *m/z* 89, butyrates). Interpreting mass spectra of ethyl esters may be confusing without accurate mass measurement because the loss of C_2H_4 can be confused with the loss of CO from a cyclic ketone.

A summary of the characteristic fragmentation of esters higher than methyl is as follows (see Figure 15.2):

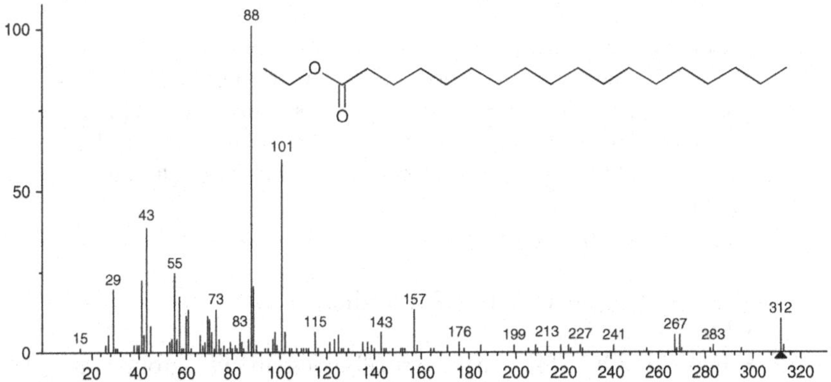

Figure 15.2 Ethyl ester of octadecanoic acid.

- $M^{+\bullet}$ peak is small.
- $[M - R']^+$ peak is observed.
- $[M - OR']^+$ peak is observed.
- $[RCOOH_2]^+$ peak is observed.
- $[R - H]^+$ or $[R - H_2]^+$ peak is observed.
- McLafferty rearrangement peak is at *m/z* 88 for ethyl esters: if the α-carbon has a methyl substitution, the major rearrangement ion will be at *m/z* 102.

D. Methyl esters of dibasic acids

Molecular ion peaks are not always observed for methyl esters of dibasic acids; however, the mass spectra of all dibasic acid methyl esters exhibit peaks representing ions of $[M - 31]^+$ and $[M - 73]^+$. By adding 31 or 73 to the appropriate ions, the molecular weight can be deduced.

E. Ethyl and higher esters of dibasic acids

1. General formula

   $ROOC(CH_2)nCOOR'$

2. Molecular ion

   A molecular ion peak is generally not observed in the mass spectra of diesters larger than diethyl malonate.

3. Fragmentation

   Characteristic losses from the molecular ions are as follows:

   $[M - (R' - H)]^+$
   $[M - (OR')]^+$
   $[M - COOR]^+$

F. Aromatic methyl esters

1. General formula

   $Ar{-}C({=}O){-}OCH_3$

2. Molecular ion

   Although the molecular ion peak is always observed, the loss of 31 ($^\bullet OCH_3$) is the most intense ion. Generally, the acid and/or protonated acid is observed. The mass spectra of *ortho*-substituents are distinguished by their intense peaks representing $[M - 32]^+$ ions, as well as $[M - 31]^+$ ions. A small peak is observed representing the $[M - 60]^+$ ion.

3. Fragmentation

   Characteristic peaks are as follows:

   - $M^{+\bullet}$
   - $[M - 31]^+$ and/or $[M - 32]^+$ (ortho effect)
   - $[M - COOCH_3]^+$
   - $[M - HCOOCH]^+$ (may be small unless an *ortho*-group is present; however, the *ortho*-group need not be a methyl group.)

G. Ethyl or higher aliphatic esters of aromatic acids

1. General formula

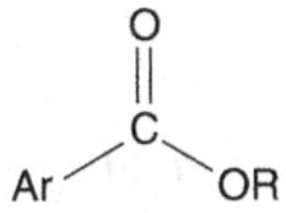

$$\mathrm{Ar{-}C({=}O){-}OR}$$

2. Molecular ion

   The molecular ion peak is observed when $R \leq C_4$. When $R \geq C_5$, the molecular ion peak is very small or nonexistent.

3. Fragmentation

   The most characteristic fragment of higher esters is the loss of the alkyl group (less one or two hydrogen atoms) through rearrangement. Characteristic peaks are as follows:

   - $M^{+\bullet}$ (small or nonexistent)
   - $[M - (R - H)]^+$ and $[M - (R - {}_2H)]^+$
   - $[ArCOO]^+$or $[ArCOO + {}_2H]^+$

   Generally the acid or protonated acid is observed. The aromatic alcohols can be differentiated by the loss of 18 Da from the molecular ion.

H. Methyl esters of cinnamic acid

1. General formula

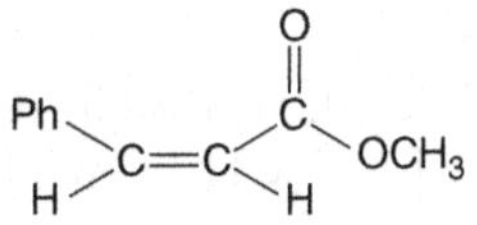

$$\mathrm{Ph(H)C{=}C(H){-}C({=}O){-}OCH_3}$$

2. Molecular ion

   The molecular peak ion is intense.

3. Fragmentation

   The esters of cinnamic acid follow the fragmentation of benzoates except there is a relatively abundant $[M - 1]^+$ peak. Prominent peaks represent the following:

   - $M^{+\bullet}$
   - $[M - 1]^+$
   - $[M - 31]^+$
   - at *m/z* 103
   - at *m/z* 77

I. Methyl esters of benzene sulfonic acids

Example 15.1

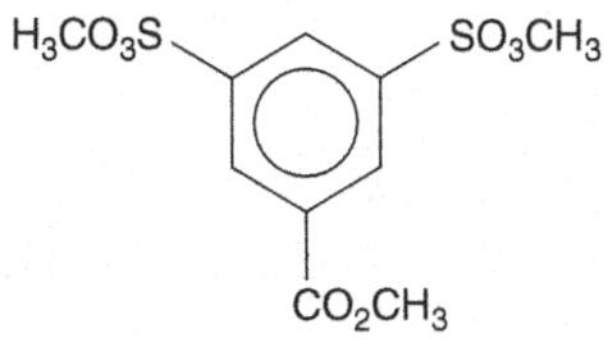

H_3CO_3S, SO_3CH_3, CO_2CH_3 (1,3,5-substituted benzene)

Prominent ions are as follows:
- $M^{+\bullet}$
- $[M - 31]^+$ ($^{\bullet}OCH_3$)
- $[M - 95]^+$ ($^{\bullet}SO_3CH_3$)
- $[M - 126]^+$ ($^{\bullet}SO_3CH_3$ + $^{\bullet}CH_3O$)
- $[M - 154]^+$ ($^{\bullet}SO_3CH_3$ + $^{\bullet}CO_2CH_3$)
- $[M - 190]^+$ ($^{\bullet}SO_3CH_3$ + $^{\bullet}SO_3CH_3$)

Example 15.2

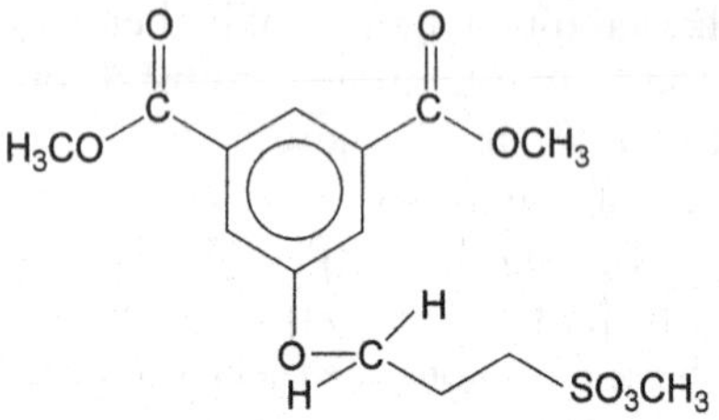

H_3CO OCH_3 SO_3CH_3

Prominent ions are as follows:
- $M^{+\bullet}$
- $[M - 31]^+$ ($^{\bullet}OCH_3$)
- $[M - 136]^+$ ($^{\bullet}CH_2{=}CHCH_2SO_3CH_3$)
- *m/z* 193 $[M - OCH_2CH_2CH_2SO_3CH_3]^+$
- *m/z* 137 $[(CH_2)_3SO_3]^+$
- *m/z* 95 $[SO_3CH_3]^+$

J. Phthalates (see Chapter 31)

K. Esters of cyano acids

The molecular ion peak is typically not observed but can be deduced by adding 31 to the highest *m/z* value peak observed (even in the case of methyl cyanoacetate). An intense peak in the mass spectra of methyl esters is at *m/z* 74, formed in a McLafferty rearrangement. Also present is a peak at *m/z* 87, but it is has a lower intensity than the peak at *m/z* 74. Characteristic losses from the molecular ion are 31, 40 (small), 59, and 73.

L. Methyl esters of aromatic cyano acids

The methyl esters of aromatic cyano acids show intense molecular ion peaks, but the intensity decreases as the length of the side chain increases. Losses from the molecular ion are 31 and 59.

REFERENCES

1. Anderson, B. A., Holman, R. T. (1974). Pyrrolidides for mass spectrometric determination of the position of the double bond in monounsaturated fatty acids. *Lipids*, 9, 185–190.

CHAPTER 16

Ethers

16.1. GC Separation of Ethers

A. Capillary columns
   1. Epichlorohydrin, propylglycidyl ether, allylglycidyl ether, phenoxy2propanone, and phenylglycidyl ether
      30-m DB-1 column, 60–175 °C at 8 °C min^{-1}.
   2. Benzoin methyl ether, benzoin propyl ether, benzoin isopropyl ether, and benzoin isobutyl ether
      30-m DB-1 column, 150–250 °C at 10 °C min^{-1}.
   3. Ethylene oxide, 2-chloroethanol, and ethylene glycol in dimethylformamide (DMF)
      30-m DB-WAX column, 60 (2 minutes)–180 °C at 15 °C min^{-1}.

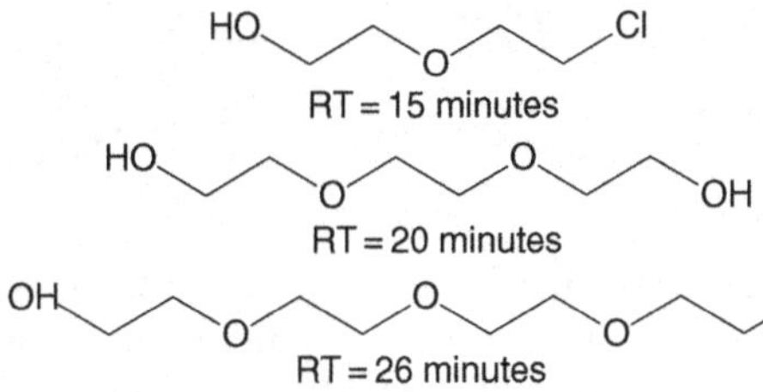

HO–CH2CH2–O–CH2CH2–Cl
RT = 15 minutes

HO–CH2CH2–O–CH2CH2–O–CH2CH2–OH
RT = 20 minutes

OH–CH2CH2–O–CH2CH2–O–CH2CH2–O–CH2CH2–OH
RT = 26 minutes

   4. 30-m DB-17 column, 75–250 °C at 6 °C min^{-1}.
   5. Dowtherm (biphenyl + biphenyl ether) and its impurities (phenol, dibenzofuran, benzophenone, terphenyl, phenoxybiphenyl, xanthone, and diphenoxybenzene)
      30-m DB-225 column, 75–215 °C at 8 °C min^{-1}. Run for 40 minutes.

16.2. Mass Spectra of Ethers

A. Unbranched aliphatic ethers
   1. General formula
      ROR′

Gas Chromatography and Mass Spectrometry
DOI: 10.1016/B978-0-12-373628-4.00016-2

2. Molecular ion
   The molecular ion peak intensity decreases with increasing molecular weight, but is still detectable through C_{16} although the intensity is low (~0.1%).
3. Fragmentation
   The initial site of the charge and radical in the molecular ion of aliphatic ethers will be associated with oxygen atom due to the loss of one of the nonbonding electrons. The intensity of the $M^{+\bullet}$ peak is low to negligible for compounds that have a $C_{\geq 4}$ chain on at least one side of the oxygen atom. Like corresponding aliphatic alcohols, as long as the number of carbon atoms is ≤3, a greater than just distinguishable $M^{+\bullet}$ peak will be present in the mass spectra of aliphatic ethers. The fragmentation of ethyl and ethyl ethers, regardless of the size of the other chain, is dominated by homolytic (radical-site-driven) cleavage. Example 16.1 is a homolytic (α-cleavage) fragmentation of an ethyl ether (Figure 16.1).

Example 16.1

Homolytic cleavage

m/z 59

and

The oxygen-substituted primary oxonium ion will then undergo a hydride-shift rearrangement fragmentation with the loss of a molecule of ethene ($H_2C{=}CH_2$) to form the ion with m/z 31 (Example 16.2).

Example 16.2

H

H_2C

CH_2

m/z 59

Hydride-shift rearrangement fragmentation

$H_2C{=}CH_2$

and

$H\overset{+}{O}{=}CH_2$

m/z 31

The mass spectra of ethyl ethers are dominated by peaks at m/z 31 and 59, both representing ions resulting from the initial fragmentation of the molecular ion (Figure 16.1). The mass spectra of methyl ethers are dominated by a peak at m/z 45. The peak with m/z 45 represents an ion with a methyl substitution on the oxygen atom of the primary oxonium ion ($CH_3–O^+{=}CH_2$), and this ion cannot undergo a hydride-shift rearrangement to form an ion with a smaller *m/z* value, thus domination of the peak at *m/z* 45 in the mass spectra of methyl ethers.

When both chains are $C_{\geq 3}$, both homolytic and heterolytic cleavages of the $M^{+\bullet}$ peak will occur; however, peaks representing heterolytic cleavage products will dominate (Figures 16.2 and 16.3). The base peak represents the ion formed by heterolytic cleavage, resulting in the loss of the largest alkoxy radical. Heterolytic cleavage is illustrated in Example 16.3.

Example 16.3

Heterolytic cleavage → *m/z* 57 and

The mass spectrum of a symmetrical ether, where the carbon chains are equal in length and are $C_{\geq 3}$, will show one peak representing heterolytic cleavage and one peak representing homolytic cleavage (Figure 16.2), whereas the fragmentation of unsymmetrical ethers will exhibit a pair of peaks representing homolytic cleavage and a pair of peaks representing ions formed by heterolytic cleavage (Figure 16.3). The rule of *loss of largest moiety* will be followed for the homolytic and heterolytic cleavages of straight-chain ethers (Figure 16.3).

The presence of peaks at *m/z* 31, 45, 59, 73, and so forth indicates the presence of ions containing an atom of oxygen; often, the ions represented by these peaks are formed by a rearrangement process. The lack of a peak corresponding to the loss of water helps distinguish the mass spectra of ethers from those of aliphatic alcohols. If the ether has a chain where $C_{\geq 4}$, it is possible to see a peak representing an $[R-(HOR')]^{+\bullet}$ ion (R is $C_{\geq 4}$) followed by a peak representing another odd electron formed by the subsequent loss of ethene ($HC_2{=}CH_2$). These peaks will have moderately low intensities.

Example 16.4

R, H, C, H, R', O, H, R, C, H, R', O, and, $\overset{+}{C}H_2$, R, C, H, H_2C_+, and, $H_2C{=}CH_2$

The mass spectra of unsymmetrical ethyl and methyl ethers that have aliphatic chains with $C_{\geq 4}$ will exhibit easily recognizable peaks representing $[M - 32]^+$ and $[M - 46]^+$ ions, respectively.

Branching at the C_1 or position of chains with $C_{\geq 4}$ can affect the appearance of spectrum of methyl and ethyl ethers. Branching at other carbon atoms on chains with $C_{\geq 4}$ has little effect on the appearance of the spectrum. The same is true for the presence of double bonds.

B. Cyclic ethers

1. General formula

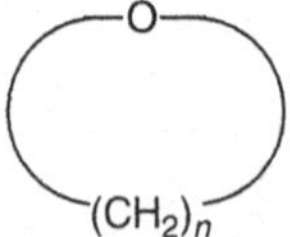

2. Molecular ion

An $M^{+\bullet}$ peak is usually present (~>25–70%) and occurs at m/z 44 (ethylene oxide), 58 (triethylene oxide), 72 (tetrahydrofuran), 86 (tetrahydropyran), and 100 (oxepane), the common unsubstituted cyclic ethers. There are also a number of different alkyl-substituted tetrahydrofurans and tetrahydropyrans.

3. Fragmentation

Losses from the molecular ions are 1 Da, 29 Da, and 30 Da with the loss of 29 (CHO) being characteristic of cyclic ethers. The intensity of the $[M-29]^+$ peak is nonexistent in the mass spectrum of oxepane; >50% in the spectra of trimethylene oxide and THF, ~10% in the spectrum of tetrahydro pyran; and >70% in ethylene oxide. The loss of 29 is also represented in the mass spectra of unsaturated cyclic ethers, such as furans and benzofurans. The $[M - 29]^+$ peak is strong in the mass spectra of THF and tetrahydropyran; ~10% in the spectra of ethyleneoxide and trimethylene oxide; and ~2% in the mass spectrum of oxepane.

C. Diaryl ethers

1. General formula

Ar–O–Ar′

2. Molecular ion

As is expected for all aromatic compounds, the $M^{+\bullet}$ peak is intense for aryl ethers.

3. Fragmentation

Peaks representing ions formed by the losses of 1 Da (H), 28 Da (CO), and 29 Da (CHO) are commonly observed in the mass spectra of diaryl ethers. In the mass spectrum of diphenyl ether, the peak at m/z 77 is intense, whereas in the mass spectra of phenyl toluoyl ethers, m/z 91 is intense. An intense peak is observed in the mass spectra of alkyl phenyl ethers at m/z 94, resulting in the formation of a phenol molecular ion through a hydride-shift rearrangement fragmentation (Example 16.5).

Example 16.5

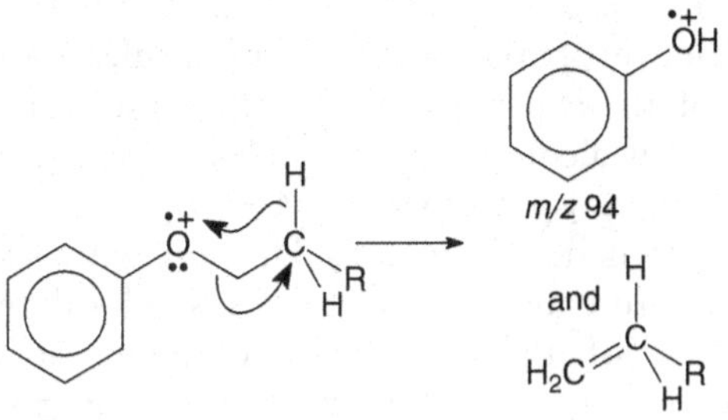

m/z 94

and

D. Benzoin alkyl ethers

1. General formula

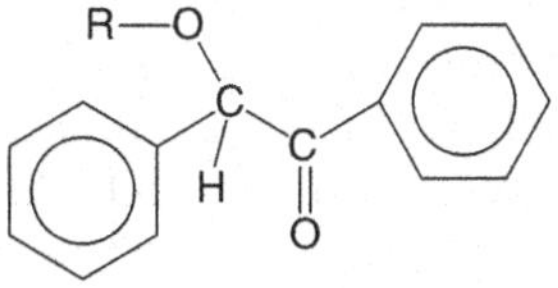

2. Molecular ion

The intensity of the $M^{+\bullet}$ peak is usually very low to nonexistent. There are spectra of three benzoin alkyl ethers in the NIST08 Mass Spectral Database: ethyl, isopropyl, and isobutyl. There are replicate spectra for the ethyl and isopropyl compounds. One of the two spectra of the ethyl compound exhibits a spectrum of the ethyl compound $M^{+\bullet}$ peak that has an intensity that is ~3% of the intensity of the base peak. The other spectrum for this compound exhibits no base peak.

3. Fragmentation

A peak representing the loss of 105 Da ($C_6H_5{}^{\bullet}C{=}O$) from the molecular ion is usually observed.

The mass spectra of all three compounds in the NIST08 Database exhibit peaks at m/z 105 that would represent a phenyl acylium ion ($C_6H_5C{\equiv}O^+$). This peak is the base peak in the isobutyl ether of benzoin; however, the relative intensity of the m/z 105 peak is less than 20% in the mass spectra of the other two compounds. There is a very intense peak at m/z 107 in the spectra of these two latter (base peak in the spectrum of the isopropyl ether and >85% relative intensity in the spectrum of the ethyl ether) compounds, which is completely missing in the spectrum of the isobutyl compound. All three spectra exhibit peaks at *m/z* 77 and 79, which have about equal intensity and range between 25 and 60 relative percent.

E. Sample mass spectra

1. In Figure 16.1, the peaks observed at *m/z* 31, 45, and 59 suggest the presence of oxygen in the molecule. By comparing the *m/z* values of these peaks with the ions listed in Appendix Q, it is evident that alcohols, ethers, and ketones are probable structures. Because there is no obvious $[M - 18]^+$ peak observed, an alcohol structure can be eliminated. The ion with *m/z* 59 indicates that the structure contains $C_2H_5O{=}CH_2$. The difference between the *m/z* value of the $M^{+\bullet}$ peak observed at *m/z* 130 and the prominent fragment ion peaks at *m/z* 59 is 71 Da, which correlates to either C_3H_7CO or C_5H_{11}. Hence, the structure for the molecule is $C_2H_5OC_6H_{13}$.
2. The $M^{+\bullet}$ peak in Figure 16.2 is clearly visible. The peaks at *m/z* 101 and 71 represent homolytic ($H_2C{=}O^+C_5H_{11}$) and heterolytic ($[C_5H_{11}]^+$) cleavages, respectively. It is clear that heterolytic cleavage dominates. The peak at *m/z* 70 represents the product of the loss of a molecule of $C_5H_{11}OH$ from the molecular ion due to a β cleavage, resulting from a γ-hydrogen-shift rearrangement (a McLafferty rearrangement). The peak at *m/z* 43 is an aliphatic ion formed by the loss of a molecule of ethene from the *n*-pentyl ion with *m/z* 71.
3. Just as was the case with the spectrum in Figure 16.2, the $M^{+\bullet}$ peak in Figure 16.3 is clearly visible in the spectrum of an unsymmetrical ether. Peaks at *m/z* 57 and 71 represent the products of heterolytic cleavage, and the peaks at *m/z* 101 and 87 represent the products of homolytic cleavage. In both cases, the ion with the smaller *m/z* value

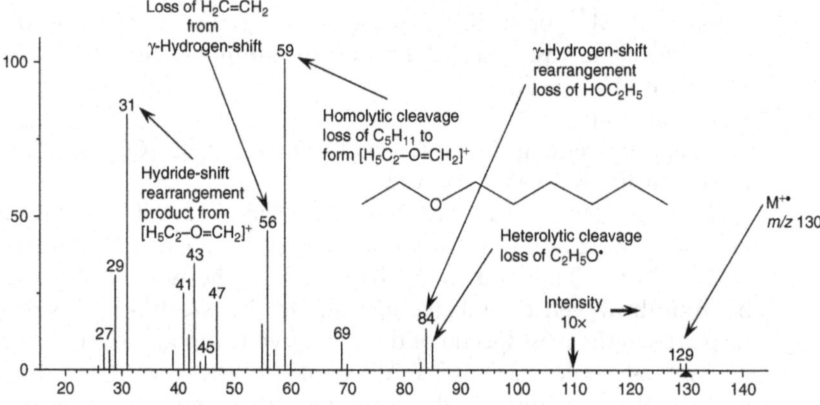

Figure 16.1 EI mass spectrum of 1-ethoxybutane.

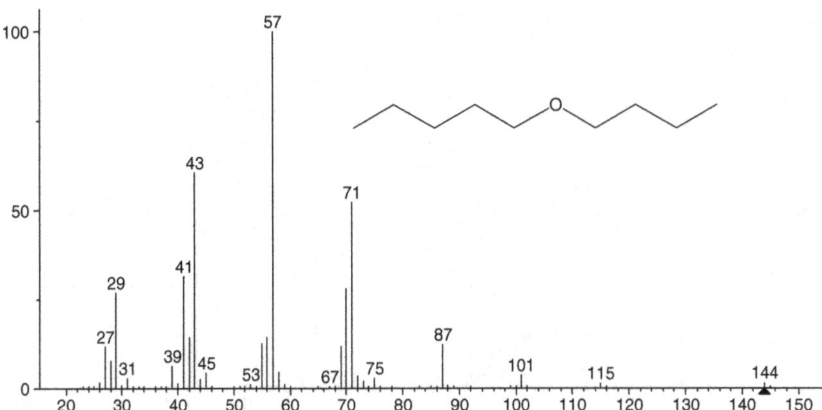

Figure 16.2 EI mass spectrum of dipentyl ether, a symmetrical ether.

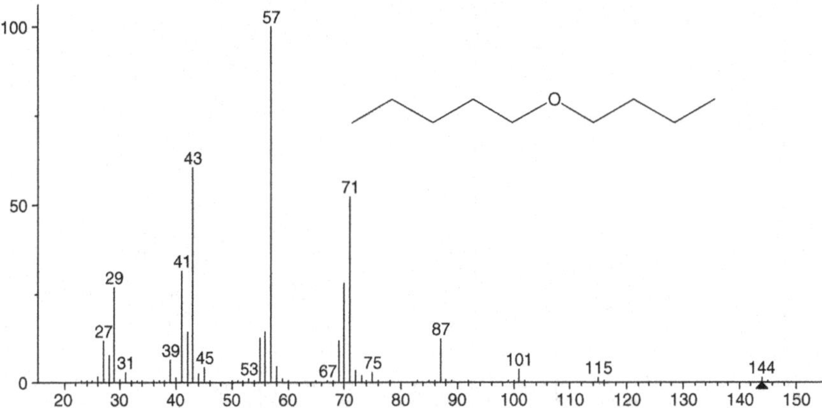

Figure 16.3 EI mass spectrum of butyl pentyl ether, an unsymmetrical ether.

is represented by the more intense of the two peaks. This is due to tendency to lose the larger moiety. Peaks representing the loss of a molecule of alcohol from both sides of the oxygen atom are also observed at *m/z* 70 and 56.

4. In Figure 16.4, the intensity of the $M^{+\bullet}$ peak at *m/z* 170 is characteristic of a $M^{+\bullet}$ peak of an aromatic compound. The characteristic losses from the molecular ions $[M - 1]^+$, $[M - 28]^+$, and $[M - 29]^+$) suggest an aromatic aldehyde, phenol, or aryl ether. An elemental composition of $C_{12}H_{10}O$ is suggested by the $M^{+\bullet}$ peak at *m/z* 170, which can be either a biphenyl or a phenylphenol ether. The simplest test to confirm the structure is to prepare a TMS derivative even though *m/z* 77 strongly indicates the diaryl ether.

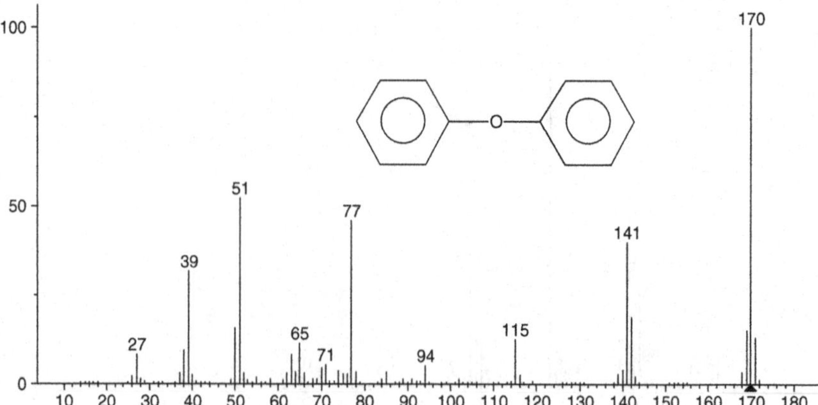

Figure 16.4 EI mass spectrum of diphenyl ether.

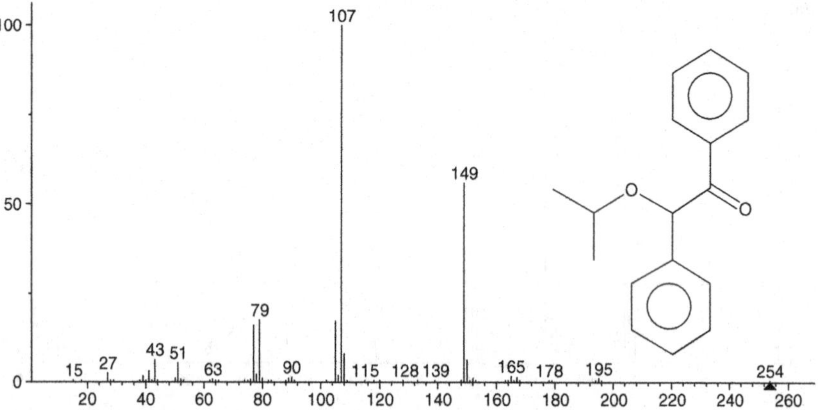

Figure 16.5 EI mass spectrum of benzoin isopropyl ether.

This will result in no change in the spectrum. Figure 16.5 is an example of the EI mass spectrum of an ether with both aliphatic and aromatic moieties.

CHAPTER 17

Fluorinated Compounds

17.1. GC Separations

A. Low-boiling fluorinated compounds,* C_1–C_2

Although capillary columns are generally preferred for most applications, porous-layer open-tubular (plot) GC columns provide the best separation of low-boiling fluorinated compounds. (Table 17.1–17.3)

1. Capillary columns
   a. CF_3Cl, CF_3CF_2Cl, CF_2Cl_2, and CHF_2Cl
      50-m Plot Alumina column (Chrompack cat. no. 4515, 0.32 mm, or cat. no. 7518, 0.53 mm), 60–200 °C at 5 °C min^{-1}.
   b. SiF_4, CH_3F, CH_3Cl, and C_2H_5F
      30-m GSQ column (J & W cat. no. 115-3432), 60–150 °C at 5 °C min^{-1}.
   c. CF_3CHClF impurities
      105-m RTX-1 capillary column at room temperature
      10-ft SP-1000 column on Carbopack B (Supelco cat. no. 1-1815 M, for the packing).
   d. O_2, N_2, CO, CF_4 (freon 14 or F-14), and CF_3CF_3 (freon 116 or F-116)
      30-m GS Molecular Sieve column at room temperature (J & W cat. no. 93802).
   e. O_2, N_2, CO, F-14, and F-116
      4-m Molecular Sieve 13× column at room temperature.
      30-m GS-Alumina column (J & W Scientific cat. no. 115-3532), 60–200 °C at 5 °C min^{-1}.
   f. CHF_3, CH_2F_2, CF_3CH_3, CHF_2Cl, CF_2Cl_2, CH_2FCl, CHF_2CH_2Cl, and $CHFCl_2$

* Fluorinated compounds are frequently referred to by a code (such as F-115). To translate this code into a molecular formula, add 90 to the number, in this example add 90 to 115. The first digit of the sum (205) is the number of carbon atoms; the second digit is the number of hydrogen atoms; and the third digit is the number of fluorine atoms; if the saturated valence is not complete, atoms of chlorine are added (i.e., C_2F_5Cl is F-115 = 90 + 115 = 205, or $C_2H_0F_5$). A four-digit number is used for unsaturated molecules. See Tables 17.2 and 17.3 for the numbering system for chlorofluorocarbons.

Gas Chromatography and Mass Spectrometry
DOI: 10.1016/B978-0-12-373628-4.00017-4

Table 17.1 Numbering system for C_1 and C_2 chlorofluorocarbons

Industrial number	Formula	Approximate boiling point (°C)	MW
11	$CFCl_3$	24	136
12	CF_2Cl_2	–30	120
13	CF_3Cl	–81	104
14	CF_4	–128	88
20	$CHCl_3$	61	118
21	$CHFCl_2$	9	102
22	CHF_2Cl	–41	86
23	CHF_3	–82	70
30	CH_2Cl_2	40	84
31	CH_2FCl	–9	68
32	CH_2F_2	–52	52
40	CH_3Cl	–24	50
41	CH_3F	–79	34
110	CCl_3CCl_3	185	234
111	$CFCl_2CCl_3$	137	218
112	$CFCl_2CFCl_2$	93	202
113	$CFCl_2CF_2Cl$	48	186
113a	CF_3CCl_3	46	186
114	CF_2ClCF_2Cl	4	170
114a	CF_3CFCl_2	4	170
115	CF_3CF_2Cl	–39	154
116	CF_3CF_3	–78	138
120	$CHCl_2CCl_3$	162	200
121	$CHCl_2CFCl_2$	117	184
121a	$CHClFCCl_3$	117	184
122	$CHCl_2CF_2Cl$	72	168
122a	$CHClFCFCl_2$	73	168
122b	CHF_2CCl_3	73	168
123	CF_3CHCl_2	27	152
123a	$CF_2ClCHClF$	28	152
123b	CHF_2CFCl_2	28	152
124	CF_3CHClF	–12	136
124a	CHF_2CF_2Cl	–10	136
125	CF_3CHF_2	–49	120
130	$CHCl_2CHCl_2$	146	166
130a	CCl_3CH_2Cl	131	166
131	$CHCl_2CHClF$	103	150
131a	$CH_2ClCFCl_2$	88	150
131b	CH_2FCCl_3	88	150
132	$CHClFCHClF$	59	134
132a	CHF_2CHCl_2	60	134

(Continued)

Table 17.1 (*Continued*)

Industrial number	Formula	Approximate boiling point (°C)	MW
132b	CF_2ClCH_2Cl	47	134
132c	$CFCl_2CH_2F$	47	134
133	$CHClFCHF_2$	17	118
133a	CF_3CH_2Cl	6	118
133b	CF_2ClCH_2F	12	118
134	CHF_2CHF_2	–20	102
134a	CF_3CH_2F	–27	102
140	$CHCl_2CH_2Cl$	114	132
140a	CH_3CCl_3	74	132
141	$CHClFCH_2Cl$	76	116
141a	$CHCl_2CH_2F$	?	116
141b	$CFCl_2CH_3$	32	116
142	CHF_2CH_2Cl	35	100
142a	$CHClFCH_2F$	?	100
143a	CF_3CH_3	–48	84
150	CH_2ClCH_2Cl	84	98
150a	CH_3CHCl_2	57	98
151	CH_2FCH_2Cl	53	82
151a	CH_3CHClF	16	82
152	CH_2FCH_2F	25	66
160	CH_3CH_2Cl	13	64
161	CH_3CH_2F	–37	48

The formula can be derived from the number by adding 90 to the industrial number. Reading the ensuing digits from right to left gives the number of fluorines, hydrogens, carbons, and double bonds. The remaining valence positions are reserved for chlorines.

Table 17.2 Numbering system for unsaturated C_2 chlorofluorocarbons

Industrial number	Formula	Approximate boiling point (°C)	MW
1110	$CCl_2{=}CCl_2$	121	164
1111	$CFCl{=}CCl_2$	71	148
1112 (trans)	$CFCl{=}CFCl$	22	132
1112 (cis)	$CFCl{=}CFCl$	21	132
1112a	$CF_2{=}CCl_2$	19	132
1113	$CClF{=}CF_2$	–28	116
1114	$CF_2{=}CF_2$	–76	100
1120	$CHCl{=}CCl_2$	88	130

(*Continued*)

Table 17.2 (*Continued*)

Industrial number	Formula	Approximate boiling point (°C)	MW
1121	$CHCl{=}CFCl$	35	114
1121a	$CHF{=}CCl_2$	37	114
1122	$CF_2{=}CHCl$	–18	98
1123	$CF_2{=}CHF$	–56	82
1130 (cis)	$CHCl{=}CHCl$	60	96
1130 (trans)	$CHCl{=}CHCl$	48	96
1130a	$CH_2{=}CCl_2$	37	96
1131	CHF-CHCl	11	80
1131 (cis)	$CHF{=}CHCl$	16	80
1131 (trans)	$CHF{=}CHCl$	–4	80
1131a	$CH_2{=}CClF$	–25	80
1132	$CHF{=}CHF$	–28	64
1132a	$CH_2{=}CF_2$	–82	64
1140	$CH_2{=}CHCl$	–14	62
1141	$CH_2{=}CHF$	–72	46
1150	$CH_2{=}CH_2$	–104	28

Table 17.3 Numbering system for cyclic chlorofluorocarbons

Industrial number	Formula	Approximate boiling point (°C)	MW
C-216	F_3C—CF_2 \ / CF_2 (three-membered ring)	–32	150
C-314	F_2C—CCl_2 / F_2C—CCl_2 (four-membered ring)	132	264
C-317	F_2C—CClF / F_2C—CF_2 (four-membered ring)	60	216
C-318	F_2C—CF_2 / F_2C—CF_2 (four-membered ring)	–6	200

25-m Poraplot Q column, 60–150 °C at 5 °C min^{-1}.

B. Medium-boiling fluorinated compounds (mostly liquids)
   1. Capillary columns
      a. Fluoroketones and fluoroethers
         30-m DB-210 column, 50–100 °C at 5 °C min^{-1} (J & W Scientific cat. no. 122-0233).
      b. C_2F_5Cl, $CFCl_3$, CF_2Cl_2, and $C_2F_3Cl_3$
         30-m DB-624 column, 40 (10 minutes)–140 °C at 5 °C min^{-1} (J & W Scientific cat. no. 125-1334).
      c. $CF_2{=}CCl_2$, $CCl_2{=}CCl_2$, CCl_3CFCl_2, $CHCl_2CF_2Cl$, CF_3CHCl_2, F-113, and F-112
         30-m DB-WAX column, 50–200 °C at 5 °C min^{-1} (J & W Scientific cat. no. 125-7032).
      d. F-11 and F-113
         30-m DB-WAX column at 50 °C.

C. Higher-boiling fluorinated compounds
   1. $C_6F_{13}CH_2CH_2OH$, $C_6F_{13}CH_2CH_2I$, $C_8F_{17}CH_2CH_2OH$ $C_4F_9CH_2CH_2OC(O)C(CH_3){=}CH_2$, and $C_6F_{13}CH_2CH_2OC(O)C(CH_3){=}CH_2$
      30-m DB-1 column, 60–175 °C at 4 °C min^{-1} (J & W scientific cat. no. 125-1032).
   2. C_2–C_{12} perfluoroalkyl iodides
      30-m DB-WAX column, 75–180 °C at 8 °C min^{-1}.

17.2. Mass Spectra of Fluorinated Compounds

A. Partially fluorinated and fluorinated aliphatic compounds
   The $M^{+\bullet}$ peak is usually not observed in the mass spectra of aliphatic fluorinated compounds (>ethane). Common losses are F, HF, or CF_3. Frequently observed peaks are at *m/z* 31 (CF), *m/z* 50 (CF_2), and *m/z* 69 (CF_3). If the peak at *m/z* 69 is intense, a CF_3 group is present. A small peak at *m/z* 51 indicates the presence of carbon, hydrogen, and fluorine. If the peak at *m/z* 51 is intense, then a CHF_2 moiety is present. The absence of a peak at *m/z* 51 and/or 47 (without chlorine) suggests a perfluorinated compound.

B. Chlorofluorocarbons
   The $M^{+\bullet}$ peak is usually not observed, but may be deduced by adding 35 to the highest *m/z* value ion observed in the mass spectrum. In the mass spectra of chlorofluorocarbons, the evidence of chlorine (or bromine) is obvious from the isotope peak patterns (X+2 patterns) observed in the fragment ions (see Chapter 20). A small peak representing a rearrangement ion at *m/z* 85 containing chlorine indicates that carbon,

fluorine, and chlorine are present even though a CF_2Cl moiety does not exist in the compound. If an *m/z* 85 peak is intense and the X+2 isotope peaks indicate the presence of chlorine, then the CF_2Cl moiety is present. Look up the abundant ions in the "structurally significant" ion tables of Appendix Q to aid in structural assignments. Also, refer to Table 17.4 for the *m/z* values of the most intense peaks, which are listed in order of decreasing intensities as many of these spectra may not be in commercial mass spectral libraries.

Table 17.4 Fluorinated compounds listed by most abundant ion

Base peak	Four next most intense peaks				Compound	Highest value *m/z* peak >1%[a]
15	34	33	14	31	CH_3F	34
15	69	47	112	31	$CH_3OCF{=}CF_2$	112
18	45	33	31	61	$CH_2FC(O)OH$	78
27	29	28	64	26	C_2H_5Cl	64
29	27	51	79	77	$CH_3CH_2CF_3$	79
29	51	79	31	50	CF_2ClCHO	114
29	69	31	100	150	$CF_3CF_2CF_2CHO$	150
30	69	50	31	64	CF_3NO	99
31	29	33	61	69	CF_3CH_2OH	83
31	29	50	69	51	$CF_3CF_2CH_2OH$	100
31	51	29	69	49	$HOCH_2(CF_2CF_2)_2H$	183
31	81	100	50	69	$CF_2{=}CF_2$	100
31	82	51	29	113	$CHF_2CF_2CH_2OH$	113
31	95	69	65	29	$C_6F_{13}CH_2CH_2OH$	364
31	107	53.5	88	57	$CF_2{=}CFCN$	107
33	51	31	32	52	CH_2F_2	52
33	69	83	51	31	CF_3CH_2F	102
33	244	163	64	83	$(C_2H_2F_3)_3PO_4$	321
39	57	108	31	38	$CF_2{=}CFCH{=}CH_2$	108
41	39	95	64	28	$CH_3C(CF_3){=}CH_2$	95
43	69	15	–	–	$(CF_3)_2CHOC(O)CH_3$	210
44	31	25	43	13	$CF{\equiv}CH$	44
45	18	51	28	44	$CF_3C(O)OH$	97
45	64	44	31	33	$CH_2{=}CF_2$	64
45	69	119	100	169	$CF_3CF_2CF_2C(O)OH$	169
45	80	82	44	26	$CH_2{=}CClF$	80
46	45	27	44	26	$CH_2{=}CHF$	46
47	27	45	26	67	CH_3CHClF	82
47	33	27	48	46	CH_3CH_2F	48
47	46	61	27	41	CH_3CHFCH_3	62

(Continued)

Table 17.4 (*Continued*)

Base peak	Four next most intense peaks				Compound	Highest value *m/z* peak >1%[a]
47	66	28	12	31	COF_2	66
47	66	33	–	–	N_2F_2	66
49	48	11	68	19	BF_3	68
49	84	86	51	47	CH_2Cl_2	84
50	52	15	49	47	CH_3Cl	50
51	33	31	52	32	CH_2F_2	52
51	64	194	196	143	$CHF_2CF_2CH_2Br$	194
51	65	47	45	27	CH_3CHF_2	66
51	67	31	50	69	CHF_2Cl	86
51	69	101	132	151	$H(CF_2)_6H$	233
51	69	101	132	151	$H(CF_2)_7H$	283
51	69	113	101	132	$CHF_2CF_2CF_2CHF_2$	151
51	69	201	127	100	$CHF_2(CF_2)_3I$	328
51	83	33	31	101	CHF_2CHF_2	102
51	83	33	64	101	$CHF_2CH_2CH_2CH_2F$	113
51	85	69	87	101	$CF_2ClCF_2CF_2CHF_2$	201
51	85	69	87	101	$CHF_2(CF_2)_3CF_2Cl$	251
51	85	69	151	87	$CHF_2CF_2CF_2Cl$	152
51	99	49	101	64	$CHF_2CF_2CH_2Cl$	151
51	101	69	31	50	CF_3CHF_2	119
51	101	85	67	31	CF_2ClCHF_2	135
51	130	132	31	111	CHF_2Br	130
52	71	33	19	14	NF_3	71
59	60	39	27	57	$CH_2=CHCH_2F$	60
59	60	39	33	57	$CH_3CF=CH_2$	60
61	96	98	26	63	$CHCl=CHCl$ (trans)	96
61	96	98	63	26	$CHCl=CHCl$ (cis)	96
61	96	98	63	26	$CH_2=CCl_2$	96
62	27	49	64	26	CH_2ClCH_2Cl	98
62	27	64	26	25	$CH_2=CHCl$	62
63	27	65	26	83	CH_3CHCl_2	98
63	82	51	31	32	$CHF=CF_2$	82
63	82	60	31	164	F_2C—S / S—CF_2	164
63	82	132	31	50	F_2C—CF_2 / CF_2	132
63	132	82	69	31	$CF_3C(S)F$	132

(*Continued*)

Table 17.4 (*Continued*)

Base peak	Four next most intense peaks				Compound	Highest value *m/z* peak >1%[a]
64	28	29	27	66	CH_3CH_2Cl	64
64	45	31	33	44	$CH_2{=}CF_2$	64
64	45	44	31	33	CHF=CHF	64
65	45	85	31	64	CH_3CF_2Cl	85
65	64	45	61	33	$CH_3CF_2CH_3$	65
67	41	54	82	56	$C_6H_{11}F$ (fluorocyclohexane)	102
67	69	31	111	32	CHFClBr	146
67	69	47	35	48	$CHFCl_2$	102
67	69	101	51	117	CF_3CHClF	136
67	83	33	51	118	$CHClFCHF_2$	118
67	86	48	69	32	SOF_2	86
67	99	83	69	79	CHClFCHClF	134
67	117	85	69	119	$CHClFCF_2Cl$	152
68	33	70	49	46	CH_2ClF	68
69	51	31	50	–	CHF_3	70
69	31	81	100	47	F_3CFC—CF_2 (epoxide, bridged by O)	147
69	31	100	50	131	F_2C—CF_2 (cyclopropane, bridged by CF_2)	150
69	31	119	100	50	$CF_3CF_2CF_3$	169
69	47	50	31	28	$CF_3C(O)F$	116
69	47	66	28	31	CF_3OCF_3	69
69	47	112	31	–	$CH_3OCF{=}CF_2$	112
69	50	25	31	34.5	CF_4	69
69	50	31	152	44	CF_3NFCF_3	152
69	51	31	151	100	$CF_3CF_2CHF_2$	151
69	51	65	77	574	$CF_3(CF_2)_7CH_2CH_2I$	574
69	51	82	151	31	CF_3CHFCF_3	151
69	58	31	108	28	CF_3CHFCN	108
69	64	114	45	95	$CF_3CF{=}CH_2$	114
69	64	133	31	45	$CF_3CH_2CF_3$	133
69	65	45	33	31	CF_3CH_3	84
69	76	50	31	38	CF_3CN	95
69	70	139	89	51	CF_3SF_3	139
69	85	47	201	119	$CF_3CF_2CCl_3$	201
69	85	147	87	97	$CF_3C(O)CF_2Cl$	147
69	85	50	87	35	CF_3Cl	85
69	85	97	31	50	$CF_3CF_2C(O)CF_2Cl$	197

(*Continued*)

Table 17.4 (*Continued*)

Base peak	Four next most intense peaks				Compound	Highest value *m/z* peak >1%[a]
69	85	116	135	31	$CF_3CFClCF_3$	185
69	89	127	70	51	CF_3SF_5	127
69	95	96	45	46	$CHF_2CF{=}CH_2$	96
69	97	50	31	147	$CF_3C(O)CF_3$	166
69	99	83	51	33	$CF_3OCH_2CF_2CHF_2$	181
69	101	169	51	150	$C_9H_2F_{18}$ (Dihydro HFP trimer)	414
69	102	63	82	32	CF_3SH	102
69	113	132	31	82	$CF_3CHFCHF_2$	133
69	113	132	82	31	$CHF_2CF_2CHF_2$	133
69	113	201	132	82	$CF_3CHFCF_2CF_3$	201
69	114	264	119	145	$CF_2{=}N(CF_2)_3CF_3$	264
69	116	147	131	–	$CF_3CF{=}CClF$	166
69	119	31	47	78	$CF_3CF_2C(O)F$	119
69	119	31	50	19	CF_3CF_3	119
69	119	31	131	100	$CF_3CF_2CF_2CF_3$	219
69	119	131	31	181	$(CF_3)_2CFCF_2CF_3$	269
69	119	169	31	47	$C_6F_{12}O_2$ (HFPO dimer)	285
69	119	169	31	100	$CF_3(CF_2)_3CF_3$	269
69	119	169	97	147	$CF_3CF_2CF_2OCF{-}C(O)F$	285
69	119	169	131	100	$CF_3(CF_2)_4CF_3$	319
69	119	169	131	219	$CF_3(CF_2)_5CF_3$	369
69	119	293	243	343	C_9F_{18} (HFP trimer B)	431
69	119	293	343	243	C_9F_{18} (HFP trimer A)	381
69	127	296	31	100	CF_3CFICF_3	296
69	129	131	148	150	CF_3Br	148
69	129	131	229	231	$(CF_3)_2CFBr$	248
69	131	31	100	31	$CF_3CF{=}CF_2$	150
69	131	119	169	50	$(CF_3)_2CF(CF_2)_2CF_3$	319
69	131	181	100	93	Perfluorinated methylcyclohexane	281
69	131	181	100	293	$C_{10}F_{17}{-}CF_3$	493
69	134	46	153	65	$CF_3N{=}SF_2$	153
69	141	47	15	–	O, $C(CF_3)_2$, O (1,3-dioxolane ring structure)	180
69	145	76	246	32	$CF_3SC(S)SCF_3$	246
69	163	–	–	–	$CF_3CCl{=}CCl_2$	198

(*Continued*)

Table 17.4 (*Continued*)

Base peak	Four next most intense peaks				Compound	Highest value m/z peak >1%[a]
69	166	147	31	131	$CF_3CCl{=}CF_2$	166
69	169	100	119	50	$CF_3CF_2CF_3$	169
69	169	119	97	100	$C_9F_{18}O_3$ (HFPO trimer)	351
69	169	127	296	177	$CF_3CF_2CF_2I$	296
69	170	63	82	31	CF_3SCF_3	170
69	181	93	200	31	$(CF_3)_2C{=}CF_2$	200
69	181	131	100	31	Perfluorinated dimethylcyclohexane	381
69	181	131	281	93	$(CF_3)_2CFCF_2CF{=}CF_2$	300
69	181	281	231	93	C_6F_{12} (HFP dimer B)	300
69	197	169	31	100	$(CF_3)_2CFC(O)CF(CF_3)_2$	197
69	202	64	133	114	CF_3SSCF_3	202
69	219	127	177	31	$F(CF_2)_4I$	327
72	46	51	39	27	$CH_2{=}CFCH{=}CH_2$	72
77	51	59	104	39	$CF_3(CH_2CHF)_2CF{=}CH_2$	206
77	78	51	59	65	$CF_3(CHFCH_2)_2CF{=}CH_2$	142
79	59	29	47	69	$F(CF_2)_8CH_2CH_3$	448
80	45	82	44	26	$CHF{=}CHCl$	80
81	15	51	101	63	$CHF_2CF_2OCH_3$	132
81	67	45	61	83	$CHFClCH_2Cl$	116
81	83	61	45	101	$CFCl_2CH_3$	101
82	63	32	31	50	CF_2S	82
82	132	84	134	31	$CFCl{=}CFCl$	132
83	51	69	33	145	$H(CF_2CF_2)_2CH_2F$	163
83	85	47	87	48	$CHCl_3$	118
83	85	69	67	31	CF_3CHCl_2	152
83	85	67	115	87	$CHClFCHCl_2$	150
83	85	69	130	199	$CHCl_2CCl_2CF_3$	199
83	85	87	95	99	$CHCl_2CHCl_2$	166
83	85	133	87	135	$CHCl_2CHF_2$	134
83	102	67	32	44	SO_2F_2	102
83	118	67	120	64	SO_2FCl	118
83	162	164	33	64	CF_3CH_2Br	162
85	69	87	185	31	$CF_3CClFCF_2Cl$	185
85	69	119	87	31	CF_3CF_2Cl	119
85	69	147	119	31	$CF_2ClCF_2CF_2CF_2Cl$	235
85	69	169	87	31	$CF_3CF_2CF_2Cl$	185
85	86	28	33	87	SiF_4	104
85	87	50	31	35	CF_2Cl_2	101
85	87	129	131	31	CF_2ClBr	145
85	87	163	50	31	$CF_2ClC(O)CF_2Cl$	198
85	117	119	47	31	$CF_2ClCF_2CCl_3$	217

(*Continued*)

Table 17.4 (*Continued*)

Base peak	Four next most intense peaks				Compound	Highest value *m/z* peak >1%[a]
85	120	122	101	69	$POClF_2$	120
85	135	87	31	137	CF_2ClCF_2Cl	151
89	70	51	35	32	SF_4	108
93	31	162	74	112	$CF_2{=}CF{-}CF{=}CF_2$	162
93	143	162	69	31	$CF_3C{\equiv}CCF_3$	162
93	143	270	74	31	F_2C — CF_2 | | FC = C—I	270
93	162	31	112	143	Perfluorocyclobutene	162
93	162	143	31	193	Perfluorocyclopentene	212
94	75	31	69	56	$CF_3C{\equiv}CH$	94
95	27	96	77	51	$CF_3CH{=}CH_2$	96
95	59	29	97	27	$CH_3CH_2CFCl_2$	130
95	61	69	130	31	$CF_3CCl{=}CH_2$	130
95	69	–	–	–	$CF_3CH{=}CHCl$	130
95	77	–	–	–	$CF_3CH{=}CH_2$	96
95	130	132	97	60	$CHCl{=}CCl_2$	130
96	70	50	75	95	Fluorobenzene	96
97	83	99	85	61	$CHCl_2CH_2Cl$	132
97	99	61	117	119	CH_3CCl_3	132
98	100	48	63	31	$CHCl{=}CF_2$	98
99	51	69	79	129	Hexfluoroisopropyl alcohol	149
99	101	49	85	79	CF_2ClCH_2Cl	134
99	129	178	127	101	$CHF_2CHClBr$	178
100	29	31	64	33	F_2C — CF_2 | | O — CH_2	111
100	51	49	45	64	CHF_2CH_2Cl	100
100	63	82	113	264	$F_2C - CF_2$ / \\ S S \\ / $F_2C - CF_2$	264
100	63	232	113	150	F_2C — CF_2 | \\ | S | / F_2C — CF_2	232
100	69	231	181	131	Perfluorodimethyl-cyclobutane	281

(*Continued*)

Table 17.4 (*Continued*)

Base peak	Four next most intense peaks				Compound	Highest value *m/z* peak >1%[a]
100	131	31	69	50	Perfluorocyclobutane	131
100	131	181	69	31	Perfluoromethylcyclobutane	231
101	51	31	111	113	CHF_2CF_2Br	180
101	83	103	85	149	$CFCl_2CHCl_2$	184
101	85	103	31	87	$CF_2ClCFClCF_2CFCl_2$	267
101	103	66	31	167	$CFCl_2C(O)CFCl_2$	195
101	103	66	35	31	$CFCl_3$	117
101	103	167	169	31	$CFCl_2CFCl_2$	167
101	133	103	67	135	$CFCl_2CHClF$	168
101	228	51	82	127	CHF_2CF_2I	228
104	85	69	50	31	POF_3	104
105	86	67	32	107	SOF_4	105
105	376	77	182	165	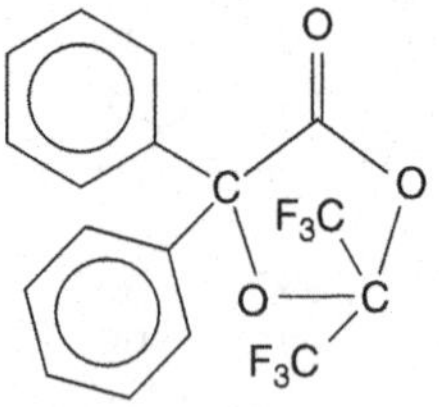 O; C; O; F3C; O; C; F3C	376
109	110	83	57	63	*o*-Fluorotoluene	110
109	244	175	194	31	1,2-Dichlorohexafluo-rocyclopentene-1	244
111	84	83	57	28	Fluoroaniline	111
111	85	–	–	–	$CHF_2CH{=}CCl_2$	146
112	64	63	92	83	Fluorophenol	112
112	64	140	125	92	*o*-Fluorophenetole	140
112	140	84	83	29	*p*-Fluorophenetole	140
113	69	31	132	82	$CF_3CH{=}CF_2$	132
113	69	63	182	31	$CF_3C(S)CF_3$	182
113	69	163	31	182	Fluorobutene	182
114	69	264	119	145	$CF_2{=}N(CF_2)_3CF_3$	264
114	116	79	44	81	$CHCl{=}CFCl$	114
115	117	101	103	79	$CFCl_2CH_2Cl$	115
116	31	66	85	118	$CF_2{=}CClF$	116
116	31	118	132	93	1,3-Dichlorohexafluo-rocyclobutane	197
117	119	121	82	47	CCl_4	117
117	119	167	165	83	$CHCl_2CCl_3$	200
117	119	167	169	47	CF_2ClCCl_3	167
117	119	201	203	199	CCl_3CCl_3	199

(*Continued*)

Table 17.4 (*Continued*)

Base peak	Four next most intense peaks				Compound	Highest value *m/z* peak >1%[a]
117	198	196	119	129	$CF_3CHClBr$	196
117	248	246	167	79	Bromopentafluorobenzene	246
118	83	33	49	120	CF_2ClCH_2F	118
118	83	33	120	49	CF_3CH_2Cl	118
119	69	129	131	31	CF_3CF_2Br	198
119	69	302	64	183	$C_2F_5SSC_2F_5$	302
119	246	127	69	177	CF_3CF_2I	246
126	75	57	76	31	$CF_2{=}CHCH{=}CF_2$	126
126	83	111	57	95	Fluoroanisole	126
127	89	108	54	129	SF_6	127
129	69	164	131	166	$CF_3CH{=}CCl_2$	164
129	131	79	81	50	CF_2Br_2	208
130	95	132	75	50	Chlorofluorobenzene	130
131	31	28	44	69	$CF_2ClCFClCF_2C(O)OH$	161
131	31	69	93	181	$CF_3CF_2CF{=}CF_2$	200
131	31	69	147	93	$CF2{=}CFCF_2Cl$	166
131	69	100	31	181	Perfluorocyclohexane	281
131	69	100	150	31	$CF_3CF{=}CF_2$	150
131	69	147	101	93	$CF_2{=}CFCF_2CFCl_2$	232
131	69	181	31	93	$CF_3(CF_2)_4CF{=}CF_2$	350
131	69	181	200	31	$CF_3CF{=}CFCF_3$	200
131	75	69	225	175	FC = CF F_2C CF_2 F_2C — CF_2	244
131	100	31	69	181	Perfluorocyclopentane	231
131	100	69	31	159	Perfluorocyclohexene oxide	231
131	113	69	31	100	Nonafluorocyclopentane	213
131	133	117	119	95	CCl_3CH_2Cl	131
131	147	69	31	149	Chloroperfluorocyclohexane	247
132	134	82	84	47	$CF_2{=}CCl_2$	132
133	117	119	135	51	CHF_2CCl_3	168
135	85	137	129	131	CF_2ClCF_2Br	214
135	101	85	103	69	CF_3CFCl_2	151
143	193	69	93	124	$CF_3C{\equiv}CCF_2CF_3$	193
145	126	96	144	51	Trifluoromethylbenzene	146
145	147	85	–	–	$CF_2ClCH{=}CCl_2$	180
147	69	149	31	182	$CF_3CCl{=}CClF$	182

(*Continued*)

Table 17.4 (*Continued*)

Base peak	Four next most intense peaks				Compound	Highest value *m/z* peak >1%[a]
147	145	31	149	79	$CFClBr_2$	189
147	216	149	69	197	$CF_3CCl{=}CFCF_3$	216
148	150	113	115	47	$CFCl{=}CCl_2$	148
151	51	69	129	131	$CF_2BrCF_2CHF_2$	230
151	117	153	119	101	CF_3CCl_3	167
151	132	69	101	201	$CF_3CCl_2CF_3$	220
155	205	224	124	69	Perfluoro-1,4-cyclohexadiene	224
161	142	111	114	162	$CF_3C_6H_4NH_2$	161
162	93	69	243	143	Perfluorocyclohexene	262
166	164	129	131	168	$CCl_2{=}CCl_2$	164
171	152	121	170	75	$CF_3C_6H_4CN$	171
173	145	189	95	75	$CF_3C_6H_4C(O)NH_2$	189
175	177	95	51	112	$CF_2BrCHFCH_2Br$	254
179	181	31	129	131	CF_2BrCF_2Br	258
179	181	85	183	216	$CF_2ClCCl{=}CCl_2$	214
185	85	131	129	69	$CF_3CFBrCF_2Cl$	229
185	117	183	119	101	CCl_3CFCl_2	183
186	117	31	93	155	Hexafluorobenzene	186
189	77	120	92	65	$CF_3C(O)NHC_6H_5$	189
194	109	196	69	85	1,2-Dichlorooctafluoro-1-hexene	294
196	69	127	177	31	CF_3I	196
197	195	147	145	31	$CF_2BrCFClBr$	274
202	55	116	31	28	Chloropentafluorobenzene	202
208	308	131	31	93	C_4F_7I	308
214	195	145	164	75	$C_6H_4(CF_3)_2$	214
227	254	127	100	31	CF_2ICF_2I	254
229	231	129	131	69	$CF_3CFBrCF_2Br$	229
234	69	133	64	165	CF_3SSSCF_3	234
263	282	213	225	163	$C_6H_3(CF_3)_3$	282

[a] Most abundant isotopes

C. Sample mass spectrum of chlorofluorocarbons

The highest *m/z* value peak observed in the mass spectrum in Figure 17.1 is at *m/z* 201. Based on the isotope pattern, the ion represented by the peak with this nominal *m/z* value contains two chlorine atoms. The $M^{+\bullet}$ peak is not observed but can be deduced by

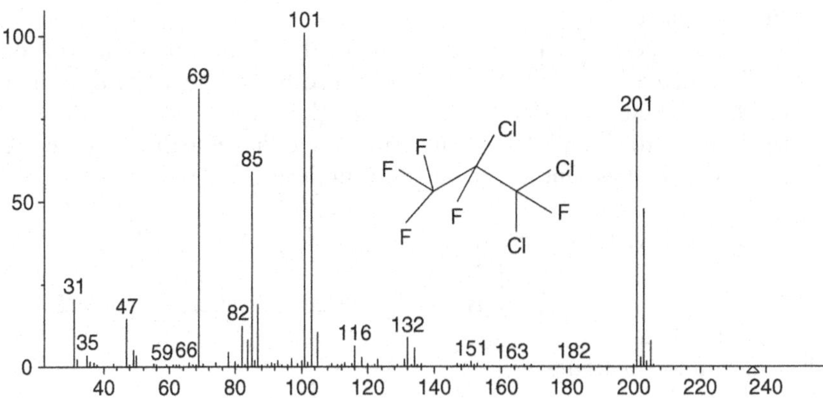

Figure 17.1 A chlorofluorocarbon. It should be noted that this spectrum is in neither the *Wiley Registry 9th Ed.* nor the *NIST08 Mass Spectral Database* and has its origin with the authors of the first edition of this book.

adding 35 to the nominal mass of the highest *m/z* value peak (*m/z* 201); therefore, it can be concluded that the molecular weight is 236 Da, and the molecule contains three atoms of chlorine. The intensity of the peak at *m/z* 101 suggests that $CFCl_2$ is present in the molecule. Notice that the peak at *m/z* 69 is more intense than the peak at *m/z* 85, indicating that a CF_3 moiety is also present in the molecule. Subtracting 69 (CF_3) and 101 ($CFCl_2$) from the deduced molecular weight of 236 Da shows that the remainder is 66 Da. The significant ion table for *m/z* 66 suggests the presence of a CFCl moiety. The deduced structure is $CF_3CFClCFCl_2$.

D. Aromatic fluorine compounds
Intense $M^{+\bullet}$ peaks are observed in the mass spectra of fluorinated benzene, ethyl benzene, toluene, and xylene. Most fluorinated aromatics molecular ions lose 19 ($^{\bullet}F$), and some lose 50 (e.g., $^{\bullet}CF_2$). The chlorofluoroaromatics can easily be identified by examining the isotope peak patterns in the vicinity of the $M^{+\bullet}$ peak.

E. Perfluorinated olefins
The $M^{+\bullet}$ peaks are usually observed in the mass spectra of perfluorinated olefins. The peak at *m/z* 31 is frequently more intense in the mass spectra of fluorinated olefins than in those of fluorinated saturated compounds.

F. Perfluorinated acids
The $M^{+\bullet}$ peaks are usually not observed in the mass spectra of perfluorinated acids but may be deduced by adding 17 ($^{\bullet}OH$) or 45 ($^{\bullet}CO_2H$) to the highest *m/z* value peak observed (greater than 1%). An intense peak at *m/z* 45 is further confirmation that the analyte is an acid.

G. Perfluorinated ketones

The $M^{+\bullet}$ peak is usually not observed in the mass spectra of perfluorinated ketones but may be deduced by adding 19 to the value of the highest *m/z* value peak observed in the mass spectrum of perfluoroacetone and 69 in the case of perfluorodiethyl ketone. A characteristic fragment ion results from α-cleavage:

Example 17.1 $CF_3C(O)$ – *m/z* 97
$CF_3CF_2C(O)$ – *m/z* 147

A peak at *m/z* 28 (CO) is usually detectable after subtracting the background.

H. Chlorofluoroacetones

1. General formula

$R_1C(O)R_2$

2. Molecular ion

The $M^{+\bullet}$ peak for chlorofluoroacetones are usually not observed but can be deduced by adding 35 to the highest *m/z* value peak observed in the spectrum. Cleavage on either side of the carbonyl group defines R_1 and R_2. By deducing the molecular weight and looking for the R_1 and R_2 groups, the particular chlorofluoroacetone can be identified. The chlorofluoroacetones and their molecular weights are given in the following list:

Compound	MW
$CF_3C(O)CF_2Cl$	182
$CF_3C(O)CFCl_2$	198
$CF_2ClC(O)CF_2Cl$	198
$CF_3C(O)CCl_3$	214
$CF_2ClC(O)CFCl_2$	214
$CFCl_2C(O)CFCl_2$	230
$CF_2ClC(O)CCl_3$	230
$CFCl_2C(O)CCl_3$	246

CHAPTER 18

Gases

18.1. GC Separations

A. Capillary columns

1. Oxygen, nitrogen, methane, and carbon monoxide
   25-m Molecular Sieve 5A Plot column at room temperature.
   (Note: This column separates oxygen and nitrogen better than the Molecular Sieve 13X column; neither column separates argon and oxygen.)
2. Nitrous oxide, carbon dioxide, and nitric oxide
   25-m GS-Q column or 25-m Poraplot Q column at room temperature.
3. Air, carbon monoxide, methane, and carbon dioxide
   25-m Carboplot OO7 column at 60 °C.
4. Carbon dioxide, carbonyl sulfide, hydrogen cyanide, propylene, and butadiene
   25-m GS-Q column or 25-m Poraplot Q column, 60–200 °C at 6 °C min^{-1}.
5. Methane, carbon dioxide, ethylene, propylene, and propane
   25-m Poraplot R column, 30–100 °C at 5 °C min^{-1}.
6. Methane, ethane, ethylene, propane, propylene, acetylene, isobutane, and *n*-butane
   25-m Poraplot Alumina column, 35–140 °C at 6 °C min^{-1}.
7. Air, hydrogen sulfide, carbonyl sulfide, sulfur dioxide, methyl mercaptan, and carbon disulfide
   25-m GS-Q column or 25-m Poraplot Q column, 60–200 °C at 8 °C min^{-1}.
8. CHF_3, CH_2F_2, CH_3CF_3, CHF_2Cl, CF_2Cl_2, CH_2FCl, CHF_2CH_2Cl, and $CHFCl_2$
   25-m Poraplot Q column, 60–150 °C at 5 °C min^{-1}.

18.2. General Information

A gas analysis usually involves some or all of the common gases: O_2, N_2, CO, CO_2, C_1–C_5 hydrocarbons, low-boiling fluorinated compounds,

Gas Chromatography and Mass Spectrometry
DOI: 10.1016/B978-0-12-373628-4.00018-6

sulfur compounds, and so forth. Separating or passing the eluate from the GC column through a sensitive thermal conductivity detector (TCD) before entering the mass spectrometer enables the qualitative and quantitative analyses of unknown gas mixtures from parts per million to percentage levels. If elaborate column switching systems are not available, two GC/MS analyses may be required on two different GC columns. For instance, CO_2 and the C_2–C_5 hydrocarbons are adsorbed on the Molecular Sieve 5A column while separating H_2, O_2, N_2, CH_4, and CO. A second GC/MS analysis is performed using GS-Q or Poraplot Q, which separates CO_2 and the C_2–C_5 hydrocarbons from composite peaks of H_2, O_2, N_2, CH_4, and CO. By performing two GC/MS analyses on two different columns, a complete gas analysis can be achieved. Remember that if a sufficient number of analyses are made, the Molecular Sieve 5A column will have to be replaced.

CHAPTER 19

Glycols

19.1. GC Separations

A. Underivatized glycols
   1. Capillary columns
      a. Propylene glycol, ethylene glycol, dipropylene glycol, diethylene glycol, triethylene glycol, and tetraethylene glycol
         30-m NUKOL column, 60–220 °C at 10 °C min^{-1}.
      b. Ethylene glycol, diethylene glycol, triethylene glycol, and tetraethylene glycol
         30-m DB-5 column, 100 (2 minutes)–200 °C at 10 °C min^{-1}.
      c. 2,3-Butanediol, 1,2-butanediol, 1,3-butanediol, and 1,4-butanediol
         30-m DB-WAX column, 50 (5 minutes)–150 °C at 15 °C min^{-1}.
      d. Isobutylene glycol, propylene glycol, dipropylene glycol, and tripropylene glycol
         30-m DB-WAX column, 180–195 °C at 2 °C min^{-1}.

19.2. Derivatization of Dry Glycols and Glycol Ethers

A. TMS derivatives
   Add 0.25 mL of TRI-Sil Z reagent to 1–2 mg of sample. Stopper and heat the mixture at 60 °C for 5–10 minutes.
B. GC separation of TMS derivatives of glycols and glycol ethers*
   1. Ethylene glycol–TMS, diethylene glycol–TMS, and glycerol–TMS
      30-m DB-1 column or 30-m DB-5 column, 60–175 °C at 4 °C min^{-1}.

*The TMS derivatives cannot be injected onto a DB-WAX column.

Gas Chromatography and Mass Spectrometry
DOI: 10.1016/B978-0-12-373628-4.00019-8

19.3. Mass Spectral Interpretation

A. Glycols and glycol ethers (underivatized)

The most intense peak observed in the mass spectra of di- and triethylene glycol is at *m/z* 45. A peak at *m/z* 31 is also expected because oxygen is present. The ion represented by the peak at *m/z* 45 is formed via heterolytic cleavage of molecular ions with the charge and radical sites on an ether–oxygen atom. The peak at *m/z* 31 represents an ion formed by a hydride-shift rearrangement fragmentation of an ion initially formed by homolytic cleavage of molecular ions with the charge and radical sites on an ether–oxygen atom (Figure 19.1). (See Chapter 16 for a detailed explanation of these ion formations.)

B. TMS derivatized glycols

The best way to identify the MWs of unknown glycols and glycol ethers is to examine the mass spectra of the TMS derivatives. The locations of the TMS derivatives are identified by plotting a mass chromatogram for *m/z* 73 and 147. The higher of the two highest *m/z* value peaks in the mass spectrum, which are 15 *m/z* units apart, is the $M^{+\bullet}$ peak of the TMS derivative. If no peaks are separated by 15 *m/z* units, add 15 to the highest *m/z* value peak observed in the mass spectrum to deduce the TMS derivative's MW.

C. Sample mass spectra

The mass spectrum in Figure 19.2 shows the presence of both *m/z* 31 and *m/z* 45 peaks, indicating that oxygen is present in the analyte. The peak at *m/z* 33 is characteristic of some hydroxy compounds and has been suggested by McLafferty [1] to occur by β-bond fragmentation accompanied by the rearrangement of two hydrogen atoms. The peak

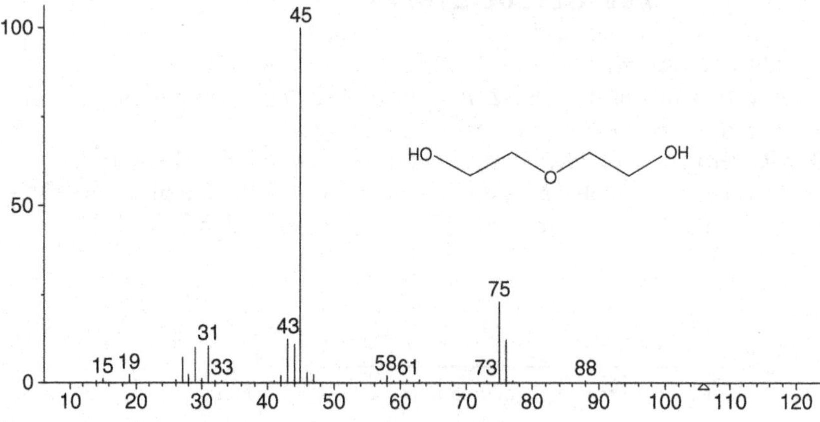

Figure 19.1 EI mass spectrum of diethylene glycol.

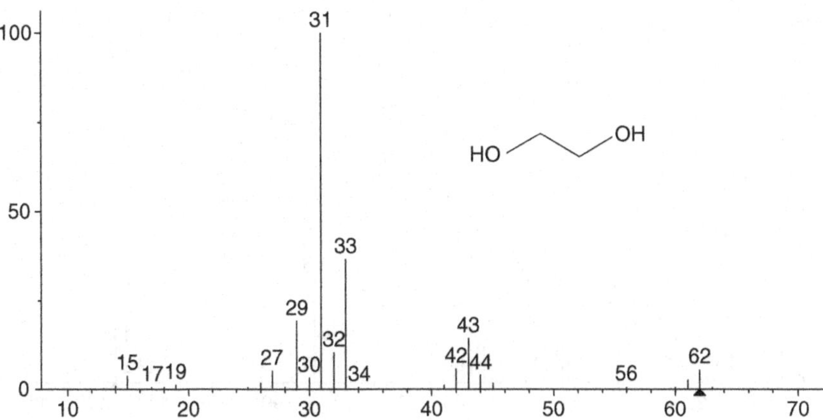

Figure 19.2 EI mass spectrum of 1,2-ethanediol.

at *m/z* 62 could represent the molecular ion of ethylene glycol. This can be confirmed by preparing the TMS derivative, which will show the presence of two hydroxy functions and confirm the MW assignment.

Figure 19.3 is the mass spectrum of propylene glycol and exhibits the presence of an abundant *m/z* 45 ion. A library search will provide strong evidence that this compound is propylene glycol. Preparation of a TMS derivative will confirm this assignment.

Figure 19.4 is the mass spectrum of 1,2-cyclohexanediol and shows an intense $M^{+\bullet}$ peak at *m/z* 116. The peaks at *m/z* 31 and 45 suggest the presence of oxygen in the analyte, and the strong fragment ion peak at *m/z* 98 ($[M - 18]^+$) is characteristic of many diols.

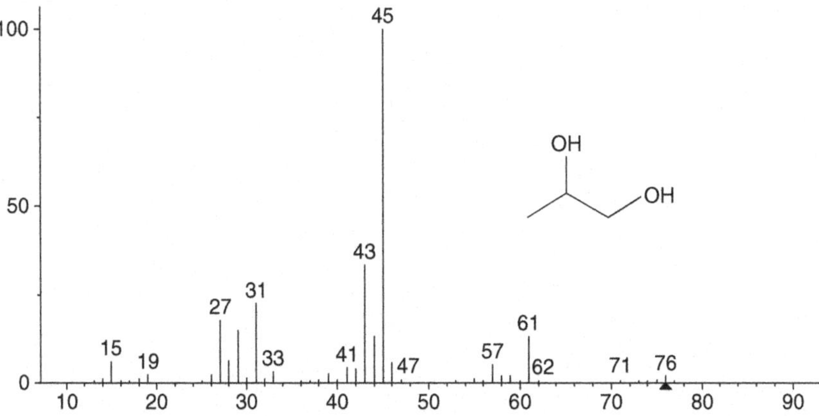

Figure 19.3 EI mass spectrum of propylene glycol.

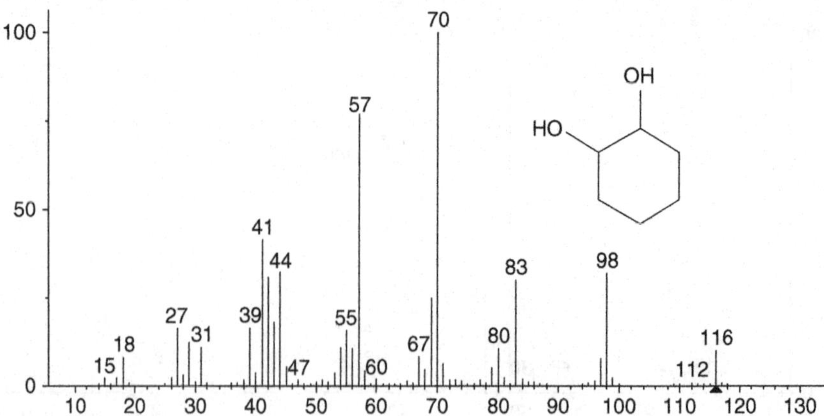

Figure 19.4 EI mass spectrum of 1,2 cyclohexanediol.

REFERENCES

1. McLafferty, F. W. (1963). *Mass Spectrsometry of Organic Ions*. New York: Academic Press.

CHAPTER 20

Halogenated Compounds (Other Than Fluorinated Compounds)

20.1. GC Separations

A. Saturated and unsaturated halogenated compounds
   1. Capillary columns
      a. Methyl chloride, vinyl chloride, methyl bromide, ethyl chloride, dichloroethanes, chloroform, carbon tetrachloride, bromochloromethane, and similar compounds
         30-m DB-624 column 35 (5 minutes)–140 °C at 5 °C min^{-1} or 25-m CP-Sil 13CB column for halocarbons, 35 (5 minutes)–80 °C at 2 °C min^{-1}.
      b. Most dichlorobutenes and trichlorobutenes
         30-m DB-WAX column, 70–200 °C at 4 °C min^{-1}.

B. Halogenated aromatics
   1. Capillary columns
      a. Chlorobenzenes
         1. 50-m DB-WAX column, 100 (8 minutes)–200 °C at 8 °C min^{-1}.
         2. *o*-, *m*-, and *p*-isomers
            50-m DB-WAX column at 125 °C.
         3. 1,3-Dichlorobenzene, 1,4-dichlorobenzene, 1,2-dichlorobenzene, 1,2,4-trichlorobenzene, and hexachlorobenzene
            30-m DB-1301 column, 75–210 °C at 10 °C min^{-1}.
      b. Chlorotoluenes
         1. Isomers of chloromethylbenzenes
            30-m CP-Sil 88 column, 50 (6 minutes)–200 °C at 4 °C min^{-1}.
      c. Chlorinated biphenyls

Molecular formula[a]	Accurate mass value
$C_{12}H_9Cl$	188.0393
$C_{12}H_8Cl_2$	222.0003
$C_{12}H_7Cl_3$	255.9613
$C_{12}H_6Cl_4$	289.9224

(*Continued*)

Gas Chromatography and Mass Spectrometry
DOI: 10.1016/B978-0-12-373628-4.00020-4

(*Continued*)

Molecular formula[a]	Accurate mass value
$C_{12}H_5Cl_5$	323.8834
$C_{12}H_4Cl_6$	357.8444
$C_{12}H_3Cl_7$	391.8054
$C_{12}H_2Cl_8$	425.7665
$C_{12}H_1Cl_9$	459.7275
$C_{12}Cl_{10}$	493.6885

[a] Not all of these isomers are completely separated by the conditions given here, but all are readily detected by plotting the masses of the molecular ions. Pesticides can interfere with polychlorinated biphenyl (PCB) analyses.

30-m DB-17 column, 150 (4 minutes)–260 °C at 4 °C min^{-1} or 30-m DB-1 column, 60 (4 minutes)–260 °C at 20 °C min^{-1}.

2. Arochlors 1016, 1232, 1248, and 1260. Arochlors have been analyzed under these GC conditions at the 50-ppm level using electron impact ionization (EI). At lower concentrations of polychlorinated biphenyls (PCBs), negative chemical ionization (CI) should be considered.
   50-m CP-Sil-88 column, 150 (4 minutes)–225 °C at 4 °C min^{-1}.

d. Chloronitrotoluenes and dichloronitrotoluenes
   30-m DB-17 column, 75–250 °C at 6 °C min^{-1}.

e. Halogenated toluene diisocyanates: chloro-, bromo-, dichloro-, and trichlorotoluene diisocyanates
   30-m DB-1 column, 70–225 °C at 4 °C min^{-1}.

f. Haloethers: bis(2-chloroethyl) ether, bis(2-chloroisopropyl) ether, and bis(2-chloroethoxy) methane
   30-m DB-1301 column, 100 (5 minutes)–130 °C at 10 °C min^{-1}, 130–250 °C at 15 °C min^{-1}.

g. Halogenated pesticides: lindane, heptachlor, aldrin, chlordane, dieldrin, DDT, and similar compounds. (See Chapter 28.)
   30-m DB-5 column, 60–300 °C at 4 °C min^{-1}. (Run for 52 °C min^{-1}.)

20.2. Mass Spectra of Halogenated Compounds (Other Than Fluorinated Compounds)

A. Aliphatic halogenated compounds

The presence of chlorine and/or bromine is easily detected by their characteristic isotopic peak patterns (see Appendix E). As in many aliphatic compounds, the intensity of the $M^{+\bullet}$ peak decreases as the size of the R group increases. For example, in the EI mass spectra of methyl chloride and ethyl chloride, the $M^{+\bullet}$ peak intensities are high, whereas in compounds with larger R groups such as butyl chloride, the molecular ion peak is relatively low or nonexistent.

The highest *m/z* value peaks observed in the mass spectra of alkyl chlorides may represent the loss of HX or X (loss of HI is seldom observed), depending on the structure of the molecule. In order to deduce the molecular ion, add the mass of X or HX to the mass at which the highest *m/z* value peak is readily observed. (Note that higher *m/z* value ions having the isotope pattern of X may be present in low abundances.) Try to select the structural type by examining characteristic low-mass fragment ions. For example, if an intense peak at *m/z* 49 is observed accompanied by a peak at *m/z* 51 that is approximately one-third its intensity of the *m/z* 49 peak, this may indicate that a terminal chlorine is present in a alkyl halide. An intense *m/z* 91 peak (with the isotope pattern for an ion with a single atom of chlorine) represents a C_6–C_{18} 1-alkyl chloride. An intense peak at *m/z* 135 (with the isotope pattern for bromine) suggests a C_6–C_{18} 1-alkyl bromide. Long-chain saturated halides may also lose an alkyl radical from the molecular ion, such as losses of 15 Da, 29 Da, 43 Da, 57 Da, 71 Da, 85 Da, and 99 Da. They can be identified as halogenated compounds, but it is difficult to deduce their molecular weights without CI or electron capture negative ionization (ECNI).

B. Sample mass spectrum of an aliphatic halogenated compound

It is obvious from the pairs of peaks at *m/z* 91 and 93 and *m/z* 105 and 107 (where the intensity of the second peak in the pair is ~33% of that of the first peak) that Figure 20.1 is a mass spectrum of a compound

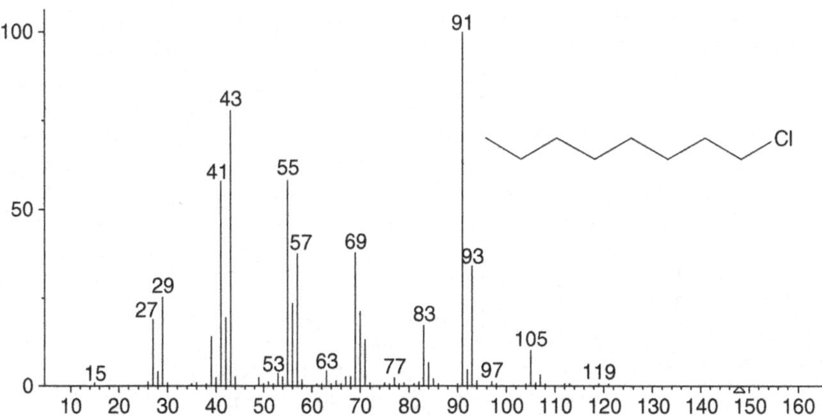

Figure 20.1 EI mass spectrum of 1-chlorooctane. The MW is 148 Da; and there is no $M^{+\bullet}$ peak, as indicated by the clear triangle.

containing at least one atom of chlorine. Although intense *m/z* 91 peaks often indicate an aromatic compound (as does a peak at *m/z* 105), the lack of an $M^{+\bullet}$ peak (all labeled peaks have an odd *m/z* value and there is no high *m/z* value peak with an even value) and the fact that there is extensive fragmentation, along with the presence of peaks two *m/z* units apart with the higher *m/z* value peak having an intensity about one-third the intensity of the higher *m/z* value peak, indicates otherwise. Unfortunately, it is not possible to know much more about the compound other than it is a 1-alkyl chloride due to the lack of an $M^{+\bullet}$ peak or any way to deduce the compound's molecular weight.

1-Alkyl halides are more easily identified than the dichloro halides. Figure 20.2 is the mass spectra of two dichlorohexanes. A very skilled interpreter of mass spectra may identify the pairs of peaks at *m/z* 125

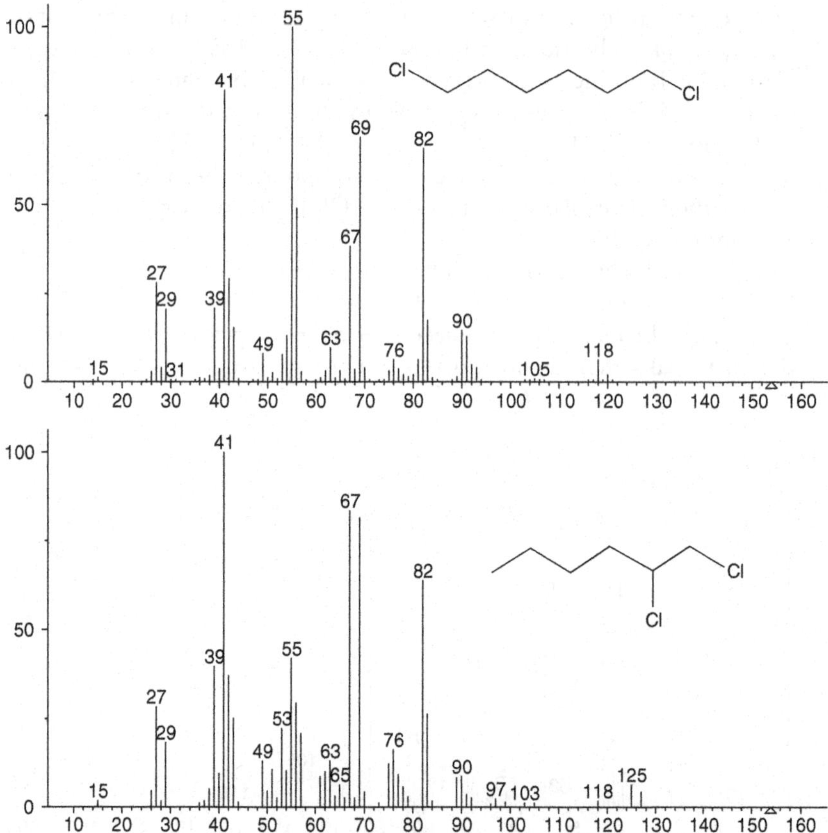

Figure 20.2 EI mass spectra of 1,6-dichlorohexane (top) and 1,2-dichlorohexane (bottom).

and 127, *m/z* 91 and 93, and *m/z* 90 and 92 in the mass spectrum of 1,2-dichlorohexane as representing ions containing an atom of chlorine or the pairs of peaks at *m/z* 118 and 120, *m/z* 91 and 93, and *m/z* 90 and 92 in the mass spectrum of 1,6-dichlorohexane as representing ions with a single atom of chlorine. A computerized library search of the two spectra would lead to a conclusion that chlorine was present, although the number of atoms might still be in question. When either of these spectra is searched against the *NIST08 Mass Spectral Database* using the *NIST MS Search Program*, hits for both compounds are present with high Match factors. This clearly shows the challenges in identifying these types of compounds based on their mass spectra alone. When the spectral data are acquired using GC/MS, the GC retention index can be used as a further indicator of identification.

C. Aromatic halogenated compounds

The $M^{+\bullet}$ peaks in the mass spectra of aromatic halogenated compounds are fairly intense. The intensity if the $M^{+\bullet}$ peak decreases with increasing length of the side chain. If an intense peak at *m/z* 91 (tropylium ion) is present, then the halogen is on the alkyl side chain rather than the ring. The loss of the halogen from the molecular ion will result in a low-intensity peak. If the halogen is on the ring, the loss of HX is favored when a CH_3 group is ortho to the halogen. This is called the ortho effect. Because of the ortho effect (see Chapter 29), it is often possible to discriminate between ortho-substitutions and meta- or para-substitutions. In the top spectrum of Figure 20.3 (the ortho-isomer of chlorotoluene), a peak is observed at *m/z* 90 (~10%), which is missing from the bottom spectrum (the para-isomer).

It is possible to distinguish the mass spectrum of a chlorotoluene from that of benzyl chloride. Examination of both spectra in Figure 20.3 shows fairly intense $[M - H]^+$ peaks (*m/z* 125 and 127) relative to the intensity of the $M^{+\bullet}$ peaks (*m/z* 126 and 128). Examination of the mass spectrum of benzyl chloride (Figure 20.4) shows a much lower intensity for the peaks representing the loss of a $^{\bullet}H$.

Even though the most probable sight of the charge and radical will be associated with halogen due to the fact that the halogen has three pairs of nonbonding electrons, the fragmentation will be dominated by benzylic cleavage (Example 20.1) resulting in the tropylium ion. The loss of $-(CH_2)_nX$ is observed if the group on the ring is $-CH_2(CH_2)_nX$. If the size of the alkyl chain with the Cl atom attached at the opposite end from the phenyl moiety is $C_{\geq 3}$, a reasonably intense peak will be at *m/z* 92 (Example 20.2) due to a β cleavage resulting from a γ-hydrogen shift (a McLafferty rearrangement fragmentation).

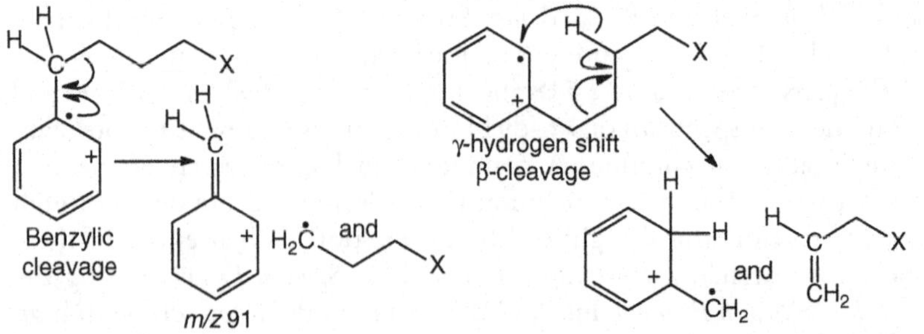

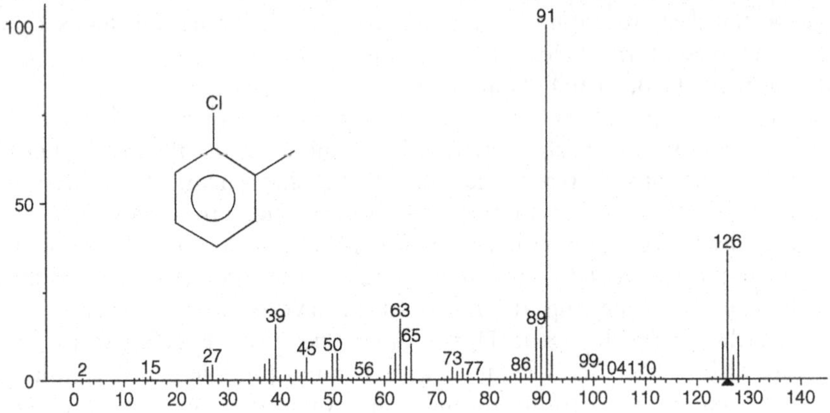

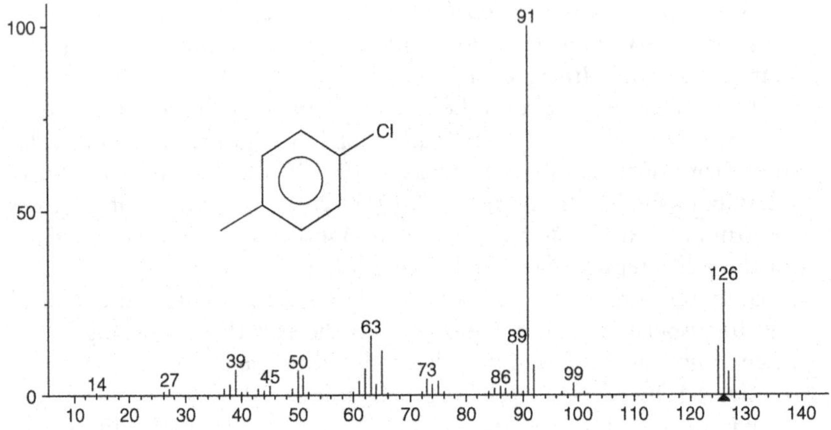

Figure 20.3 EI mass spectra of 1,2-dichlorobenzene.

D. Sample mass spectrum of an aromatic halogenated compound
Figure 20.5 is an example of a mass spectrum of an aromatic dichlorocompound. The intensity of the $M^{+\bullet}$ peak indicates that an aromatic compound is present. The isotope peak pattern is that of an ion

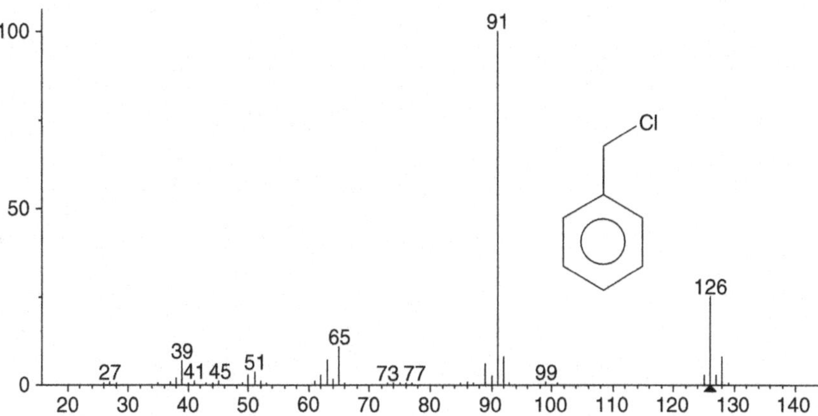

Figure 20.4 EI mass spectrum of benzyl chloride.

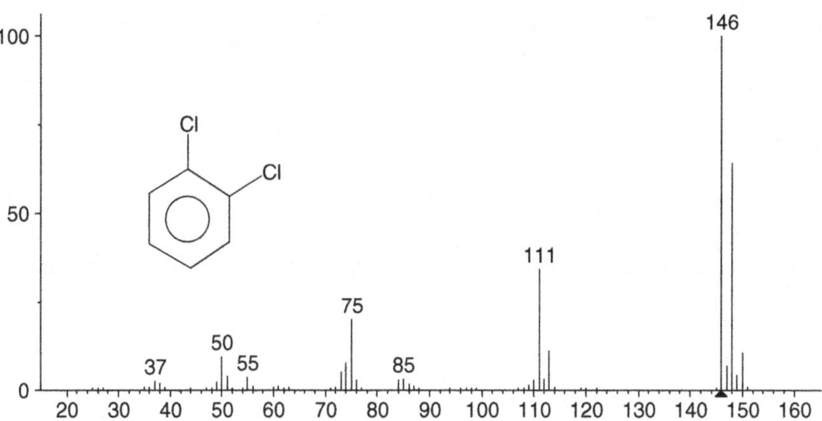

Figure 20.5 EI mass spectrum of *o*-dichlorobenzene.

with two atoms of chlorine; subtracting 70 mass units from the *m/z* value of the $M^{+\bullet}$ peak gives 76, which can be reconciled with the formula C_6H_4. It should be noted that no ortho effect is observed in the mass spectrum of *o*-dichlorobenzene. There is no difference in any of the mass spectra of the three regioisomers of dichlorobenzene.

CHAPTER 21

Hydrocarbons

21.1. GC Separation of Hydrocarbons

A. Saturated and unsaturated aliphatic hydrocarbon compounds
   1. C_1–C_5 hydrocarbon compounds
      a. 50-m Al_2O_3/KCl Plot column,
         40 (1 minute)–200 °C at 10 °C min^{-1} or 60–200 °C at 3 °C min^{-1}.
      b. 30- to 50-m GS-Q column or Poraplot Q column,
         35–100 °C at 10 °C min^{-1}, 100 °C (5 minutes); 100–200 °C at 10 °C min^{-1}; or 40 (2 minutes)–115 °C at 10 °C min^{-1}.
   2. C_4 hydrocarbon isomers
      2-m Picric acid column on Carbopak C, 30–100 °C at 2 °C min^{-1}.
   3. C_4–C_{12} hydrocarbon compounds
      100-m SQUALANE column, 40–70 °C at 0.5 °C min^{-1}.
   4. Natural gas
      30-m GS-Q column at 75 °C.
   5. C_5–C_{100} hydrocarbon compounds
      30-m SIM DIST-CB Chrompack column, 100–325 °C at 10 °C min^{-1}; injection port at 325 °C.

B. Low-boiling aromatic hydrocarbon compounds
   1. Benzene, toluene, ethylbenzene, *p*-xylene, *m*-xylene, *o*-xylene
      30-m CP CW57 CB column, 50–200 °C at 5 °C min^{-1}.
   2. Benzene, toluene, ethylbenzene, *p*-xylene, *m*-xylene, *o*-xylene, butylbenzene, styrene, *o*-diethylbenzene, *m*-diethylbenzene, and *p*-diethylbenzene
      25-m DB-WAX column, 50 (2 minutes)–180 °C at 5 °C min^{-1}.

C. Polynuclear aromatic hydrocarbon compounds
   1. Naphthalene, acenaphthane, acenaphthene, fluorene, phenanthrene,* anthracene,* fluoranthene, pyrene, benzanthracene, benzophenanthrene, and benzpyrene
      30-m DB-1 column, 120 (6 minutes)–275 °C at 10 °C min^{-1}.
   2. Phenanthrene and anthracene and other polynuclear aromatic compounds
      30-m DB-1301 column, 100–260 °C at 5 °C min^{-1}.

---

* Not separated using the DB-1 column

Gas Chromatography and Mass Spectrometry
DOI: 10.1016/B978-0-12-373628-4.00021-6

D. Gasoline

Gasoline contains over 250 components of a mixture of C_4–C_{12} hydrocarbons, which vary in concentration from batch to batch. Some of these components are isobutane, *n*-butane, isopentane, *n*-pentane, 2,3-dimethylbutane, 3-methylpentane, *n*-hexane, 2,4-dimethylpentane, benzene, 2-methylhexane, 3-methylhexane, 2,2,4-trimethylpentane, 2,3,4-trimethylpentane, 2,5-dimethylhexane, 2,4-dimethylhexane, toluene, 2,3-dimethylhexane, ethylbenzene, methylethylbenzenes, *m*-xylene, *p*-xylene, *o*-xylene, trimethylbenzenes, naphthalene, methylnaphthalenes, and dimethylnaphthalenes.

1. 30-m DB-1701 column, 35 (10 minutes)–180 °C at 6 °C min^{-1}.
2. 30-m DB-1 column, 35 (10 minutes)–200 °C at 6 °C min^{-1}.

21.2. Mass Spectra of Hydrocarbon Compounds

A. *n*-Alkanes

The intensity of the $M^{+\bullet}$ peaks decreases with increasing chain length but is still detectable at C_{40}. In contrast to methyl-substituted branched alkanes, the loss of a methyl group is not favored for *n*-alkanes. This is because the methyl radical is not very stable. Usually the first fragment ion peak below the $M^{+\bullet}$ peak represents an $[M-29]^+$ ion (loss of a $^\bullet CH_2CH_3$). Compounds of $C_{\geq 4}$ show a base peak at *m/z* 43 or 57. Alkanes yield a series of ions represented by peaks differing by 14 *m/z* units (e.g., 43, 57, 71, 85, etc.). It is important to note that the successively smaller *m/z* value peaks represent ions that are formed by successively larger alkyl radicals; not be the sequential loss of a $-CH_2-$ unit from the ion with a 14 *m/z* unit previously higher value. Straight-chain aliphatic hydrocarbon ions cannot lose a $-CH_2-$ unit, which would be a diradical. It should also be noted that fragment ions of the form $^+C_{m+2m+1}$ where *m* is >½ the total number of carbon atoms in the original molecule are formed exclusively by fragmentation of the molecular ion. The fragment ions having *m* values of <½ the total number of carbon atoms in the original molecule are formed by fragmentation of the molecular ion and by secondary fragmentations of ions with higher *m/z* values. This is partially the reason for the high abundance of C_3, C_4, C_5, and C_6 ions.

B. Sample mass spectrum of an alkane

Figure 21.1 is an example of a C_{14} aliphatic hydrocarbon. The $M^{+\bullet}$ peak is observed along with peaks representing the $[M-29]^+$ fragment ion and the low mass ions with *m/z* 43, 57, and so forth separated by 14 *m/z* units.

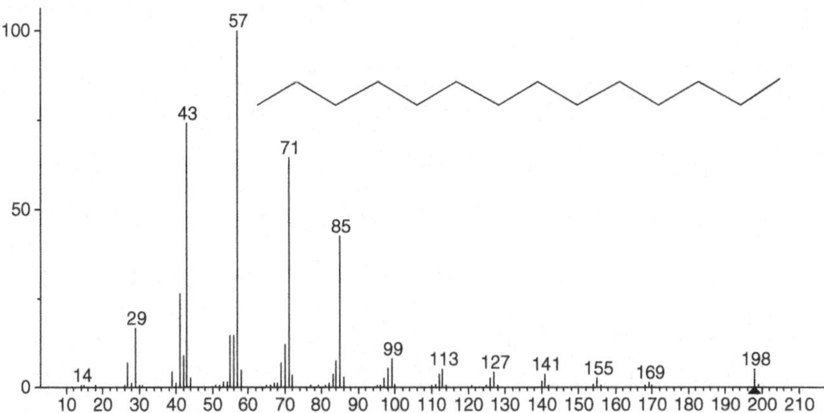

Figure 21.1 EI mass spectrum of *n*-tetradecane.

C. Branched alkanes
   The intensity of the $M^{+\bullet}$ peak decreases with increased branching; therefore, the $M^{+\bullet}$ peak may be nonexistent. The loss of 15 Da from the molecular ion indicates a methyl side chain. The mass spectra of branched alkanes are dominated by the tendency for fragmentation at the branch points, and hence the location of the branch point can be determined based on the *m/z* values of the three peaks representing the more stable secondary carbenium ions (see Watson and Sparkman) [1].
D. Sample mass spectrum of a branched alkane
   Figure 21.2 is the mass spectrum of methyl-substituted straight-chain alkane. The peaks representing the $[M - CH_3]$ ion are obvious at *m/z* 407. Even though methyl radicals are not very stable and their loss from hydrocarbon chains is not favored, when a methyl group is a *special methyl*, it will be lost; and being a substituent on a chain makes the methyl moiety in this case a *special methyl.* The peaks at *m/z* 197 and 251 represent secondary carbenium ions that retain the methyl moiety. Unlike the other peaks in the mass spectrum, these peaks do not conform to a continuing decrease in intensity with increasing *m/z* values. The high intensity of the peaks 1 *m/z* unit lower is due to the propensity for the secondary carbenium ions to lose a hydrogen radical. The peaks at *m/z* 169 and 224 represent primary carbenium ions formed by cleavage of the bond between the secondary carbon atom of the branch point and the first carbon atom of either of the two long chains attached to this secondary carbon atom.

   Figure 21.3 is the mass spectrum of another branched hydrocarbon. Note the absence of the $M^{+\bullet}$ peak, which is not uncommon for long-chain branched aliphatic compounds. The spectrum exhibits the continuous

Figure 21.2 EI mass spectrum of 13-methylnonacosane.

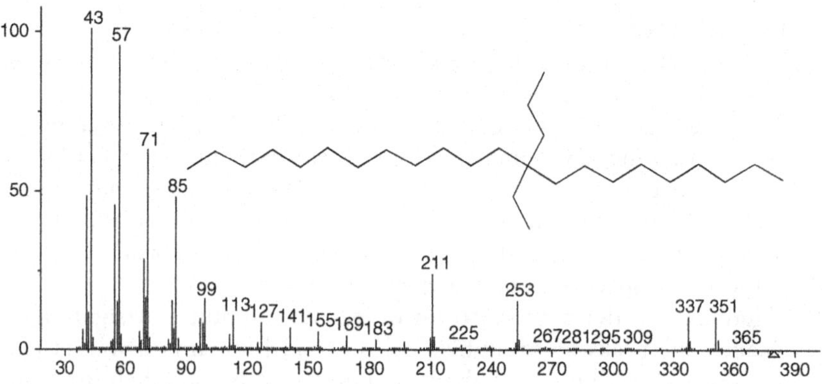

Figure 21.3 EI mass spectrum of 10-ethyl-10-propyldocosane.

periodicity of peaks of decreasing intensity every 14 *m/z* units from left to right toward the position of where the $M^{+\bullet}$ peak would be if it were present. This rhythm is broken by four peaks that have what appear to be abnormally high intensities (*m/z* 211, 253, 337, and 351). These peaks reflect that there are two branches on an aliphatic hydrocarbon backbone, and they are attached to the same carbon atom. The intensities of these peaks are high because fragmentation of any of the carbon–carbon bonds associated with the tertiary carbon at the branch point will lead to very stable tertiary carbenium ions. These peaks are not accompanied by peaks of high intensities that are 1 *m/z* unit lower because the available hydrogen moiety associated with secondary carbenium ions is not present on tertiary carbenium ions.

E. Cycloalkanes

The intensity of the $M^{+\bullet}$ peak is greater in the mass spectra of cycloalkanes than it is in the spectra of straight-chain alkanes containing the same number of carbon atoms. Fragmentation of the ring is usually characterized by the loss of 28 *m/z* unit and 29 *m/z* unit. The tendency to lose alkenes, such as $H_2C{=}CH_2$, produces mass spectra that contain a greater number of ions with even *m/z* values than are seen in the mass spectra of straight- or branched-chain hydrocarbons. A saturated ring with an aliphatic side chain favors cleavage at the bond connecting the ring to the rest of the molecule. Compounds containing cyclohexyl rings fragment forming ions with *m/z* 83, 82, and 81, corresponding to ring fragmentation and the loss of one and two hydrogen atoms (Figure 21.4).

F. Alkenes

$M^{+\bullet}$ peaks are usually intense for low-molecular-weight compounds. Alkyl cleavage with the charge remaining on the unsaturated portion is very often the base peak. A series of fragment ions with *m/z* 41, 55, 69, 83, and so forth are characteristic. Methods are available to locate the position of the double bond in aliphatic compounds [2]. In the case of a 1-alkene, successively larger olefin molecules will be lost from the molecular ion to form even-electron fragment ions that are represented by peaks at even *m/z* values (Figure 21.5).

G. Alkylbenzenes

Alkylbenzenes have $M^{+\bullet}$ peaks at the following *m/z* values: 92, 106, 120, 148, and so forth. The $M^{+\bullet}$ peak intensity decreases with increasing alkyl chain length but can be detected up to at least C_{16}. Characteristic fragment ions are observed at *m/z* 39, 51, 65, 77, and 91.

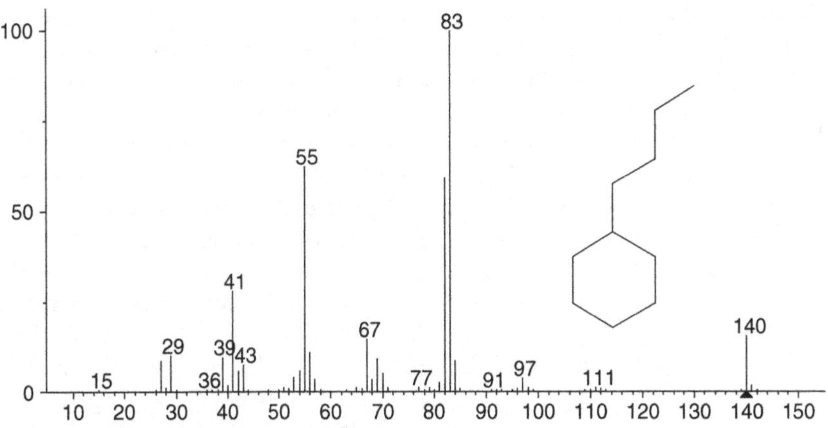

Figure 21.4 EI mass spectrum of *n*-butylcyclohexane.

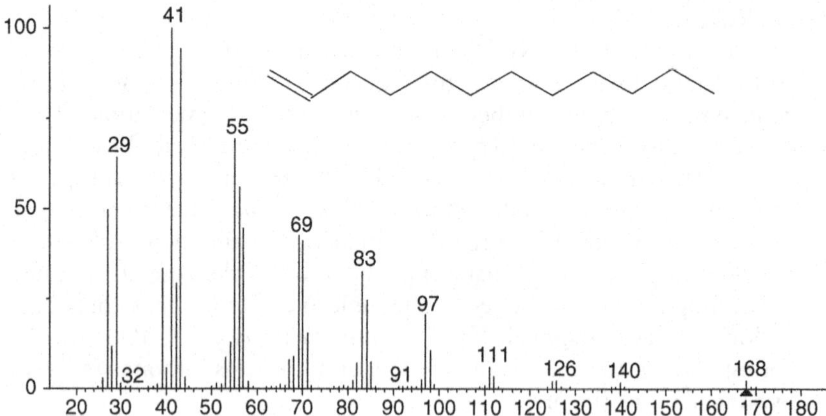

Figure 21.5 EI mass spectrum of 1-dodecene.

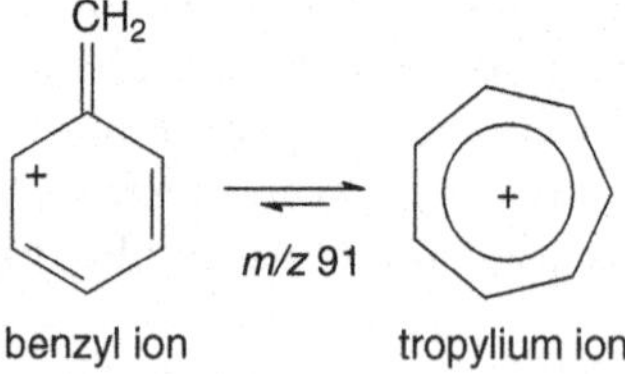

Benzylic cleavage is the dominant driving force for the fragmentation of alkylbenzenes. With the charge and radical sites associated with the aromatic ring, a bond (other than the bond to the ring) on the benzylic carbon atom will cleave due to the pairing of an electron in that bond with the ring radical site. The resulting benzyl ion then forms the much more stable tropylium ion. Dialkyl-substituted benzenes do not exhibit an ortho effect (Chapter 29). The larger of the two alkyl groups is lost in the formation of a substituted tropylium ion (Figure 21.6). An excellent method for identifying alkylbenzenes was developed by Meyerson [3] and should be consulted.

Regioisomers of dialkyl-substituted benzenes such a xylene cannot be differentiated by mass spectrometry. Their EI mass spectra are essentially identical. This mass spectrum is somewhat unique and can be identified as that of a xylene; the three regioisomers are then differentiated by their retention indices. Ethylbenzene has a very similar mass spectrum to that of the xylenes. The single difference is that the mass spectra of the xylenes exhibit an $[M - 1]^+$ peak, which does not have a significant intensity in the mass spectrum of ethylbenzene. The loss of a methyl radical as opposed to the loss of a hydrogen radical from the benzylic

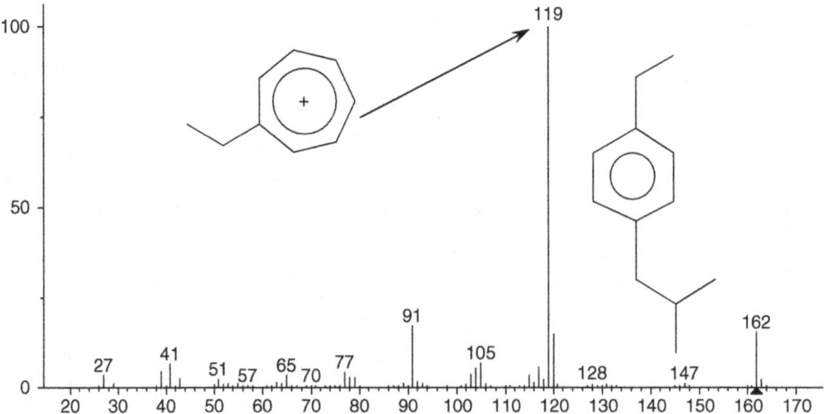

Figure 21.6 EI mass spectrum of 1-ethyl-4-(2-methylpropyl)benzene. The peak at *m/z* 119 represents an ethyl-substituted tropylium ion due to the loss of the larger isopropyl moiety.

carbon of the molecular ion is favored in the fragmentation of ethylbenzene; therefore, the lack of an $[M - 1]^+$ peak. In the molecular ion fragmentation of xylene, benzylic cleavage will only result in the loss of a hydrogen radical, because the only substitution on the two benzylic carbon atoms is hydrogen; thus, the $[M - 1]^+$ peak.

Benzylic Carbon Atoms

H_3C CH_3 H C—H CH_3

Xylene Ethylbenzne

H. Polynuclear Aromatic Hydrocarbons

Unsubstituted polynuclear aromatic hydrocarbons show intense $M^{+\bullet}$ peaks. The mass spectra of these compounds will also exhibit peaks representing double-charge ions in the mass spectrum of chrysene (Figure 21.7) where there are peaks at *m/z* 113 and *m/z* 114. These peaks represent ions that have a mass of 226 Da and 228 Da, respectively, but that have two charge states. Peaks representing multiple-charge ions in EI mass spectra are rare, but they are found in the spectra of aromatic hydrocarbons. The peak at *m/z* 39 in the mass spectrum of benzene represents a molecular ion with two charges, not a fragment ion.

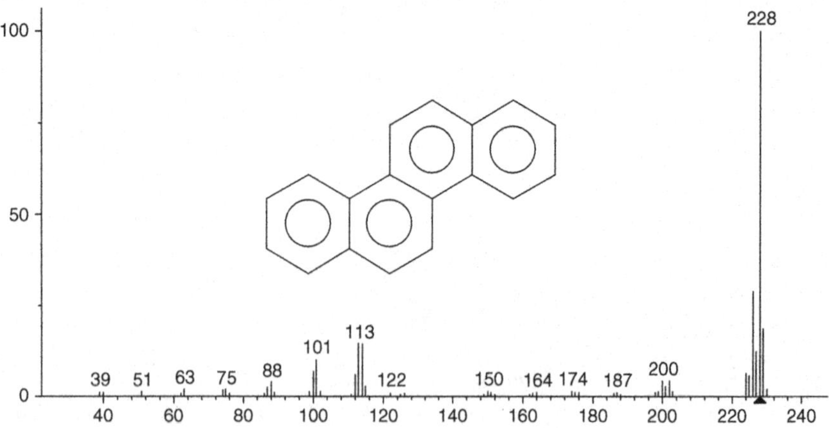

Figure 21.7 EI mass spectrum of chrysene. The peaks at *m/z* 113 and 114 represent double-charge ions that have respective masses of 226 Da and 228 Da.

The aklylated polynuclear aromatics and the alkylated benzenes fragment similarly:

R
CH_2
Loss of•R
CH_2
m/z 141

Characteristic fragment ions of alkylnaphthalenes are *m/z* 141 and 115. Using the GC conditions previously outlined in this chapter, most of the molecular ions of the following compounds can be found by plotting the accompanying accurate mass values (listed in order of elution):

Compounds	Accurate *m/z* Value
Naphthalene	128.0626
Acenaphthene	154.0783
Fluorene	166.0783
Phenanthrene	178.0783
Anthracene	178.0783
Fluoranthene	202.0783
Pyrene	202.0783
Benzanthracene	228.0939
Chrysene	228.0939
Benzpyrene	252.0939

If all isomers are present, identification is straightforward; but if only one isomer is present, standards may have to be injected into the GC-MS to obtain retention times under the GC conditions used.

REFERENCES

1. Watson, J. T., Sparkman, O. D. (2007). *Introduction to Mass Spectrometry: Instrumentation, Applications and Strategies for Data Interpretation*. Chichester, U.K.: Wiley.
2. Schneider, B., Budzikiewicz, H. (1990). A facile method for the location of a double bond in aliphatic compounds. *Rapid Commun. Mass Spectrom.*, 4, 550.
3. Meyerson, S. (1955). Correlations of alkylbenzene structures with mass spectra. *Appl. Spectrosc.*, 9, 120.

CHAPTER 22

Isocyanates

22.1. GC Separations

A. Toluene diisocyanates (TDI), xylene isocyanates, chloro-TDI, bromo-TDI, dichloro-TDI, and trichloro-TDI: 30-m DB-1 column, 70–225 °C at 4 °C min^{-1}.
B. *m*-Phenylene-diisocyanate, toluenediisocyanate, xylenediisocyanate, butylated hydroxytoluene, 5-chlorotoluenediisocyanate, and methylene-*bis*-(4-cyclohexylisocyanate): 30-m DB-1 column, 100–300 °C at 8 °C min^{-1}.
C. Xylene diisocyanates: 30-m DB-1 column, 70–235 °C at 4 °C min^{-1}.

22.2. Mass Spectral Interpretation

A. Mass spectra of aliphatic isocyanates
   1. General formula: RNCO.
   2. Molecular ion: The $M^{+\bullet}$ peaks of aliphatic isocyanates are observed up about C_8.
   3. Fragmentation: Characteristic fragments of aliphatic isocyanates include the following: *m/z* 56 (CH_2NCO), *m/z* 70 (CH_2CH_2NCO), *m/z* 84 ($CH_2CH_2CH_2NCO$), and so forth. Even though the charge and radical are associated with the isocyanate, the radicals are lost from the other end of the molecular ion (Figure 22.1).
B. Mass spectra of aromatic isocyanates
   1. General formula: ArNCO.
   2. Molecular ion: The $M^{+\bullet}$ peaks of the aromatic isocyanates and diisocyanates are usually observed depending on the length of the alkyl groups on the ring. As seen in Figures 22.2–22.6, when the isocyanate moiety is attached to the aromatic ring, the intensity of the $M^{+\bullet}$ peak is very high. The isocyanate moiety is attached to a benzyl carbon; the intensity of the $M^{+\bullet}$ peak is respectable but not as great as when it is attached directly to the ring.
   3. Fragmentation: Losses from the molecular ion include 28 (CO), 29 (H + CO), and 55 (CO + HCN). Sometimes loss of hydrogen is observed, particularly if there is a methyl group on the ring. There

Gas Chromatography and Mass Spectrometry
DOI: 10.1016/B978-0-12-373628-4.00022-8

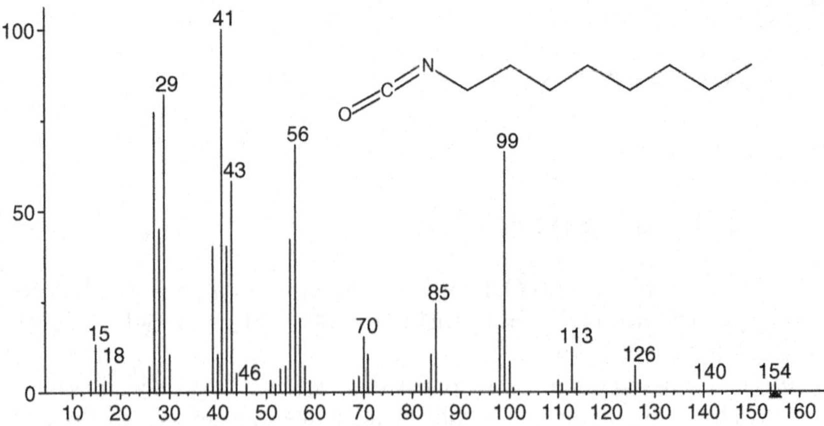

Figure 22.1 EI mass spectrum of 1-octylisocyanate.

does not appear to be an ortho effect for the fragmentation of alkyl-substituted aromatic isocyanates (Figures 22.5 and 22.6).

C. Sample mass spectra

Figures 22.2 and 22.3 are the mass spectra of aromatic isocyanates. Losses of 28 Da (CO) and 29 Da (H + CO) are evident in both spectra. The intensity of the peak at *m/z* 91 ($[M - NCO]^+$) in Figure 22.2 indicates the benzyl structure shown.

Figure 22.4 is an aromatic diisocyanate. The observed successive losses of 28 Da and 56 Da are similar to the losses found with quinone or anthraquinone.

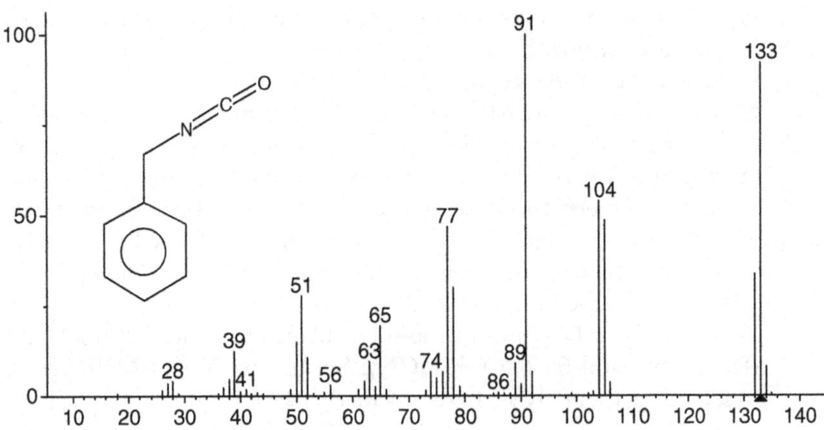

Figure 22.2 EI mass spectrum of benzyl isocyanate.

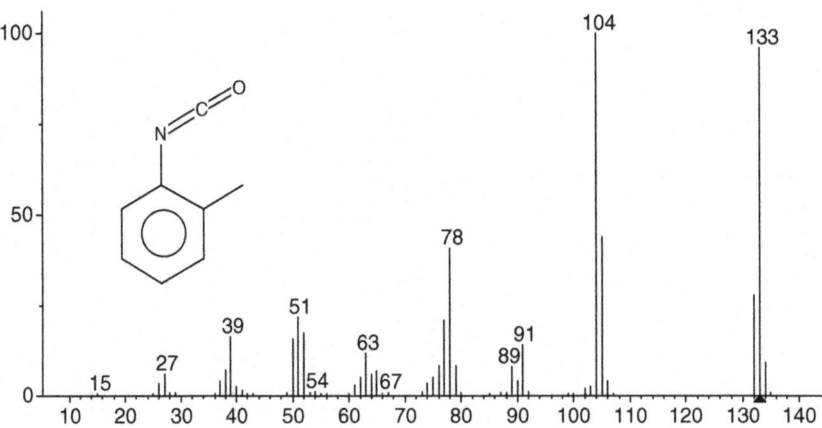

Figure 22.3 EI mass spectrum of 1-isocyanate-2-methyl benzene.

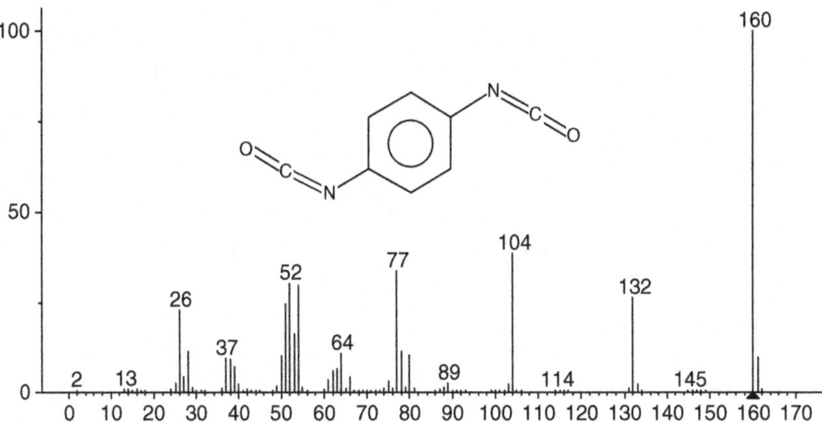

Figure 22.4 EI mass spectrum of 1,4-diisocyanate benzene.

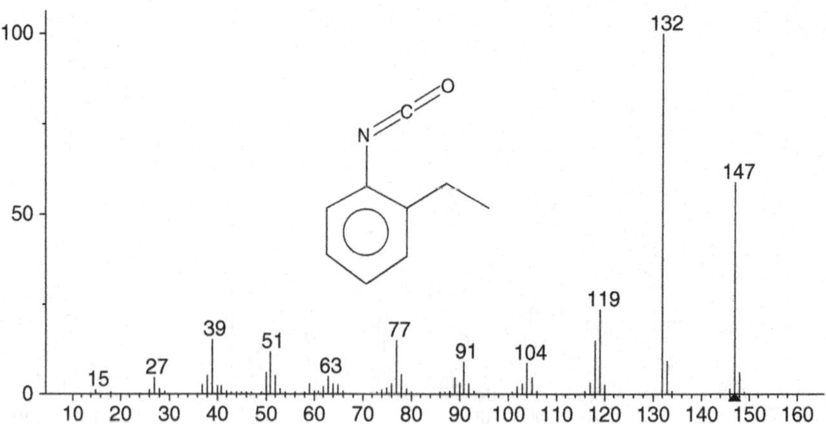

Figure 22.5 EI mass spectrum of *o*-ethylphenyl isocyanate.

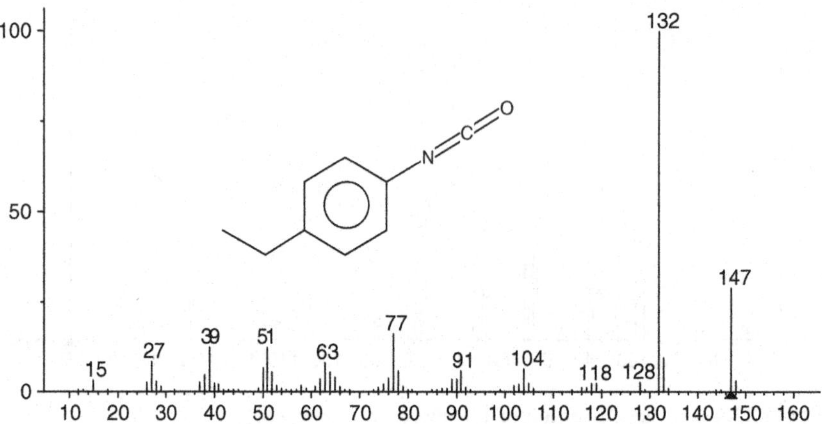

Figure 22.6 EI mass spectrum of *p*-ethylphenyl isocyanate.

CHAPTER 23

KETONES

23.1. GC SEPARATION OF KETONES

A. Capillary columns
   1. Most ketones from acetone to 3-octanone: 50-m DB-5 column, 40 (3 minutes) to 250 °C at 10 °C min^{-1} (nonselective).
   2. TMS derivatives of multifunctional ketones: 30-m DB-210 column (selective for ketones), 40–220 °C at 6 °C min^{-1}.
   3. Cyclohexanone, cyclohexanol, cyclohexenone, dicyclohexyl ethercyclohexyl valerate, cyclohexyl caproate, valeric acid, and caproic acid: 30-m FFAP column, 60–200 °C at 6 °C min^{-1}.

23.2. DERIVATIVES OF KETONES

A. Methoxime derivatives

   This derivative is useful for determining the presence and number of keto groups as well as for protecting the ketone from enolization. Some diketones that polymerize readily, such as 2,3-butanedione, should be freshly distilled prior to the methoxime derivatives being prepared.
   1. Preparation of methoxime derivatives: Add 0.5 mL of MOX reagent to the sample. Heat at 60 °C for 3 hours. Evaporate the reaction mixture to dryness with clean dry nitrogen. Dissolve in minimum amount of ethyl acetate. Some solids will not dissolve.

B. Other derivatives

   The ketone group can be reduced to an alcohol that can then be silylated. This procedure has been used to identify the keto group in carbohydrates.
   1. Reduction and preparation of derivatives: Concentrate the aqueous mixture to 0.5 mL. Add 20 mg of sodium borohydride dissolved in 0.5 mL of ion-exchange water. Let this solution stand at room temperature for 1 hour. Destroy the excess sodium borohydride by adding acetic acid until gas evolution stops. Evaporate the solution to dryness. Add 5 mL of methanol and evaporate again to dryness. Prepare the TMS derivative by adding 250 μL of MSTFA reagent and then heat at 60 °C for 5 minutes.

Gas Chromatography and Mass Spectrometry
DOI: 10.1016/B978-0-12-373628-4.00023-X

23.3. Mass Spectra of Ketones

A. Aliphatic ketones

1. General formula: RC(O)R′.
2. Molecular ion: The molecular weight of aliphatic ketones can often be determined from its prominent $M^{+\bullet}$ peak. In general, the intensity of the $M^{+\bullet}$ peaks of ketones is greater in the spectra of C_3–C_8 compounds than in those of C_9–C_{11} compounds. A $M^{+\bullet}$ peak is usually observed for methoxime derivatives.
3. Fragmentation: The unsymmetrical ketones usually yield four major fragment ion peaks resulting from cleavage on either side of the carbonyl group: R, $RC{\equiv}O^+$, R'^+, and $R'C{\equiv}O^+$. The aliphatic ions are due to heterolytic cleavage and the acylium ions are due to homolytic cleavage. The peaks representing the acylium ions are usually more intense than representing the alkyl ions. A peak representing the loss of 31 Da from the molecular ion of methoxime derivatives is usually observed

Homolytic cleavage

Loss of CO

Loss of CO

Heterolytic cleavage

4. Characteristic fragment ions: The mass spectra of aliphatic ketones can also exhibit intense peaks that represent ions that are formed through a β cleavage resulting from a γ-hydrogen shift (the McLafferty rearrangement fragmentation*) if one of the branches is $C_{\geq 3}$. The intensities of these peaks are far less for C_3 compounds than for those where $C_{>3}$. If there are two branches and both are $C_{\geq 3}$, the product of the first McLafferty rearrangement will undergo a second fragmentation to produce another odd-electron ion, usually having an m/z value of 58. This is why a peak at m/z 58 is considered to be an indicator of an aliphatic ketone. If the carbonyl carbon is C_4, the peak at m/z 58 may not be very intense. Another odd-electron ion peak (has an even m/z value) at a higher m/z value indicates that

*The γ-hydrogen shift-induced β cleavage was named after Fred McLafferty by Carl Djerassi because of the extensive studies of this reaction that were carried out by McLafferty; however, the process was actually first reported by Nicholson [1].

the ketone has two branches with $C_{\geq 3}$. This peak represents the ion formed by the rearrangement fragmentation occurring within the larger of the two branches. The size of the remaining branch can be determined by subtracting 43 from the *m/z* value of this higher *m/z* value odd-electron ion peak.

If the carbonyl is on C_3, the McLafferty rearrangement fragmentation product will be represented by a peak at *m/z* 72. Methyl ketones produce an abundant ion with *m/z* 43. Low-intensity peaks at *m/z* 31, 45, 59, 73, and so forth reveal oxygen in the unknown and are especially useful in distinguishing ketone spectra from isomeric paraffin spectra.

B. Sample mass spectrum of aliphatic ketones

In the mass spectrum of 2-hexanone (Figure 23.1), the $M^{+\bullet}$ peak is apparent at *m/z* 100, which can be confirmed by preparing the methoxime derivative. The compound type is verified by the presence of peaks at *m/z* 43 and 58.

C. Cyclic ketones

1. General formula:

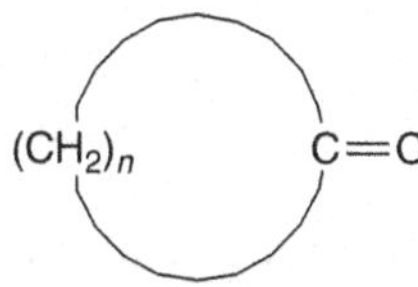

2. Molecular ion: $M^{+\bullet}$ peaks of cyclic ketones are relatively intense. Characteristic fragment ions of cyclic ketones are represented by peaks at *m/z* 28, 29, 41, and 55. Cyclic ketones also lose CO and/or C_2H_4 (*m/z* 28) from the molecular ion (C_6 and higher). Lower

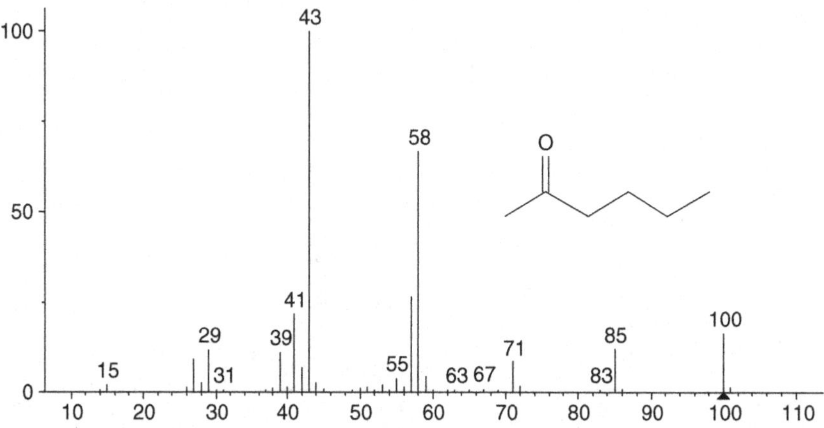

Figure 23.1 EI mass spectrum of 2-hexanone.

intensity peaks representing ions corresponding to loss of H_2O are frequently observed. Ketosteroids are a special class of cyclic ketones and have intense $M^{+\bullet}$ peaks.

D. Sample mass spectrum of cyclic ketones

Figure 23.2 is a mass spectrum of a typical cyclic ketone. This spectrum has peaks representing 28 Da and 29 Da from the molecular ion, which is characteristic of cyclic ketones. Also, the loss of 18 Da from the molecular ion is frequently observed.

E. Aromatic ketones

1. General formula:

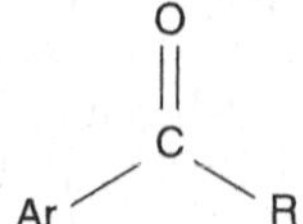

2. Molecular ion: The $M^{+\bullet}$ peak is always present.
3. Fragmentation: Fragmentation occurs on both sides of the carbonyl group. For example, in acetophenone, the most intense peaks are observed at *m/z* 77, 105, and 120 (see Figure 23.3). Peaks at *m/z* 39, 50, and 51 also suggest the presence of an aromatic ring. Aromatic compounds such as quinone, tetralone, and anthraquinone lose CO.

F. General comments

The fragmentation of ketones has been widely studied and reported in the literature. One detailed explanation of the fragmentation of aliphatic ketones can be found in Sparkman et al. [2]

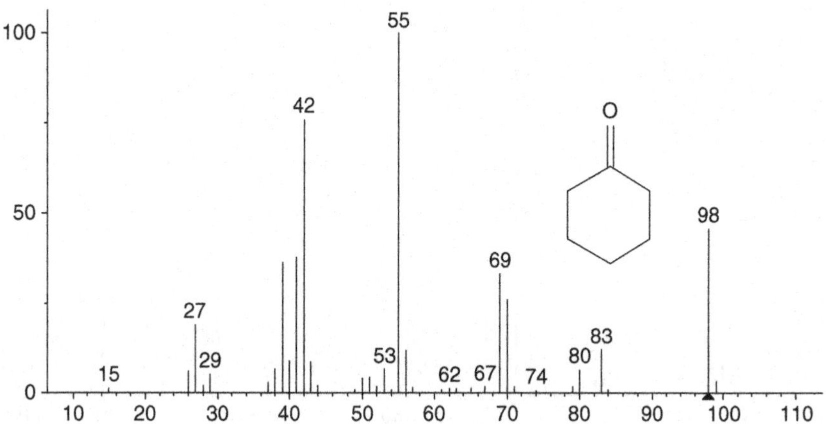

Figure 23.2 EI mass spectrum of cyclohexanone.

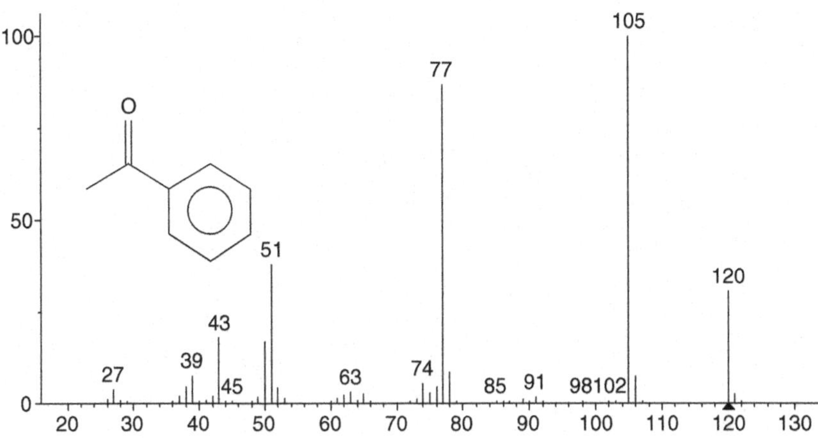

Figure 23.3 EI mass spectrum of actophenone.

REFERENCES

1. Nicholson, A. J. C. (1954). The photochemical decomposition of the aliphatic methyl ketones. *Trans. Faraday Soc.,* 50, 1067-73.
2. Sparkman, O. D., Jones, P. R., Curtis, M. Anatomy of an ion's fragmentation after electron ionization. *Current Trends in Mass Spectrometry.* Iselin, NJ, USA: Advanstar; Part I, May, 2009, 18-29. Part II, July 2009, 12-19. Supplement to: LC–GC and Spectroscopy.

CHAPTER 24

NITRILES

24.1. GC SEPARATION OF NITRILES

A. General
   1. Capillary columns
      a. Ethyl succinonitrile, 1,1-cyanophenylethane, 1,2-cyanophenylethane, methylglutaronitrile, and adiponitrile: 30-m DB-210 column, 50–225 °C at 10 °C min^{-1}.
      b. Dicyanobutenes: 30-m DB-210 column, 50–225 °C at 6 °C min^{-1}.
      c. Isoxazole, fumaronitrile, propanedinitrile, and ethylene nitrile: 30-m DB-WAX column, 100–225 °C at 8 °C min^{-1}.

24.2. MASS SPECTRA

A. Aliphatic mononitriles
   1. General formula: RCN.
   2. Molecular ion: The aliphatic mononitriles may not show molecular ions when R is greater than C_2.
   3. Fragmentation: The saturated aliphatic mononitriles with molecular weights greater than 69 are characterized by intense peaks at *m/z* 41, 54, 68, 82, 96, 110, 124, 138, 152, 166, and so forth. Aliphatic nitriles undergo the McLafferty rearrangement producing an ion with *m/z* 41.
      The aliphatic mononitriles may not show $M^{+\bullet}$ peaks, but $[M-1]^+$, $[M-27]^+$, or $[M-28]^+$ peaks are usually observed. Sometimes a loss of 15 Da may also be observed. If CH_3 is replaced by CF_3, as in the case of $CF_3CH_2CH_2CN$, a fluorine is first lost from the molecular ion followed by the loss of HCN (from the $[M-F]^+$ ion). This influence of the CF_3 group diminishes as the alkyl chain length increases.

B. Sample mass spectrum of an aliphatic mononitrile
   The mass spectrum in Figure 24.1 has intense peaks at *m/z* 41, 54, 68, and so forth, which suggests an alkyl nitrile. The highest *m/z* value peak at *m/z* 96 represents the $[M-1]^+$ ion. Also observed are peaks representing $[M-15]^+$, $[M-27]^+$, and/or $[M-28]^+$ ions.

Gas Chromatography and Mass Spectrometry
DOI: 10.1016/B978-0-12-373628-4.00024-1

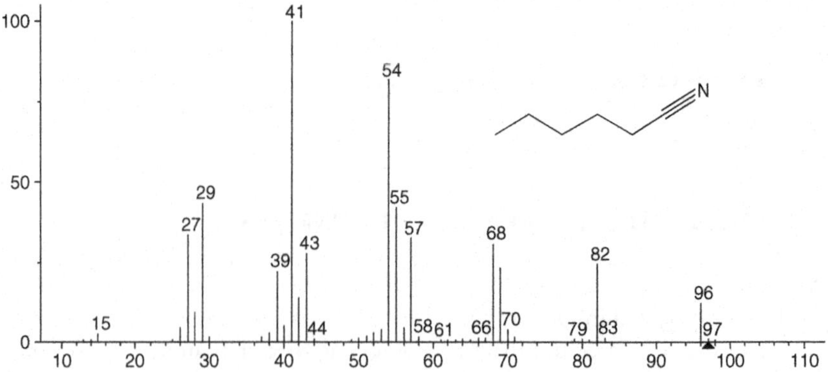

Figure 24.1 EI mass spectrum of hexanenitrile.

C. Aliphatic dinitriles
   1. General formula: $NC(CH_2) \times CN$.
   2. Molecular ion: The $M^{+\bullet}$ peaks of aliphatic dinitriles with molecular weights greater than 80 are usually not observed, but their molecular weights can be deduced by adding 40 Da (CH_2CN) to the highest *m/z* value peak of reasonable intensity.
   3. Fragmentation: The highest two *m/z* value peaks usually represent the $[M-40]^+$ ion and a less-abundant $[M-28]^+$ ion. Intense peaks at *m/z* 41, 54, and 55 also should be present (see Figure 24.2). Dinitriles also may have peaks representing ions with *m/z* 82, 96, 110, 124, 138, 152, and so forth. Except for adiponitrile and methylglutaronitrile, the peak at *m/z* 68 has a very low intensity in the mass spectra of aliphatic dinitriles. Adiponitrile may be distinguished from methylglutaronitrile by the relative intensities of

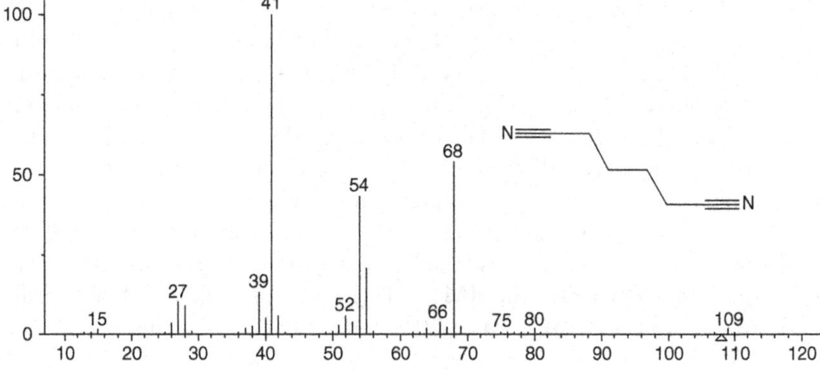

Figure 24.2 EI mass spectrum of hexanedinitrile.

peaks at *m/z* 41 and 68. If the peak at *m/z* 41 is of greater intensity than the peak at *m/z* 68, the mass spectrum suggests adiponitrile. Conversely, if the intensity of the peak at *m/z* 68 is greater than that of the peak at *m/z* 41, the mass spectrum represents methylglutaronitrile.

D. Aromatic nitriles and dinitriles
   1. General formula: ArCN and $Ar(CN)_2$.
   2. Molecular ion: The mass spectra of aromatic nitriles and dinitriles show intense $M^{+\bullet}$ peaks, consistent with the mass spectra of aromatic compounds.
   3. Fragmentation: The loss of 27 Da (HCN) from the molecular ion of aromatic and dinitriles is characteristic. The tolunitriles show a loss of hydrogen as well as the loss of HCN and H_2CN. If the methyl group is replaced by a CF_3 group, the loss of 19 (F) and 50 (CF_2) Da is represented by a very intense peak, whereas the intensity of the peak representing the loss of 27 Da is practically nonexistent. If the nitrile group is on the side chain rather than on the aromatic ring, the loss of 27 Da (HCN) will occur (Figure 24.3).

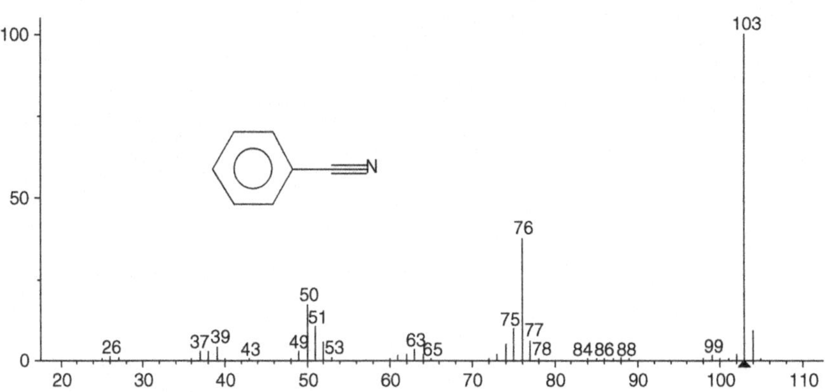

Figure 24.3 EI mass spectrum of benzonitrile.

CHAPTER 25

NITROAROMATICS

25.1. GC SEPARATION OF NITROAROMATICS

A. Capillary columns
   1. Aniline, nitrobenzene, phenylenediamine isomers, nitroaniline isomers, azobenzene, and azoxybenzene: 30-m DB-1 column, 75–250 °C at 10 °C min^{-1}.
   2. Dichlorobenzene, aminotoluene, nitrotoluene isomers, diaminotoluene, and dinitrotoluene isomers: 30-m DB-17 column, 100–250 °C at 8 °C min^{-1}.
   3. Nitroanilines: 30-m DB-17 column, 75–250 °C at 10 °C min^{-1}.
   4. Nitrophenols, nitroaminotoluenes, *N,N*-dimethylnitroanilines, chloronitrotoluenes, chloronitroanilines, nitronaphthalenes, dinitrotoluenes, dinitroanilines, dinitrophenols, dinitrochlorobenzenes, dichloronitrotoluenes, chlorodinitroanilines, dinitronaphthalenes, and trichloronitrotoluenes: 30-m DB-17 column, 75–250 °C at 10 °C min^{-1}.

B. Dinitrobenzenes
   30-m DB-17 column at 225 °C.

C. Nitrodiphenylamines
   30-m DB-225 column or 30-m CP-Sil 43CB column, 75–225 °C at 10 °C min^{-1}.

D. Di(4-nitrophenyl)ether and 1,2-di(4-nitrophenyl)ethane
   30-m DB-17 column, 100–275 °C at 6 °C min^{-1}.

25.2. MASS SPECTRA OF NITROAROMATICS

A. Nitroaromatics
   1. General formula: ArNO.
   2. Molecular ion: The mass spectra of nitroaromatic compounds are characterized by intense $M^{+\bullet}$ peaks.
   3. Fragmentation: Some or all of the following peaks representing fragment ions are observed: $[M - 16]^+$ (loss of O), $[M - 17]^+$ (loss

Gas Chromatography and Mass Spectrometry
DOI: 10.1016/B978-0-12-373628-4.00025-3

of OH from *ortho*-isomers), $[M - 30]^+$ (loss of NO), $[M - 46]^+$ (loss of NO_2), $[M - 58]^+$ (loss of NO + loss of CO), and $[M - 92]^+$ (loss of NO_2 + second molecule of NO_2).

B. Nitrotoluenes (MW = 137 Da)

The mass spectrum of the *o*-nitrotoluene isomer is easy to identify because of the loss of OH from the molecular ion. The mass spectra of all the nitrotoluene isomers exhibit a peak representing the loss of 30 Da from the molecular ion. The *m*- and *p*-isomers can be distinguished from each other by the intensities of the peak at *m/z* 65 relative to the intensity of the $M^{+\bullet}$ peak, particularly if both isomers are present (see Figure 25.1).

C. Nitroanilines (MW = 138 Da)

The mass spectra of all three isomers are different. The spectrum of the *o*-isomer has a peak representing the loss of 17 Da from the $M^{+\bullet}$ peak (loss of OH, a low-intensity peak). The mass spectra of the *m*- and *p*-isomers have peaks representing the loss of 16 Da from the molecular ions and can be distinguished by comparing the relative intensities of the *m/z* 65 fragment ion peaks versus their $M^{+\bullet}$ peaks. The *m/z* 65 peak is the most intense peak in the mass spectrum (the base peak) of the *m*-isomer (see Figure 25.2), whereas the $M^{+\bullet}$ peak at *m/z* 138 is the base peak in the mass spectrum of the *p*-isomer.

D. Nitrophenols (MW = 139 Da)

o-Nitrophenol loses OH from the molecular ion. The mass spectra of both the *m*- and the *p*-isomers exhibit peaks representing the loss of 16 Da, 30 Da, and 46 Da from the molecular ion. The *m*- and *p*-isomers can be distinguished from each other by the high intensity of the peak at *m/z* 109 in the mass spectrum of the *p*-isomer and its low intensity in the mass spectrum of the *m*-isomer (see Figure 25.3).

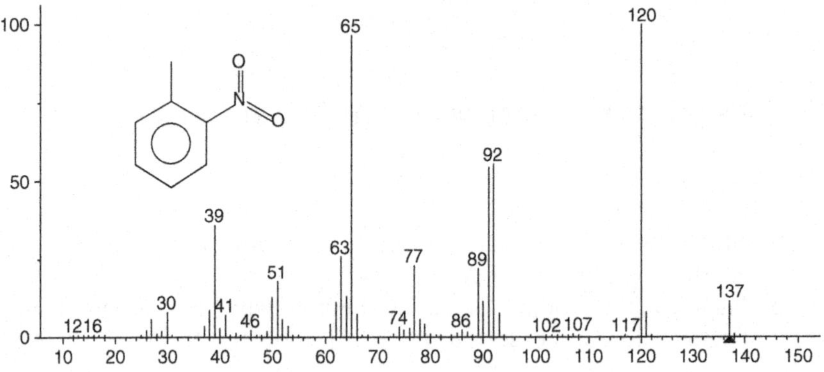

Figure 25.1 EI mass spectrum of *o*-nitrotoluene.

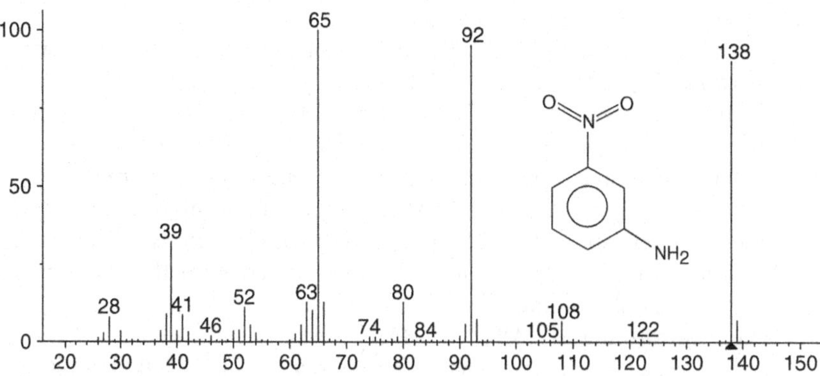

Figure 25.2 EI mass spectrum of *m*-nitroaniline.

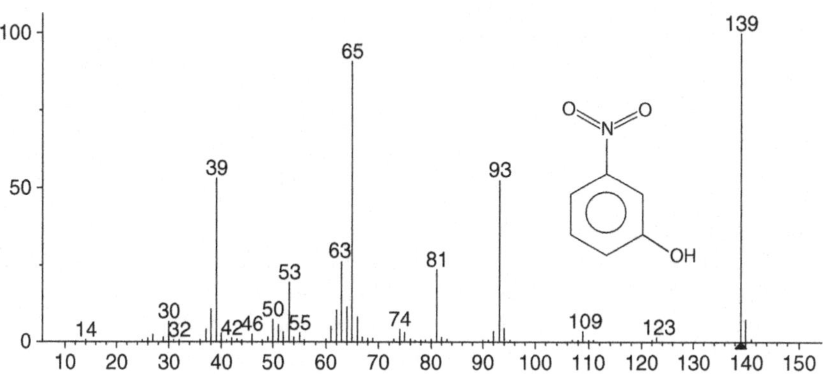

Figure 25.3 EI mass spectrum of 3-nitrophenol.

E. Nitroaminotoluenes (MW = 152 Da)
   The mass spectra of nitroaminotoluenes have intense $M^{+\bullet}$ peaks and a characteristic peak at *m/z* 77 representing a fragment ion. Other intense peaks that are frequently observed are at *m/z* 135, 107, 106, 104, 79, and 30. The peak representing the loss of 17 Da is especially intense when the methyl and nitro groups are ortho to each other.
F. *N,N*-dimethylnitroanilines (MW = 166 Da)
   Again, the *o*-isomer of *N,N*-dimethylnitroaniline loses OH from the molecular ion resulting in an intense peak at *m/z* 149. Other abundant ions in the mass spectra include *m/z* 119, 105, 104, 77, and 42. In the mass spectrum of the *p*-isomer, the peak at *m/z* 136 is reasonably abundant.

G. Dinitrobenzenes (MW = 168 Da)
Some intense peaks in the mass spectra of the dinitrobenzene isomers are at *m/z* 30, 168, 75, 50, 76, 122, and 92. The *m*- and *p*-isomers can be distinguished from one another by the ratio of intensities of the peaks at *m/z* 76 and 75. In the mass spectrum of the *m*-isomer, the peaks at *m/z* 76 and 75 have approximately equal intensities, whereas in the *p*-isomer, the peak at *m/z* 75 is more intense than the peak at *m/z* 76. The *o*-isomer can be distinguished from the other isomers by the greater intensities of the peaks at *m/z* 63 and 64 relative to the intensity of the peaks at *m/z* 76 and 75.

H. Chloronitrotoluenes (MW = 171 Da)
Intense peaks in the mass spectra occur at *m/z* 171, 141 ($[M - NO]^+$), 125 ($[M - NO_2]^+$), and 113 ($[(M - (NO + CO))]^+$). If chlorine (35 Da and 37 Da) is lost from the molecular ion, then the chlorine atom is located on the alkyl group.

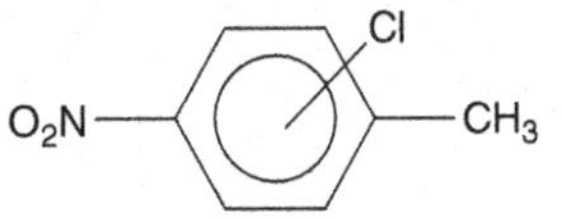

Cl
O_2N — CH_3 O_2N — CH_2Cl

$[M - Cl]^+$ $[M - Cl]^+$

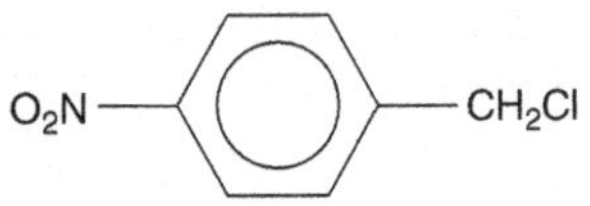

(Not observed) (Intense peak)

I. Nitronaphthalenes (MW = 173 Da)
The most intense peaks in the mass spectra of the 1- and 2-nitronaphthalenes are at *m/z* 127, 115, and 173. The 1-isomer can be distinguished from the 2-isomer by the presence of peaks at *m/z* 145 and 143. The intensity of the peak at *m/z* 115 is greater relative to the base peak in the spectrum of the 1-isomer than it is in the spectrum of the 2-isomer.

J. Dinitrotoluenes (MW = 182 Da)
The $M^{+\bullet}$ peaks are detectable in the mass spectra of all isomers except the 2,3-isomer. The 3,4- and 3,5-isomers have the most intense $M^{+\bullet}$ peaks and the 2,6-isomer has the least intense (2%). Because the nitro groups of the 2,3- and 2,6-isomers are ortho to the methyl group, OH is readily lost at the expense of the molecular ion. If the $M^{+\bullet}$ peak (*m/z* 182) is intense, the unknown mass spectrum is either the 3,4- or the 3,5-isomer. If the *m/z* 165 peak is very intense, then the mass spectrum represents the 2,3-, 2,4-, 2,5-, or 2,6-isomer.

K. Dinitroanilines (MW = 183 Da)
The $M^{+\bullet}$ peak is the base peak in the mass spectra of the dinitroanilines. Other important peaks are observed at *m/z* 153, 137, 107, and 91. The intensity of the peak at *m/z* 137 is very low in the mass spectrum of the 2,6-isomer, which is also characterized by a peak representing the loss of 17 Da from the molecular ion.

L. Dinitrophenols (MW = 184 Da)
Dinitrophenols are characterized by abundant $M^{+\bullet}$ peaks. The mass spectrum of the 2,4-isomer has intense fragment ion peaks at *m/z* 154, 107, 91, and 79. None of these peaks are intense in the mass spectra of the 2,6- or 2,5-isomers. The mass spectrum of the 2,6-isomer has a peak at *m/z* 126, which is not present in the spectrum of the 2,5-isomer. All of the isomers have peaks at *m/z* 63 and 46.

M. Dinitrochlorobenzenes (MW = 202 Da)
Intense peaks are observed at *m/z* 30, 202, 75, 110, 63, 74, 50, and 109. Most of the isomers can be distinguished by the peaks at *m/z* 186, 172, and 156. For example, all of these peaks are present in the mass spectra of the 2,4-isomer, none in the 3,4-isomer, and 186 (small peak) and 156 in the 2,5-isomer.

N. Dichloronitrotoluenes (MW = 205 Da)
A mass spectrum representing a dichloronitrotoluene is indicated by the presence of a $M^{+\bullet}$ peak at an odd *m/z* value, an X+2 isotope pattern indicating the presence of two atoms of chlorine, and peaks representing losses of 30 Da and 46 Da from the molecular ion. Again, when the chlorine atoms are on the benzene ring, the loss of chlorine from the molecular ion does not occur. An $[M - Cl]^+$ peak indicates that at least one of the chlorines is on the alkyl group.

O. Nitrodiphenylamines (MW = 214 Da)
Intense peaks that are characteristic of the mass spectra of nitrodiphenylamines are at *m/z* 214, 184, and 164.

P. Chlorodinitroanilines (MW = 217 Da)
The mass spectra (Figure 25.4) of chlorodinitroanilines are characterized by peaks at *m/z* 217, 201 (low intensity), 187 ($[M - NO]^+$), 171 ($[M - NO_2]^+$), 141 ($[M - (NO_2 + NO)]^+$), and 125 ($[M - 2(NO_2)]^+$).

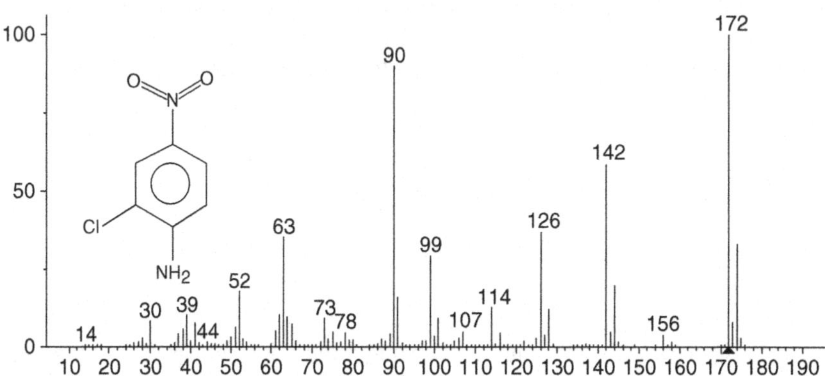

Figure 25.4 EI mass spectrum of 2-chloro-4-nitroanilines.

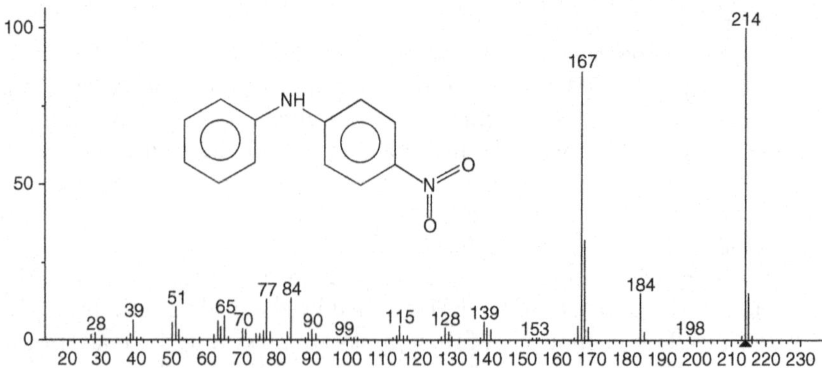

Figure 25.5 EI mass spectrum of *o*-nitrodiphenyl amine.

Q. Dinitronaphthalenes (MW = 218 Da)
The mass spectra (Figure 25.5) of all the isomers show intense $M^{+\bullet}$ peaks and intense peaks at *m/z* 126 and 114, which are characteristic of the naphthalene moiety. In the mass spectra of the 2,3- and 1,8-dinitro isomers, the intensity of the *m/z* 126 peak is very low. The 2,3-isomer has an intense peak at *m/z* 127.

R. Trichloronitrotoluenes (MW = 239 Da)
The presence of three chlorine atoms is easily determined by the X+2 isotope pattern. The odd molecular weight shows the presence of nitrogen. The loss of 30 and 46 Da from the molecular ions shows the presence of the nitro group.

S. Sample mass spectra
In Figure 25.2 (*m*-nitroaniline), the intensity of the $M^{+\bullet}$ peak at *m/z* 138 suggests an aromatic compound. Examining the losses from the molecular ion ($[M - 46]^+$ and $[M - 30]^+$) shows that it is a nitro compound. Because the molecular ion is of even mass, an even number of nitrogen atoms must be present. Looking up the correlation for *m/z* 92 in Appendix Q for possible structures, the following is suspected:

NH_2

The structure is a nitroaniline. Generally, the *o*-isomer can be detected by the losses of both 16 Da and 17 Da from the molecular ion, whereas the *m*- and *p*-isomers lose only 16 Da. This ortho effect applies when amino, hydroxyl, and methyl groups are ortho to the nitro group. In this example, only a 16-Da loss is observed. This loss along with a peak at *m/z* 65 (the base peak in the mass spectrum) suggests an *m*- or *p*-isomer. The $M^{+\bullet}$ peak appears at *m/z* 139 in Figure 25.3. The intensity of this peak suggests an aromatic compound and the odd *m/z* value for the ion suggests that the analyte has an odd number of nitrogen atoms. Peaks representing $[M - 46]^+$, $[M - 30]^+$, and $[M - 16]^+$ suggest a nitroaromatic compound. Looking up the correlation for *m/z* 93 in Appendix Q suggests a nitrophenol. The presence or absence of an OH group can be determined by preparing a TMS derivative.

CHAPTER 26

Nitrogen-Containing Heterocyclic Compounds

26.1. GC Separations of Nitrogen-Containing Heterocyclic Compounds

A. Capillary columns
   1. Pyridines, 2,3,4-methylpyridines (picolines), and aniline: 30-m Carbowax column + KOH on Carbopack B, 75–150 °C at 3 °C min^{-1}.
   2. Quinolines and acridines: 30-m DB-5 column, 75–275 °C at 10 °C min^{-1}.
   3. Pyrroles, indoles, and carbazoles: 30-m DB-17 column, 75–275 °C at 10 °C min^{-1}.
   4. Benzonitrile, aniline, nitrobenzene, benzo-*p*-diazine, biphenyl, azobenzene, and dibenzoparadiazine: 30-m DB-225 column, 75–215 °C at 10 °C min^{-1}.
   5. Phenazine and anthracene: 30-m DB-1 column, 100–225 °C at 6 °C min^{-1}.
   6. Imidazoles and benzimidazoles: 30-m DB-5 column, 75–250 °C at 10 °C min^{-1}.
   7. Creatinine–TMS and uric acid–TMS: 30-m DB-1 column, 100–250 °C at 10 °C min^{-1}.

26.2. Mass Spectra of Nitrogen-Containing Heterocyclics

A. Mass spectra of pyridines, quinolines, and acridines
   1. Molecular ion: The $M^{+\bullet}$ peak is usually intense except when long-chain alkyl groups are attached to the ring.
   2. Fragment ions: Nitrogen-containing heterocyclics such as pyridines, quinolines, and acridines lose HCN and/or HCN + H from their molecular ions.
   3. Examples
      a. Pyridines

Gas Chromatography and Mass Spectrometry
DOI: 10.1016/B978-0-12-373628-4.00026-5

The mass spectra of methylpyridines (picolines) and dimethylpyridines (lutidines) have prominent peaks at *m/z* 65 and 66. Aniline can be distinguished from picolines by the peak at *m/z* 78 in the mass spectra of picolines. If the alkyl group, R, is attached to the carbon atom adjacent to the nitrogen atom, RCN easily can be lost. The mass spectra of alkylpyridines are characterized by peaks at *m/z* 65, 66, 78, 92, 106, and so forth.

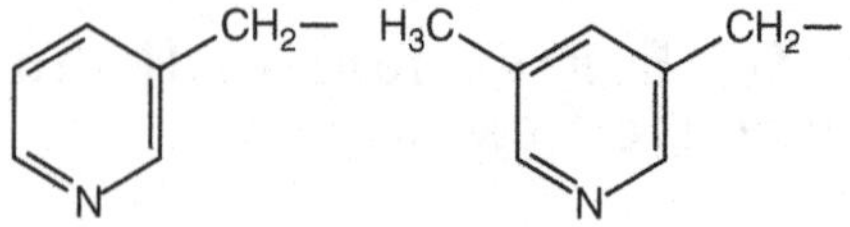

CH_2— H_3C CH_2—

N N

m/z 92 *m/z* 102

b. Quinoline (benzpyridine)

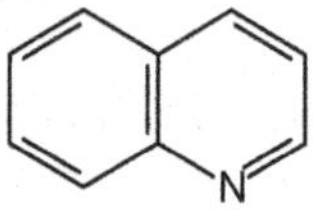

N MW = 129 Da

Some characteristic peaks in the mass spectra of quinolines include *m/z* 102, 128, 156, and so forth. Again, as in alkylpyridines, RCN is lost if the alkyl group is attached to the carbon atom adjacent to the nitrogen atom.

c. Acridine (dibenzpyridine)

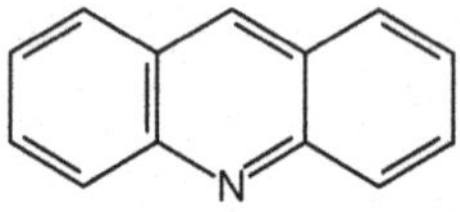

N MW = 179 Da

The $M^{+\bullet}$ peak is the most intense peak (the base peak) in the mass spectrum. Characteristic fragment ion peaks in the mass spectrum are observed at *m/z* 89, 90, and 151.

B. Mass spectra of pyrroles, indoles, and carbazoles
   1. Molecular ion: Molecular ions of these heterocyclics are usually abundant except when long chains or tertiary alkyl groups are attached to the ring.
   2. Fragment ions: Alkyl groups attached to a carbon atom of the ring (C-alkyl) fragment beta to the ring. A test for the presence or absence of the C-alkyl pyrrole is to prepare a TMS derivative. If the TMS derivative cannot be formed, an alkyl or other group is attached to the nitrogen atom. *N*-alkyl-substituted pyrroles will undergo a β-cleavage analogist to a benzylic cleavage to form an ion with *m/z*

80 ($C_4H_4^{+}N{=}CH_2$). Provided that there are a sufficient number of C atoms in the alkyl moiety, the molecular ion will undergo an γ-hydrogen shift-induced β cleavage resulting in the loss of an olefin and an ion with *m/z* 81. Both peaks at *m/z* 80 and 81 have significant intensities; the *m/z* 81 peak will be the base peak, and the peak at *m/z* 80 will be more intense (~80% of the base peak) than the $M^{+\bullet}$ peak, which will have an intensity of ~50%.

3. Examples
   a. Indole (benzopyrrole)

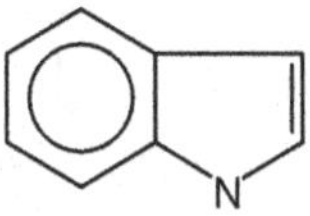

MW = 117 Da

The presence of indole and carbazole can also be determined by their TMS derivatives. The mass spectrum of indole–TMS has intense peaks at *m/z* 189, 174, and 73.

   b. Carbazole (dibenzopyrrole)

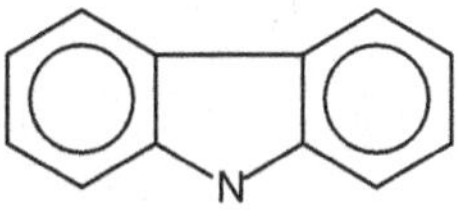

MW = 167 Da

The mass spectrum of carbazole–TMS has intense peaks at *m/z* 239 (100%), 224 (58%), and 73 (58%). Underivatized alkyl pyrroles have characteristic ions at *m/z* 80, 94, 108, and so forth, whereas alkyl indoles have peaks representing characteristic fragment ions at *m/z* 103, 115, 144, and so forth.

C. The mass spectra of pyrazines, quinoxalines, and phenazines
1. Molecular ions: $M^{+\bullet}$ peaks are observed.
2. Fragment ions: Characteristic fragment ions involve loss of a nitrogen and the adjacent carbon atom (RCN).
3. Examples
   a. Pyrazines:

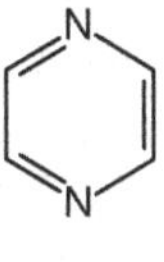

MW = 80 Da

Alkyl pyrazines lose RCN from their molecular ions when the alkyl group is attached to the carbon adjacent to the nitrogen atom.

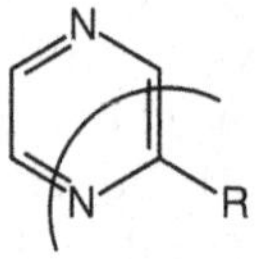

b. Quinoxaline

The mass spectrum of quinoxaline has an intense $M^{+\bullet}$ peak at *m/z* 130.

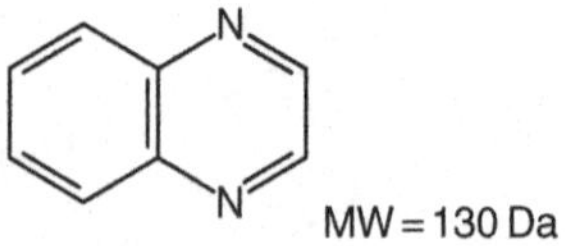

MW = 130 Da

The loss of HCN from the molecular ion yields an abundant ion represented by a peak at *m/z* 103. A second loss of HCN results in an abundant ion represented by a peak at *m/z* 76. The molecular ion of dimethylquinoxaline loses acetonitrile:

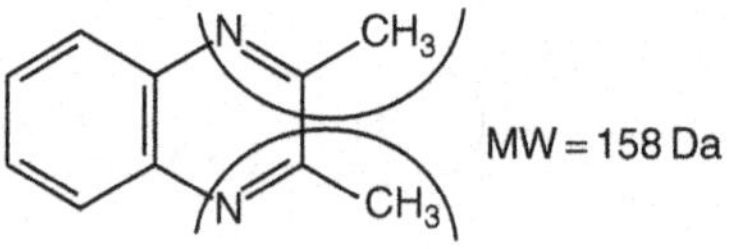

MW = 158 Da

This loss is followed by another loss of acetonitrile, resulting in an ion represented by an intense peak at *m/z* 76.

c. Phenazine (dibenzopyrazine)

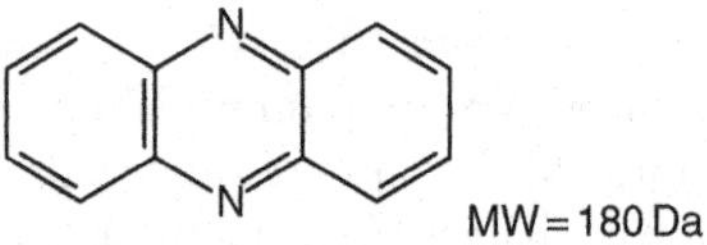

MW = 180 Da

The mass spectrum of phenazine is characterized by peaks at *m/z* 180, 179, 153, 90, and 76.

D. Mass spectra of imidazoles and benzimidazoles

1. Molecular ions: Intense $M^{+\bullet}$ peak.
2. Fragment ions: All nitrogen-containing heterocyclics lose HCN and/ or HCN + H from their molecular ions.

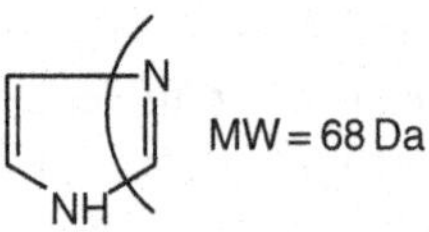

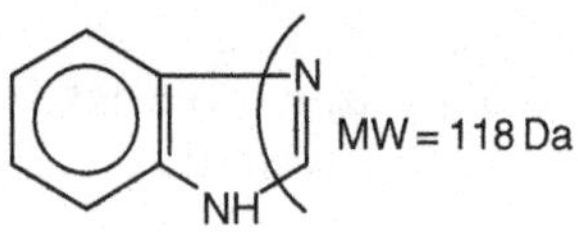

MW = 68 Da MW = 118 Da

Imidazole has an intense $M^{+\bullet}$ peak at *m/z* 68, but the molecular ion loses H and HCN to yield abundant ions with *m/z* 67 and 41, respectively. Benzimidazole loses HCN from the molecular ion (see Figure 26.1). A characteristic fragment ion for alkylbenzimidazoles where the alkyl has three or more carbon atoms is represented by a peak at *m/z* 132. The ion is an odd electron produced by a γ-hydrogen shift-induced β cleavage resulting in the loss of a 1,2-olefin. There will also be a peak at *m/z* 131 representing an even-electron ion formed by a single-bond-fragmentation analogist to a benzylic cleavage.

E. Creatinine–TMS (MW = 329 Da) and uric acid–TMS (MW = 456 Da)
The mass spectra of these nitrogen-containing heterocyclic compounds are characterized by peaks representing the following ions:
Creatinine–TMS *m/z* 115, 73, and 329
Uric acid–TMS *m/z* 73, 456, and 441.

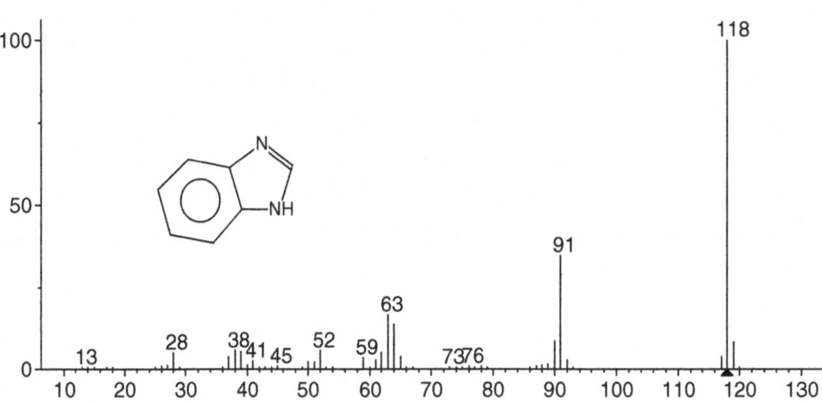

Figure 26.1 EI mass spectrum of 1,3-benzimidazole.

CHAPTER 27

NUCLEOSIDES (TMS DERIVATIVES)

Nucleosides consist of a purine or a pyrimidine base and a ribose or a deoxyribose sugar connected via a β-glycosidic linkage. These compounds are associated with structures of RNA (ribose sugars) and DNA (deoxyribose sugars). The compounds are very polar and their analysis by GC/MS is only possible when they have been derivatized. It is possible to get fairly respectable spectra when the pure compounds are introduced into an EI source using a direct insertion probe. The structures of the nucleosides of most biological interest are shown in Figure 27.1.

27.1. DERIVATIZATION

Add 0.25 mL of DMF (*N,N*-dimethylformamide) and 0.25 mL of TRI-Sil TBT reagent to the sample in a screw-cap septum vial. If TRI-Sil TBT reagent is not readily available, add 0.25 mL of acetonitrile and 0.25 mL of BSTFA reagent instead. Heat at 60 °C for at least 1 hour for ribonucleosides or for a minimum of 3 hours for deoxyribonucleosides. After cooling to room temperature, inject 1–2 μL of the reaction mixture directly into the gas chromatograph. The resulting derivatives have been reported to be stable for weeks if tightly capped and refrigerated [1].

27.2. GC SEPARATION OF DERIVATIZED NUCLEOSIDES

A. Capillary column
   1. 2′-Deoxyuridine, thymidine, 2′-deoxyadenosine, 2′-deoxycytidine, 2′-deoxyguanosine: 10–30-m DB-17 column, 100–275 °C at 10 °C min^{-1}; injection port at 280 °C.

Gas Chromatography and Mass Spectrometry
DOI: 10.1016/B978-0-12-373628-4.00027-7

2′-Deoxyadenosine
dA

2′-Deoxyguanosine
dG

Thymidine
dT

2′-Deoxycytidine
dC

2′-Deoxyuridine
dU

Figure 27.1 Structures of the five biologically active nucleosides.

27.3. Mass Spectra of TMS Nucleosides [2]

Display a mass chromatogram of *m/z* 103 to determine the elution time of the TMS–nucleosides. Next, determine the molecular weight by identifying the $M^{+\bullet}$ peak (which is usually observed) associated with an $[M - 15]^+$ peak and often with $[M - 90]^+$, $[M - 105]^+$, and $[M - 203]^+$ peaks. The $M^{+\bullet}$ peak of the TMS derivatives of ribonucleosides is 88 *m/z* units higher than the TMS derivatives of the deoxyribonucleosides. If the difference between the *m/z* value of the $M^{+\bullet}$ peak and the base peak is 260, the sugar portion is deoxyribose. A difference of 290 represents an *o*-methylribose, and a difference of 348 Da suggests a ribose.

Nucleoside–TMS	MW	*m/z* Values that indicate the base	Base
2′-Deoxyuridine	444	169, 184	Uracil
Thymidine	458	183	Thymine (5-methyluracil)
2′-Deoxyadenosine	467	192, 207	Adenine
2′-Deoxycytidine	443	168, 183	Cytosine
2′-Deoxyguanosine	555	280, 295	Guanine

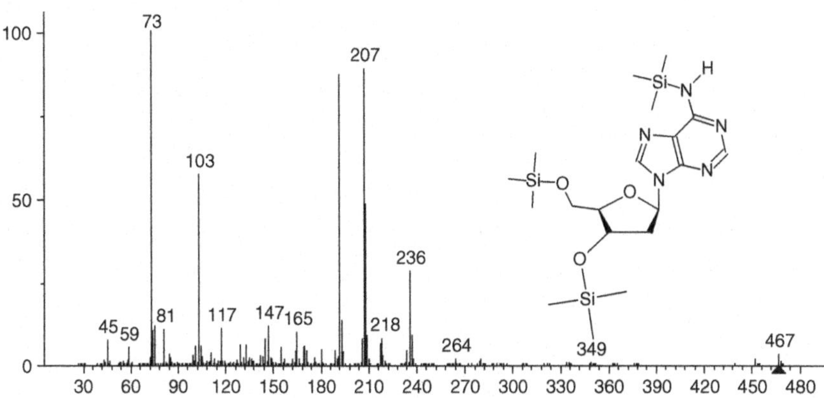

Figure 27.2 TMS derivative of 2′-deoxyadenosine, EC: $C_{19}H_{37}N_5O_3Si_3$, MW 467 Da.

If necessary, the nucleosides can be hydrolyzed to the sugar and the base by heating in formic acid [2]. Kresbach et al. [3] have used pentafluorobenzylation combined with electron capture negative ionization for the detection of trace amounts of nucleobases. For fragmentation patterns of TMS 2′-, 3′-, and 5′-deoxynucleosides, see Reimer et al. [4] (Figure 27.2).

REFERENCES

1. Schram, K. H., McCloskey, J. L. (1979). In: Tsuji K, ed. *GLC and HPLC of Therapeutic Agents*. New York: Marcel Dekker.
2. Crain, P. F. (1990). In: McCloskey, J. A., ed. *Methods in Enzymology*, Chapter 43 (Vol. 193). San Diego, CA: Academic Press.
3. Kresbach, G. M., Annan, R. S., Saha, M. G., Giese, R. W., Vouros, P. (1988). *Mass Spectrometric and Chromatographic Properties of Ring-Penta Fluorobenzylated Nucleobases used in the Trace Detection of Alkyl DNA Adducts*. Proceedings of the 36th Annual Conference of the American Society for Mass Spectrometry, San Francisco, June 5–10.
4. Reimer, M. L. J., McClure, T. D., Schram, K. H. (1988). *Investigation of the Fragmentation Patterns of the TMS Derivatives of 2'-, 3'-, and 5'-Deoxynucleosides*. Proceedings of the 36th Annual Conference of the American Society for Mass Spectrometry, San Francisco, June 5–10.

CHAPTER 28

Pesticides

28.1. Chlorinated Pesticides

A. GC separations

1. Lindane, heptachlor, aldrin, a- and g-chlordane, dieldrin, DDT, and similar compounds.
   a. 30-m CP-Sil 8 CB* column, 60–300 °C at 4 °C min^{-1}. Run for 52 minutes.
   b. 30-m DB-5 column, 50 (2 minutes)–140 °C at 20 °C min^{-1}; 140–300 °C at 4 °C min^{-1}.*
   c. 15–30-m DB-608 column, 140 (2 minutes)–240 °C at 10 °C min^{-1}; 240 (5 minutes)–265 °C at 5 °C min^{-1}.
   d. 50-m CP-Sil 88 column, 60–225 °C at 20 °C min^{-1}.

B. Pesticide extraction procedures

1. For pesticide extraction procedures pertaining to food samples, refer to U.S. Government Manuals on pesticide residue analysis [1].
2. For pesticide extraction from aqueous samples [2].
3. The Environmental Protection Agency (EPA) has prepared a manual of pesticide residue analysis dealing with samples of blood, urine, human tissue, and excreta, as well as water, air, soil, and dust. *Manual of Analytical Methods*, edited by Thompson, JF, Quality Assurance Section, Chemistry Branch EPA, Environmental Toxicology Division, Pesticides Health Effects Research Laboratory, Research Triangle Park, NC 27711.
4. Pesticide bulletins are available from Supelco, Inc., Supelco Park, Bellefonte, PA 16823.

C. Structure of common chlorinated pesticides and abundant ions

The following values reported for the MW of these chlorinated pesticides are actually the *nominal mass* values for these molecules. The *nominal mass* of Cl is 35. The *atomic mass* (weight) of Cl is 35.453; therefore, the integer *molecular weight* for lindane is 291 Da. Integer masses reported in mass spectrometry are based on *nominal mass*

* The DB-5 column may be used but does not provide enough GC resolution if metabolites are present.

Gas Chromatography and Mass Spectrometry
DOI: 10.1016/B978-0-12-373628-4.00028-9

because mass spectrometry involves the masses of the isotopes of the elements, not their atomic weights.

1. Lindane

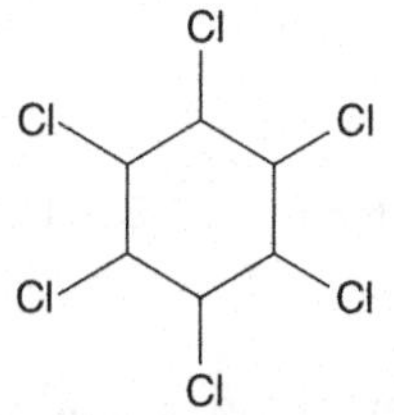

$C_6H_6Cl_6$ (MW = 288 Da)

2. Heptachlor

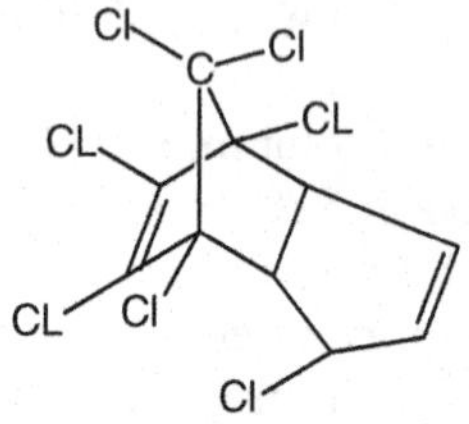

$C_{10}H_7Cl_7$ (MW = 370 Da)

3. Aldrin

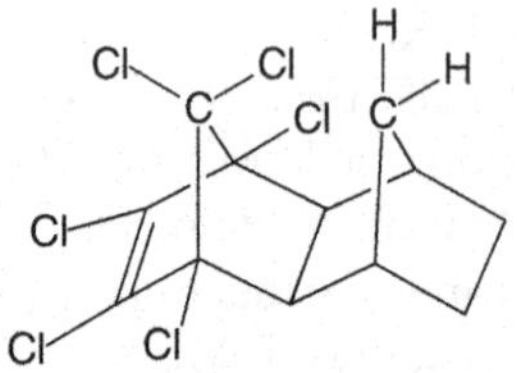

$C_{12}H_8Cl_6$ (MW = 362 Da)

4. Chlordane

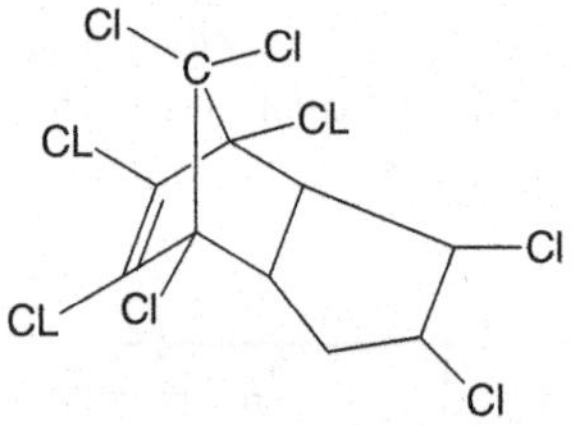

$C_{10}H_6Cl_8$ (MW = 406 Da)

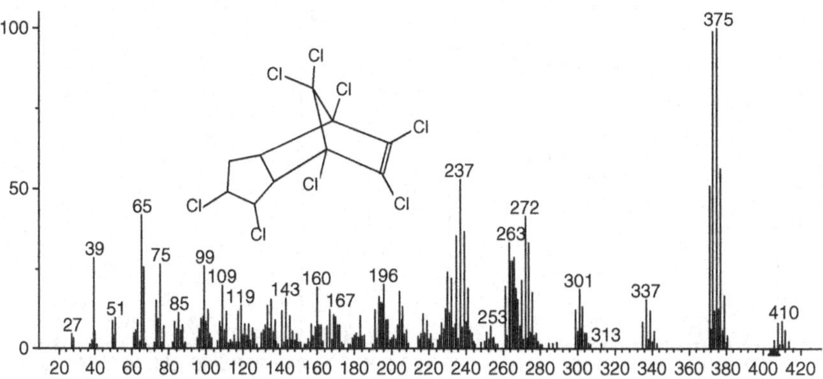

Figure 28.1 EI mass spectrum of chlordane.

The $M^{+\bullet}$ peak for chlordane is discernable in Figure 28.1. Note the X+2 isotope pattern for eight chlorine atoms. The base peak at *m/z* 373 is the X+2 isotope peak with a nominal *m/z* value of 371, which represents the ion formed by the loss of a $^{\bullet}$Cl from the molecular ion.

5. Dieldrin

$C_{12}H_8Cl_8O$ (MW = 378 Da)

6. DDT

$C_{14}H_8Cl_5$ (MW = 352 Da)

The $M^{+\bullet}$ peak is apparent in the mass spectrum of DDT at *m/z* 352 (Figure 28.2) with the classic X+2 isotope pattern for an ion with five

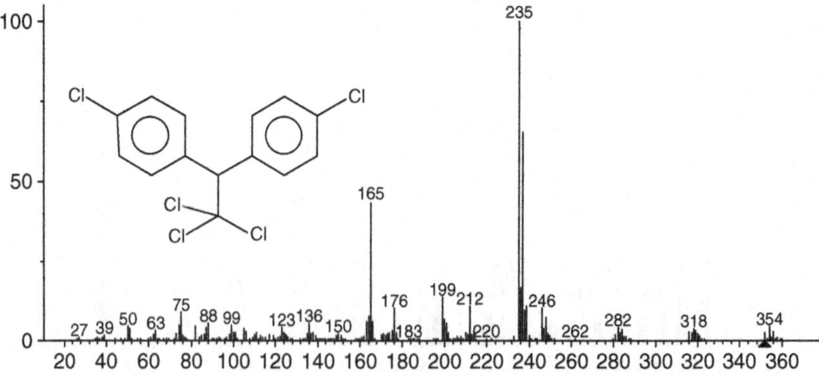

Figure 28.2 EI mass spectrum of DDT.

atoms of chlorine (see Appendix E). The major fragment ion due to the loss of CCl_3 is represented by the peak at *m/z* 235.

7. Methoxychlor

CH3 CH3 O O H C C Cl Cl Cl

$C_{16}H_{15}Cl_3O_3$ (MW = 344 Da)

28.2. Organophosphorus Pesticides

A. GC separations

1. Diazinon, malathion, dimethoate, trichlorofon,* and so forth. 30-m DB-5, SPB-5, or Supelco PTE-5 column, 100–300 °C at 4 °C min^{-1} or 150 (3 minutes)–250 °C at 5 °C min^{-1}.
2. Dichlorovos, phorate, dimethoate, diazinon, disulfoton, methylparathion, malathion, parathion, azinphos-methyl, azinphos-ethyl, and so forth. 50-m CP-Sil 13CB column, 75–250 °C at 10 °C min^{-1}.

---

* Trichlorofon loses HCl from the molecular ion producing a spectrum identical to that of dichlorovos. These pesticides can be differentiated by preparing a TMS derivative and/or their retention times.

B. Structures of common organophosphorus pesticides and *m/z* values representing the most abundant ions

1. Dichlorovos*

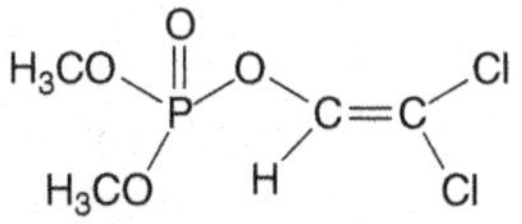

$C_4H_7Cl_2O_4P$ (MW = 220 Da)

2. Trichlorofon*

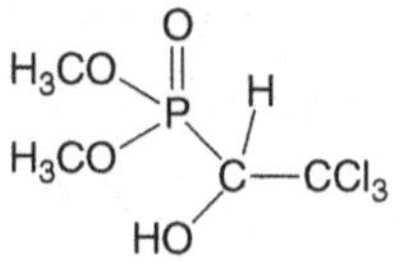

$C_4H_8O_4Cl_3P$ (MW = 256 Da)

3. Phorate (thimet)

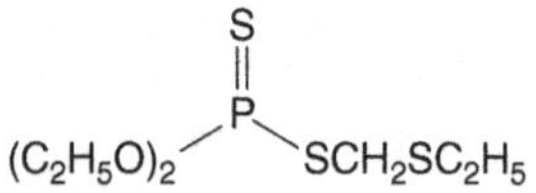

$C_7H_{17}O_2PS_3$ (MW = 260 Da)

4. Dimethoate

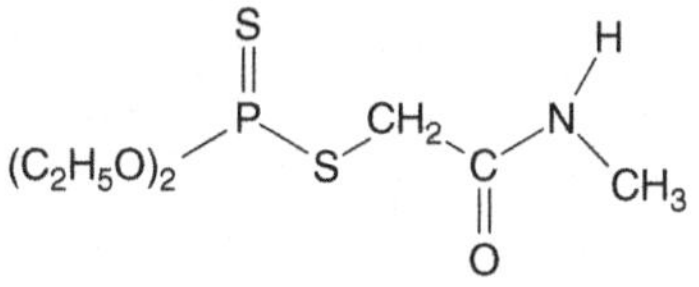

$C_5H_{12}NO_3PS_2$ (MW = 229 Da)

5. Diazinon (dimpylate)

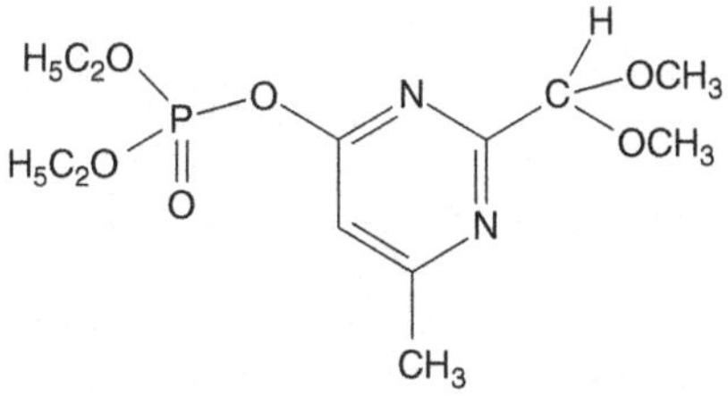

$C_{12}H_{21}N_2O_3PS$ (MW = 304 Da)

---

* Trichlorofon loses HCl from the molecular ion producing a spectrum identical to that of dichlorovos. These pesticides can be differentiated by preparing a TMS derivative and/or their retention times.

6. Disulfoton

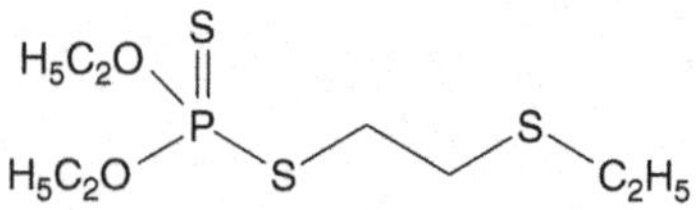

$C_8H_{19}O_2PS_3$ (MW = 274 Da)

7. Methyl parathion

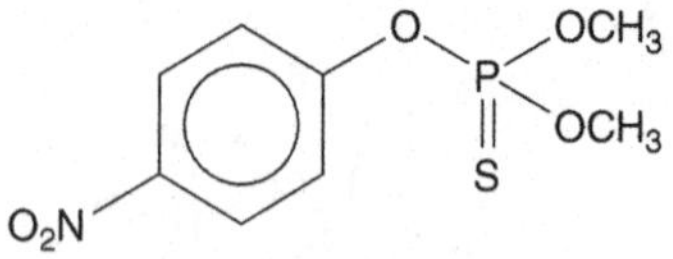

$C_8H_{10}NO_5PS$ (MW = 263 Da)

8. Malathion

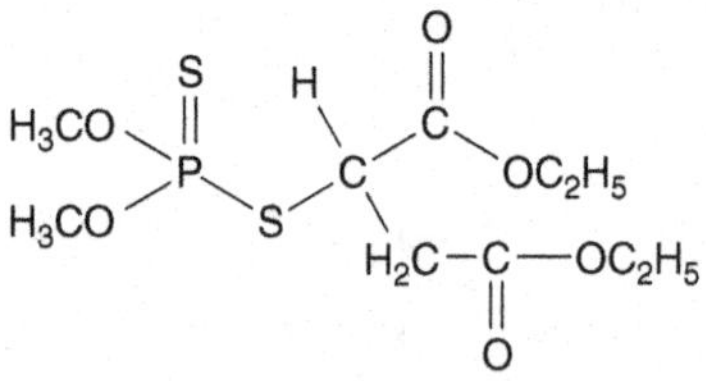

$C_{10}H_{10}O_6PS_2$ (MW = 330 Da)

9. Parathion

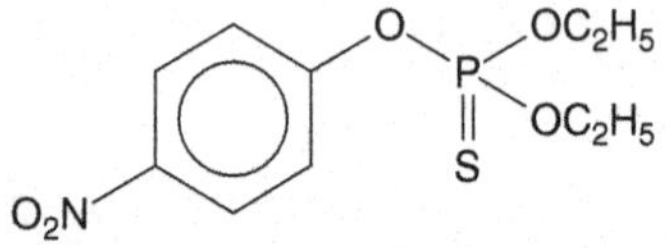

$C_{10}H_{14}NO_5PS$ (MW = 291 Da)

10. Azinphos-methyl

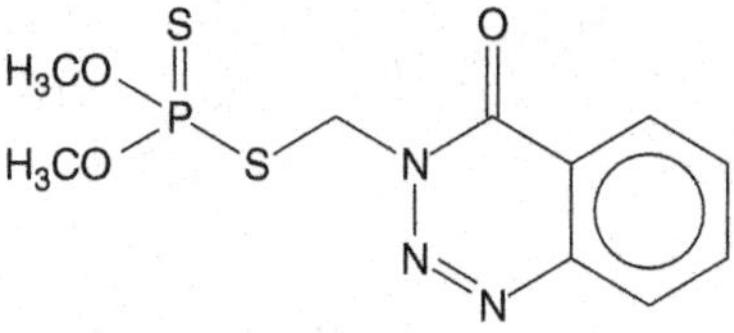

$C_{10}H_{12}N_3O_3PS_2$ (MW = 317 Da)

11. Azinphos-ethyl

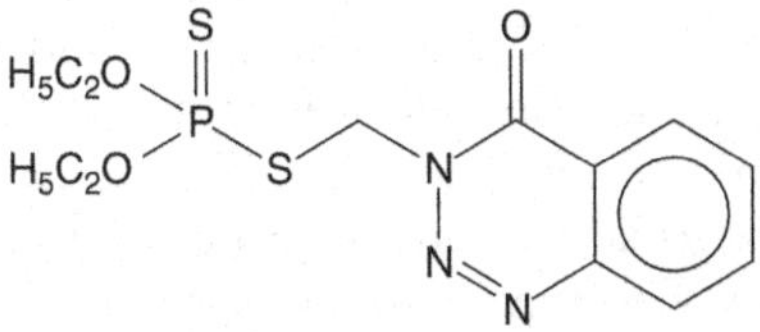

$C_{12}H_{16}N_3O_3PS_2$ (MW = 345 Da)

28.3. Mass Spectra of Pesticides

Pesticides have widespread health and environmental considerations. GC/MS analyses of this class of compounds are extensive for a wide variety of matrices from soil and water to commodities and living plants to biological fluids such as blood plasma and urine. The challenge often is how to get the pesticide from the sample into a form that can be injected into the GC. The mass spectra of this class of compounds can be very complex due to the often high number of atoms of X+2 elements that are present and due to the fact that the EI mass spectra of many pesticides do not exhibit a $M^{+•}$ peak. The significance of the analyses of pesticides by GC/MS is evident by the large number of EI mass spectral databases of these compounds.

Some of these databases are only available in a hardcopy format, although many of the actual spectra are now included in the *NIST08 Mass Spectra Database* and/or the *Wiley Registry of Mass Spectra*, 9th ed. The authors of the first edition of this book recommended *Mass Spectrometry of Pesticides and Pollutants*, (Safe S, Hutzinger O. Boca Raton, FL: CRC Press; 1973, with two additional printings in 1976, and one each in 1977, 1979, and 1980, indicating its popularity; ISBN: 08493-5033-6; contains 275 spectra). Unfortunately, this book has been out of print for many years and finding copies is very difficult. Another hardcopy-only publication are the first and second editions of *Handbook of Mass Spectra of Environmental Contaminants* (Hites RA. CRC Press: 2nd ed., 1992, ISBN:0873715349; 533 spectra (reviewed *JASMS* 5:598); 1st ed., 1985, ISBN:084930537; 394 spectra). In addition to these three hardcopy-only collections, Agilent Technologies has three different electronic databases of pesticide EI mass spectra for use with their *GC/MS ChemStation*: *Stan Pesticide Mass Spectral Library* (340 spectra); the *Agilent Pesticide RTL* (retention time locking) *Library* (926 spectra including spectra of endocrine disrupters); and the *Japanese Positive List Pesticide RTL Library* (431 spectra). John Wiley & Sons has announced *Mass Spectra of Pesticides 2009*, a database available in electronic format for the *NIST11 MS Search Program* and other proprietary instrument manufacturers' data systems that contains 1,238 spectra and is authored by Rolf Kühnle. The Wiley database includes structures; however, none of the Agilent databases do.

The challenge in a GC/MS pesticide analysis can be in achieving the desired limit of detection or limit of quantitation. It may be necessary to use electron capture negative ionization (ECNI), which will be highly specific for the halogenated target analytes and will ignore the presence of nonhalogenated matrix compounds. The importance of the role of retention indices (RI values) in the confirmation of the identity of pesticides should not be forgotten.

If pesticide analyses are a major function of the laboratory, a mass spectral database should be created of pure standards and standards at different concentration levels (down to the desired limit of detection) in usually encountered matrices using the instrument(s) that is(are) normally used for these analyses. When using EI with sample having complex matrices, the use of AMDIS for data processing can be very helpful.

A list of the most abundant ions for a few pesticides is found in Table 28.1. The use of this list and/or a computer library search will often be sufficient to

Table 28.1 Ions for identifying pesticides

Base peak	Four next most intense peaks				Compound	Highest m/z peak >1%
66	263	265	79	261	Aldrin	362
75	121	260	97	–	Phorate (thimet)	260
77	32	160	93	76	Azinphos-methyl	317
79	82	263	81	277	Dieldrin	378
81	100	61	60	59	Methyldemeton	230
87	75	55	–	–	Aldicarb	190
87	93	125	229	–	Dimethoate	229
88	89	97	274	–	Disulfoton	274
97	197	199	314	–	Chlorpyrifus	349
100	272	274	65	270	Heptachlor	370
109	79	185	145	–	Dichlorovos	220
109	79	185	145	–	Trichlorofon	256
109	81	149	99	–	Paraoxon	275
109	97	137	291	–	Parathion	291
109	263	125	–	–	Methyl parathion	263
110	152	81	–	–	Propoxur	209
132	160	77	–	–	Azinphos-ethyl	345
173	127	125	93	158	Malathion	330
179	137	152	199	304	Diazinon (dimpylate)	304
181	183	109	217	–	Lindane	288
227	274	308	–	–	Methoxychlor	344
235	237	165	236	239	DDT	352
373	375	377	–	–	Chlordane	404

identify many of the most encountered commercial pesticides. Also, see Chapters 20, 29, and 30.

In addition to compounds listed here, there may be many others. Many of the other compounds contain Br instead of Cl. It is important to be careful in the evaluation of X+2 isotope peak patterns.

REFERENCES

1. Food and Drug Administration (1994). *Pesticide Analytical Manual*, Volumes I and II, Washington, DC: U.S. Department of Health and Human Services, Food and Drug Administration, U.S. Government Printing Office.
2. Eisert, R., Leusen, K. (1995). Determination of pesticides in aqueous samples by solid-phase microextraction coupled to gas chromatography-mass spectrometry. *J. Am. Soc. Mass Spectrom.*, 6, 11-19.

CHAPTER 29

PHENOLS

29.1. GC SEPARATIONS OF UNDERIVATIZED PHENOLS AND DIHYDROXYBENZENES

A. Capillary columns
   1. General conditions for the separation of phenols: 30-m DB-5 column, 80–150 °C at 8 °C min^{-1}.
   2. Phenol, 2,6-xylenol, *m*-, *p*-, and *o*-cresols, 2,4,6-trimethylphenol, 2,5-xylenol, 2,4-xylenol, 2,3,6-trimethylphenol, 2,3-xylenol, 3,5-xylenol, and 3,4-xylenol: 25-m DB-1701 column, 75 (2 minutes)–140 °C at 4 °C min^{-1}.
   3. Phenol, 2,4,6-trichlorophenol, *p*-chloro-*m*-cresol, 2-chlorophenol, 2,4-dichlorophenol, 2,4-dimethylphenol, 2-nitrophenol, 4-nitrophenol, 2,4-dinitrophenol, and pentachlorophenol: 25-m DB-1701 column, 40 (2 minutes) to 210 °C at 6 °C min^{-1}.
   4. Dowtherm impurities: benzene, phenol, naphthalene, and dibenzofuran from diphenyl and diphenyl ether. (Diphenyl and diphenyl ether are not separated under these conditions.): 30-m DB-225 column, 75–215 °C at 8 °C min^{-1}. Run 40 minutes.
   5. Dihydroxybenzenes, catechol, and resorcinol: 30-m DB-WAX column, 150–230 °C at 10 °C min^{-1}.

29.2. DERIVATIZATION OF PHENOLS AND DIHYDROXYBENZENES

A. Add 250 μL of MSTFA or TRI–Sil/BSA formula D to approximately 1 mg of sample in a septum-stoppered vial. (TRI–Sil/BSA Formula D contains DMF, which may interfere in the GC separation of some low-boiling TMS derivatives.) Heat at 60 °C for 15 minutes.

Gas Chromatography and Mass Spectrometry
DOI: 10.1016/B978-0-12-373628-4.00029-0

29.3. GC Separations of Derivatized Phenols and Dihydroxybenzenes

A. Capillary columns
  1. Catechol–TMS, resorcinol–TMS, and hydroquinone–TMS (MW = 254 Da). (The presence of phosphoric acid interferes with this separation.): 30-m DB-1 column, 60 (5 minutes) to 250 °C at 4 °C min^{-1}.
  2. TMS derivatives of phenols and dihydroxybenzenes: 30-m DB-1 column, 60 (5 minutes) to 250 °C at 4 °C min^{-1}.

29.4. Mass Spectra of Phenols

A. Underivatized phenols
  1. Molecular ion: The $M^{+\bullet}$ peaks are very intense in phenol, methylphenol, and dimethylphenol, as would be predicted for aromatic compounds.
  2. Fragment ions: The losses of 28 Da and 29 Da from the molecular ions are characteristic. Methylphenol can be distinguished from dimethylphenol by comparing the $[M - 1]^+$ and the $[M - 15]^+$ peaks. Methylphenol has an intense $[M - 1]^+$ peak, whereas the $[M - 15]^+$ peak for dimethylphenol is more intense. Methylphenol can be distinguished from benzyl alcohol by the presence of an *m/z* 107 peak in the mass spectrum of methylphenol, which is characteristic of alkylphenols.

B. Sample mass spectrum of an underivatized phenol
The mass spectrum in Figure 29.1 shows a $M^{+\bullet}$ peak at *m/z* 122. The intense fragment ion peak at *m/z* 107 can be either of the following structures as suggested by Appendix Q; (R+dB = 4, see Chapter 5):

CH / OH — I; HO, CH_2 — II; H_3C, OH — III

Losses of 28 Da (CO) and 29 Da (HCO) from the molecular ion suggest that the hydroxyl group is attached to the benzene ring, thus eliminating structure I. Unlike the molecular ions of aliphatic alcohols, the molecular ions of phenols do not lose a molecule of water.

By preparing a TMS derivative, the presence of the hydroxyl group can be established. However, like the lack of differences in the spectra of the underivatized dimethylphenol, the mass spectra of the TMS derivatives

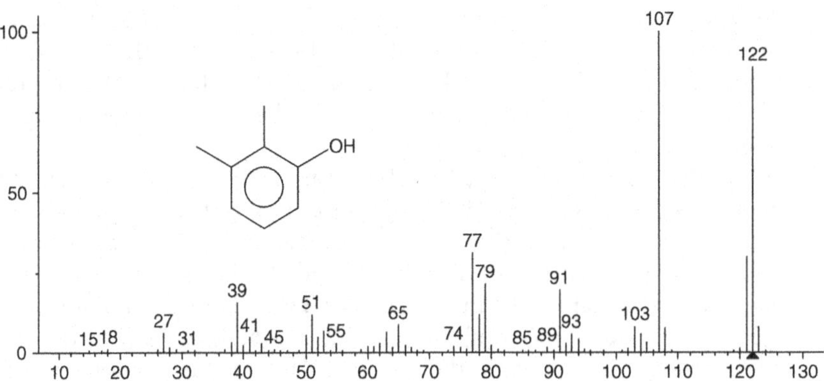

Figure 29.1 EI mass spectrum of 2,3-dimethylphenol.

will not allow for the determination of the position of the aromatic substitution. The regioisomers of dimethylphenol cannot be determined using only mass spectrometry. Differentiating these regioisomers in a GC/MS analysis may be possible using retention indices; however, just as ethylbenzene can be differentiated from the xylene regioisomers (Chapter 21), the spectrum of 1-phenylethanol can be differentiated from the spectra of dimethylphenol regioisomers in a way other than the fact that there is little or no ion current represented by peaks corresponding to $[M - 28]^+$ and $[M - 29]^+$. The intensity of the $[M - 28]^+$ and $[M - 29]^+$ peaks is very low in the spectra of all of the regioisomers of dimethylphenol (Figure 29.1). The intensity of the peak at *m/z* 77 in all the spectra of 1-phenylethanol in the *NIST08 Mass Spectral Database* is significantly less than the intensity of the peak at *m/z* 79 (Figure 29.2); whereas in all the spectra of the various regioisomers of

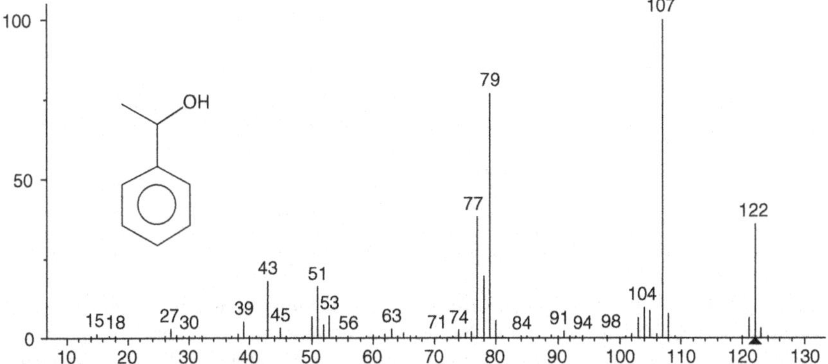

Figure 29.2 EI mass spectrum of 1-phenylethanol.

dimethylphenol, the intensity of the peak at *m/z* 77 is greater than the intensity of the peak at *m/z* 79—not by as much as the difference in intensities of the two peaks with these *m/z* values in the spectra of 1-phenylethanol but by enough that the difference is obvious (Figure 29.2).

Another very important point about the mass spectra of substituted phenols is the ortho effect. When the position of certain substituents are ortho to a group with a labile hydrogen, a rearrangement fragmentation will occur with the loss of a molecule and the formation of an odd-electron ion. Compare the mass spectra of 2,6-dichlorophenol (top) and 4,5-dichlorophenol (bottom) in Figure 29.3. The loss of HCl from the molecular ion followed by the loss of CO from the resulting ion is

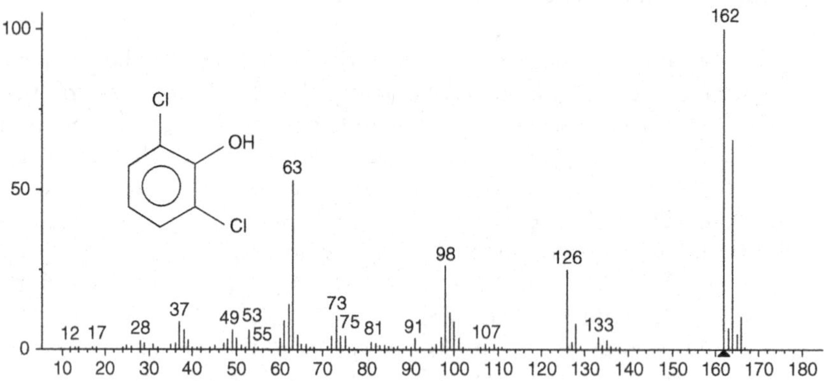

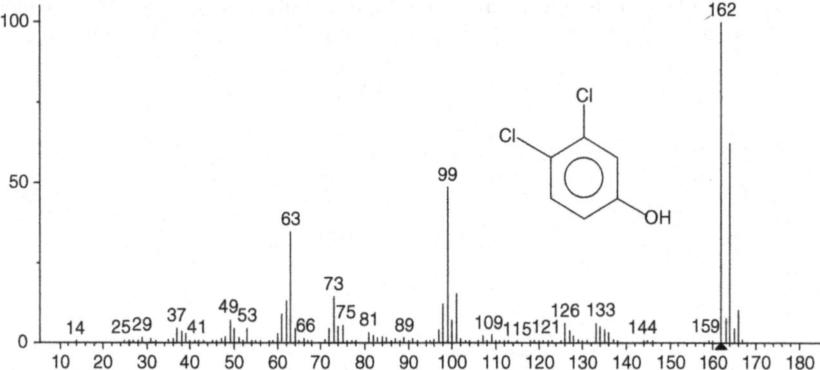

Figure 29.3 (Top) EI mass spectrum of 2,6-dichlorophenol; (bottom) EI mass spectrum of 4,5-dichloropheonol. Notice the presence of the peaks representing odd-electron fragment ions that have retained the Cl atom in the top spectrum that are missing from the bottom spectrum. These peaks represent $[M - HCl]^{+}$ and $[(M - HCl) - CO]^{+}$ ions.

obvious in the 2,6-isomer's mass spectrum. The missing evidence of this same fragmentation is just as obvious in the mass spectrum of 4,5-isomer.

This ortho effect results in a mass spectral differentiation of *ortho*-isomers from *meta*- and *para*-isomers.

C. Phenols as TMS derivatives
   1. Molecular ion: Intense $M^{+\bullet}$ peaks are observed.
   2. Fragment ions: Benzyl alcohol–TMS can be distinguished from a cresol–TMS because benzyl alcohol has its base peak at *m/z* 91. The *o*-, *m*-, and *p*-cresol derivatives can be distinguished from each other and from benzyl alcohol by the relative intensities of the peaks at *m/z* 91.

Derivative–TMS	Abundant ions
Phenol–TMS	151, 166
Cresol–TMS	165, 180, 91
Benzyl alcohol–TMS	91, 165, 135, 180
Xylenol–TMS (six isomers)	179, 194, 105

If a hydroxylated phenol group is suspected after an initial analysis without derivatization, prepare the TMS derivative and reanalyze to determine the presence and number of hydroxyl groups.

29.5. Aminophenols

See Chapter 11.

29.6. Antioxidants*

A. GC separation of antioxidants
   1. 2,6-Di-*tert*-butyl-4-methylphenol (BHT), 2,4-di-*tert*-butylphenol, and 6-*tert*-butyl-2,4-dimethylphenol: 30-m Supelcowax-10 column, 50–200 °C at 5 °C min^{-1}.

B. Mass spectra of antioxidants

*For further information on polymer additives including antioxidants, see Cortes et al. [1] and Asamoto et al. [2] in Chapter 31.

Antioxidant	*m/z* values of intense peaks
2,6-Di-*tert*-butyl-*p*-cresol (BHT) (IONOL)	205, 57, 220
2,4-Di-*tert*-butyl phenol	191, 57, 192, 206
6-*Tert*-butyl-2,4-dimethylphenol	163, 135, 178

CHAPTER 30

Phosphorus Compounds

30.1. GC Separations

A. Capillary columns
   1. Hexamethylphosphoramide, pentamethylphosphoramide, tetramethylphosphoramide, trimethylphosphoramide, and the metabolite that was postulated to be

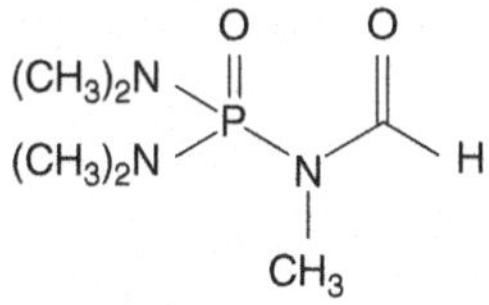

      30-m FFAP-DB column, 80–220 °C at 12 °C min^{-1}, or 30-m CW20M column, 60–200 °C at 10 °C min^{-1}.
   2. and $PHOCH_2CH_2P(CH_3)_2$

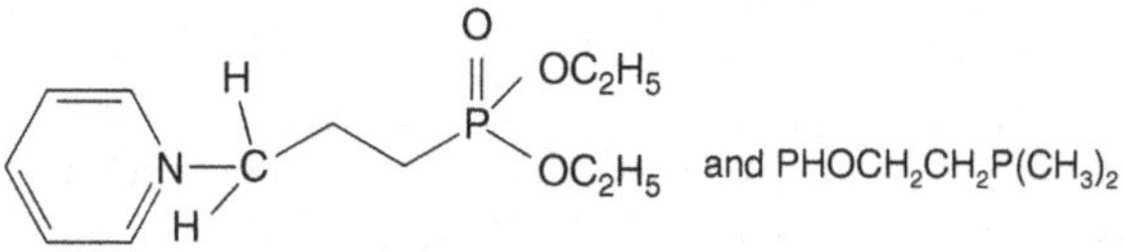

      30-m DB-1 column, 70–220 °C at 10 °C min^{-1}.
   3. Phosphorus pesticides (see Chapter 28): Diazinon, malathion, and similar compounds.
      a. 30-m DB-5 column, 150 (3 minutes) and 250 °C at 5 °C min^{-1} or 100–300 °C at 4 °C min^{-1}.
      b. 50-m CP-Sil 13CB column, 75–250 °C at 10 °C min^{-1} or
      c. 30-m CP-Sil 7 column, 190–240 °C at 4 °C min^{-1}.

Gas Chromatography and Mass Spectrometry
DOI: 10.1016/B978-0-12-373628-4.00030-7

30.2. Mass Spectra of Phosphorus Compounds

A. Alkyl phosphites and alkyl phosphonates

1. General formulas:

$$H{-}P(=O)(OR)_2 \quad \text{and} \quad R^1{-}P(=O)(OR_2)_2$$

2. Characteristic fragment ions

$$R^1{-}P(=O)(OR_2)_2 \longrightarrow \left[R^1{-}P(OH)_3\right]^+$$

If R^1 = H, m/z 83 is observed.
If R^1 = CH_3, m/z 97 is observed.
If R^1 = C_2H_5, m/z 111 is observed.
If R^1 contains a γ-hydrogen, then the McLafferty rearrangement occurs.

B. Alkyl phosphates

1. General formula: $(RO)_3PO$.
2. Characteristic fragment ion peaks and their relative intensities.

$(C_2H_5O)_3$ PO	>	$(C_2H_5O)_2$ $P(OH)_2$	>	$C_2H_5OP(OH)_3$	>	$P(OH)_4$
m/z 182		m/z 155		m/z 127		m/z 99

C. Phosphoramides (hexamethylphosphosphoramide)

1. Formula: $[(CH_3)_2N]_3PO$.
2. Molecular ion: The $M^{+\bullet}$ peak is easily observed.
3. Characteristic fragment ion: A very intense fragment ion peak is observed at m/z 135 corresponding to $[(CH_3)_2N]_2PO$ (Figure 30.1).

D. Mass spectra of phosphorus pesticides (see Chapter 28)

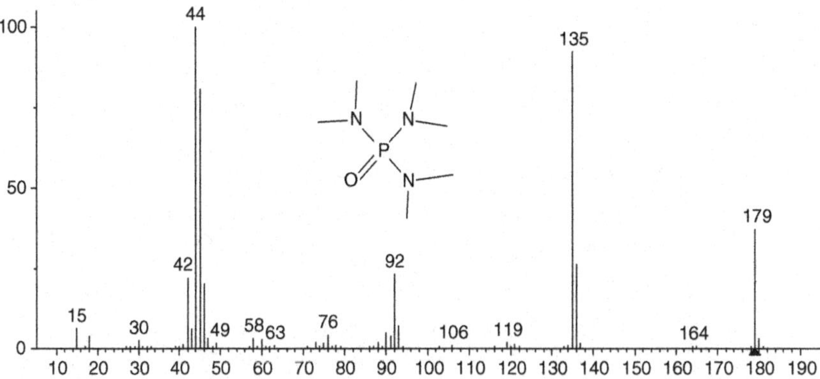

Figure 30.1 EI mass spectrum of hexamethylphosphoramide.

CHAPTER 31

Plasticizers and Other Polymer Additives (Including Phthalates)

31.1. GC Separations

A. Capillary columns

1. Triethylcitrate (MW = 276 Da), dibutylphthalate (MW = 278 Da), dibutylsebacate (MW = 314 Da), acetyltributylcitrate (MW = 402 Da), trioctylphosphate (MW = 444 Da), di-(2-ethylhexyl)adipate (MW = 370 Da), di-(2-ethylhexyl)phthalate (MW = 390 Da), and di-(*n*-decyl)phthalate (MW = 444 Da): 15-m DB-1 column, 75–275 °C at 10 °C min^{-1}.
2. Dimethylphthalate (MW = 194 Da), diethylphthalate (MW = 222 Da), butylbenzylphthalate (MW = 312 Da), di-(2-ethylhexyl)phthalate (MW = 390 Da), and dioctylphthalate (MW = 390 Da): 15-m DB-1 column, 150–275 °C at 15 °C min^{-1}.
3. BHT (MW = 220 Da), Tinuvin P (MW = 225 Da), acetyltributylcitrate (MW = 402 Da), butylbenzylphthalate (MW = 312 Da), stearamide (MW = 283 Da), eurcylamide (MW = 337 Da), and Irganox 1076 (MW = 530 Da): 15-m DB-1 column, 150–300 °C at 15 °C min^{-1}.
4. Dimethyl terephthalate impurities: acetone (MW = 58 Da), benzene (MW = 78 Da), toluene (MW = 92 Da), xylene (MW = 106 Da), methylbenzoate (MW = 150 Da), CHO–$(C_6H_4)CO_2CH_3$ (MW = 164 Da), NC–$(C_6H_4)CO_2CH_3$ (MW = 161 Da), methyltoluate (MW = 150 Da), *p*-nitromethylbenzoate (MW = 181 Da), dimethylterephthalate (MW = 194 Da), methyl DMT (MW = 208 Da), $CH_3OC(O)$–C_6H_4–$CH(OCH_3)_2$ and isomer (MW = 210 Da), and $CH_3OC(O)$–C_6H_4–C_6H_3–$(CO_2CH_3)_2$ (MW = 328 Da): 30-m DB-1 column, 100–250 °C at 10 °C min^{-1}.

Gas Chromatography and Mass Spectrometry
DOI: 10.1016/B978-0-12-373628-4.00031-9

31.2. Mass Spectra

Phthalate Isophthalate Terephthalate

A. Phthalates, isophthalates, and terephthalates

If the MW is greater than 190 Da and there is an intense peak at *m/z* 149, this suggests an ester of a benzene dicarboxylic acid where R^1 or R^2 or both are larger than CH_3. If *m/z* 149 is the most intense peak (the base peak), the mass spectrum probably represents a phthalate where both R^1 and R^2 are an ethyl group or larger (see Figure 31.1). The most intense peaks in the mass spectra of isophthalates and terephthalates represent the $[M - OR]^+$ ion.

m/z 149

The following peaks are important peaks to examine (if present):

- $M^{+\bullet}$ peak (small or nonexistent; when R is C_5 or higher, there is a better chance of seeing the $M^{+\bullet}$ peak).

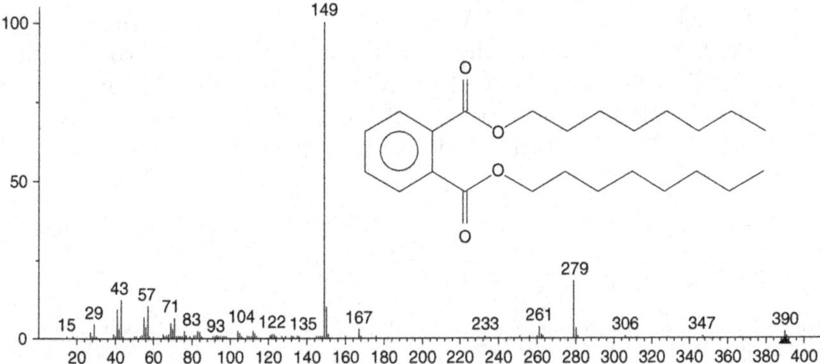

Figure 31.1 EI mass spectrum of di-*n*-octylphthalate.

- $[M - (R - H)]^{+\bullet}$ or $[M - (R - 2H)]^{+}$.
- $[M - OR]^{+}$ can be identified by associating with a discernible peak 18 *m/z* units higher (which represents $[M - (R - 2H)]^{+}$). The *m/z* value of the $[M - OR]^{+}$ peak can be used to determine the molecular weight of the analyte.*
- R^{+} (may be no more than 5%).

1. Dimethyl phthalate (194 Da): *m/z* 119 (0.5%), *m/z* 120 (1.9%), *m/z* 136 (6%) = $[M - (COOCH_3)]^{+}$, *m/z* 163 (100%) = $[M - (OCH_3)]^{+}$, and *m/z* 194 (11%) = $M^{+\bullet}$.
2. Dimethyl isophthalate (194 Da): *m/z* 119 (4%), *m/z* 120 (6%), *m/z* 136 (24%), *m/z* 163 (100%), and *m/z* 194 (24%).
3. Dimethyl terephthalate (194 Da): *m/z* 119 (14%), *m/z* 120 (23%), *m/z* 135 (20%), *m/z* 163 (100%), and *m/z* 194 (25%).
4. Diethyl phthalate (222 Da): *m/z* 149 (100%) = $[M - (OC_2H_5 + C_2H_4)]^{+}$, *m/z* 177 (28%) = $[M - (OC_2H_5)]^{+}$, *m/z* 194 (20%) = $[M - (C_2H_4)]^{+}$, and *m/z* 222 (3.4%) = $M^{+\bullet}$.
5. Diethyl terephthalate (222 Da): *m/z* 149 (45%), 166 (28%) = $[M - 2(C_2H_4)]^{+}$, *m/z* 177 (100%), *m/z* 194 (21%), and *m/z* 222 (11%).
6. Di-*n*-propylphthalate (250 Da): *m/z* 43 (6.6%), *m/z* 149 (100%), *m/z* 209 (8%), and *m/z* 250 (0.33%) = $M^{+\bullet}$.
7. Di-*n*-butylphthalate (278 Da): *m/z* 57 (5%), *m/z* 149 (100%), *m/z* 223 (5.6%), and *m/z* 278 (0.78%) = $M^{+\bullet}$.
8. Di-*n*-octylphthalate (*m/z* 390): *m/z* 57 (36%), *m/z* 71 (23%), *m/z* 149 (100%), *m/z* 167 (34%), and *m/z* 390 (1%) = $M^{+\bullet}$.

B. Other additives

Excellent discussions of the mass spectra of polymer additives are given by Cortes et al. [1] and Asamoto et al. [2].

REFERENCES

1. Cortes, H. J., Bell, B. M., Pfeiffer, C. D., Graham, J. D. (1989). Multidimensional Chromatography using On-line Coupled Microcolumn Size Exclusion Chromatography-Capillary Gas Chromatography-Mass Spectrometry for Determination of Polymer Additives. *J. Microcolumn. Sep.*, 1(6), 278–288.
2. Asamoto, B., Young, J. R., Citerin, R. J. (1990). Laser desorption Fourier-transform ion cyclotron resonance mass spectrometry of polymer additives. *Anal. Chem.*, 62, 61–70.

* The $[M - OR]^{+}$ ion has the structure of $C_6H_4(CO_2)OR$. The $C_6H_4(CO)_2O$ moiety has a mass of 148 Da. Subtract this value from the *m/z* value of the $[M - OR]^{+}$ ion (261 in the example of di-*n*-octylphthalate or 261 − 148 = 113). Add this value to the *m/z* value of the $[M - OR]^{+}$ ion and add 16 for the second OR oxygen (261 + 113 + 16 = 390); this results in the MW of the analyte.

CHAPTER 32

Prostaglandins (MO–TMS Derivatives)

Prostaglandins are members of a group of lipid compounds that are derived from fatty acids enzymatically. They have important functions in the bodies of animals. The name prostaglandin is derived from the prostate gland. When prostaglandin was first isolated from seminal fluid in 1935 by the Swedish physiologist Ulf von Euler [1] and independently by M.W. Goldblatt [2], it was believed to be part of the prostatic secretions. In fact, prostaglandins are produced by the seminal vesicles. It was later shown that many other tissues secrete prostaglandins for various functions.

Every prostaglandin contains 20 carbon atoms, including a 5-carbon ring. They are mediators and have a variety of strong physiological effects, such as regulating the contraction and relaxation of smooth muscle tissue. Although they are technically hormones, they are rarely classified as such.

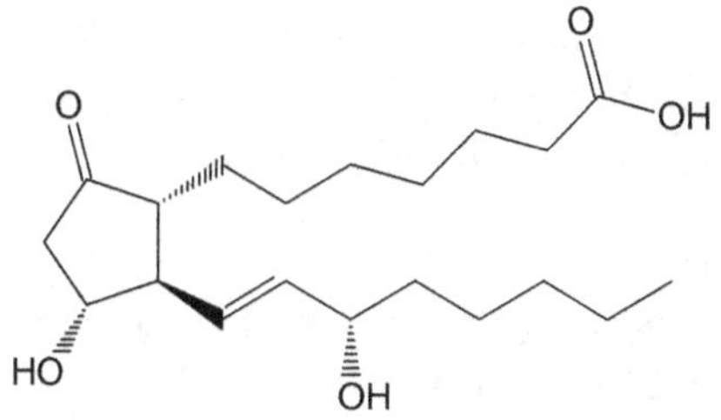

Prostaglandin A's (PGAs) and prostaglandin B's (PGBs) may be formed by the dehydration of PGEs. When the free-acid-form PGA_1 was analyzed by introduction on a direct insertion probe using EI mass spectrometry, the mass spectrum exhibited a $M^{+\bullet}$ peak (*m/z* 336) that was ~30% of the intensity of the base peak, which was at *m/z* 247, which could be the result of the loss of 71 ($[M - C_5H_{11}]^+$) from the molecular ion followed by the loss of a molecule of water. There was an $[M - H_2O]^+$ peak at *m/z* 318 (~90% of the intensity of the $M^{+\bullet}$ peak) and a peak at *m/z* 265 ($[M - C_5H_{11}]^+$, ~40% of the base peak). The problem with these free acids and GC/MS is that the acids will not fly through the GC. Even when the acid OH is converted to a methyl ester, the remaining hydroxyls will not allow for good chromatography. The prostaglandins are usually derivatized to form a methyl oxime of the keto group and TMS adducts of the three hydroxyls.

Gas Chromatography and Mass Spectrometry
DOI: 10.1016/B978-0-12-373628-4.00032-0

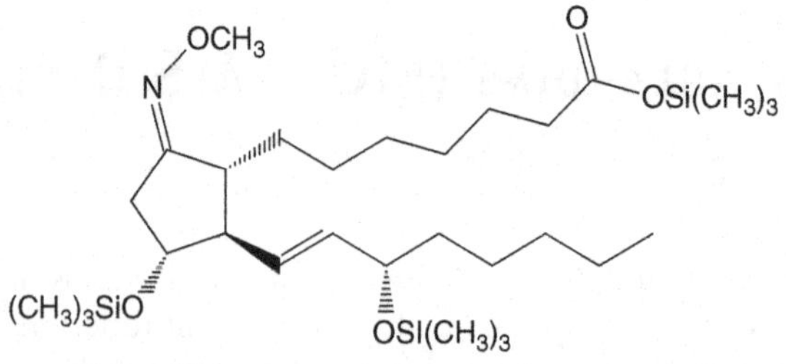

32.1. Derivatization (MO–TMS)

Evaporate the sample extract to dryness with clean dry nitrogen. Store the dried extract at dry-ice temperatures until the sample is derivatized (otherwise the PGEs and the hydroxy-PGEs may convert to PGAs and PGBs). Add 0.25 mL of MOX reagent to the dried extract and let the reaction mixture stand at room temperature overnight. Evaporate to dryness with clean dry nitrogen. Add 0.25 mL of MSTFA or BSTFA reagent. Let the reaction mixture stand at room temperature for at least 2 hours. Evaporate the excess reagent with flowing nitrogen gas. Dissolve the residue in a minimum amount of hexane for GC/MS analysis.

32.2. GC Separation of Derivatized Prostaglandins

A. PGA_1–MO–TMS, PGB_1–MO–TMS, PGB_2–MO–TMS, PGE_1–MO–TMS, PGE_2–MO–TMS, PGF_1a–TMS, 6–KETOPGF_1–MO–TMS, PGF_2a–TMS, and thromboxane B_2–TMS

1. 30-m DB-17 column, 175–270 °C at 8 °C min^{-1}; injection port at 280 °C.
2. 30-m DB-1701 column, 100 (2 minutes) to 275 °C at 5 °C min^{-1}; injection port at 280 °C.
3. 30-m DB-225 column, 100–225 °C at 15 °C min^{-1}; injection port at 280 °C.*

---

*These GC conditions are suitable for analyzing many prostaglandins, thromboxanes, leukotrienes, and other metabolites of arachidonic acid, such as the hydroxyeicosatetraenoic (HETE) acids. However, the 5-, 12-, and 15-HETE isomers are difficult to separate using GC methods. Sometimes the methyl ester–TMS derivatives provide a better GC separation; or for ketoprostaglandins, the MO–methyl ester–TMS derivatives often give a better separation.

32.3. Mass Spectra of MO–TMS Derivatives of Prostaglandins

A. Molecular ion
   The $M^{+\bullet}$ peaks of the MO–TMS derivatives are usually of low intensity but are detectable (see Figures 32.1 and 32.2).

B. Characteristic fragment ions
   Characteristic fragment ion peaks represent $[M - 15]^+$ and $[M - 31]^+$ ions, with the peak representing the $[M - 31]^+$ ion being more intense. The $[M - 31]^+$ peak is intense only for ketoprostaglandins. By plotting mass chromatograms of the following fragment ions, the presence or absence of a particular prostaglandin can be determined even though complete GC resolution may not be obtained.

Prostaglandin	Characteristic abundant ions
PGA_1–MO–TMS (MW = 509 Da)	*m/z* 388, 419, 438, 478
PGB_1–MO–TMS (MW = 509 Da)[a]	*m/z* 388, 438, 478, 494
PGB_2–MO–TMS (MW = 507 Da)	*m/z* 386, 476, 492, 507
PGE_1–MO–TMS (MW = 599 Da)	*m/z* 426, 478, 528, 568, 584

[a] PGB_1 can be distinguished from PGA_1 by the more intense *m/z*-478 ion relative to the other fragment ions in the mass spectrum of PGB_1–MO–TMS.

Prostaglandin	Most abundant ions
PGE_2–MO–TMS (MW = 597 Da)	*m/z* 436, 566, 582, 476
PGF_1a–TMS (MW = 644 Da)[a]	*m/z* 367, 368, 438, 483
6-KETOPGF_1–MO–TMS (MW = 687 Da)	*m/z* 436, 476, 566, 656
PGF_2a–TMS (MW = 642 Da)	*m/z* 391, 462, 481, 552
Thromboxane B_2–TMS (MW = 658 Da)	*m/z* 211, 301, 387
12-HETE–TMS (MW = 464 Da)	*m/z* 213, 324, 449

[a] In the mass spectra of TMS derivatives of nonketoprostaglandins, the molecular ion is usually not observed. Generally, the fragment ions are [M – 15]+, [M – 71]+, [M – 90]+, [M – 161]+, and [M – 180]+.

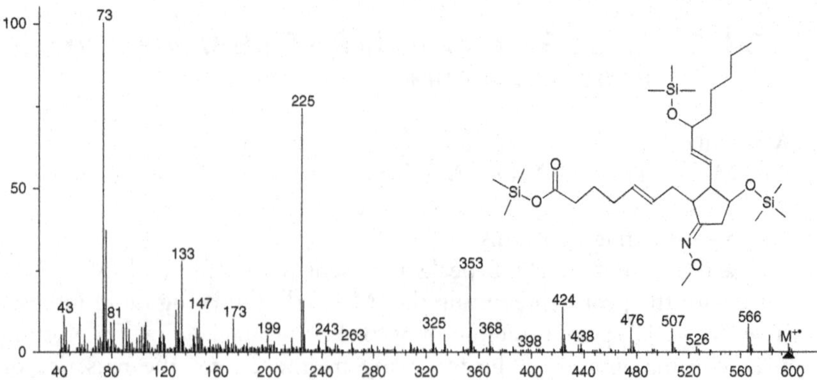

Figure 32.1 PGE$_2$–methoxime–diTMS.

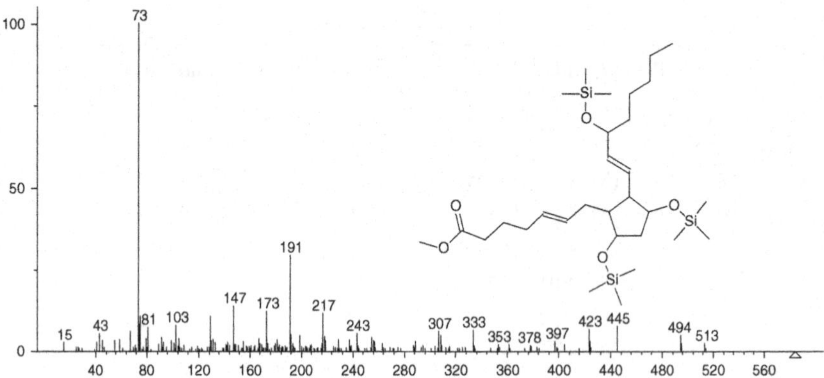

Figure 32.2 PGF$_2$–methyl–triTMS.

REFERENCES

1. Von Euler, U. S. (1935). Über die spezifische blutdrucksenkende Substanz des menschlichen Prostata-und Samenblasensekrets. *Wien. Klin. Wochenschr.*, 14(33), 1182–3. http://www.springerlink.com/content/g602m231xpw85226/fulltext.pdf
2. Goldblatt, M. W. (1935). Properties of human seminal plasma. *J. Physiol.*, 84(2), 208–18.

CHAPTER 33

Solvents and Their Impurities

There are generally two types of analyses that are requested with reference to solvents. The first is the identification of residual solvents in products and the second is the identification of impurities in common industrial solvents. Certain GC conditions have been found to separate most of the common solvents. Always examine the mass spectra at the front and back of the RTIC chromatographic peaks to determine if they are homogeneous. Also remember that isomers may not be detected by this approach if they are not separated. AMDIS can be very beneficial for the data analysis of these types of samples.

33.1. GC Separations of Industrial Solvent Mixtures

A. Capillary columns
   1. Acetone, THF, methanol, ethyl acetate, isopropyl acetate, isopropyl alcohol, ethanol, *n*-propyl alcohol, toluene, and 2-methoxyethanol
      a. 30-m DB-FFAP column or 30-m HP-FFAP column, 60–200 °C at 6 °C min^{-1}.
      b. 25-m DB-WAX column or 25-m CP-WAX 52CB column, 50 (2 minutes) to 65 °C at 1 °C min^{-1}; 65–150 °C at 10 °C min^{-1}.
   2. Ethanol, acetonitrile, acetone, diethyl ether, pentane, ethyl acetate, and hexane: 25-m Poraplot Q column, 60–200 °C at 6 °C min^{-1}.
   3. Benzene, propyl acetate, allyl acetate, 1-pentanol, cyclohexanone, cyclohexanol, dicyclohexyl ether, cyclohexyl valerate, butyric acid, valeric acid, caproic acid, 1,5-pentanediol, dicyclohexyl succinate, and dicyclohexyl glutarate: 30-m DB-FFAP column, 60–200 °C at 6 °C min^{-1}.

33.2. GC Separations of Impurities in Industrial Solvents

A. Capillary columns
   1. Diaminotoluene impurities: dichlorobenzene, toluidine, nitrotoluene isomers, and dinitrotoluene isomers: 30-m DB-17 column, 100–250 °C at 8 °C min^{-1}.

Gas Chromatography and Mass Spectrometry
DOI: 10.1016/B978-0-12-373628-4.00033-2

2. 3,4-Dichloroaniline impurities: aniline, chloroaniline, dichloroaniline isomers, trichloroaniline isomers, tetrachloroaniline, tetrachloroazobenzene, and pentachloroazobenzene: 30-m DB-1 column, 100–280 °C at 10 °C min^{-1}.
3. Diethylene glycol impurities: acetaldehyde, 2-methyl-1,3-dioxolane, dioxane, and diethylene glycol: 30-m Nukol column, 50–220 °C at 8 °C min^{-1}.
4. *N,N*-dimethylacetamide (DMAC) impurities: *N,N*-dimethylacetonitrile, *N,N*-dimethylformamide, DMAC, *N*-methylacetamide, and acetamide: 60-m DB-WAX column, 60–200 °C at 7 °C min^{-1}.
5. Ethanol impurities: ethyl acetate, methanol, *n*-propyl alcohol, isobutyl alcohol: 25-m CP-WAX 57 column for glycols and alcohols (Chrompack cat. no. 7615), 50 (5 minutes) to 180 °C at 8 °C min^{-1}.
6. Isopropyl alcohol impurities: ethanol, *n*-propyl alcohol, and *t*-butyl alcohol: 30-m NUKOL column, 45–200 °C at 6 °C min^{-1}.
7. *t*-Butyl alcohol, *n*-butyraldehyde, methyl ethyl ketone, 2-methylfuran, tetrahydrofuran, 4-methyl-1,3-dioxolane, 2-methyl THF, 3-methyl THF, and tetrahydropyran: 30-m DB-1 column at 60 °C.
8. Toluene impurities: benzene, ethylbenzene, and *o*-, *m*-, and *p*-xylenes: 25-m DB-WAX column, 50 (2 minutes) to 180 °C at 5 °C min^{-1}.

33.3. Mass Spectra of Solvents and Their Impurities

Solvents and their impurities represent a wide class of compound types; therefore, a discussion of common mass spectral features is meaningless. However, most of the mass spectra for all solvents can be found in the *NIST08 Mass Spectral Database* and/or the *Wiley Registry of Mass Spectra*, 9th ed. The Demo Database containing mass spectra of 2,378 compounds has spectra for most of the common solvents.

One important point regarding solvent impurities is that some distributors of high-purity solvents like reagent-grade dichloromethane will add compounds (like cyclohexanone) for various reasons (such as antioxidation). These added compounds can interfere with a GC/MS analysis and every effort possible must be made to be aware of such *contamination*.

CHAPTER 34

STEROIDS

34.1. GC SEPARATION OF UNDERIVATIZED STEROIDS

A. Animal or plant steroids
   1. 30-m DB-17 column, 225–275 °C at 5 °C min^{-1}; injection port at 290–300 °C. (Dissolve the dried extract in the minimum amount of methylene chloride or toluene.)

34.2. DERIVATIZATION OF STEROIDS

A. Preparation of TMS Derivatives: Add 0.25 mL of TRI-Sil TBT reagent (the only reagent that was found to react with the hydroxyl group in position 17) and 0.25 mL of pyridine to the dried extract. Cap tightly and heat at 60 °C for 1–12 hours or overnight.

B. Preparation of Methoxime (MO)–TMS Derivatives (especially for hydroxyketosteroids, which may decompose under the given GC conditions unless the MO–TMS derivatives are prepared): Add 0.25 mL of MO hydrochloride in pyridine to the dried extract. Let stand for 3 hours at 60 °C or overnight at room temperature. Evaporate to dryness with clean dry nitrogen. Add 0.25 mL of TRI-Sil TBT and 0.25 mL of pyridine to the dried reaction mixture. Cap tightly and heat at 60 °C for a few hours or overnight at room temperature.

34.3. GC SEPARATION OF DERIVATIZED STEROIDS

A. TMS derivatives
   1. 30-m DB-17 column, 225–275 °C at 5 °C min^{-1}.
   2. 15-m CP-Sil 5 column or 15-m CP-Sil 8 column, 200–300 °C at 15 °C min^{-1}.

B. MO–TMS derivatives
   1. 30-m DB-17 column, 100–200 °C at 15 °C min^{-1}, then 200–275 °C at 5 °C min^{-1}; injection port temperature at 280–300 °C.

Gas Chromatography and Mass Spectrometry
DOI: 10.1016/B978-0-12-373628-4.00034-4

2. 30-m CP-Sil 5 column 100–210 °C at 20 °C min^{-1}, then 210–280 °C at 2 °C min^{-1}.

34.4. Mass Spectra of Underivatized Steroids

A. Molecular ion

Because of low volatility, most steroids are derivatized before analysis by GC/MS. $M^{+\bullet}$ peaks are usually observed for steroids sufficiently volatile to be analyzed underivatized by GC/MS (see Figure 34.1). Some important steroids in urine include estrone, estradiol, estriol, pregnanediol, and 17-ketosteroids, which can be analyzed by GC/MS as the TMS or the MO–TMS derivatives. The plant steroids such as camposterol, ergosterol, stigmasterol, cholestanol, and sitosterol are generally analyzed as the TMS derivatives.

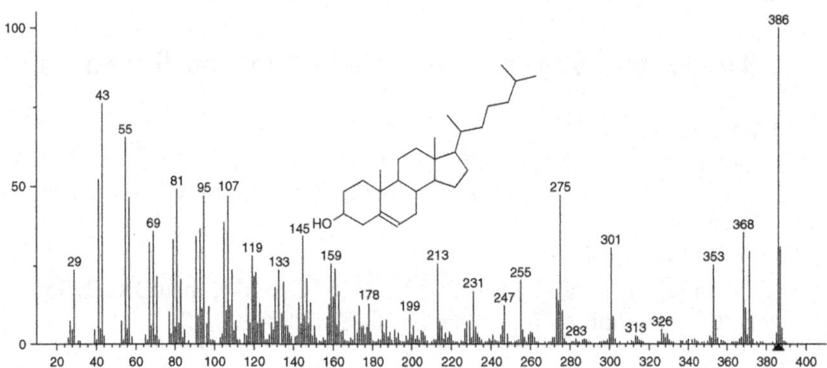

Figure 34.1 EI mass spectrum of cholesterol.

B. Characteristic fragment ions

The most characteristic cleavage is the loss of carbons 15, 16, and 17, together with the side chain and one additional hydrogen. It is possible to determine the elemental composition of the side chains of steroids by the difference in the mass between the molecular ion and an intense peak more than 15 *m/z* units below the $M^{+\bullet}$ peak. Typically, this ion corresponds to $[M - (R + 42)]^+$ where 42 Da is the mass of the C_3H_6 moiety. If there is no side chain, as in the cases of estrone, estradiol, and estriol, then the substitution on the D-ring can be determined using the following losses from the molecular ions:

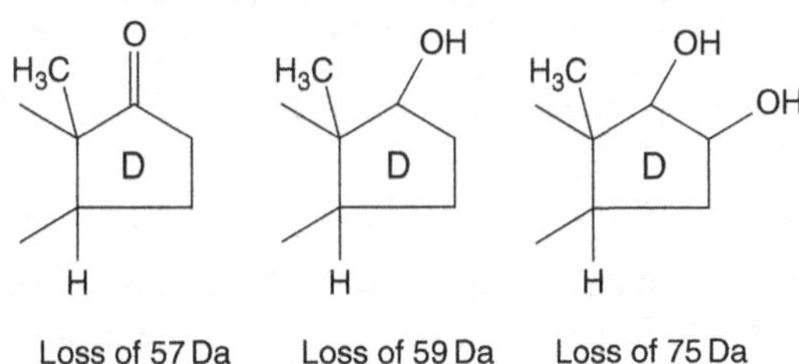

34.5. Mass Spectra of TMS Derivatives of Steroids

Plot a mass chromatogram for *m/z* 73 to determine if a TMS derivative was prepared and which GC peak(s) to examine.

A. Molecular ion

Generally, the $M^{+\bullet}$ peak is observed, but not always, as in the case of pregnanediol–TMS (MW = 464 Da). Again, if two high *m/z* value peaks are not observed that are 15 units apart, then add 15 to the highest *m/z* value peak observed to deduce the *m/z* value of the molecular ion.

B. Fragment ions

Common losses from the molecular ions include 90 Da and 105 Da. See Figure 34.2 for the mass spectrum of the TMS derivative of cholesterol.

34.6. Mass Spectra of MO–TMS Derivatives

A. Molecular ion

The $M^{+\bullet}$ peaks of MO–TMS derivatives are generally more intense than those of the TMS-only derivatives.

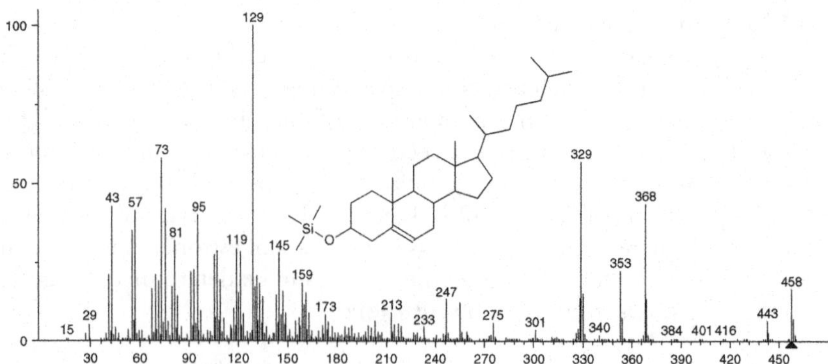

Figure 34.2 EI mass spectrum of cholesterol–TMS.

B. Fragment ions

The $[M - 31]^+$ fragment ion peak is characteristic and is more intense than the $[M - 15]^+$ peak. If two high *m/z* value peaks are observed that are 16 units apart, add 31 to the *m/z* value of the more intense peak and 15 to the highest *m/z* value peak to deduce the *m/z* value of the molecular ion of the derivative. By subtracting 29 Da for each keto group and 72 Da for each hydroxyl group, the original MW of the steroid can be determined. For example, in the case of androsterone-MO-TMS, the $M^{+\bullet}$ peak is at *m/z* 391. By subtracting 101 (29 + 72) Da, the original MW of 290 Da is deduced.

CHAPTER 35

SUGARS (MONOSACCHARIDES)

35.1. GC SEPARATION OF DERIVATIZED SUGARS

A. Monosaccharides

1. Preparation of the TMS derivative: Add 0.5 mL of TRI-Sil Z reagent (trimethylsilylimidazole in pyridine) to 1–5 mg of the sample. (This derivatizing preparation does not react with amino groups and tolerates the presence of water.) Heat in a sealed vial at 60 °C until the sample is dissolved. An alternate method is to let the reaction mixture stand at room temperature for at least 30 minutes (or overnight). This procedure is not appropriate for amino sugars.
2. GC separation of TMS derivatives: arabinose, fucose, xylose, mannose, galactose, α-glucose, and β-glucose: 30-m DB-1 column, 60–250 °C at 8 °C min^{-1}.

B. Amino sugars (glucosamine and galactosamine as TMS derivatives)

1. Preparation of TMS derivative: To derivatize the amino sugars as well as the non-amino sugars, substitute TRI-Sil TBT or TRI-Sil/BSA (Formula P) reagent for TRI-Sil Z and follow the procedure given in Section 35.1.A.1.
2. GC separations of amino sugars as TMS derivatives: 30-m DB-1701 column, 70–250 °C at 4 °C min^{-1}. (Depending on the amounts present, complete GC separation may not be achieved.)

C. Sugar alcohols (as acetates)

1. Preparation of the acetate derivative: Evaporate the extract to dryness. Add 50 μL of three parts acetic anhydride and two parts pyridine. Heat at 60 °C for 1 hour. Evaporate to dryness with clean dry nitrogen and dissolve the residue in 25 μL of ethyl acetate.
2. GC separation of the acetate derivatives: Rhamnitol, fucitol, ribitol, arabinitol, mannitol, galacitol, glucitol, and inositol: 30-m CP-Sil 88 column at 225 °C for 60 minutes.

D. Reduced sugars (as acetates)

1. Preparation of the acetate derivative: Concentrate the aqueous mixture of saccharides to approximately 0.5 mL in a 20–50 mL container. Reduce the saccharides by adding 20 mg of sodium borohydride that has been dissolved carefully into 0.5 mL of water and let the reducing mixture stand at room temperature for at least

Gas Chromatography and Mass Spectrometry
DOI: 10.1016/B978-0-12-373628-4.00035-6

1 hour. Destroy the excess sodium borohydride by adding acetic acid until the gas evolution stops. Evaporate the solution to dryness with clean nitrogen. Add 10 mL of methanol and evaporate the solution to dryness. Acetylate with 0.5 mL (three parts acetic anhydride and two parts pyridine) overnight. Evaporate to a syrupy residue and add 1 mL of water. Evaporate again to dryness to remove the excess acetic anhydride. Dissolve the residue in 250 μL methylene chloride.

2. GC separation of reduced and acetylated sugars: 30-m CP-Sil 88 column at 225 °C.

35.2. MASS SPECTRAL INTERPRETATION

A. Mass spectra of TMS derivatives of sugars

1. Molecular weight

   In general, to deduce the molecular weight of the TMS derivative of sugars, add to 105 the highest mass observed (loss of $(CH_3)SiOH + {}^{\bullet}CH_3$).

   - 333.1374 is the highest mass observed for arabinose, ribose, ribulose, xylose, lyxose, and xylulose.
   - 347.1530 is the highest mass observed for fucose and rhamnose.
   - 435.1875 is the highest mass observed for sorbose, allose, altrose, galactose, gulose, idose, and mannose.

2. Fragmentation

   DeJongh et al. [1] have described the fragmentation of the TMS derivatives of sugars. Comparison of GC retention times together with the mass spectra is sufficient to identify the sugars. The mass spectra suggest certain structural features. For instance, peaks at *m/z* 191, 204, and 217 suggest a TMS hexose. If the peak at *m/z* 204 is more intense than the peak at *m/z* 217, the hexose is the pyranose form; but if the *m/z* 217 peak is most intense, then it is a furanose (Figure 35.1). Aldohexoses can be differentiated from ketohexoses by the peaks at *m/z* 435 for aldohexoses versus *m/z* 437 for ketohexoses.

m/z	Ion structure
191	$(CH_3)_3SiOCH{=}OSi(CH_3)_3]^+$
204	$(CH_3)_3SiOCH{=}CHOSi(CH_3)_3]^{+\bullet}$
217	$(CH_3)_3SiOCH{=}CH{-}CH{=}OSi(CH_3)_3]^+$

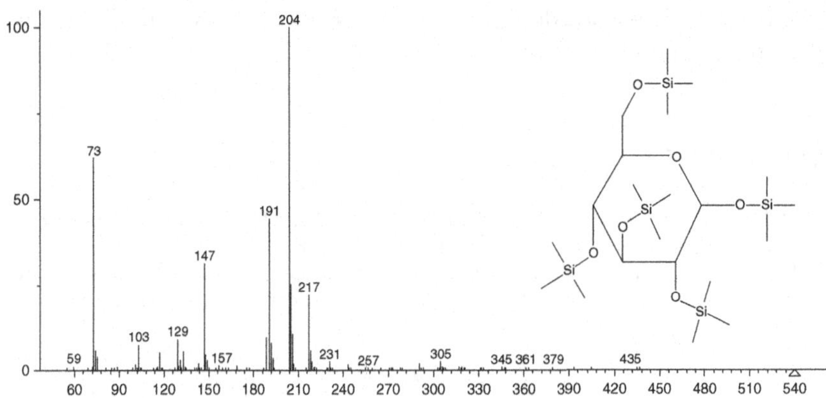

Figure 35.1 EI mass spectrum of TMS–α-glucose.

B. Sample mass spectrum

C. Mass spectra of amino sugars as TMS derivatives

1. Molecular ion: By plotting mass chromatograms for the *m/z* value of the $M^{+\bullet}$ peak and the accurate *m/z* value 362.1639, the presence of these amino sugars can be determined.

	Highest *m/z* value peak observed
Glucosamine	393.1643
Galactosamine	467.2375

2. Fragmentation: The TMS derivatives of amino sugars have their base peak at *m/z* 131.0765.

D. Mass spectra of sugar alcohol as acetates

1. Molecular ion: Chemical ionization using ammonia as reagent gas establishes the molecular weights of sugar acetates.
2. Fragmentation: Biemann et al. [2] have shown the fragmentation of alditol acetates. The acetates have intense ions at *m/z* 43, 103, and 145. They also appear to lose 42 Da, 59 Da, 60 Da, and 102 Da from their molecular ions

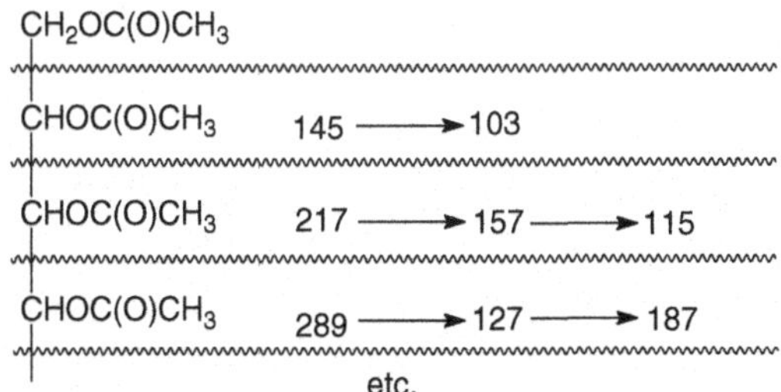

The mass spectrum of alditol acetates are easy to interpret as they fragment at each C–C bond as shown.

REFERENCES

1. DeJongh, D. C., Radfon, T., Hribar, J. D., et al., (1969). Analysis of derivatives of carbohydrates by GC/MS. *J. Am. Chem. Soc.*, 91, 1728.
2. Biemann, K., Schnoes, H. K., McCloskey, J. A. (1963). Applications of mass spectrometry to structure problems. Carbohydrates and their derivatives. *Chem. Ind. (London)*, 448, 449.

CHAPTER 36

SULFUR COMPOUNDS

36.1. GC SEPARATIONS

A. Capillary columns

1. Hydrogen sulfide, carbonyl sulfide, sulfur dioxide, and methylmercaptan: 25-m GS-Q column or 25-m Poraplot Q column, 50–120 °C at 5 °C min^{-1}.
2. Hydrogen sulfide, methane, carbon dioxide, ethane, and propane: 25-m Poraplot Q column or 25-m GS-Q column at 60 °C.
3. Carbon dioxide, carbonyl sulfide, hydrogen cyanide, propylene, butadiene, and furan: 25-m GS-Q column or 25-m Poraplot Q column, 60–200 °C at 6 °C min^{-1}.
4. Carbonyl sulfide, carbon disulfide, tetrahydrofuran, and toluene: 25-m DB-1 column, room temperature to 200 °C at 10 °C min^{-1}.
5. Methane, ethene, ethane, propene, acetaldehyde, methyl formate, butene, acetone, furan, dimethyl sulfide, isoprene, isobutyraldehyde, diacetyl, methylfuran, and isovaleraldehyde: 25-m Poraplot Q column or 25-m GS-Q column, 60–200 °C at 8 °C min^{-1}.
6. 2-Propanethiol, 1-propanethiol, 1-methyl-1-propanethiol, 2-methyl-1-propanethiol, 1-butanethiol, 1-pentanethiol, allyl sulfide, propyl sulfide, and butyl sulfide: 25-m GS-Q column or 25-m Poraplot Q column, 150–230 °C at 10 °C min^{-1}.
7. *n*-Propyl disulfide, and *n*-propyl trisulfide: 25-m DB-FFAP column at 100 °C.
8.

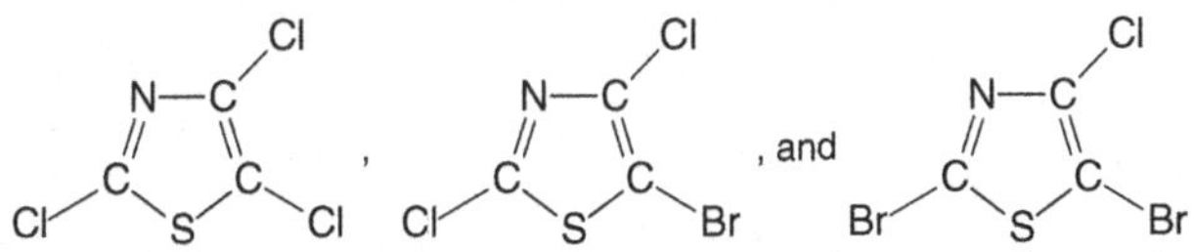

   2,4,5-trichlorothiazole (Cl, Cl, Cl), 2,4-dichloro-5-bromothiazole (Cl, Cl, Br), and 4-chloro-2,5-dibromothiazole (Br, Cl, Br)

   25-m DB-210 column at 120 °C.
9. Phenanthrylene sulfide and pyrene: 30-m DB-210 column, 150–225 °C at 5 °C min^{-1}.

Gas Chromatography and Mass Spectrometry
DOI: 10.1016/B978-0-12-373628-4.00036-8

36.2. Mass Spectra of Sulfur Compounds

A. Aliphatic thiols (mercaptans)

1. General formula: RSH.
2. Molecular ion: The presence of sulfur can be detected by the ^{34}S isotope (4.4% per atom of S) and the large mass defect of sulfur in accurate mass measurements. In primary aliphatic thiols, the $M^{+\bullet}$ peak's intensities range from 5% to 100% of the base peak.
3. Fragment ions: Loss of 34 (H_2S) *m/z* units from the molecular ion of primary (1°) thiols is characteristic. In the mass spectra of secondary (2°) and tertiary (3°) thiols, a loss of 33 rather than a loss of 34 is observed from the $M^{+\bullet}$ peak. If *m/z* 47 and 61 are reasonably represented by reasonably intense peaks and the *m/z* 61 peak is approximately 50% of the intensity of the *m/z* 47 peak, then the mass spectrum represents a primary thiol (see Figure 36.1). If the intensity of the peak at *m/z* 61 is much less than 50% of the intensity of the peak at *m/z* 47, then the mass spectrum represents a secondary or tertiary thiol. Also, remember that saturated sulfur compounds generally undergo α cleavage (breaking of the C_1–C_2 bond) to lose the largest alkyl group—analogous to the fragmentation of aliphatic alcohols.

B. Thioethers (sulfides)

1. General formula: R–S–R′.
2. Molecular ion: $M^{+\bullet}$ peaks usually have reasonably intensities. The mass spectra of cyclic thioethers exhibit intense $M^{+\bullet}$ peaks and an intense fragment ion peak due to a double homolytic cleavage (e.g., *m/z* 60 $[CH_2\text{–}S\text{–}CH_2]^+$) (See Scheme 36.1).
3. Fragmentation: The mass spectra of thioethers can be distinguished from those of 1°, 2°, and 3° thiols by the absence of the losses of

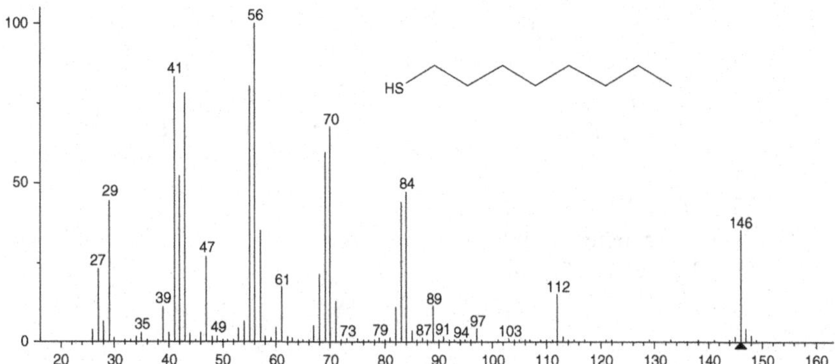

Figure 36.1 EI mass spectrum of 1-octanethiol.

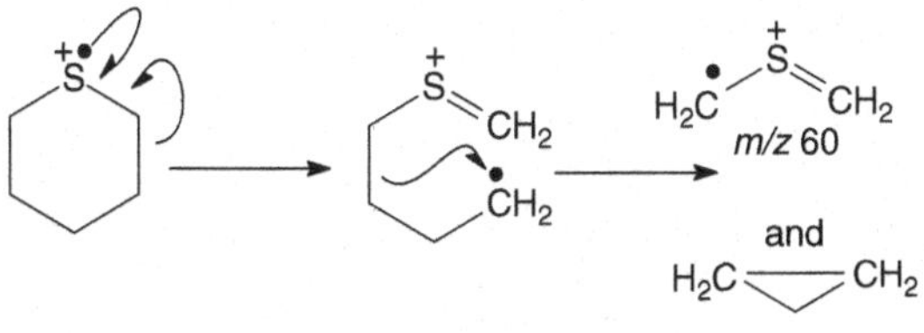

Scheme 36.1

either 33 Da or 34 Da from molecular ions. Fragmentation also occurs alpha to the sulfur atom with dominant loss of the larger alkyl group. Fragmentation can occur on either side of the sulfur atom with the rearrangement of a hydrogen atom. Fragmentation of thioethers is analogous to the fragmentation of aliphatic ethers (Chapter 16).

C. Aromatic thiols
   1. General formula: ArSH.
   2. Molecular ion: The mass spectra of aromatic thiols exhibit intense $M^{+\bullet}$ peaks as would be expected for the mass spectra of any aromatic compound (see Figure 36.2).
   3. Fragmentation: Characteristic losses from the molecular ion include 26 Da, 33 Da, and 44 Da (CS analogous to the loss of CO from the molecular ions of phenols). Ions characteristic of the phenyl group are also observed. Like the mass spectra of the corresponding aminophenols, there is an observed ortho effect. However, the spectrum of the *meta*-isomer differs considerably from the spectra of both the *para*- and *ortho*-isomers by a significant increase in the intensity of the peaks at *m/z* 80 and 81 and in the fact that the intensity of these two peaks are about equal in the spectrum of the

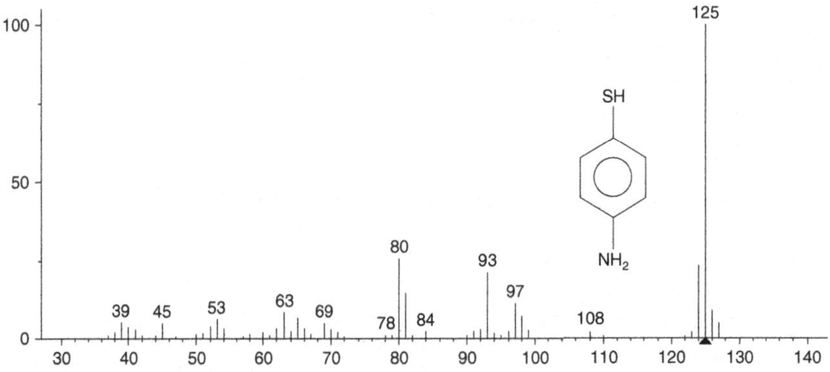

Figure 36.2 EI mass spectrum of 4-aminothiophenol.

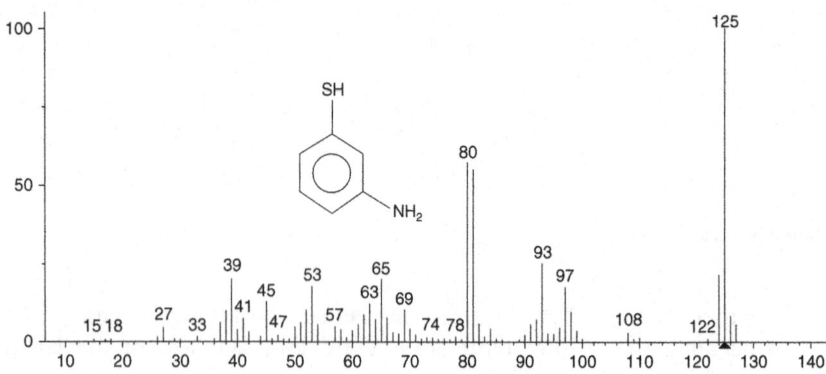

Figure 36.3 EI mass spectrum of 3-aminothiophenol.

meta-isomer, whereas the intensity of the *m/z* 80 peak is clearly dominant in the spectra of the other two isomers. This difference in the intensities of the peaks at *m/z* 80 and 81 along with retention index data may be sufficient for an unambiguous identification of the *meta*-isomer (see Figure 36.3).

D. Thioesters
   1. General formula: RC(O)SR′.
   2. Molecular ion: The $M^{+\bullet}$ peak in the mass spectra of thioesters is more intense than that of the corresponding oxygenated esters.
   3. Fragmentation: Peaks representing R^+ and RCO^+ are frequently the most intense observed. Peaks representing $RCOS^+$, SR'^+, and R'^+ are also usually observed.

E. Isothiocyanates
   1. General formula: RNCS.
   2. Molecular ion: $M^{+\bullet}$ peaks are usually observed.
   3. Fragmentation: A peak representing a characteristic fragment ion is observed at *m/z* 72 ($H_2C{=}N{=}C{=}S$). For R > *n*-C_5, the peak at *m/z* 115 is prominent.

SECTION 3

APPENDICES

APPENDIX A

Definitions of Terms Related to Gas Chromatography

Adjusted retention time (t'_R): The retention time for an analyte (t_R) minus that of an unretained compound (t_0): $t'_R = t_R - t_0$.
Adjusted retention volume (V'_R): The retention volume for an analyte (V_R) minus the retention volume of an unretained compound (V_0): $V'_R = V_R - V_0$.
Bleed (**a.k.a. column bleed**): An appearance of a background signal from a chromatographic system caused by the stationary phase or contamination of the inlet system. Column bleed usually increases with increasing column temperature.
Capacity factor (k'): Expresses the retention of a compound as compared to an unretained compound, such as methane: $k' = (t_R - t_0)/t_0$.
Capillary column: A tube usually made of deactivated fused silica that is not filled with packing material, typically of 0.15–0.53 mm i.d. and 10–60 m in length. The inner walls are coated with a liquid stationary phase and the outer walls are coated with a polyimide coating to reduce fragility. These are wall-coated open-tubular (WCOT) columns. Less common are porous-layer open-tubular (PLOT) columns in which a porous material rather than a liquid stationary phase is deposited on the inner walls.
Conditioning: Maintaining a column with a flow of carrier gas (mobile phase) at the maximum expected operating temperature of the column for some period of time (often overnight). The purpose is to remove excessive stationary phase and solvent used in the manufacturing process and to minimize bleed.
Carryover: Appearance of peaks in a chromatogram representing substances that were not in the sample but are residuals from an earlier analysis. See the table on GC troubleshooting in Appendix M.
Dead volume: The total volume in the system that is not swept by the carrier gas.
Derivatization: Chemical reaction of a compound that yields a product that is more volatile, more stable, and has improved GC behavior over the original substance. The derivative might also improve the mass spectral characteristics of the product; i.e., the product might have a more complex and unique mass spectrum than the original compound.

Gas Chromatography and Mass Spectrometry
DOI: 10.1016/B978-0-12-373628-4.00037-X

Efficiency: Degree of band broadening for a given retention time. Efficiency is expressed as the number of theoretical plates, N, or as the height equivalent to a theoretical plate, HETP. Also see **Theoretical plates.**

Gas holdup: V_0 is the volume of carrier gas that passes through the column to elute an unretained substance, such as argon or methane. The time required is t_0.

Height equivalent to a theoretical plate (HETP): A measure of the efficiency of a column usually expressed in millimeters.

HETP = L/N where L is the length of a column in millimeters and N is the number of theoretical plates. The reciprocal of HETP is also used to describe efficiency and is often expressed by the term "plates per meter."

Kováts retention indices: A method for reporting the retention time of a compound relative to the retention times of *n*-alkanes that were separated under the same chromatographic conditions (see The stationary phase under Section 2.3 in Chapter 2).

Linear velocity ***(u)***: The column length in centimeters divided by the number of seconds required for the carrier gas to pass through the column (see 2.3.3 Carrier Gas Considerations in Chapter 2). In GC/MS, the letter **u** is also used as the symbol for the unified atomic mass unit (see the term **dalton** (**Da**) in Appendix B).

McReynolds' constants: A system of classification of stationary phases that compares the retention of 10 compounds of different polarity to a reference nonpolar stationary phase (squalene) (see The stationary phase under Section 2.3 in Chapter 2).

Resolution: The degree to which two peaks are separated:

$$R = 2[(t_2 - t_1)/(w_1 + w_2)]$$

where t_2 and t_1 are the retention times of the two peaks, and w_1 and w_2 are the widths of the two peaks.

Retention time (**t_R**): The time required for a substance to pass through the column and be detected.

Retention volume (**V_R**): The volume of carrier gas that passes through the column to elute a substance.

Selectivity: The characteristics of the stationary phase that determine how far the peak maxima of two components will be separated in time.

Theoretical plates (**N**): A measure of the efficiency of the column or sharpness of the peaks. The more theoretical plates a column has, the narrower the peaks will be.

$$N = 5.555\left(\frac{t_R}{W_{1/2}}\right)^2$$

where $W_{1/2}$ is the width of the peak at half-height.

Van Deemter equation: An equation relating efficiency (HETP in millimeters) to linear flow velocity in a chromatographic column. The efficiency is expressed as the height equivalent to a theoretical plate:

$$\mathrm{HETP} = A + \frac{B}{u} + Cu$$

where A, B, and C are constants, and u is the linear velocity of the carrier gas. The first term refers to the tortuosity of the path of the analytes as they move through the packing material; therefore, A is dropped for open-tubular columns (see 2.3.3 Carrier Gas Considerations in Chapter 2).

APPENDIX B

Definitions of Terms Related to Mass Spectrometry

For a complete list of definitions, see Sparkman OD. *Mass Spectrometry Desk Reference*, 2nd ed. Pittsburgh, PA: Global View Publishing; 2006.

Accurate mass: More correctly, measured accurate mass is an experimentally determined mass of an ion, a radical, or a molecule that allows the elemental composition to be deduced. For ions with a mass ≤200, a measurement to ±5 ppm is sufficient for the determination of the elemental composition. The term *measured accurate mass* is used when reporting the mass to some number of decimal places (usually a minimum of three). A measured mass should be reported with a precision of the measurement. This term should not be used to describe the mass calculated from published tables.

Alpha (α) **cleavage** (**a special form of homolytic cleavage**): A fragmentation (homolytic cleavage) that results from one of the pair of electrons between the atom attached to the atom with the odd electron and an adjacent atom that pairs with the odd electron. After fragmentation, the atom that contains the charge when the ion is formed retains the charge. A radical is lost as a result of the fragmentation. This fragmentation is homolytic cleavage because it involves the movement of a single electron (see *exact mass*).

amu: See *dalton.*

Atmospheric pressure chemical ionization (**APCI**): Chemical ionization at atmospheric pressure.

Atomic weight (**a.k.a. atomic mass**): The weighted average of the mass of the naturally occurring isotopes of an element. The integer atomic mass of oxygen was an absolute value of 16 by definition. The chemical atomic mass scale made the determination of the atomic mass of newly discovered elements easy by forming their oxides. The atomic mass is used in the calculation of the average mass. Also see *nominal mass* and *monoisotopic mass.*

Average molecular weight (**M_r**): The mass of a molecule calculated using the atomic weight of the elements. Also see *molecular weight.*

Background: A signal in the mass spectrum due to electrical noise of the instrument or due to chemical noise from instrument contamination, column bleed, or sample matrix.

Gas Chromatography and Mass Spectrometry
DOI: 10.1016/B978-0-12-373628-4.00038-1

Base peak: The peak in the mass spectrum with the greatest intensity.
Benzylic cleavage: A fragmentation that takes place at the carbon atom attached to a phenyl group. When the phenyl group is C_6H_5, the benzylic cleavage will result in a benzyl ion with a formula of $C_6H_5 = CH_2^+$, which can be isomeric with the tropylium ion. Benzylic cleavage is a special case of homolytic cleavage and is due to the loss of a pi (π) electron, which places the site of the charge and the radical on the phenyl ring.

Collisionally activated dissociation (**CAD**): The same process as collision-induced dissociation (CID).
Collision-induced dissociation (**CID**): An ion/neutral process in which the projectile ion impacts a target neutral species. Part of the translational energy of the ion is converted to internal energy causing subsequent fragmentation.
Chemical ionization (**CI**): The formation of new ionized species when gaseous molecules interact with ions. This process may involve the transfer of an electron, proton, or other charged species between the reactants in an ion/molecule reaction. CI refers to positive ions, and negative CI is used for the formation of negative ions through ion/molecule reaction. The process of negative chemical ionization (NCI) is different than that of electron capture negative ionization (ECNI). See *electron capture ionization* and *negative chemical ionization*.
dalton (**Da**): A unit of mass on the atomic scale that is equal to 1/12 the mass of an atom of ^{12}C, the most abundant naturally occurring isotope of carbon. A dalton is a synonym for a unified atomic mass unit (*u*). Prior to the establishment of the carbon-12 standard for mass on the atomic scale, two different definitions of oxygen-16 standard were used and the symbol for atomic mass was *amu*. This symbol is now considered obsolete.
Dark matter: A mass spectral experiment results in charged particles (ions) and neutral particles (that are produced by the fragmentation of the initially formed ion). These neutral particles (radicals and molecules) are referred to as the *neutral losses* from the originally formed ion. These neutral losses are the dark matter of the mass spectrum. The dark matter is determined by subtracting an *m/z* value of a peak (usually representing a fragment ion) that is lower than that of another peak (usually the $M^{+\bullet}$ peak).

Detection limit: The detection limit is the smallest amount of analyte that is put into the mass spectrometer which provides a signal that can be distinguished from background noise.

EIC: See *mass chromatogram*.

Electron capture negative-ion (ECN) detection, electron capture negative ionization (ECNI), or resonance electron capture ionization (RECI): Refers to producing negative ions by the reaction of thermal electrons with molecules in a mass spectrometer. Thermal electrons are low-energy electrons that have average kinetic energies which are the same order-of-magnitude as the thermal energy of the molecules that they are intended to ionize. These terms are the correct terms to describe the formation of negative ions when an analyte molecule captures a low-energy electron. This process, which was first reported as a GC/MS technique by Ralph Dougherty and later by Don Hunt and George Stafford, usually happens under high-pressure conditions that are similar to those used in conventional CI. This is probably the reason that this technique is often incorrectly referred to as negative chemical ionization (NCI); see *chemical ionization* and *negative chemical ionization*. The resulting negative molecular ions are symbolized as $M^{-\bullet}$, which indicates an odd-electron negative ion. The results of electron capture/ionization are as follows.

Resonance electron capture

$$AB + e^{-}(\sim 0.1\text{eV}) \rightarrow AB^{-\bullet}$$

Dissociative electron capture

$$AB + e^{-}(0-15\text{eV}) \rightarrow A^{\bullet} + B^{-}$$

Ion-pair formation that results from electron capture

$$AB + e^{-}(>10\text{eV}) \rightarrow A^{-} + B^{+} + e$$

Electron ionization (EI): Ionization of analyte molecules in the gas phase (10^{-1}–10^{-4} Pa) by electrons accelerated between 50 V and 100 V. The original standard was 70 V. Ionization by electrons is according to the following reaction: $M + e^{-} \rightarrow M^{+\bullet} + 2e^{-}$. The term *electron impact* is obsolete and should not be used.

Electron volt (eV): The unit of energy used to describe the fragmentation of an ion. An electron volt is a unit of energy that is the work done on an electron when passing through a potential rise of 1 V. 1 eV = 1.602×10^{-19} J. The energy of the electron beam in EI mass spectrometry is expressed in electron volt. In modern instruments, the ionization energy standard for EI mass spectrometry is 50–100 eV (i.e., electrons are accelerated between 50 V and 100 V in the ion source). In the early days of mass spectrometry, the standard was set at 70 eV; and standard EI is still said to be carried out at 70 eV resulting in 70-eV spectra. One electron volt is equivalent to 23 kcal of energy.

Electrostatic analyzer (**ESA**): A velocity-focusing device for producing an electrostatic field perpendicular to the direction of ion travel. Ions of a given kinetic energy are brought to a common focus. This analyzer is used in combination with a magnetic analyzer to increase resolution and mass accuracy.

Elemental composition (**EC**): A chemical formula stating the number of atoms of the various elements in a molecule, radical, or ion.

Exact mass: More correctly, the calculated exact mass is the mass determined by summing the mass of the individual isotopes that compose a single ion, a radical, or a molecule based on a single mass unit being equal to 1/12 the mass of the most abundant naturally occurring stable isotope of carbon. The exact mass values used for the isotopes of each element are recorded in tables of isotopes. The mass (atomic mass) of an element that appears in the periodic table is a weighted average of these exact mass values of the naturally occurring stable isotopes of that element. If the mass is calculated with the exact mass value of the most abundant naturally occurring stable isotope of each element in the ion, radical, or molecule, then the *calculated exact mass* is the same as the *monoisotopic mass*.

Extracted ion chromatogram or **profile** (**EIC** or **EIP**): See *mass chromatogram*.

eV: See *electron volt*.

FC-43: See *perfluorotributylamine*.

Field ionization (**FI**): A soft ionization technique sometimes used in GC/MS that results in a high abundance of low-energy molecular ions. Ionization of an analyte molecule in the vapor phase takes place in an electrical field (10^7–10^8 V cm^{-1}) maintained between two sharp points or edges of two electrodes.

Fragmentation pattern: The result of the decomposition of an ion (usually the molecular ion in GC/MS) to produce a series of fragment ions with specific abundances. This pattern is displayed as a mass spectrum and is a mass spectral graphic presentation of the analyte.

Fragment ion: An ion formed from a charged species representing an intact molecule. The formation of a fragment ion is unimolecular and results in a charged species and a neutral, which is either a radical or a molecule.

GC/MS interface: An interface between a gas chromatograph and a mass spectometer that provides a continuous introduction of eluant gas to the mass spectometer ion source.

Gamma (γ)**-hydrogen shift-induced beta** (β) **cleavage** (**a.k.a. the McLafferty rearrangement**): A rearrangement reaction that was originally described by an Australian chemist, A.J.C. Nicholson (Nicholson 1954), but was named after F.W. McLafferty (McLafferty 1993) because of the extent to which he studied and reported the reaction in a wide variety of compound types. An odd-electron fragment ion is formed by the loss of a molecule. This fragment results from a γ-hydrogen shift to an unsaturated group such as a carbonyl (when the site of the odd electron and the charge is on the oxygen atom). The γ-hydrogen shift causes the radical site to move to the carbon atom that originally contained the γ-hydrogen. This new location of the radical site initiates a cleavage reaction that causes the fragmentation of the carbon–carbon bond that is beta to the unsaturated group and the loss of a terminal olefin.

Heterolytic (a.k.a. charge-site-driven or inductive, i) cleavage: A fragmentation that results from the pair of electrons between the atom attached to the atom with the charge and an adjacent atom that moves to the site of the charge. This fragmentation involves the movement of the charge site to the adjacent atom. A radical is lost as a result of the fragmentation. The movement of a pair of electrons is symbolized by a double-barbed arrow

Homolytic cleavage: A fragmentation that results from one of a pair of electrons between two atoms moving to form a pair with the odd electron. After fragmentation, the atom that contains the charge when the ion is formed retains the charge. A radical is lost as a result of the fragmentation. This reaction involves the movement of a single electron and is symbolized by a single-barbed arrow, the so-called "fishhook" convention.

Hydride shift: The movement of a hydrogen proton with the two electrons that attach it to an adjacent atom. This movement results in a bond being broken and a new bond being formed.

A hydride shift

Hydrogen shift: The movement of a hydrogen atom, usually in response to a radical site. A hydrogen shift results in a hydrogen proton and one of the two electrons in the bond between the hydrogen and an adjacent atom moving away from the ion to another location on the ion. A bond is broken, and a bond is formed.

Ion abundance: Refers to the number of ions present. Ions are found in mass spectrometers and not on mass spectra. Ion abundance is *not* a synonym for peak intensity.
Ion cyclotron resonance (ICR) analyzer: A device to determine the mass-to-charge ratio of an ion in a magnetic field by measuring its cyclotron frequency.
Ionization: A process that produces an ion from a neutral atom or molecule.
Ion series: A series of peaks in an EI mass spectrum that indicate structural moieties such as a series of peaks spaced 14 *m/z* units apart, indicating the presence of an alkyl moiety. Each of the peaks represents carbenium ions that differ in structure by a single $-CH_2-$ group. The starting *m/z* value of the series can be a clue to the compound type; e.g., *m/z* 31 with a succession of peaks, each 14 *m/z* units higher than the previous, indicates substitute aliphatic alcohols. An ion series can be useful in determining the type of a compound. Another ion series is that associated with an aromatic compound (39, 51, 65, 77, 91, etc.).
Ion trap analyzer (a.k.a. quadrupole ion trap, QIT): A mass-resonance analyzer that produces a three-dimensional rotationally symmetric quadrupole field capable of storing ions at selected mass-to-charge ratios.
Isotope: An atom of an element that differs from other atoms of that element in the number of neutrons in its nucleus.
Isotope peak: A peak that represents an ion that has the same elemental composition as the monoisotopic ion but has one or more atoms of one or more of its elements that are a higher mass isotope. Also see *monoisotopic peak.*
Isotopic ion: An ion containing one or more of the less abundant naturally occurring isotopes of the elements that make up the structure.
MS/MS (mass spectrometry/mass spectrometry): A process by which a stable ion is collisionally activated so that it will undergo fragmentation. This is done to gain specificity or to learn more about the structure of an ion of a specific *m/z* value. MS/MS is sometimes referred to as *tandem mass spectrometry.*
MW: See *molecular weight.*
Magnetic analyzer: A direction-focusing device that produces a magnetic field perpendicular to the direction of ion travel. All ions of a given momentum with the same mass-to-charge ratio are brought to a common focus.
Mass chromatogram (a.k.a. reconstructed or **extracted ion current (RIC** or **EIC) chromatogram** or **profile)**: A display of a data set from a GC-MS prepared by plotting the ion current from an individual ion, a small series of *m/z* values, or the sum of a series of noncontiguous *m/z* values versus the record (spectrum) number.
Mass defect: The difference between the integer mass of an atom, ion, radical, or molecules and its exact mass. Mass defect due to hydrogen can be a factor for ions that have a mass >500 Da. A negative value for mass defect due to multiple Cl and/or Br atoms can also be a factor.

Mass discrimination: The attribute of an *m/z* analyzer resulting in the transition or detection of ions of one *m/z* value more efficiently than another. An instrument can exhibit either *high-* or *low-mass discrimination.* Mass discrimination can be due to instrument design or the contamination condition of the instrument.

Mass spectrometer: An instrument used to separate ions according to their mass-to-charge ratios and to determine the abundance of ions.

Mass spectrum (also known as spectrum): The data display from a mass spectrometer. The mass spectrum can be displayed graphically or as numerically as a series of *m/z* values paired with an absolute or relative intensity.

Mass-to-charge ratio (*m/z*): An abbreviation for division of the observed mass of an ion by the number of charges the ion carries. Thus, $C_6H_6^{+\bullet}$ has *m/z* 78, but $C_6H_6^{2+\bullet}$ has *m/z* 39.

Metastable ion: An ion that dissociates into a production and a neutral product during the flight from the ion source to the detector. Metastable ions can be observed when they occur in a field-free region of the mass spectometer.

millibar (mbar): See *pascal.*

Molecular ion: An ion formed by the addition ($M^{-\bullet}$) or removal ($M^{+\bullet}$) of an electron from a molecule without fragmentation. All molecular ions are odd-electron ions; i.e., they have an odd number of electrons.

Molecular weight (MW): In mass spectrometry, this term is used to refer to the nominal mass or the monoisotopic mass of a molecule. It is not based on the atomic weights of the elements but rather on the masses of the most abundant naturally occurring isotopes of the elements. Also see *average molecular weight.* In this book, the symbol "MW" means the nominal mass of a molecule.

Monoisotopic mass: The mass of an ion calculated using the exact mass of the most abundant isotope of each element in the formula (e.g., C = 12.0000; O = 15.9949).

Monoisotopic peak: A peak representing an ion where only a single isotope of any element is present, and the isotope for each element is the most abundant naturally occurring isotope.

Negative chemical ionization (NCI): Formation of an ion with a negative charge using negatively charged reagent ions. The resulting negative ions are usually adducted ions like RCl^{-}. See *chemical ionization.*

Neutral loss: Describes the nonionic product of a fragmentation. The neutral losses are inferred by the mass spectrum rather than explicitly stated. The mass of a neutral loss (radical or molecule) is determined by the subtraction of the *m/z* values of two related mass spectral peaks.

Nominal mass: The mass calculated for an ion when using the integer mass values of the most abundant naturally occurring isotope of each element in the formula (e.g., C = 12, O = 16, S = 32, etc.).

Nominal mass peak: A monoisotopic peak reported as an integer mass based on the nominal masses of all the atoms comprising the ion.
PFK: See *perfluorokerosene*.
pascal (Pa): The derived SI unit of pressure, equal to 1 newton (the force required to accelerate a standard kilogram at a rate of $1\,\mathrm{m\,sec}^{-2}$ over a frictionless surface) per square meter. Many mass spectrometry applications refer to pressure in torr. 1 Torr = ~133 Pa. The unit millibar (mbar, 10^{-3} bar) is equal to exactly 100 Pa.
Peak: Perhaps the most confusing term in GC/MS. There are chromatographic peaks; or, to be more specific, there are reconstructed chromatographic peaks; and there are mass spectral peaks. In GC/MS, chromatographic peaks are usually depicted as vertical lines; however, the chromatographic peak has a width (representing the energy dispersion of a monoisotopic ion, the minimum and maximum *m/z* values of ions that cannot be resolved by the instrument, or both) and a height, which is directly proportional to the ion current being represented. Care must always be taken when using the word *peak* to make sure it is clear whether a *chromatographic peak* or a *mass spectral peak* is meant.
Peak intensity: Refers to the height of a peak on a mass spectrum. The peak intensity may be expressed in terms of signal strength or relative intensity (relative to the base peak or a specified peak on the mass spectrum). Peak intensity is *not* a synonym for ion abundance. Peaks have intensities, and ions have abundances.
Precursor ion: An ion that undergoes fragmentation to produce a product ion and a neutral product in an MS/MS experiment. The term *precursor ion* should not be used as a synonym for molecular ion. The term *parent ion* should not be used as a synonym for precursor ion.
Perfluorokerosene (PFK): A volatile mixture of aliphatic hydrocarbons where all the hydrogen atoms have been replaced with fluorine atoms used for tuning and calibration of the *m/z* scale of double-focusing mass spectrometers. Ions formed by PFK are used in peak matching for accurate mass assignments.
Perfluorotributylamine: A volatile organic compound used in EI for tuning to calibrate the *m/z* scale of quadrupole and quadrupole ion trap mass spectrometers. It has the following structure:

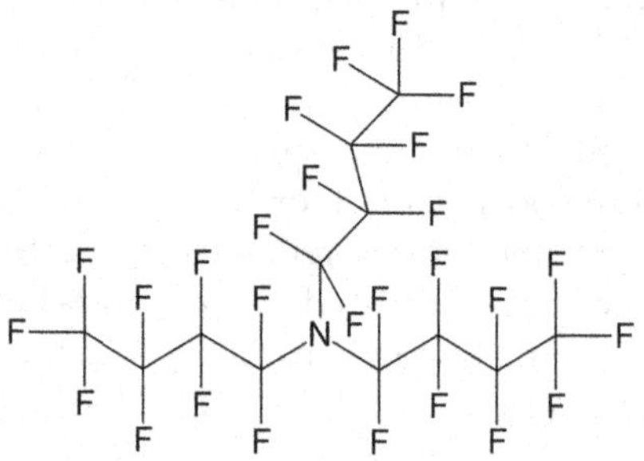

Product ion: An ion related to a precursor ion by a process such as fragmentation in an MS/MS experiment.
QIT: See *ion trap analyzer.*
QMF: See *quadrupole m/z analyzer.*
Quadrupole *m/z* analyzer: A mass filter (QMF) that creates a quadrupole field with dc and rf components so that only ions of a selected *m/z* value are transmitted to the detector.
RIC: See *mass chromatogram.*
Radical ion (**odd electron ion**): An ion containing an unpaired electron that is both a radical and an ion.
Rearrangement ion: A dissociation product ion in which atoms or groups of atoms have transferred positions in the original ion during the fragmentation process. This usually involves the loss of a molecule.
Reconstructed ion chromatogram or **profile** (**RIC**): See *mass chromatogram.*
Resolution: The difference in *m/z* values of two ions that can just be separated (Δm). This is based on the same definition used for resolving power (see *resolving power*).
Resolving power (**10% valley definition**): Let two monoisotopic peaks of equal height in a mass spectrum at masses m and (m − Δm) be separated by a valley that at its lowest point is 10% of the height of the peaks. The resolving power (R) is equal to m/Δm.
Resolving power (**FWHM**): It is not always possible to have two monoisotopic peaks of equal intensity; therefore, R can be calculated by measuring the full width of a monoisotopic peak at half its maximum in *m/z* units. R calculated in this way is usually 2× the value obtained using the 10% valley method.
Scan: A term, more often than not, inappropriately (QIT and time-of-flight (TOF) mass spectrometers are not scanning instruments) used as a synonym for *spectrum* or *mass spectrum.*
Selected ion monitoring (**SIM**): Describes the operation of a mass spectometer in which the ion currents of one or several selected *m/z* values are recorded, rather than the entire mass spectrum.
Selected reaction monitoring (**SRM**): A process by which the ion current of a product ion of a specific *m/z* value produced from a precursor of a specified *m/z* value is monitored. This process is used to gain specificity for analytes in complex matrices.
Spectral skewing: A term used to describe the phenomenon of changes in relative intensities of mass spectral peaks between different spectra during an acquisition of GC/MS data due to the changes in concentration of the analyte in the ion source as the chromatographic component elutes. This phenomenon is not observed in ion trap (quadrupole or magnetic) or time-of-flight (TOF) mass spectrometers. The TOF-MS records all ions sent down the flight tube in a single

pulse. The ion trap mass spectrometer records all the ions that have been stored in the trap.
Tandem mass spectrometry: See *MS/MS.*
Time-of-flight (**TOF**) **analyzer**: A device that measures the flight time of ions having a given kinetic energy through a fixed distance. The flight times of ions are proportional to *m/z* values.
torr (**Torr**): See *pascal.*
Total ion current (**TIC**): The sum of all the separate ion currents contributed by the ions that make up the spectrum.
Unified atomic mass unit (***u***): See *dalton.*
Unimolecular ion decomposition: Refers to the fact that the decomposition of an ion under EI conditions is due solely to the ion's internal energy. The pressure in the ion source is sufficiently low so as to prevent bimolecular or other collision reactions (the collision of an ion resulting in either a fragmentation or an increase in internal energy that will result in fragmentation).

APPENDIX C

Atomic Masses and Isotope Abundances and Other Information for the Determination of an Elemental Composition from Isotope Peak Intensity Ratios

Table C.1 Types, symbols, integer and exact masses, and natural isotopic abundance ratios for elements most often encountered in organic mass spectrometry

Type	Element	Symbol	Integer mass[a]	Exact mass[b]	Abundance	X+1 factor[c]	X+2 factor[d]
X	Hydrogen	H	1	1.0078	99.99		
		D or ^{2}H	2	2.0141	0.01		
X+1	Carbon	^{12}C	12	12.0000	98.91		
		^{13}C	13	13.0034	1.1	$1.1n_C$	$0.0060n_C^2$
X+1	Nitrogen	^{14}N	14	14.0031	99.6		
		^{15}N	15	15.0001	0.4	$0.37n_N$	
X+2	Oxygen	^{16}O	16	15.9949	99.76		
		^{17}O	17	16.9991	0.04	$0.04n_O$	
		^{18}O	18	17.9992	0.20		$0.20n_O$
X	Fluorine	F	19	18.9984	100		
X+2	Silicon	^{28}Si	28	27.9769	92.2		
		^{29}Si	29	28.9765	4.7	$5.1n_{Si}$	
		^{30}Si	30	29.9738	3.1		$3.4n_{Si}$
X	Phosphorus	P	31	30.9738	100		
X+2	Sulfur	^{32}S	32	31.9721	95.02		
		^{33}S	33	32.9715	0.76	$0.8n_S$	
		^{34}S	34	33.9679	4.22		$4.4n_S$
X+2	Chlorine	^{35}Cl	35	34.9689	75.77		
		^{37}Cl	37	36.9659	24.23		$32.5n_{Cl}$
X+2	Bromine	^{79}Br	79	78.9183	50.5		
		^{81}Br	81	80.9163	49.5		$98.0n_{Br}$
X	Iodine	I	127	126.9045	100		

(*Continued*)

Gas Chromatography and Mass Spectrometry
DOI: 10.1016/B978-0-12-373628-4.00039-3

Table C.1 (*Continued*)

[a] The integer mass of the most abundant (This may not always be the lowest mass naturally occurring stable isotope of the element, as is the case with the elements in this table. The lowest mass isotope is usually not the most abundant for the transition elements.) naturally occurring stable isotope of an element is the element's *nominal mass*. The nominal mass of an ion is the sum of the nominal masses of the elements in its elemental composition (e.g., $C_3H_6O^{+\bullet}$ has a nominal mass of 58).

[b] The exact mass of the most abundant (This may not always be the lowest mass naturally occurring stable isotope of the element, as is the case with the elements in this table. The lowest mass isotope is usually not the most abundant for the transition elements.) naturally occurring stable isotope of an element is the element's *monoisotopic mass*. The monoisotopic mass of an ion is the sum of the monoisotopic masses of the elements in its elemental composition (e.g., $C_3H_6O^{+\bullet}$ has a monoisotopic mass of 58.0417). The difference between the integer mass of an isotope and that isotope's exact mass is the isotope's *mass defect*.

[c] Assume X = 100%; X represents the relative intensity of the first peak in a cluster of peaks corresponding to isotopic variants of a given ion.

[d] The factor is multiplied by the number (*n*) of atoms of the element present to determine the magnitude of the intensity contribution for a given isotope. For example, the contribution at X+1 due to ^{15}N for an ion containing three nitrogens would be 0.37 × 3 = 1.11 relative to 100 at X.

See Appendix F for details on assigning elemental composition.

Rings Plus Double Bonds

This formula is only valid when the elements are at their lowest valence state. Double bonds associated with nitrogen and phosphorus at a valence of 5 or sulfur at a valence of 6 are not determinable by this formula.

For the general elemental composition: $C_xH_yN_zO_n$ (halogens treated as hydrogen; Si as C; P as N; and S as O). Elements with the same valence are treated the same.

$$R{+}dB = x - \tfrac{1}{2}y + \tfrac{1}{2}z + 1$$

If the calculation ends in ½, the ion is an even-electron ion. If the calculation ends in an integer, the ion is an odd-electron ion.

APPENDIX D

X+1 and X+2 Values for Ions Containing Atoms of C and H Based on Isotope Contributions

Table D.1 X+1 and X+2 values for ions containing atoms of C and H based on isotope contributions

C number	X+1	X+2	C number	X+1	X+2
C_1	1.1	0.01	C_{21}	23.1	2.6
C_2	2.2	0.02	C_{22}	24.2	2.9
C_3	3.3	0.05	C_{23}	25.3	3.2
C_4	4.4	0.10	C_{24}	26.4	3.4
C_5	5.5	0.15	C_{25}	27.5	3.7
C_6	6.6	0.22	C_{26}	28.6	4.1
C_7	7.7	0.29	C_{27}	29.7	4.4
C_8	8.8	0.38	C_{28}	30.8	4.7
C_9	9.9	0.49	C_{29}	31.9	5.0
C_{10}	11.0	0.60	C_{30}	33.0	5.4
C_{11}	12.1	0.73	C_{31}	34.1	5.8
C_{12}	13.2	0.86	C_{32}	35.2	6.1
C_{13}	14.3	1.0	C_{33}	36.3	6.5
C_{14}	15.4	1.2	C_{34}	37.4	6.9
C_{15}	16.5	1.3	C_{35}	38.5	7.3
C_{16}	17.6	1.5	C_{36}	39.6	7.8
C_{17}	18.7	1.7	C_{37}	40.7	8.2
C_{18}	19.8	1.9	C_{38}	41.8	8.7
C_{19}	20.9	2.2	C_{39}	42.9	9.1
C_{20}	22.0	2.4	C_{40}	44.0	9.6

The isotope contributions from other elements are additive in experimental data (e.g., the X+1 contributions from N, O, S, etc. are combined with the above-listed X+1 contribution from carbon; similarly, the X+2 contributions from elements that have X+2 isotopes are combined with the X+2 contribution from carbon).

Gas Chromatography and Mass Spectrometry
DOI: 10.1016/B978-0-12-373628-4.00040-X

APPENDIX E

Isotope Peak Patterns for Ions Containing Atoms of Cl and/or Br

Table E.1 Numeric values for chlorine and bromine isotopic abundance ratios

Cl/Br	X	X+2	X+4	X+6	X+8	X+10
Cl	100	32.5	–	–	–	–
Cl_2	100	65.0	10.6	–	–	–
Cl_3	100	97.5	31.7	3.4	–	–
Cl_4	76.9	100	48.7	10.5	0.9	–
Cl_5	61.5	100	65.0	21.1	3.4	0.2
Cl_6	51.2	100	81.2	35.2	8.5	1.1
ClBr	76.6	100	24.4	–	–	–
Cl_2Br	61.4	100	45.6	6.6	–	–
Cl_3Br	51.2	100	65.0	17.6	1.7	–
$ClBr_2$	43.8	100	69.9	13.7	–	–
Cl_2Br_2	38.3	100	89.7	31.9	3.9	–
Cl_3Br_2	31.3	92.0	100	49.9	11.6	1.0
$ClBr_3$	26.1	85.1	100	48.9	8.0	–
Cl_2Br_3	20.4	73.3	100	63.8	18.7	2.0
Br	100	98.0	–	–	–	–
Br_2	51.0	100	49.0	–	–	–
Br_3	34.0	100	98.0	32.0	–	–
Br_4	17.4	68.0	100	65.3	16.0	–

Note: all intensities are normalized to the base peak in the cluster whether or not the X peak is the base peak.

Gas Chromatography and Mass Spectrometry
DOI: 10.1016/B978-0-12-373628-4.00041-1

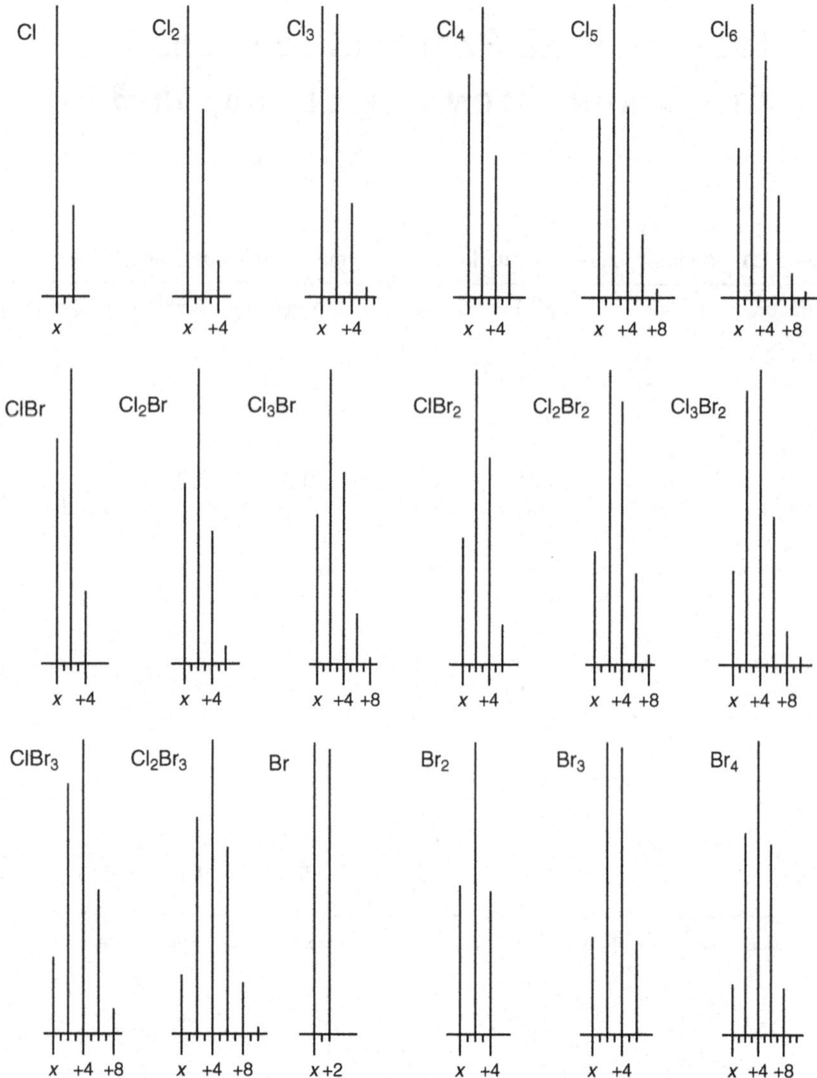

Figure E.1 Graphical representation of relative isotope peak intensities for any given ion containing the indicated number of chlorine and/or bromine atoms. Note: each intensity is two mass units apart.

APPENDIX F

Steps to Follow in the Determination of an Elemental Composition Based on Isotope Peak Intensity Ratios

The element types (X, X+1, and X+2) are found in *Appendix C*.

The precision with which the intensity of the peaks can be measured in most GC/MS instruments is $\pm 10\%$. This precision deteriorates as the intensity of the peaks gets closer to the detection limit. This means that accounting for $\pm 10\%$ of the initial percent intensity of an isotope peak will signify that the elemental composition due to that peak has been rationalized. Once a remainder of $\pm 10\%$ has been reached in any one of the following steps, move to the next step even though there is sufficient intensity and mass remaining to add atoms of another element. To do otherwise can lead to overinterpretation.

It is also important that the mass must be reconciled to zero. There can be $\pm$ remainders for the rationalization of the X+1 and X+2 peak intensities, but the mass must reconcile to zero.

1. **Determine the nominal *m/z* peak in a cluster of peaks**
   This is the lowest *m/z* value peak in a cluster of peaks where all the peaks at higher *m/z* values can be rationalized as isotope peaks. When the ion contains multiple atoms of the X+2 elements Cl and/or Br, the nominal *m/z* peak may not be the base peak in the cluster. The X+2 peak may be the base peak. The nominal *m/z* peak is designated as the X peak.
2. **Assign the X+2 elements with the exception of oxygen (Si, S, Cl, and Br)**
   The X+2 elements are assigned before the X+1 elements because the X+2 isotope peak patterns, when atoms of Cl and/or Br are present, are very obvious; and the X+2 elements of Si and S have contributions at X+1, which could alter the intensity of the X+1 peak due to the number of carbon and/or nitrogen atoms present.
3. **Assign the X+1 elements (C and N)**
   Knowing whether the nominal mass peak represents a fragment ion or an $M^{+\bullet}$ or a protonated molecule (in the case of chemical ionization) based on the peak's *m/z* value will allow the determination of whether or not the peak represents an ion that contains an odd number of nitrogen atoms.

Gas Chromatography and Mass Spectrometry
DOI: 10.1016/B978-0-12-373628-4.00042-3

If the peak is known to present an ion with an odd number of nitrogen atoms, initially attempt a solution that has a single atom of nitrogen atoms; and then assign the number of carbon atoms. If it is not known that the ion contains an odd number of nitrogen, begin by assigning the number of atoms of carbon and fill in with nitrogen if necessary. Do not forget to consider the presence of two atoms of ^{13}C and the contribution to the intensity of the X+2 peak (see *Appendix D*).

4. **Assign the number of atoms of oxygen**
   The contribution of the presence of ^{18}O to the intensity of the X+2 peak is only 0.2%; therefore, it is necessary to account for the presence of two ^{14}C atoms before accounting for the atoms of oxygen. To do otherwise could lead to an incorrect value for the number of atoms of oxygen. It is also important to remember that due to the low contribution of ^{18}O represented by the X+2 peak that the number of atoms of oxygen can be in error by ± 1.
5. **Assign the X elements (F, P, I, and H)**
   Use these elements to reconcile the remaining mass to zero. Even though hydrogen has two naturally occurring stable isotopes, the abundance of deuterium compared to the abundance of protonium is so low that hydrogen can be considered as an X element for ions normally encountered in GC/MS. H, F, and I all have a valence of 1. Phosphorus has a valence of 3; therefore, the assignment of phosphorus can sometimes be tricky. After the X elements have been assigned, the value left in the mass column must be zero.
6. **Calculate the number of rings and/or double bonds (*R*+*dB*) based on the elemental composition**
   One way of determining whether or not the proposed elemental composition is reasonable is to determine the number of rings plus double bonds that would be present in an ion with the proposed elemental composition. The general formula for the number of rings and/or double bonds is as follows:

$$R+dB = \#\mathrm{C} - \tfrac{1}{2}\#\mathrm{H} + \tfrac{1}{2}\#\mathrm{N} + 1$$

where C is the number of atoms of carbon and any other elements with a valence of 4 (Si); H is the number of atoms of hydrogen and any other elements with a valence of 1 (F, Cl, Br, I); and N is the number of atoms of nitrogen or any other elements with a valence of 3 (P).

Rings and double bonds associated with oxygen and sulfur are taken into account by the formula, but the number of atoms of these two elements is not a part of the calculation.

If the *R*+*dB* calculation ends in a whole number, the ion is an odd-electron ion like a $M^{+\bullet}$.

If the $R+dB$ calculation ends in a ½, the ion is an even-electron ion like a fragment ion or a protonated molecule (in the case of chemical ionization). The number of rings and/or double bonds for a fragment ion is reached by dropping the ½ (rounded down); in the case of a protonated molecule, the calculated value is rounded up to the nearest whole number.

Sulfur and phosphorus have normal valences of 2 and 3, respectively; however, sulfur can also have valences of 4 and 6 and phosphorus can have a valence of 5. If any of the higher valence states exists for either of these two elements in the configuration of an elemental composition, any rings or double bonds associated with these elements will not be accounted for in the $R+dB$ calculation.

The $R+dB$ value must *not* be a negative number. A phenyl ring accounts for a total of four rings + double bonds.

7. **Propose a possible structure**
   Once an elemental composition has been reached that results in a reasonable number of rings and/or double bonds, propose one or more structures to see if the results are reasonable.
8. Whenever possible, compare the elemental composition reached, using accurate mass measurements with what is determined using isotope peak intensity ratios and vice versa.

An example of an elemental composition determination based on the intensity of isotope peaks is as follows:

Step 1: Determine the nominal m/z value peak; m/z 149

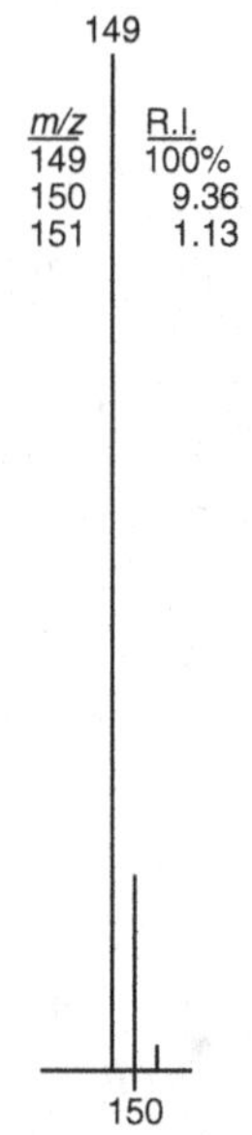

Step 2: Determine number X+2 elements except O; None, because X+2 = 1.13

Step 3: Determine number X+1 elements; start with C because X representing $OE^{+\bullet}$ or EE^{+} is unknown.

Based on X+1 value of 9.36, #C atoms is 8 or 9. Choose 8 because X+2 for 8 or 9 C atoms is far less than 1.13%; O must be contributing to X+2 and O has a contribution at X+1, which means X+1 is not all due to C.

	Mass	X+1	X+2
	149	9.36	1.13
8C × 12 =	96	8.8	0.38
	53	0.56	0.75

There is enough of a remainder in the Mass and the X+1 columns to consider an atom of N; however, 0.56% is well within ±10% of 9.36. Therefore, it is time to move to Step 4.

Step 4: Assign the number of atoms of oxygen. 0.75 rounds nicely to 0.8, which would indicate 4 atoms of O; however, there are only 48 units of mass remaining. Therefore, the maximum number of O atoms is 3.

$$3O \times 14 = \frac{48}{5} \quad \frac{0.12}{0.44} \quad \frac{0.60}{0.15}$$

The X+2 remainder of 0.15 is >10% of 1.13; however, at the low initial level of 1.13, this remainder is reasonable, especially with only 5 remaining in the Mass column.

Step 5: With only 5 remaining in the Mass column, it can only be accounted for with 5 atoms of hydrogen, resulting in an elemental composition of $C_8H_5O_3$.

The odd nominal *m/z* value, the elemental composition with no N atoms, and the *R*+*dB* calculation are all in agreement for an EE^{+} ion with no atoms of N.

Step 6: Calculate the number of rings plus double bonds.

R+*dB* = 8 C atoms − 5 H atoms/2 + 0 N atoms/2 + 1 = 6½

Ending in ½ shows that the ion is a fragment ion. Round to 6 to get the actual number of rings and/or double bonds.

Step 7: Propose a structure: A phenyl group will account for 6 carbon atoms and 1 ring and 3 double bonds. This leaves 2 atoms of carbon, 3 atoms of oxygen, 2 rings and/or double bonds, and 5 atoms of H. Two carbonyl groups account for 2 double bonds, the remaining atoms of carbon, and 2 of the atoms of oxygen. Four hydrogen atoms will be required for the four positions not occupied by the 2 carbonyl groups. This leaves 1 atom of oxygen and 1 atom of hydrogen, which could constitute the structure shown below.

This is a common ion representing phthalate contaminations found in GC/MS data files.

A second example is as follows:

Step 1: Determine the nominal *m/z* value peak; *m/z* 155

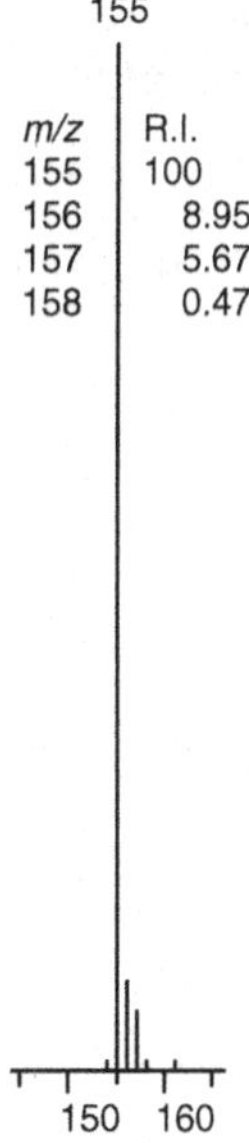

Step 2: Determine the number of X+2 elements except O.

The X+2 value is too high to represent contributions from 2 atoms of ^{13}C and/or atoms of ^{18}O. It is too low to represent ^{37}Cl or ^{81}Br. This means that the X+2 value is probably due to either S or Si with some added contribution from O.

Si has a significant X+1 contribution (5.2% per atom). If only 1 atom of Si was present, ~2.3% of the X+2 value would have to be due to C and O and this does not seem possible. If 2 atoms of Si were present, there would not be agreement at X+1; therefore, the most likely candidate is S and maybe O.

	Mass	X+1	X+2
	155	8.95	5.67
1S × 32 =	32	0.80	4.40
	123	8.15	1.27

The 1.27 remainder at X+2 is >10% of 5.67; however, X+2 from two atoms of ^{13}C and from ^{18}O have yet to be taken into account.

Step 3: Determine the number of X+1 elements; start with C because X representing $OE^{+\bullet}$ or EE^{+} is unknown.

Based on the adjustment of the X+1 value due to a contribution from S, it appears that there are 7 atoms of C.

$$7C \times 12 = \frac{84}{39} \quad \frac{7.70}{0.45} \quad \frac{0.29}{0.98}$$

There is enough of a remainder in the Mass and the X+1 columns to consider an atom of N; however, 0.45% is well within ±10% of 8.95. Therefore, it is time to move to Step 4.

Step 4: Assign the number of atoms of oxygen. 0.98 indicates as many as 5 atoms of O; however, there are only 39 units of mass remaining. Therefore, the maximum number of O atoms is 2.

$$3O \times 14 = \frac{32}{7} \quad \frac{0.08}{0.37} \quad \frac{0.40}{0.58}$$

The X+2 remainder of 0.58 is only slightly >10% of the original 5.67; however, this remainder is reasonable, especially with only 7 remaining in the Mass column. At 0.45, the remainder at X+1 is well within the ±10% guideline.

Step 5: With only 7 remaining in the Mass column, it can only be accounted for with 7 atoms of hydrogen, resulting in an elemental composition of $C_7H_7SO_2$.

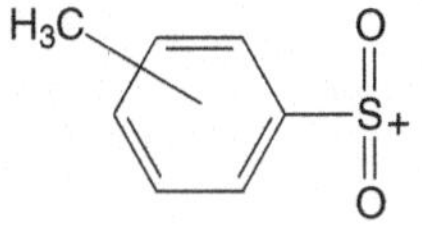

Step 6: Calculate the number of rings plus double bonds.
$R+dB = 7$ C atoms − 7 H atoms/2 + 0 N atoms/2 + 1 = 4½
Ending in ½ shows that the ion is a fragment ion. Round to 4 to get the actual number of rings and/or double bonds.

Step 7: Propose a structure: A phenyl group will account for 6 carbon atoms and 1 ring and 3 double bonds. This leaves 1 atom of carbon, 2 atoms of oxygen, 1 atom of S, and 7 atoms of H. There could be a methyl group attached to the ring. There could also be a SO_2 attached to the ring. The proposed structure has a total of 6 $R+dB$, but this is not a problem because this proposed structure of the sulfur atom has a valence of 6, not the lowest valence state of 2.

APPENDIX G

Derivatization in GC/MS

Reasons for Derivatization

Derivatization is used in GC and GC/MS for several reasons:

- Some highly polar compounds require derivatization in order to pass through a GC column. These compounds are not volatile enough for GC analysis without derivatization. Examples are amino acids and sugars.
- There are somewhat less polar compounds that can be determined by GC without derivatization; however, they are subject to absorption on active sites in the GC system. The chromatographic peaks represented by these compounds may exhibit tailing, especially as the column ages, and the limits of detection will be relatively high due to absorption. Examples are free fatty acids, phenols, and some drugs of abuse (morphine). Derivatization will eliminate these problems.
- Compounds may be derivatized to introduce fluorinated functional groups which significantly increase volatility. For example, conversion of the hydroxyl groups in high-molecular-weight compounds, such as disaccharides, to trifluoroacetyl esters will facilitate their analysis by GC. Fluorinated derivatives also make it possible to use GC electron capture detection, rather than flame ionization detection, and electron capture negative ionization (ECNI) in GC/MS, both of which provide for lower limits of detection through increased sensitivity and specificity.
- Some compounds that are to be determined by GC/MS fragment into ions that are not very useful for identification or quantitation. In the example shown in Figure G.1, underivatized amphetamine yields an EI mass spectrum with only two peaks, one at *m/z* 44 and the other at *m/z* 91. The $M^{+\bullet}$ (molecular ion) peak for this compound can be considered to be nonexistent. Complex matrices will produce ions with the same *m/z* values as those representing the analyte, making the analyte somewhat difficult to identify and/or quantitate. When amphetamine is derivatized with 4-carbethoxyhexafluorobutyrl chloride [1], not only is the

Gas Chromatography and Mass Spectrometry
DOI: 10.1016/B978-0-12-373628-4.00043-5

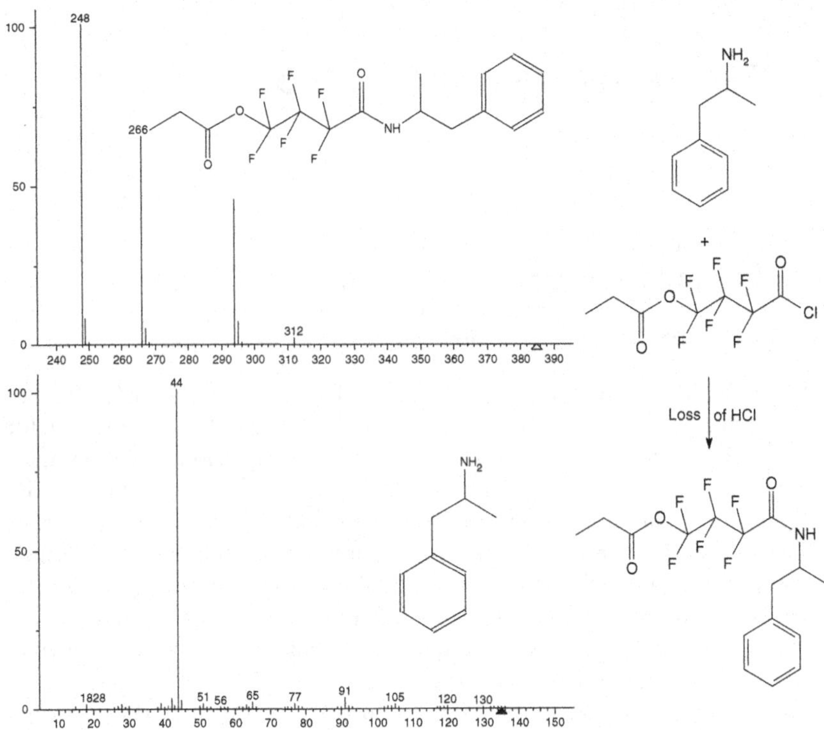

Figure G.1 The reaction scheme for the conversion of amphetamine to the 4-carbethoxyhexafluorobutyrl derivative (nominal mass 385 Da) (right), the mass spectra of amphetamine (lower left), and the derivatized compound (upper left). Whereas the spectrum of the underivatized amphetamine exhibits only two peaks at *m/z* 44 and 91, the derivatized compound exhibits a mass spectrum with three peaks representing unique ions at *m/z* 248, 266, and 294.

amphetamine retained longer on the column to assure separation from most compounds in the matrix, but also three unique ions appear in the mass spectrum.

Functional Groups and Their Derivatives

Compounds can be derivatized if they contain reactive functional groups such as hydroxyl, carboxyl, keto, sulfhydryl, amines, and amides. These functional groups are converted to various different types of ethers and esters. Some of these groups are summarized in Table G.1.

Table G.1 A list of many of the functional groups that can be derivatized for GC and GC/MS analyses. The reagents listed are all commercially available from Pierce*, Regis Technologies, Inc., and many chromatography supply companies. General procedures for derivatization are packaged with the reagents.

Functional group	Procedure	Reagent	Derivative increase in MW/functional group		Notes
Hydroxyl 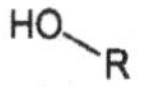 Alcohols	Silylation	BSA BSTFA BSTFA + TCMS	TMS (trimethylsilyl) ethers 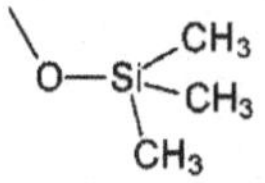	72	Most often-used derivative with good thermal stability and poor hydrolytic stability; TCMS is a catalyst
		HMDS			Weak donor usually used with TCMS
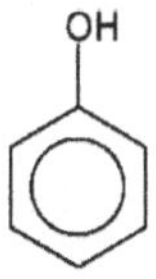		MSTFA MSTFA + TCMS			
		TMCS			Weak donor usually used with HMDS; can be used with salts
or 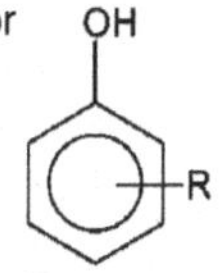 Phenols		TMSI Tri-Sil® Reagents			
		MTBSTFA	TBDMS (*tert*-butyldimethylsilyl) ethers	114	More stable than TMS; good MS fragmentation patterns
		MTBSTFA + TBDMCS	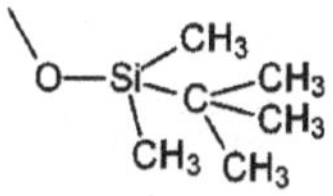		TBDMCS is a catalyst
	Acylation	MBTFA TFAA TFAI	TFA (trifluoroacetate) esters 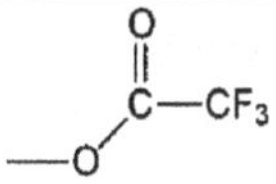	96	Good for trace analysis with ECD
		PFPI PFAA	PFP (pentafluoropropionate) esters 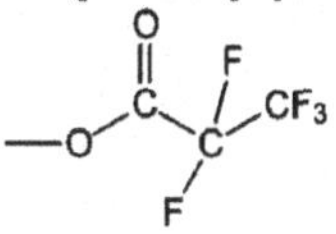	146	Good for trace analysis with ECD
		HFBI HFAA	HFB (heptafluorobutyrate) esters 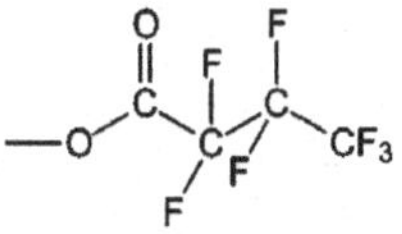	196	Good for trace analysis with ECD
	Alkylation	PFBBr	Pentafluorobenzylether 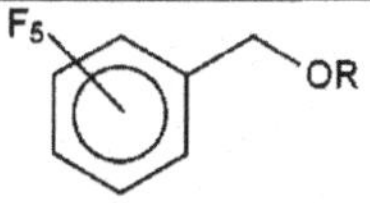	180	With aldoxides only (R–O$^-$)
		MethElute™ Reagent	Methyl ether 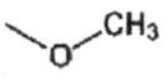	14	Derivatization in injector especially for specific drugs

(*Continued*)

Table G.1 *(Continued)*

Functional group	Procedure	Reagent	Derivative increase in MW/functional group		Notes
Carboxyl	Silylation	BSA BSTFA BSTFA + TCMS MSTFA TMCS	TMS (trimethylsilyl) esters	72	Easily formed, generally not stable, analyzed quickly Can be used with some salts
		TMSI Tri-Sil® Reagents			
		MTBSTFA	TBDMS (*tert*-butyldimethylsilyl) esters	114	More stable than TMS derivatives
		MTBSTFA + TBDMCS			TBDMCS is a catalyst
	Alkylation	PFBBr	Pentafluorylbenzyl esters	180	Used with ECD
		BF_3-Methanol	Methyl esters	14	Best for large samples of fatty acids
		Methy-8® Reagent MethElute™ Reagent			Fatty acids and amino acids Derivatization in injector especially for specific drugs
Amides Primary Secondary	Silylation	BSTFA + TCMS MSTFA + TCMS Tri-Sil® Reagents	TMS (trimethylsilyl) amides *from primary amides* *from secondary amides*	72	Difficult to form due to steric hindrance TCMS used as a catalyst
Amines Primary	Silylation	BSTFA + TCMS	TMS (trimethylsilyl) amines	72	Primary and secondary amines, TCMS used as a catalyst
		MSTFA + TCMS	*from primary amines*		Reaction by-products more volatile

(Continued)

Table G.1 (*Continued*)

Functional group	Procedure	Reagent	Derivative increase in MW/functional group		Notes
Secondary		Tri-Sil® Reagents	from secondary amines		
		MTBSTFA + TBDMCS	TBDMS (*tert*-butyldimethylsilyl) amines; from primary amines; from secondary amines	114	Difficult to form, very stable TBDMCS aids derivatization
	Acylation	MBTFA TFAA TFAI	Trifluoroacetamides; from primary amines; from secondary amines	96	
		PFAA	Pentafluoropropionamides; from primary amines; from secondary amines	146	Good for ECD detection
		HFBA	Heptafluorobutyamides; from primary amines; from secondary amines	196	Good for ECD detection

(*Continued*)

Table G.1 *(Continued)*

Functional group	Procedure	Reagent	Derivative increase in MW/functional group		Notes
	Alkylation	MethElute™ Reagent	Methyl amines R–N(H)–CH_3 R–N(CH_3)–CH_3 *from primary amines* R–N(R′)–CH_3 *from secondary amines*	14	Derivatization in injector especially for specific drugs
Keto R–C(=O)–R′	Methylation	MOX reagent	Methoxime R–C(=$NOCH_3$)–R′	29	Used for steroids and α-keto acids to prevent formation of multiple derivatives

*Pierce Chemical is now a division of Thermo Scientific. Their Web site (www.piercenet.com) contains general informatio about the derivatizing reagents. The first edition of this book included part numbers for these reagents, but part numbers hav been removed in this edition due to the fact that the reagents are available from many vendors.

USEFUL EQUIPMENT FOR DERIVATIZATION

Normally in the sample cleanup process, the analytes to be derivatized are in a solvent after a liquid–liquid extraction or after passing through a solid-phase extraction cartridge. Derivatization usually involves two steps:

1. Removal of the solvent by evaporation under a gentle stream of clean dry nitrogen gas.
2. Addition of the derivatization reagent (sometimes in a solvent such as acetonitrile or pyridine) to the dry sample. In this step, the mixture is usually heated for about 20 minutes.

For the evaporation, a dry block heater with an evaporator is recommended.* Dry block heaters require aluminum blocks with wells to accommodate culture tubes or vials of a specific size. These aluminum blocks can also accommodate a thermometer. The evaporation is usually conducted while heating to 25–35 °C. To minimize analyte losses, it is desirable to use silanized vials and culture tubes.**

*Available from Pierce, a division of Thermo Scientific, www.piercenet.com

**The user may prepare silanized glassware with dimethyldichlorosilane, but the glassware is commercially available from scientific supply houses and is strongly recommended.

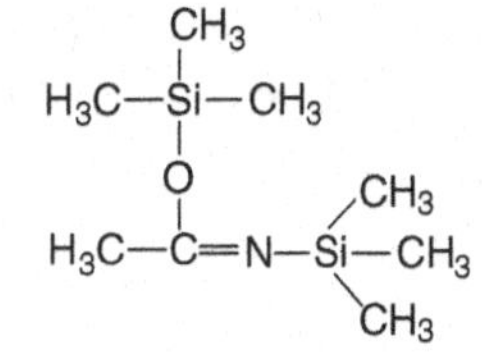

BSA
(*N,O*-Bis(trimethylsilyl)acetamide)

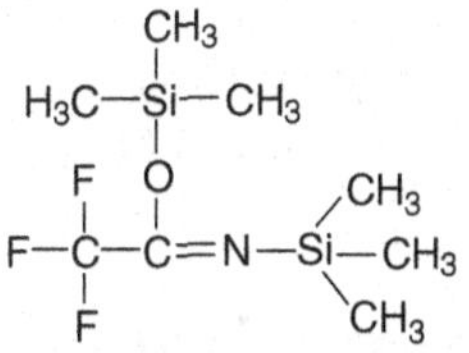

BS TFA
(*N,O*-Bis(trimethylsilyl)trifluoroacetamide)

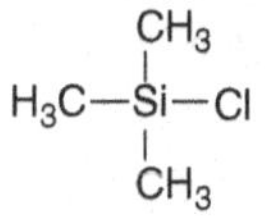

TMCS
(Trimethylchlorosilane)

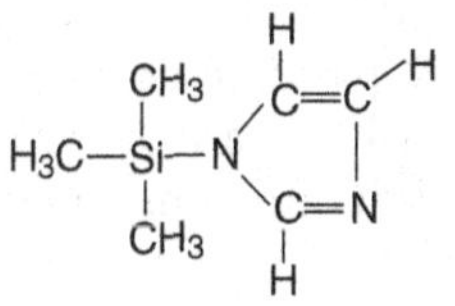

TMSIM
(Trimethylsilylimidazole)

Figure G.2 Structures of some of the more common silyl-derivatizing reagents.

For the derivatization, it is convenient to have a second dry block heater. The reason is that the derivatization reaction is usually carried out at a higher temperature (60–80 °C) than the evaporation. In most cases, the derivatized compound can be injected directly onto the column. If the column contains polar functional groups that react with the derivatizing agent (such as polyethylene glycol or "wax" columns), then the derivatizing reagent must be removed by evaporation.

The alkylating derivatizing reagent (MethElute™) normally reacts very rapidly with the analyte and does not require an incubation period; therefore, this reagent is mixed with the analyte and injected directly into the injector, where the derivatization takes place.

Sample Derivatization Procedures

The following are suggested procedures for derivatizing compounds with hydroxyl, carboxyl, amino, or keto functional groups. Derivatization of a compound with one functional group is normally a simple procedure. One possible problem is steric hindrance; when silylating functional groups where steric hindrance is a factor, the silylation procedures below should be modified by using a silylating agent that contains a catalyst such as BSTFA with TMCS. If steric hindrance is a factor, longer reaction times or higher temperatures might give better results.

Observe safety precautions, including eye protection, when following these procedures and handling these reagents. All derivatization procedures and evaporation of solvents should be carried out in a fume hood.

Hydroxyl Groups

TMS Procedure

1. Add 100 μL MSTFA or 100 μL BSTFA to 1–5 mg dry sample.
2. Add 50 μL pyridine or other silylation-grade solvent such as acetonitrile.
3. Cap the vial and mix well.
4. Heat at 60 °C for 5 minutes.
5. Cool the reaction mixture and inject 1–2 μL into the GC/MS system.

TBMS Procedure

1. Add 100 μL of acetonitrile or other silylation-grade solvent to 1–5 mg dry sample.
2. Add 100 μL MTBSTFA reagent to the sample and heat at 60 °C for 15 minutes.
3. Cool the reaction mixture and inject 1–2 μL into the GC/MS system.

Acetyl Procedure

1. Add 40 μL of silylation-grade pyridine and 60 μL acetic anhydride to the dry sample in a septum vial.
2. Heat at 60 °C for at least 30 minutes.
3. Evaporate to dryness with clean dry nitrogen.
4. Dissolve the residue in 25 μL ethyl acetate.
5. Inject 1–2 μL into the GC/MS system.

Carboxylic Acids

TMS Procedure

Use MSTFA, BSTFA, or Tri-Sil/BSA in pyridine. For keto acids, the methoxime derivative should be prepared first. (See the procedure in Chapter 7, Section 7.3.B.)

1. Add 200 μL of MSTFA, BSTFA, or Tri-Sil/BSA to 1–5 mg of dry sample.
2. Cap tightly and heat at 60 °C for 15 minutes.
3. Inject 1–2 μl into the GC/MS system.

TBDMS Derivative Procedure for Acids

The TBDMS derivative has been used for low-molecular-weight acids such as formic and for dicarboxylic acids such as itaconic, citraconic, and mesaconic where a 30-m 50% trifluoropropyl 50% methyl polysiloxane (DB-210, RTX-200, or HP-210) column programmed from 60 °C to 210 °C at 10 °C per minute separates the latter acids as the TBDMS derivatives.

1. Add 50 µL of MTBSTFA reagent to 1–5 mg of dry sample. Solvents that may be used (if necessary) are pyridine, acetonitrile, tetrahydrofuran (THF), and *N,N*-dimethylformamide (DMF).
2. Allow the reaction mixture to react at room temperature for 30 minutes for low-boiling acids. For high-boiling acids, heat at 60 °C for 5–20 minutes.
3. Inject 1–2 µL into the GC/MS system.

Methyl Ester Procedures

Anhydrous Methanol/H_2SO_4 Procedure

This is an old method, but it is found to be best for trace components. A byproduct is dimethyl sulfate (MW = 126 Da), which shows ions at *m/z* 95, 96, and 66.

1. Add 250 µL of methanol and 50 µL concentrated sulfuric acid to 1–5 mg of dry acid sample.
2. Cap the vial and heat at 60 °C for 45 minutes.
3. Cool the reaction mixture and add 250 µL of distilled water with a syringe.
4. Add 500 µL of chloroform or methylene chloride.
5. Shake the mixture for 2 minutes.
6. Inject 1–2 µL of the chloroform or methylene chloride (bottom layer) into the GC/MS system.

Anhydrous methanolic HCl Procedure

1. Add 250 µL of 3N methanolic HCl to 1–5 mg of dry sample.
2. Cap tightly with a teflon-lined cap and heat at 60 °C for 20 minutes.
3. Cool the reaction mixture and neutralize carefully before sampling.*
4. Inject 1–2 µL into the GC/MS system.

* HCl does not appear to adversely affect the GC columns like other mineral acids.

BF_3/Methanol Procedure

1. Add 250 µL of BF_3/methanol reagent to 1–5 mg of dry sample.
2. Cap tightly and heat at 60 °C for 20 minutes.
3. Cool in an ice-water bath and add 2 mL of distilled water.
4. Extract within 5 minutes with 2 mL of methylene chloride.
5. Extract with another 2 mL of methylene chloride.
6. Combine the extracts and evaporate the methylene chloride with clean dry nitrogen to approximately 100 µL, or evaporate to dryness and add 100 µL of methylene chloride.
7. Dry the extract by adding a small amount of anhydrous sodium sulfate.

Methyl-8® Procedure

1. Add 100 µL of Methyl-8® reagent to 1–5 mg of dry sample.
2. If necessary, add 100 µL of pyridine, chloroform, methylene chloride, THF, or DMF.
3. Cap tightly and heat at 60 °C for 15 minutes.
4. Inject 1–2 µL into the GC/MS system.

A Method for Determining the Number of –COOH Groups in a Molecule

The TMS derivative of an acid can be converted to the methyl ester using anhydrous methanolic HCl. First, obtain a mass spectrum of the TMS derivative of the acid, and then evaporate the TMS reaction mixture to dryness with clean dry nitrogen. Add 250 µL of anhydrous methanolic HCl and heat at 60 °C for 20 minutes. Many TMS derivatives of acids are converted to methyl esters at room temperature after 20 minutes. If the sample is rerun as the methyl ester, the number of carboxyl groups can be determined by the mass differences before and after making the methyl ester from the TMS derivative.

A Method for Determining the Active Hydrogen Content of a Molecule in a Single GC/MS Analysis

This method is especially valuable for identifying metabolites and other trace biological materials. A small portion of the sample is dissolved in the minimum amount of solvent, and the solution is divided into two equal portions. The usual TMS derivative is prepared using the TMS procedure provided above, using BSA reagent for one portion. The other portion is used to prepare the deuterated derivative using BSA–D_{18}*. After the reactions are completed, the two portions are mixed and a 1–2-μL sample is injected immediately into the GC/MS system. In general, the front of the GC peak contains the deuterated derivative, whereas the back of the GC peak contains the regular (D_0) TMS derivative. The mass difference between the MW of the D_0-TMS and the MW of the D_9-TMS derivatives is used to determine the number of TMS groups present. For each 9-Da difference, there will be one TMS group. The deuterated TMS derivative will have a fragment ion at *m/z* 82, whereas the D_0-TMS derivative will have a fragment at *m/z* 73. To obtain the MWs, add 18 (CD_3) to the highest mass ion of substantial intensity (excluding isotope peaks) for the D_9-TMS derivative and add 15 (CH_3) for the D_0-TMS derivative. (Note: an $M^{+\bullet}$ peak of low intensity may be present.)

Primary and Secondary Amines

The amine group is the most difficult [2] to silylate, and a silylating reagent should be used that contains trimethylchlorosilane, which acts as a catalyst. Even though amines are difficult to silylate, primary amines can form both the di-TMS and the mono-TMS derivatives. The amount of each and the rate of formation is solvent dependent. Polar solvents (e.g., acetonitrile) favor the formation of di-TMS derivative; whereas in neutral solvents (e.g., dichloromethane), the formation of the mono-TMS derivative is favored. The derivative should always be handled in such a way as to avoid water because the *N*-TMS is very hydrolytically unstable.

The following procedures are for amines with at least 3 carbons or boiling points >60 °C because they all involve derivatizing the dry sample. Amines with lower boiling points would evaporate during the drying process. Volatile amines can pass through a polar GC column without derivatization. For low-molecular-weight amines that require derivatization in order to yield a mass spectrum that is useful for identification (as in the example of amphetamine above), consult the literature.

TMS Procedure

1. Add 50 μL MSTFA + TCMS or 100 μL BSTFA + TCMS to 1–5 mg dry sample.
2. Add 50 μL dichloromethane.

* Available from CDN Laboratories, www.CDNisotopes.com

3. Cap the vial and mix well.
4. Heat at 60 °C for 5 minutes or let stand at room temperature for 10–20 minutes.
5. Cool the reaction mixture and inject 1–2 μL into the GC/MS system.

TBMS Procedure

1. Add 100 μL of acetonitrile or other silylation-grade solvent to 1–5 mg dry sample.
2. Add 100 μL MTBSTFA + TBDMCS reagent to the sample and heat at 60 °C for 15 minutes.
3. Cool the reaction mixture and inject 1–2 μL into the GC/MS system.

Trifluroacetyl Procedure

1. Add 200–500 μL of toluene containing 0.05 M triethylamine to the dry sample.
2. Add 50 μL trifluoroacetic anhydride (TFAA), cap, and mix well.
3. Heat for 5 minutes at 45 °C.
4. Cool to room temperature.
5. Add 400–1,000 μL 5% sodium bicarbonate solution.
6. Mix by vortexing until the top layer is clear.
7. Centrifuge.
8. Inject 1–2 μL of the top layer. Do not inject any of the bottom layer.

Acetyl Procedure

1. Add 40 μL of silylation-grade pyridine and 60 μL of acetic acid to the dry sample in a septum vial.
2. Heat at 60 °C for at least 30 minutes.
3. Evaporate to dryness with clean dry nitrogen.
4. Dissolve the residue in 25 μL acetonitrile or pyridine.
5. Inject 1–2 μL into the GC/MS system.

Methyl-8® Procedure (Primary Amines Only)

1. Add 50 μL of Methyl-8® reagent and 50 μL acetonitrile to <0.1 mg of dry sample.
2. Heat at 60 °C for 30 minutes.
3. Inject 1–2 μL into the GC/MS system.

Method to Determine the Number of –NH or NH_2 Groups

1. Analyze the sample first as the TMS derivative.
2. Add 250-μL MBTFA to the TMS reaction mixture.
3. Heat at 60 °C for 30 minutes to convert the TMS groups of –NH or –NH2 to trifluoroacetyl groups.

Keto Groups

Methoxime Procedure

1. Add 0.5 mL MOX reagent to the dry sample.
2. Heat at 60 °C for 3 hours.
3. Evaporate the reaction mixture to dryness with clean dry nitrogen.
4. Dissolve the residue in the minimum amount of ethyl acetate. Do not add the solvent if the TMS derivative is to be prepared as follows:

TMS Derivative of Methoxime Product

Add 250 μL of MSTFA reagent to the dry methoxime derivative and let stand for 2 hours at room temperature.

When derivatizing compounds with multiple functional groups such as carbohydrates, keto acids, and amino acids or with single functional groups with more than one active hydrogen (amines), it is important to verify that only one derivatized compound results. For these reasons, when considering derivatization of complex compounds, it is strongly advised that the references[3–6] be consulted.

REFERENCES

1. Czarny R. J. Hornbeck C. L. (1989). Quantitation of methamphetamine and amphetamine in urine by capillary GC/MS. Part II, derivatization with 4-carbethoxyhexaflurobutyryl chloride. *J Anal Toxicol.* 13, 257–262.
2. Kataoka H. (1996). Review: derivatization reactions for the determination of amines by gas chromatography and their applications in environmental analysis. *J. Chromatogr. A,* 733(1, 2), 19–34.
3. Knapp D. R. (1979). *Handbook of Analytical Derivatization Reactions.* New York: Wiley-Interscience, ISBN:047103469X.
4. Blau, K. Halket, J, eds. (1993). *Handbook of Derivatives for Chromatography.* 2nd ed. New York: Wiley, ISBN:047192699X

5. Blau K, King G, eds. (1978). *Handbook of Derivatives for Chromatography*. 1st ed. London: Heyden, ISBN:0855012064.
6. Zakin, V. Halket J. A. (2009). *Handbook of Derivatives for Mass Spectrometry*. Chichester, UK: IM Publications, ISBN: 9781901019094.

APPENDIX H

Points of Comparison of LC/MS vs GC/MS

There is no such thing as a universal analytical technique, especially when it comes to the analysis of a mixture of organic compounds or even a simple organic compound in a complex organic matrix. Both LC/MS and GC/MS can be used. The following is a list of various chromatography/mass spectrometry systems. Bolded items are considered to be advantages over the other type of system.

LC/MS (Electrospray and APCI)	GC/MS (EI and CI)
Instrument purchase price with HPLC system: $120,000–$250,000++.	**Instrument purchase price with GC: $75,000–$150,000.**
HPLC system is often not included with instrument; some vendors rely on a second vendor or user choice for HPLC system.	**Gas chromatograph always included with instrument.**
Nearly all commercial transmission quadrupole systems have MS/MS option.	Most GC/MS manufacturers now have a tandem quadrupole instrument. There is a third-party add-on for the Agilent GC-MS.
System can be operated for continuous-flow liquid samples without HPLC; solid-phase extraction, dialysis, direct-process stream sampling, and other liquid sampling techniques may be substituted for HPLC.	Continuous-flow samples can only be introduced via GC. Some EI/CI instruments originally designed for GC/MS have had the GC replaced with a membrane introduction for continuous-flow introduction.
No commercial options for static direct-sample introduction; i.e., sample must always be in solution.	**Nearly all systems have direct-sample probe option available for static introduction of solid and neat liquid samples.**

(*Continued*)

Gas Chromatography and Mass Spectrometry
DOI: 10.1016/B978-0-12-373628-4.00044-7

(*Continued*)

LC/MS (Electrospray and APCI)	GC/MS (EI and CI)
High volume of solvent vapor places a high demand on vacuum and ventilation system; multipump, multistage vacuum systems increase cost and maintenance requirements.	**Relatively low gas load into the MS requires less pumping efficiency in vacuum system; lower initial cost and lower maintenance costs for vacuum system.**
Columns are moderately efficient; column lifetime can be considerably shortened by contamination and precipitation; expensive to replace.	**Extremely efficient capillary columns; contamination issues are minimal; column lifetime easily extended by removal of the first few centimeters.**
EI applicable only to compounds that ionize in solution; APCI more broadly applicable but still requires a polar functional group.	**EI will ionize nearly anything**; CI similar requirements to LC/MS APCI.
Typical LC/MS spectra are relatively featureless; not likely to provide enough data for identification of unknowns beyond an elemental composition.	**Typical EI spectra are information-rich and are commonly used for unambiguous identification of unknowns.**
Spectral libraries must be created by the user for a particular model of instrument; limited commercial availability of libraries.	**Spectral libraries for EI are well established and commercial availability is very good; libraries are usable on all manufacturers' instruments.**
Electrospray and APCI are gentle ionization techniques resulting in little or no fragmentation; in-source CAD and/or MS/MS can be used to generate fragment ions; degree of fragmentation fully controllable by the user.	The high energy of the EI process can result in complete sample fragmentation with loss of molecular ion peak. CI can be used in this case but with severe loss of spectral information (MW only).
Typical LC/MS spectra can be complicated by adduct ions, clustering, multimer formation.	**GC/MS EI and CI spectra generally very clean and easily interpreted (provided initial chromatographic separation is good).**

(*Continued*)

(*Continued*)

LC/MS (Electrospray and APCI)	GC/MS (EI and CI)
Electrospray subject to ion suppression effects from solvent systems and sample constituents. APCI is more rugged.	**Ion suppression effects not a concern with EI source**; ion/molecule reactions may complicate CI.
Samples may range from relatively volatile small molecules (>100 Da) to extremely large macromolecules (>100 kDa) with electrospray; APCI limited to volatile samples <2,000 Da.	Samples limited to highly volatile to moderately volatile molecules (~10–800 Da). Higher masses can be analyzed when volatile analytes contain extensive number of F atoms. Pyrolysis front-ends applicable to some polymers and macromolecules.
Polar/ionic samples work extremely well in both electrospray and APCI; relative volatility of samples is irrelevant.	Polar/ionic samples may be problematic in GC injectors and columns; volatility differences may result in sampling discrimination; these types of compounds may require derivatization.
Thermally labile compounds well suited to electrospray analysis because they must be ionizable in solution; may be analyzed by APCI if high temperature not required for volatility.	Compounds exhibiting thermal lability at 200 °C and higher cannot be analyzed unless they are derivatized.
Derivatives not required for good ionization efficiency.	Derivatives often required to improve sample volatility.
No phase change required for liquid samples; "dirty" liquid samples, even serum, can often be directly injected; matrix components easily diverted prior to MS.	Sample preparation required for liquid samples to facilitate phase change when injected. Serum and other "dirty" samples require extensive sample prep prior to GC injection.
Liquid handling systems can fully automate processes such as combinatorial chemistry, followed by sample preparation and injection onto LC-MS.	Difficult to reliably integrate GC/MS injection with extensive liquid handling and liquid sample prep automation.
Mobile phases must be prepared carefully; hazardous solvent waste disposal required.	**No mobile-phase issues (He or H_2 only)**.

(*Continued*)

(*Continued*)

LC/MS (Electrospray and APCI)	GC/MS (EI and CI)
Pump maintenance (seals, valves, etc.) must be carried out to ensure retention time precision; interface must be periodically clean, especially if involatile mobile phase is being used; **mass analyzer remains relatively uncontaminated for long periods of use**.	**No moving parts; low maintenance requirement; parts such as inserts and septa more easily replaced than pump parts**; mass analyzer subject to contamination.
Due to solution chemistry consideration, more highly skilled operator required.	**Operator can be more easily trained to run the instrument**.
Cannot be used for the analysis of permanent gases or nonpolar compounds.	**Works as well with permanent gases as it does with solutions. Ideally suited for nonpolar compounds**.
Cannot be used with SPME or headspace analyses.	**Works well with SPME and headspace**.

APPENDIX I

List of Available EI Mass Spectral Databases

NIST/EPA/NIH Mass Spectral Database (NIST08)

220,435 Spectra of 192,262 Compounds Includes ~28 K+ Selected Replicate Spectra. The *NIST08 Database* is provided with electronic structures in a format optimized for the NIST Mass Spectral (MS) Search Program.

In addition, the *NIST08 Database* includes 292,924 Kovâts retention index (KRI) values for 42,888 compounds (21,847 of which there are mass spectral data) in the Database on nonpolar and polar columns. Each RI value has the complete literature citation (including title) from which it was extracted and all the GC conditions under which it was measured. The NIST MS Search Program now has a routine that will estimate the KRI for compounds in the Database where there are no previous literature values. This routine will estimate a KRI for any structure displayed by the NIST MS Search Program. There are 14,802 CAD spectra of 5,308 different ions (both positive ions and negative ions) acquired using triple-quadrupole mass spectrometers and quadrupole ion trap instruments.

Provided by all mass spectrometry instrument manufacturers as an option for use with their proprietary library search programs and for the NIST Mass Spectral Search Program for Windows™ – includes Prof. Henneberg's Industrial Chemical Collection and Complete Collections of Chemical Concepts, Georgia & Virginia Crime Lab, TNO Flavor & Fragrances, AAFS Toxicology Section – Drug Lib., Assoc. of Official Racing Chemists, St. Louis Univ. Urinary Acids, VERIFIN & CBDCOM Chemical Weapons, Japan AIST/NIMC Collection, Russian Academy of Sciences: Institute of Petrochemical Synthesis, Eastman Chemical Company, and Military Institute of Chemistry and Radiometry (Poland).

NIST Retention Index Database with Search Program

The NIST Retention Index Database with a Search Program is also provided by a number of NIST software distributors. The Search Program offers far more flexibility in searching the RI Database and displaying results.

Wiley/NIST Registry of MS Data, 9th Ed.

668,092 Spectra + Optional Additional 152,590 NIST Spectra from *NIST08*—820,682 Spectra.

Distributed by John Wiley & Sons through Mass Spectrometry Instrument Manufacturers and directly to end users. JW&S distributes the Wiley

Gas Chromatography and Mass Spectrometry
DOI: 10.1016/B978-0-12-373628-4.00045-9

9 & 9N in the NIST format along with the NIST MS Search Program. The additional NIST spectra contain an important set of spectra collected under close control of tune conditions by the U.S. EPA. The Wiley/NIST (Wiley 9N) is offered by most instrument manufacturers. The Wiley 9N is provided with electronic structures from Wiley.

Maurer/Pfleger/Weber Mass Spectral and Gas Chromatography Data of Drugs, Poisons, Pesticides, Pollutants and Their Metabolites, 2007 Ed., with Two-Volume Text, 3rd Ed.
7,840 Spectra and RI values: >2,300 acetylated, >1,000 methylated, >700 trimethylsilylated, >400 trifluoroacetylated, and >200 each of pentafluoropropinylated and heptafluorobutylated derivatives. Unlike previous versions, this version contains full spectra and structures in electronic format. This is the fourth electronic release. Previously known as the Pfleger/Maurer/Weber Library. Data in the electronic format is available with a 2-volume text from John Wiley & Sons and their distributors. The 2-volume text is also available without the electronic version of the data.

Ion Trap Terpene Mass Spectral Library, 4th Ed.
EI spectra acquired on an HP5970 of 2,205 Terpenes include hardcopy printout of all spectra with structures. This database is available for most commercial data systems. It is a companion to Identification of Essential Oil Components by Gas Chromatography/Mass Spectrometry from Allured Publishing. Electronic structures are not available. This database is not provided in the NIST MS Search Program format; however, the Allured Web site contains the following statement, "The HP [sic] Chemstation [sic] library can be converted to the NIST library format by using Lib2NIST, a free download program from http://chemdata.nist.gov."

History of the Admas Mass Spectral Library

Ion Trap Terpene MS Library
570 EI Spectra of Terpenes acquired on an ITD700.

Spectra acquired pre-1989 on the Finnigan ITD700 (Pre-AGC and Re-Axial Modulation) and provided with the Finnigan & Varian Ion Trap Data Systems at no charge.

A separate book containing the spectra, Identification of Essential Oil Components by Ion Trap Mass Spectrometry, was published by John Wiley Sons © 1989.

Ion Trap Terpene MS Library, 2nd Ed.
1,252 EI Spectra of Terpenes acquired on an ITD700 © 1995 available in most commercial data systems' formats.

Provided as a companion to Identification of Essential Oil Components by Gas Chromatography/Mass Spectrometry from Allured Publishing, Carol Stream, IL, USA.

Terpene MS Library, 3rd Ed.
EI Spectra of 1,607 Terpenes acquired on an HP5970 © 2004 available in most commercial data systems' formats.

Provided as a companion for most commercial data systems to Identification of Essential Oil Components by Gas Chromatography/Quadrupole Mass Spectrometry from Allured Publishing, Carol Stream, IL, USA.

Terpene MS Library, 4th Ed.

EI Spectra of ~2,200 Terpenes acquired on an HP5970 © 2007 available in most commercial data systems' formats.

Provided as a companion for most commercial data systems to Identification of Essential Oil Components by Gas Chromatography/Quadrupole Mass Spectrometry, 4th ed., from Allured Publishing, Carol Stream, IL, USA.

All the books that accompanied the electronic versions of the four editions of this database contained structures that illustrated chirality; however, there have never been any electronic structures provided.

ChemStation Pesticide Mass Spectral Library (340 EI Spectra)

340 EI Spectra available from Agilent Technologies as an option for use with their Proprietary Library Search Program (The Stan Pesticide Library, G1038A).

Other Agilent GC/MS ChemStation Mass Spectral Databases

Agilent has a series of mass spectral databases that are part of their retention time locking (RTL) system. These databases are provided in the GC/MS ChemStation and a NIST AMDIS format. The ChemStation format databases can be converted to the NIST MS Search Program format.

MS LIBRARY	APPLICATION AREA	NUMBER OF SPECTRA (approx.)	PRODUCT NUMBER
Fiehn GC/MS Metabolomics RTL Library	Dr. Oliver Fiehn UC Davis	1,050 Spectra of ~700 metabolites	G1676AA
Agilent Pesticides RTL Library	Pesticides and endocrine disrupters	926	G1672AA
Forensic Toxicology RTL Library	Forensic	723	G1675AA
Hazardous Chemicals RTL	Environmental	731	G1671AA
Environmental Semi-VOAS RTL	Environmental	273	G1677AA
Indoor Air Toxics RTL Library	Environmental	171	G1673AA
Japanese Positive List Pesticide RTL	Pesticides	431	G1675AA

John Wiley and Sons offer a number of Specialty Mass Spectral Databases that contain spectra that are not in the *Wiley Registry of Mass Spectral Data*.

Mass Spectra of Designer Drugs 2000
Dr. Peter Roesner
Regional Department of Criminal Investigation Kiel; Kiel, Germany
ISBN: 978-3-527-32392-0
Contains 11,011 spectra of 9,436 compounds and 3,892 Kováts indices.
The Designer Drugs database first appeared in 2003. It had no electronic structures. The fourth edition came just a year after the third edition. It is not clear whether Wiley plans additional editions on a yearly basis; however, there has been some talk about a possible subscription. Along with the 2007 (third) edition, there was a two-volume set of books (copyright 2007, ISBN: 978-3-527-30798-2). There was not a new volume printed for the 2008 edition or the 2009 edition. Wiley has stated that the spectra added to the 2007 edition for the 2008 edition will be made available in hardcopy. Unlike the previous three versions, Designer Drugs 2008 and Designer Drugs 2009 contain structures in electronic format. These structures are provided in the NIST MS Search Program format and the Agilent PBM format. The 2009 edition contains all the spectra from the previous editions.

Mass Spectra of Volatile Compounds in Food, 2nd Ed.
Central Institute of Nutrition and Food Research (TNO)
A.J. Zeist, The Netherlands, ISBN: 0-471-44056-6

Mass Spectra of Flavors & Fragrances of Natural & Synthetic Compounds (FFNSC 1.3)
Luigi Mondello
ISBN: 978-0-470-42521-3
Contains 1,831 mass spectra, RI values, and structures.

Mass Spectra of Pesticides 2009
Rolf Kühnle
ISBN: 978-3-527-32488-0
Contains 1,238 mass spectra.

Mass Spectra of Androstanes, Oestrogens and Other Steroids 2008
Professor H.L.J. Makin
St. Bartholomew's and the Royal London School of Medicine and Dentistry, London, U.K.
ISBN: 978-3-527-32193-3
Contains 3,722 electron ionization mass spectra, and structures.
Mass Spectra of Androgens, Estrogens, and other Steroids database was published in 2005 (prepared in 2004) containing ~3 K spectra. A book entitled *Mass Spectra and GC Data of Steroids: Androgens and Estrogens* authored by HLJ

Makin, DJH Trafford, and J Nolan with data on "nearly 2,500 spectra" was published in 1998 (ISBN: 3-527-29644-1). This database has structures in the NIST MS Search Program format, but not in the Agilent format.

Mass Spectra of Pharmaceuticals and Agrochemicals 2006
Rolf Kühnle
ISBN: 978-3-527-31615-1
Contains 4,563 mass spectra.

Mass Spectra of Organic Compounds
Dr. Alexander Yarkov
Chemical Block, Moscow, Russia, ISBN: 0-471-66773-0
Contains 37,055 mass spectra.

Mass Spectral Collection: Geochemicals, Petrochemicals and Biomarkers, 2nd Ed.
J.W. de Leeuw
Netherlands Institute of Sea Research, Texel, The Netherlands, ISBN: 0-471-64798-5
Contains 1,100 mass spectra.

Other Mass Spectral Databases

Georgia State Crime Laboratory Toxicological Mass Spectral Library
Approximately 1,000 Spectra
Included in the NIST MS Database—Hardcopy of spectra along with NMR, IR, and UV available as a 7-volume set from CRC Press—Available to Law Enforcement Agencies in electronic format for use with the Agilent GC/MS ChemStation on request.

In addition to the databases listed above, there is an AAFS (American Association for Forensic Scientist) Toxicological EI Mass Spectral Library containing 2,300+ spectra. This database was developed in Canada and is available on the Web at http://www.ualberta.ca/~gjones/mslib.htm as a free download.

NIST08 Demo
The National Institute of Standards and Technology (NIST) provides all of its software and a demo database of EI mass spectra at http://chemdata.nist.gov. Select the link labeled NIST/EPA/NIH Mass Spectral Library. As you scroll down the page, you will see a graphic of the *NIST MS Search Program's* main display. Under the graphic is a line of text which reads "A NIST08 demo version may be downloaded here." The word *here* is the link to download all the items listed below

- Version 2.0d of NIST MS Search
- MS Interpreter
- Isotope/Formula Calculator
- ~2,400 spectra from NIST05 Database

- Current Version of AMDIS
- On request, Interfaced to ChemStation
- ***Free Download***

Notes on Mass Spectral Databases

A database containing mass spectra of all the compounds you regularly analyze should be developed on the instrument(s) used in your laboratory. If you use different types of analyzers (i.e., QIT, TOF, QMF, and double-focusing), a separate database should be maintained for each instrument type. The database should contain multiple spectra of each compound over the expected concentration range. As much information as can be provided about a compound should be included with each spectrum along with a chemical structure that shows chirality where appropriate.

APPENDIX J

INFORMATION REQUIRED FOR REPORTING A GC/MS ANALYSIS

The amount of information appearing in journal articles regarding the acquisition of data by GC/MS is often insufficient for those wishing to reproduce the method to do so.

Mass spectrometry is increasingly being used by chromatographers who are used to treating whatever is on the exit end of the column as a detector. This means that a great deal of the mass spectrometry particulars necessary to reproduce an analysis is often omitted and/or not recorded in the analyst's notebook. When reporting the instrument's settings and condition for a publication, a tabular format is best. Separate tables should be created for the mass spectrometer (MS) and the gas chromatograph (GC).

It is important to remember that whereas most GC/MS systems come with a specific GC mated to a specific MS, some manufacturers have different model GCs as options. A good example of combinations of GCs and mass spectrometers that can be confusing is the case of the Agilent (then Hewlett–Packard) 5890 GC and 5972 MS followed by the 6890 GC and 5972 MS. There was also a 6820 GC with a 5973 MS and a 6890 GC with a 5973 MS.

As a general rule of thumb, any value that can be set in a method editor on a computerized data system or any value that can be read during acquisition should be reported.

The following is a list of required information:

Mass Spectrometer—manufacturer, model, and type: **e.g., Varian, 4000, External Ionization Quadrupole Ion Trap, Walnut Creek, CA, USA**.

Note: Reporting the mass spectrometry instrument must include relevant details. An example of a potential lack of information in reporting just the manufacturer and model name (number) is seen for instruments that are available with various pumping system options. It is important to include the specifics of the pumping system with the instrument's description. The Agilent 5973 has been sold with an oil diffusion pump, a $70\,L\,sec^{-1}$ turbomolecular pump, and a $250\,L\,sec^{-1}$ turbomolecular pump. Both the $70\,L\,sec^{-1}$ and the $250\,L\,sec^{-1}$ turbomolecular pump instruments offer chemical ionization; however, CI performance on the $70\,L\,sec^{-1}$ system is far inferior to the performance of the $250\,L\,sec^{-1}$ system.

Gas Chromatography and Mass Spectrometry
DOI: 10.1016/B978-0-12-373628-4.00046-0

Some manufacturers offer various types of electronic upgrades to a given model of an instrument. If such an upgrade has been installed on an instrument, the details should be included with the instrument's description.

Type of sample introduction: e.g., GC, direct insertion probe, desorption CI or FD probe, flow (etc.); specifically, all makes and models of valves, third-party devices, (etc.).

Sign of ions/type of ionization: e.g., positive EI, positive CI, FI, negative ECI, negative (etc.).

Ionization energy: In the case of EI, CI, and ECI GC/MS (where a beam of electrons is directly or indirectly involved with the ionization), the electron energy should be reported; e.g., for an EI analysis: 70 eV.

Relevant voltages: Most instruments report various voltages for lens, detectors, etc. All of these should be reported. The labels of these voltages vary from manufacturer to manufacturer. In the case of an MS/MS analysis, the collision energy should be reported.

Reagent gas and pressure (where appropriate): In the case of CI or ECNI where a reagent gas such as methane is used, the gas and its pressure in the ion source should be reported. If the pressure in the ion source cannot be determined, some determinable value should be reported such as the gauge pressure on the supply size of the gas cylinder.

Gases used: Any gas such as a GC carrier or reagent gas should be reported as to type, source (gas generator, boil-off of a liquid, etc.), temperature, and flow rate and/or pressure. If an MS/MS experiment is being carried out, the collision gas and its pressure should be reported (i.e., Ar at a pressure of 3.0×10^{-1} Pa).

Relevant pressures: e.g., pressure in the ion source, analyzer, interface in LC/MS systems, detector region of the analyzer, (etc.).

***m/z* scale calibration**: The name of the manufacturer's tune file, the material used for calibration (e.g., PFTA, PFK). In the case of an accurate mass measurement, the reference *m/z* value and compound should be reported.

Type of data acquisition: e.g., continuous acquisition of spectra or selected ion monitoring (SIM), alternating continuous scan/SIM, MS/MS mode (fixed-precursor-ion analysis, fixed-product-ion analysis, or common-neutral-loss analysis), selected reaction monitoring (SRM), (etc.).

Data acquisition range: e.g., *m/z* 20–350, SIM ions, SRM transitions, precursor ions, (etc.).

Data acquisition rate: Some instruments do not allow the direct setting of the acquisition rate (spectra per second or time for the acquisition of a single spectrum). These values are determined by setting the number of spectra to be averaged before saving a spectrum to a data file. For example, the Agilent MSD ChemStation for the 5973 allows the setting of the **start mass** and **end mass** on the **scanning mass range**

tab of the **edit scan parameters** dialog box and the **sampling rate (2^*n*)** on the **threshold & sampling rates** tab. Changing the value in any of these three fields will change the value reported in the Scans/Sec field of the Summary Settings pane of the dialog box. The value entered in the **sampling rate (2^*n*)** is the power to which 2 is raised to determine the number of spectra that are averaged before a spectrum is stored. Both this number and the *data acquisition rate* (Scans/Sec) should be reported.

The Varian quadrupole ion trap mass spectrometer's data system has entry fields for the starting and ending *m/z* values for the data acquisition. It also has an entry field for the number of spectra to be acquired in 1 second. This causes the number of *microscans* (spectra averaged before a spectrum is stored) to change. Again, both the spectra per second and the number of microscans should be reported as the *data acquisition rate.*

In the case of an SIM or SRM analysis, the dwell time for each ion should be specified along with the time these ions are monitored during the analysis.

Temperature of the ion source: e.g., 200 °C.

Temperature of the analyzer (mass spectrometer): e.g., 150 °C.

Other relevant temperatures: e.g., temperatures of the GC column transfer line, APCI interface (which is an ion-source temperature).

Mass spectrometer's resolving power (or resolution): In the case of most GC/MS analyses using transmission quadrupole or QIT mass spectrometers, the resolution is unit (ions differing by 1 *m/z* unit can be separated) even though the data system can report to the nearest 0.05–0.01 *m/z* unit. This difference in resolution and reported value should be noted. In some cases, using a transmission quadrupole in the SIM mode, the instrument can be set to monitor the nearest 0.05 *m/z* unit. These used values should be reported in the acquisition range portion of the method. In the case of data acquired under high resolving power conditions, all relevant information should be reported; e.g., reported resolving power and definition used (10,000 based on 10% valley definition or 20,000 based on FWHM definition), the reference compound (and *m/z* value used, if appropriate), the determined accuracy of mass measurements (usually in ppm).

Type of ion detection device: e.g., electron multiplier, microchannel plate, photomultiplier, (etc.).

Electron multiplier (or ion detection device) voltage: e.g., EM voltage = 1,500. In the case of an electron multiplier, the gain should be reported even though, in most cases, it is 10^5.

Solvent delay: The delay between the time the analysis was started (usually from the sample introduction time) and the time mass spectral acquisition begins. This is mainly used in GC/MS to make sure that the solvent elutes the column before the ionization filament and the EM are turned on.

Data acquisition time: The amount of time that data are acquired.

Gas Chromatograph—manufacturer, model, and type: **e.g., Agilent, 6890, 220 Volt, Little Falls, Delaware, USA.**

Injector type: e.g., split/splitless, cold-on-column, large volume, (etc.). All pressure used with the injector should be reported.

Injection type (do not confuse with injector type): e.g., split, splitless, (etc.). If a split injection is used, the split ratio should be stated. If a splitless injection is being made, the split delay and the purge flow to split vent flow should be reported.

Injector temperature: e.g., 250 °C. If a subambient condition is being used, what is the coolant?

GC column: e.g., column length (m), column diameter (mm; i.e., 0.32 mm or μm; i.e., 320 μm, sometimes called 320 microns [sic]) stationary phase (DB™-5MS), thickness of stationary phase in open-tubular capillary columns (μm; 0.25 μm, sometimes called 0.25 microns [sic]), column's material of construction (fused silica).

Carrier gas, linear velocity, flow rate: e.g., He, 32 cm sec^{-1}, 1 mL min^{-1}. There may be other parameters that are specific to an instrument make or an injector type.

Column oven temperature details: Initial temperature and hold time; rate of rise (°C min^{-1}) to next temperature; hold time for second plateau; rate of rise to next plateau, (etc.). If the initial temperature is a result of a cryogen, give particulars of the cryogen (CO_2 or LN_2) flow rates, (etc.).

Other detectors: If other GC detectors such as the FID or ECD are used, report all relevant data such as temperature, gases, flow rate, (etc.).

Note: When a **chromatographic device** is used, the sample size and method of introduction should be specified. In the case of GC sample introduction, the use of slow or fast injections should be specified. If an **autosampler** is used, all settable parameters associated with the autosampler should be reported. The sample solvent should always be reported along with the exact or approximate analyte concentration.

Data Analysis: **Manufacturer, software name, and version number; e.g., Thermo Electron, Xcaliber™, Version 1.4, San Jose, CA, USA.**

Additional data analysis software including version number or publication date and where acquired: e.g., MS Tools, v.5.14, purchased from ChemSW, Fairfield, CA, USA.

Databases used in identifications including version number and where acquired: e.g., NIST Mass Spectral Database (NIST02), purchased from Scientific Instrument Services, Ringoes, NJ, USA.

Third-Party Software for Use with GC/MS

There are a number of programs available to facilitate the analysis of data and to help in the design of an analysis. Some of these programs must be purchased, and some are available as free downloads.

National Institute of Standards & Technology (NIST) Software & Databases

The National Institutes of Standards and Technology (NIST) has a number of programs that are available at no charge (**FREE DOWNLOADS**) and can be used in the analysis of GC/MS data. All of these programs can be accessed from the top menu at http://chemdata.nist.gov.

Free Downloads from NIST

One important download from the NIST Web site is the Demo version of the NIST/EPA/NIH Mass Spectral Database, **NIST08**. The Demo version of NIST08 has spectra of 2,378 compounds. Many of these spectra include retention index values and citations (described below under the heading **NIST Databases that Must Be Purchased**). The Demo version includes a completely functioning copy of the **NIST Mass Spectral (MS) Search Program, AMDIS** (see Chapter 5); the **Mass Spec Interpreter** (a program that has many features to aid in the interpretation of EI–$M^{+\bullet}$ and CI–MH^{+} mass spectra; it will match mass spectral peaks to parts of molecules when the spectrum is associated with a structure); and a separate isotope/formula calculator. The included **MS Search Program** allows for the building of User libraries and can be interfaced with Agilent ChemStation. To do the latter, macros will need to be installed in the ChemStation's program directory. These macros can be obtained by contacting the author at ods@csi.com.

One very important feature of the **NIST MS Search Program** is that it will predict a nonpolar retention index (*n*-alkane scale) for any spectrum

Gas Chromatography and Mass Spectrometry
DOI: 10.1016/B978-0-12-373628-4.00047-2

(including User Library spectra) that has an associated structure. User libraries can be built with associated structures.

AMDIS (Automated Mass spectral Deconvolution and Identification System) can not only be used to view data files for most any commercially available GC/MS data system; but it also does an excellent job of chromatographic peak deconvolution, which provides for background-limited spectra for library searches in the **MS Search Program** or interpretations.

NIST also has a **Mass Spectrum Digitizer Program**, which uses graphic presentation of mass spectra that are in a JPG format and converts them to a tabular text format that is suitable for direct use in a User library in the **NIST MS Search Program**. Spectra can be taken from the literature that is available in a PDF format, put into a JPG format, and made available for library searching.

The NIST Chemdata Web site is also the source of program upgrades, user manuals, and other valuable information such as that about the NIST **WebBook**, a Web site with a diverse collection of data including electron ionization (EI) mass spectra of many compounds (not as inclusive as the NIST08 Database). Under the heading **NIST MS Search Program** (second link) on the Chemdata Home Page is a link to **Lib2NIST**, which is a utility for making copies of netCDF and Agilent ChemStation mass spectral databases to a NIST format. This is a very valuable utility.

NIST Databases That Must Be Purchased

NIST also has two Databases available for purchase: the NIST/EPA/NIH Mass Spectral Database (**NIST08**) and **the NIST GC Retention Index (RI) Database** with its own Search Program. A new version of the NIST/EPA/NIH MS Database is published about every 3 years. The number in the name is the publication year. The next update is scheduled for 2011 (**NIST11**). The **NIST08 DB** contains 219,744 spectra of 191,436 compounds; replicate spectra are maintained in a separate file so that the combination or only the main library can be searched by itself, either with or without user libraries. **NIST08** also includes all the RI records that are in the **NIST GC RI Database**; however, the **NIST MS Search Program** does not allow the interrogation of the **RI DB** to near the same extent as it can be interrogated by the **RI Search Program. NIST08** also includes an MS/MS Database of about 14 K spectra of ~5.3 K different ions in both positive and negative modes. Unlike the Demo version of **NIST08**, the full version can be searched in a way that the Hit Lists will contain spectra of derivatives and stereoisomers of identified compounds.

The **NIST RI Database** contains 292,924 records for 42,888 compounds, 21,847 of which have mass spectra (Figure K.1). The records in the

2. Value: 1626.1 iu
Column Type: Capillary
Column Class: Standard nonpolar
Active Phase: DB-1
Column Length: 30 m
Carrier Gas: He
Column Diameter: 0.26 mm
Phase Thickness: 0.25 um
Data Type: Normal alkane RI
Program Type: Ramp
Start T: 120 °C
End T: 280 °C
Heat Rate: 8K/min
Start Time: 1 min
End Time: 22 min
Source: Manca, D; Ferron, L; Weber, J-P **A System for Toxicological Screening by Capillary Gas Chromatography with Use of Drug Retention Index Based on Nitrogen-Containing Reference Compounds** *Clin. Chem.*, *35*(4), **1989**, 601–607.

Figure K.1 Example of a Retention Index record with GC method and citation.

RI Database have been taken from the literature and include not only the reported retention index value but also the complete GC method along with the literature citation including the tile.

The **RI Database Search Program** allows for multiple search options and types, including RI ranges as well as by compound name (DB includes multiple synonyms), formal (elemental composition), molecular weight (nominal mass), and Chemical Abstracts Service registry number (CASrn) (Figure K.2). The search results can be constrained. This **Database** and its accompanying **Search Program** are ideal when a new compound or new compound type needs to be analyzed by GC. Just look up the compound or a similar compound, and a method is readily available. One user needed a method for the analysis of organo-tin compounds. A Sequential Search using Sn > 0 as a Constraint resulted in 32 Hits. The **Program** also features a Structure Similarity Search. If a chromatographic peak appears between peaks representing two known compounds, look up the RI values for those known compounds and do an RI Search over the RI range to see if there is a candidate for the unknown; or, in the case of regioisomers, which isomer elutes in what order. The **NIST RI Database** and its accompanying **Search Program** are also ideal for use with GC systems as well as GC/MS systems.

The **NIST Mass Spectral** and the **GC Retention Index Databases** are only available through **NIST** authorized distributors. The distributors offering the **RI Database** are listed at the bottom of the Web page for the **Database** and **RI Search Program**. There is a link on the Chemdata Web page to a list of distributors of the **MS Database.**

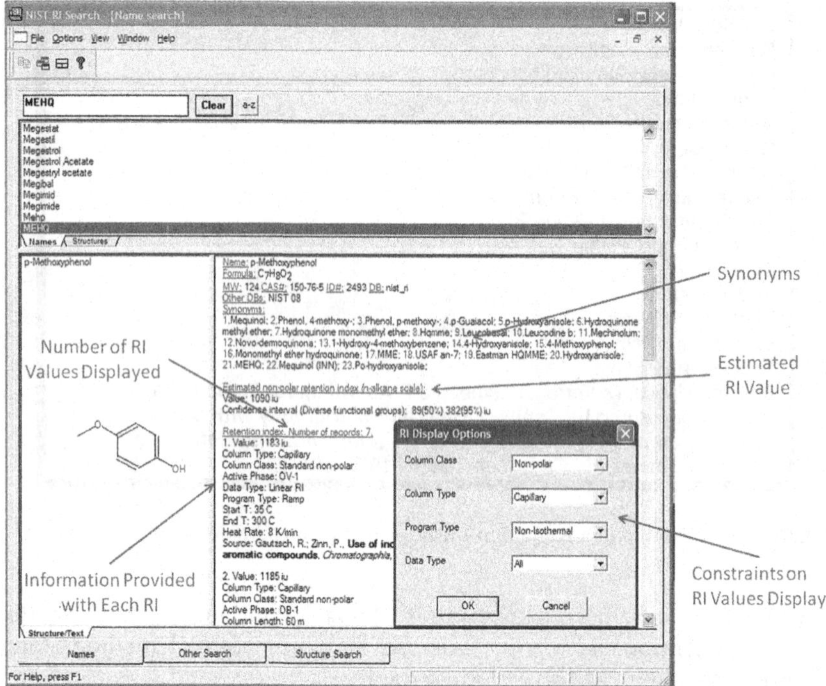

Figure K.2 NIST RI Search Program display of the results of a Name Search.

Another Free Program for Download

Many manufacturers do not allow for the use of their data analysis software on more than a single computer. **AMDIS** can be used for multiple workstations to perform data manipulations; however, there is another program that has more traditional features—the **Wsearch32** (http://www.wsearch.com.au). This program's display of a reconstructed total ion current (RTIC) chromatogram and mass spectrum is shown in Figure K.3. The program allows for the display of mass chromatograms and can have multiple Windows. Wsearch32 has its own library search routine that will allow you to use mass spectral databases formatted for Agilent ChemStation Probability Based Matching (PBM) search. It will also allow you to interface to the **NIST MS Search Program**.

Be careful if you interface to the **NIST MS Search Program**; the **Wsearch32** *Library Search Preferences* are set to *Overwrite* the **NIST** *Spec List* by default. You need to change this setting to *Append* unless you want your *Spec List* wiped out each time you send a spectrum to the **NIST MS Search Program**.

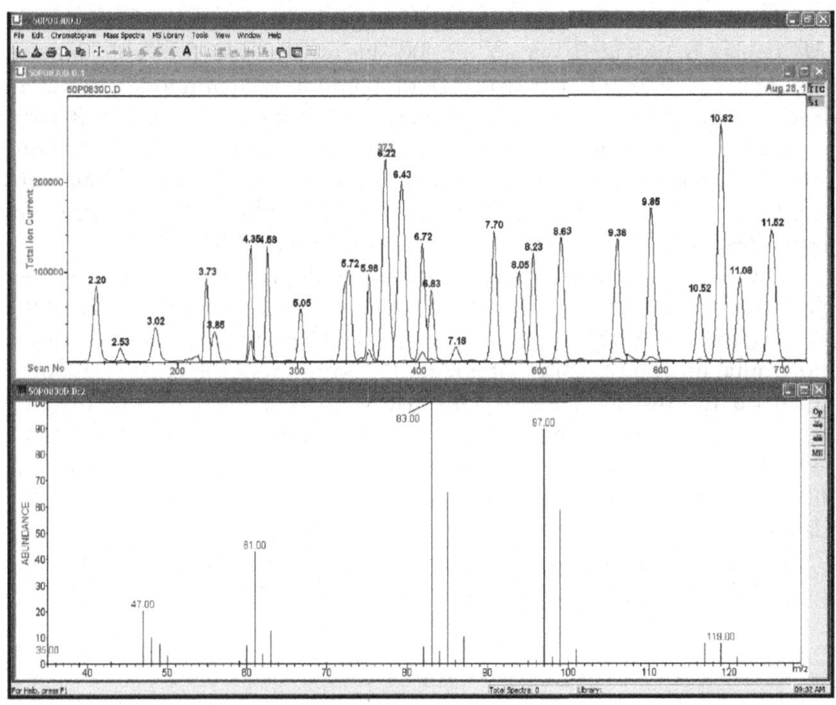

Figure K.3 RTIC chromatogram (top) and mass spectrum of scan 373 (bottom) using Wsearch32.

MassFinder for Data Analysis of GC/MS Data Files

This is a commercially available software package for the visualization, evaluation, and presentation of GC/MS GC data in the Agilent ChemStation MS (/*.D/data.ms), Agilent ChemStation GC (/*.D/*.CH), MassSpec/Finnigan (*.mss), GCQ or ITS40 (*.ms, includes older Varian Saturn data files), Varian SMS (*.sms), JEOL Shrader (*.lrp), PE TurboMass (*RAW/_functns.inf) <nominal spectra>, Thermo Finnigan XCalibur (*.RAW) <converter>, Shimadzu (*.gqd) <converter>, and the JCAMP-DX import (*.jdx). Many mass spectrometry acquisition systems can export JCAMP-DX data files which are compatible with *MassFinder*. *MassFinder* can also use mass spectral libraries in its own proprietary format MassFinder (*.mfl), the JCAMP-DX (*.jdx/*.jdl). *NIST MS Search Program* format (*.msp) <only small user libraries>, and Agilent *ChemStation* MS format<conversion necessary>. *MassFinder* can also call the *NIST MS Search Program.*

The software is targeted toward research in flavors and fragrances. *MassFinder* is provided with a ~2,000-spectra database (available only in the *MassFinder* proprietary format) of compounds commonly found in essential oils, particularly monoterpenes, sesquiterpene hydrocarbons, oxygenated sesquiterpenes, diterpenes, and related aromatic and aliphatic constituents such as esters and lactones. Each entry consists of the mass spectrum, substance name, molecular formula, graphical chemical structure (emf files), and RI on nonpolar stationary phases (DB 5).

The RTIC and mass chromatograms are displayed on the left as they would appear from a strip chart recorder. The spectrum in the middle of the center display of three spectra represents the spectrum at the position of the horizontal line on the chromatogram. The position of the horizontal line is fixed on the display, and the chromatogram is scrolled up or down by placing the Mouse pointer on the chromatogram and using the Mouse wheel. The top spectrum is the next lower numbered spectrum in the data file, and the bottom spectrum is the next higher numbered spectrum in the data file. See Figure K.4.

The three spectra on the right are spectra from the library (either the one provided with the program or a user-generated library). The middle

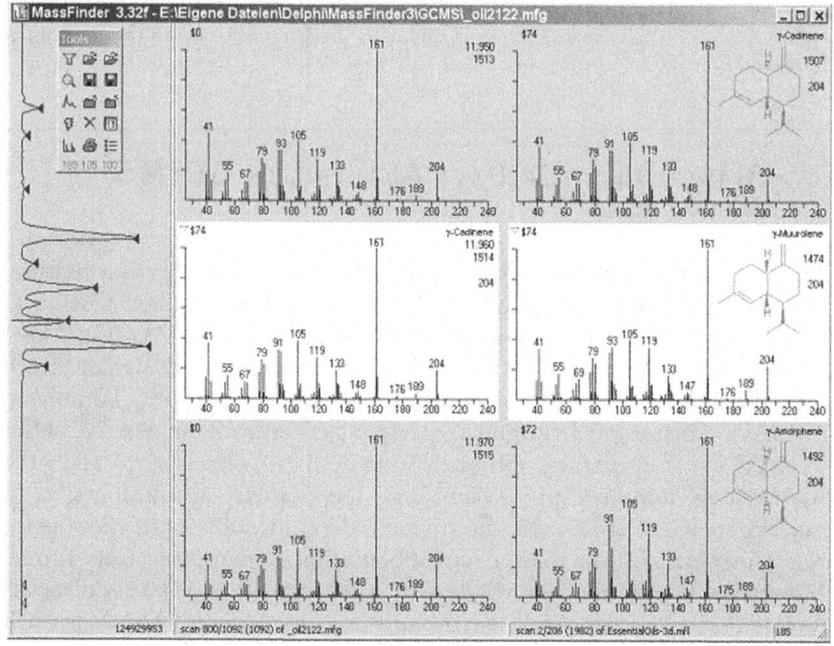

Figure K.4 A typical display of the MassFinder software.

spectrum is identified as the second-best library match, and the top spectrum is the best library match. The library match is based on the mass spectral data and retention-time data.

A detailed review of this program and database is available in Sparkman, OD "Review MassFinder 3" (A Software Program Including a Collection of EI Mass Spectra of Essential Compounds) *J. Am. Soc. Mass Spectrom.* 2007;18(6):1137–1144. *MassFinder* is a product of Dr. Hochmuth, Scientific Consulting, Hamburg, Germany; and more information is available at http://www.massfinder.com.

Cerno Bioscience's *MassWorks*™ for Accurate Mass Assignments of Integer *m/z* Value Data

Another interesting GC/MS utility (which must be purchased) is **Cerno Bioscience's** *MassWorks* software. Many GC/MS data systems allow data to be acquired in the profile mode as opposed to the centroid mode. In the profile mode, as many as 10 or 20 data points are acquired and stored (*m/z*-intensity pairs) for each integer *m/z* value, whereas in the centroid mode, the same number of points is acquired; however, only the single fractional *m/z* value of the most intense measurement is stored, along with the intensity, for each integer *m/z* value. *MassWorks* will use the data from a profile-mode acquisition of a substance that produces ions of known elemental compositions (like PFTBA) and thus know exact *m/z* values to calibrate the system. This calibration is then used with the profile-mode data of integer mass assigned peaks to obtain an accurate mass measurement.

The software can also be used to deconvolute mass spectral multiple peaks such as two peaks both having an integer *m/z* value of 57, but one represents the $C_4H_9^+$ ion (exact mass 57.0704 Da) and the other represents the $H_5C_2{-}C{\equiv}O^+$ ion (exact mass 57.0340 Da). Not only can the software show that two peaks are present, but it can also provide accurate-enough mass values to allow for unambiguous elemental compositions and determine the abundance of the two ions. It should be noted that the size of data files acquired using the profile mode are considerably larger than those acquired using the centroid mode. Data acquisition is not compromised because the same number of data points is acquired whether using the centroid or profile mode—the difference is in what is stored.

The authors have successfully used this software with data acquired on an Agilent 5975 transmission quadrupole mass spectrometer under GC/MS conditions. There have been reports by **Cerno** that the software works with other manufacturer's GC/MS systems just as well. For more information about *MassWorks* or to purchase this program, go to http://www.cernobioscience.com.

Mass Spec Tools™ from ChemSW

MS Tools is a collection of programs that has utilities for all types of GC/MS but is more applicable in some respects to data acquired with high resolving power instruments such as TOF or double-focusing mass spectrometers. These programs are for displaying and processing mass spectra and related data and for performing common calculations related to mass spectrometry. The display and processing functions are intended to make use of accurate mass data, but integer mass spectra can also be displayed.

The Elemental Composition Workshop and the **SearchFromList** programs provide functions for examining mass spectra and assigning elemental compositions to mass spectral peaks or searching user-defined databases by using exact masses and isotopic abundances. These programs are not intended to replace the normal display and processing functions of a mass spectrometer data system for calibration, processing full GC/MS data sets, and so forth. Instead, the programs provide a means for identifying and labeling components in individual mass spectra exported in a text format such as JCAMP-DX (*.jsp, *.jdx, etc.), NIST (*.msp), and Shrader (*.txt).

The **Mass Spec to Metafile** program creates vector graphic (metafile) displays from mass spectra or mass chromatogram files saved in text format. The output can be imported into reports, presentations, or graphic editors for preparing publication-quality figures. There are 12 different programs in this suite.

1. **Mass Spec Periodic Table**: Displays a periodic table with the information that is relevant to mass spectrometry. Clicking on an element produces a graphical display of the isotopic abundances for that element, along with a table of exact masses and abundances for each element (see Figure K.5).
2. **Isotope Ratio Calculator**: More than just an exact-mass isotope ratio calculator, this program provides a means for simulating the isotopic abundances and masses of average mass and the appearance of single-charge or multiple-charge isotopic clusters at any given resolving power. Overlapping isotopic clusters can be displayed, and isotopic enrichment can be modeled through user-defined isotopes. The Program's display is shown in Figure K.6.
3. **Mass Series Calculator**: Useful for calculating reference tables based on polymers or clusters with repeating subunits. The program can also calculate ionic clusters and multiple-charge ions (including multimers). The Program's display is shown in Figure K.7.
4. **Elemental Composition Workshop**: Calculate elemental compositions from exact masses, isotopic peak intensities, or both. The program can predict the presence or absence and approximate number of certain

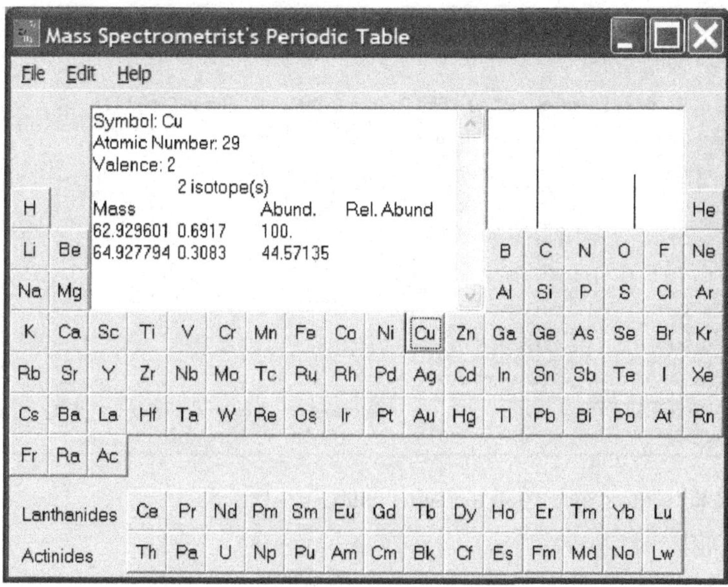

Figure K.5 Mass Spec Tools Periodic Table for mass spectrometrists.

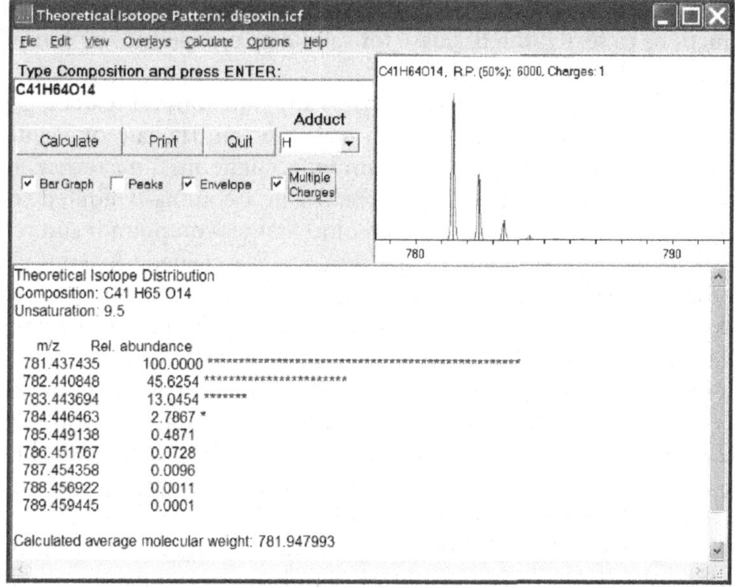

Figure K.6 Mass Spec Tools utility to calculate resolving power.

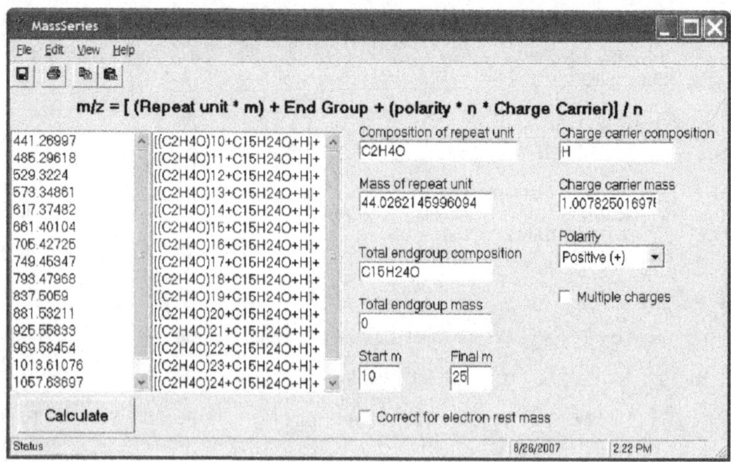

Figure K.7 Mass Spec Tools Ion Series utility.

X+2 elements (S, Si, Cl, and Br) from the intensity of the isotope peaks. An optional function filters out unreasonable compositions based on heuristic rules. Neutral loss compositions and ion correlation functions are supported. Individual masses can be processed or an entire spectrum. The results can be printed, saved to a file, or cut and pasted into a report. If the **NIST Mass Spectral Database** is installed, the program provides functions to search the database for candidate compounds by formula or name (see Figures K.8 and K.9).

5. **SearchFromList**: This program has many functions for searching accurate mass spectra against a custom mass spectral database or identifying components in a complex spectrum by accurate mass measurements. A list of target compounds can be created in a comma-delimited text or *Microsoft*® *Excel*™ format. Both absolute (target compound) and relative (e.g., metabolite or adduct) mass lists can be created. A series can be defined to search a spectrum for polymer repeat units, clusters, adducts, alkyl series, or silyl derivatives. A measured spectrum can be searched against a mass spectral database consisting of text-format mass spectra stored in a specified directory on the disk. Additional functions include spectrum comparison, spectrum addition and subtraction, and *Kendrick Mass Defect* analysis. An optional *Microsoft Access*™ database can be used to enter or look up detailed information about a given mass spectrum in a created mass spectral database (see Figure K.10).
6. **Mass Spec to Metafile**: This program takes as input a text-format mass spectrum (profile or bar graph) or chromatogram and creates a vector graphic (metafile) that can be cut and pasted into a report or presentation

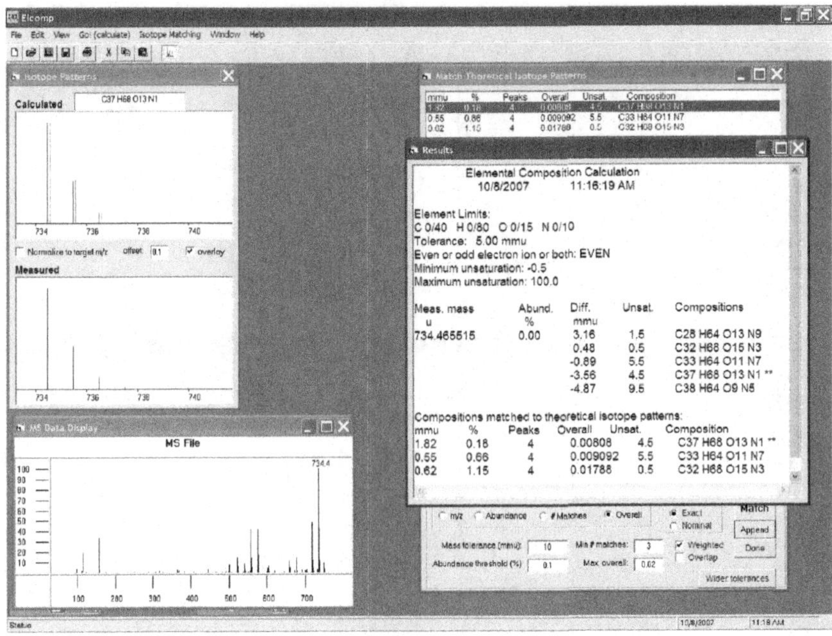

Figure K.8 Display of Elemental Composition Workshop after determining a composition based on isotope peak intensities and accurate mass data.

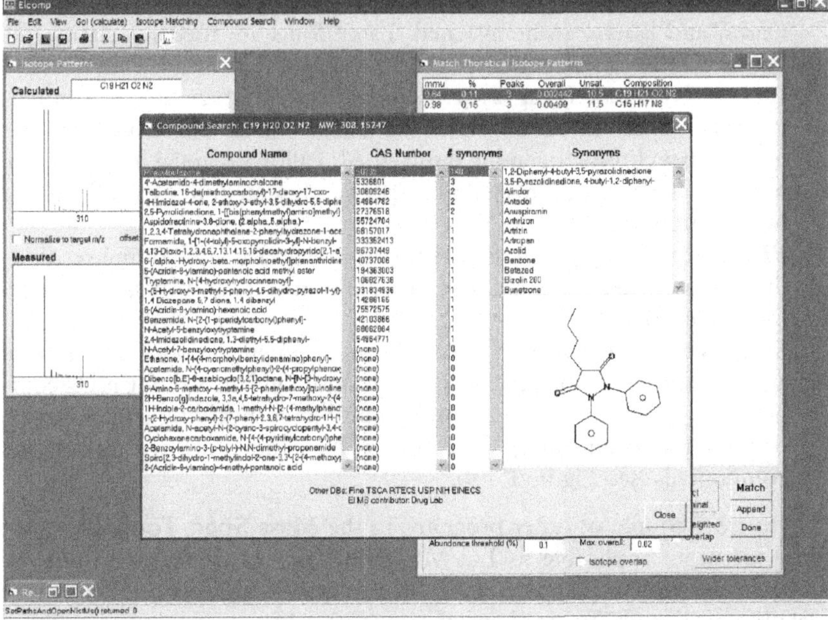

Figure K.9 The Library Search overlay of the Elemental Composition Workshop utility.

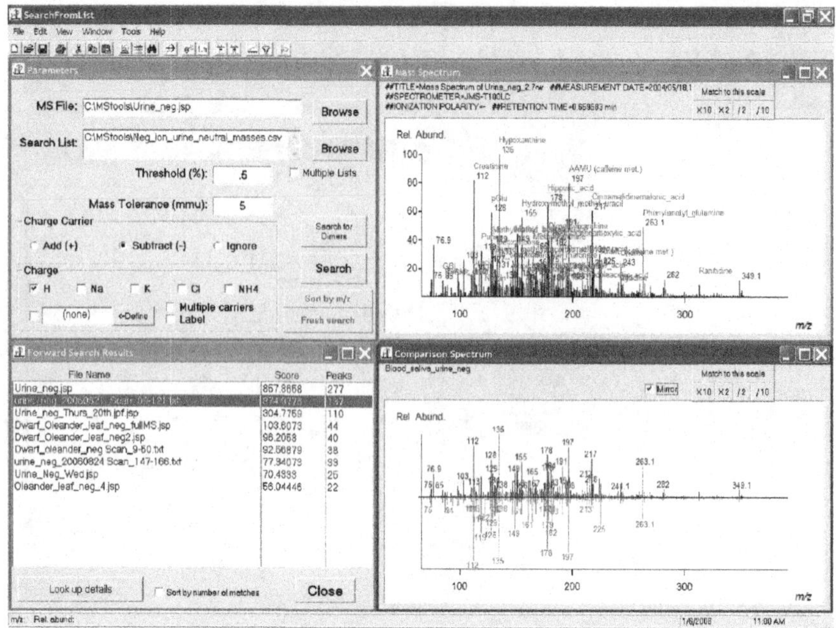

Figure K.10 Results of a search of a mass spectrum of an unknown substance against a generated database using the Search From List program.

for further editing. The *m/z* axis range, the spacing between *m/z* tic marks, and labels can be assigned. Large fonts are used for clarity (see Figure K.11).

7. **Interactive Peak Integrator**: Permits manual integration of chromatographic peaks in RTIC chromatograms or mass chromatograms for text-format chromatographic displays. Also allows manual integration of mass spectral peaks for text display spectra presented in the profile mode (see Figure K.12).
8. **PPMCalc**: This program allows for an interactive view of the relationship between parts per million, resolving power, and mass differences (see Figure K.13).
9. **Subtract Composition**: Subtract one elemental composition from another and view the resultant composition and masses for each composition. Useful for thinking through problems involving fragmentation or finding the remaining "core" group for compounds with polymeric components (see Figure K.14).

There are a couple of other programs in the **Mass Spec Tools** suite, but they are more applicable to LC/MS. The program is available from the publisher ChemSW (Fairfield, CA). To purchase **MS Tools** or obtain more details, go to http://www.chemsw.com/13057.htm.

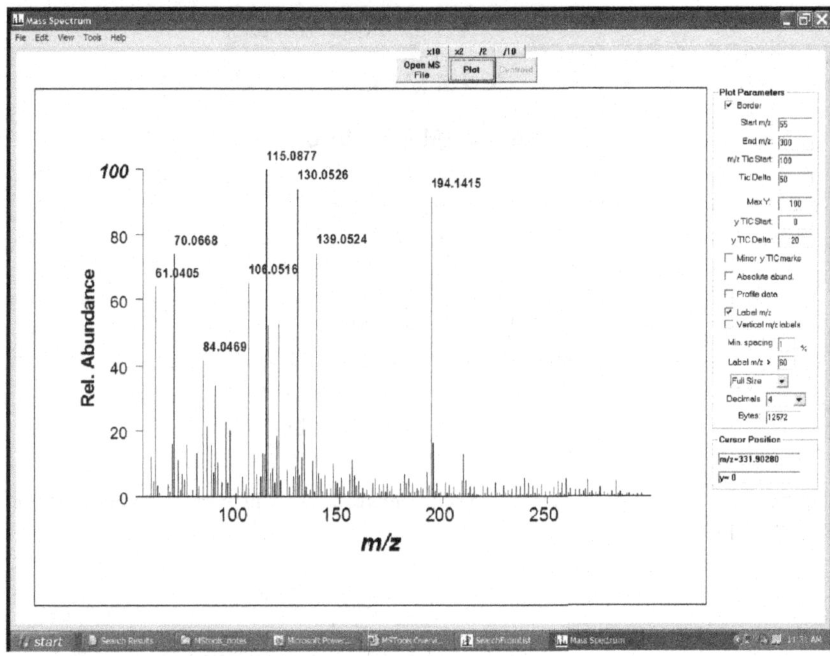

Figure K.11 Utility of customizing the display of a mass spectrum using the Mass Spec to Metafile program. Chromatograms can also be edited with this program.

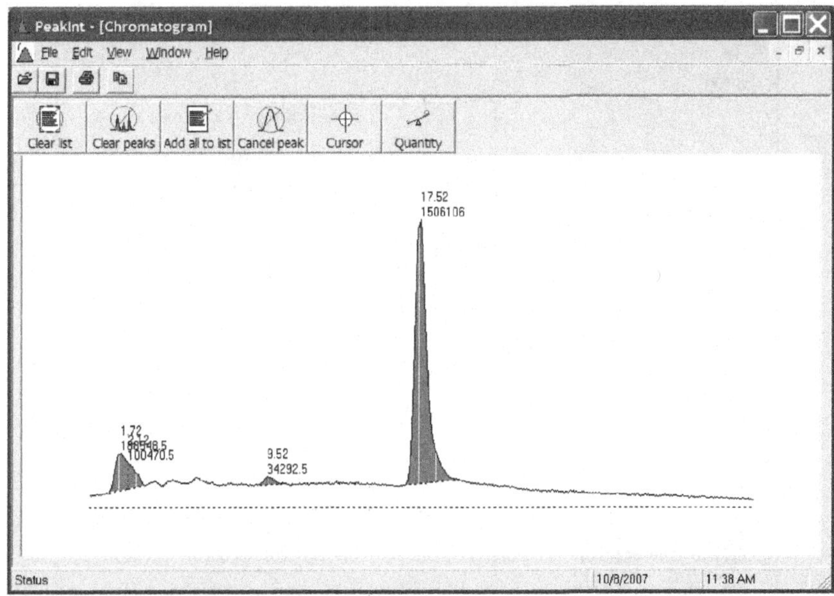

Figure K.12 Display from Interactive Peak Integrator Program.

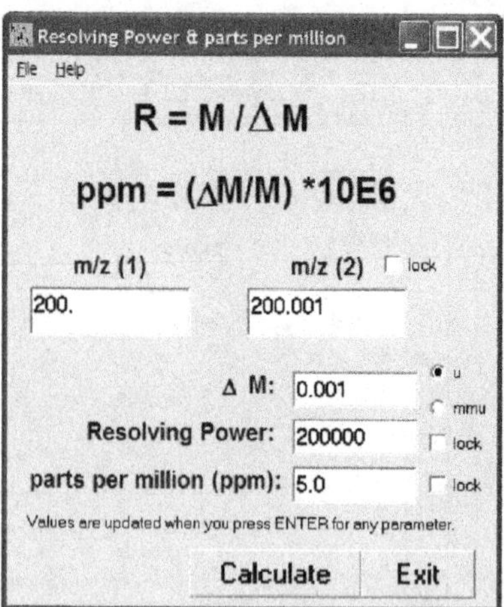

Figure K.13 Mass Spec Tools Conversion between resolving power (R) and parts per million (ppm).

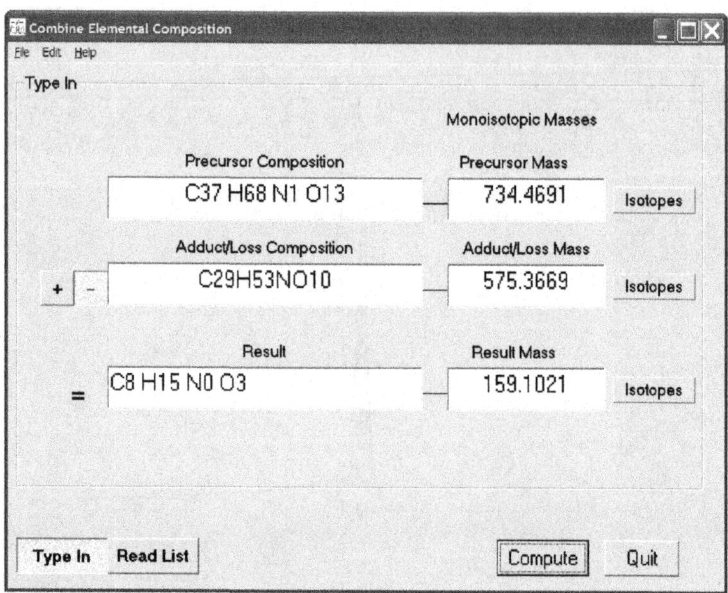

Figure K.14 Subtract Composition program's display.

Advanced Chemistry Development, Inc. (ACD/Labs)

ACD/Labs is a scientific software company that publishes a number of programs for gas chromatography, for mass spectrometry, and to aid in the manipulation of data for the identification of organic compounds. These programs include **ACD/GC Simulator, ACD ChromManager, ACD/MS Fragmenter, ACD/SpecManager Analytical Data Management System**, and **ACD/ChemSketch**. All ACD/Labs software is described in detail at http://www.acdlabs.com.

1. **ACD/GC Simulator** models a hypothetical gas chromatogram for specified compounds under the specified chromatographic conditions on the basis of their calculated boiling points. Also, **ACD/GC Simulator** predicts the retention time for a new compound if the chromatographic experimental data are available. Based on two experiments for the same compounds under different conditions with unsatisfactory results, the optimal conditions for the best separation can be found. And moreover, after the optimal conditions are found, the retention time for novel compounds under the new conditions can be predicted. **ACD/GC Simulator** lets the chromatogram be directly transferred to **ACD/ChromManager** for databasing, analyzing, and processing. Conversely, GC chromatograms can be transferred from **ACD/ChromManager** to **ACD/GC Simulator** for optimization and prediction (see Figure K.15).
2. **ACD/ChromManager** is a software designed to process, database, and manage chromatographic separations such as from GC or GC/MS. The program can also be used for chromatograms resulting from HPLC, LC/UV (DAD or PDA), and CE. A complete array of processing tools can be applied to raw chromatographic data, structures can be assigned to peaks, method information can be added, and chromatograms and methods can be stored in a relational database. Separation methods can be retrieved from this database through searching by method parameters, as well as chemical structure, substructure, and structural similarity.
3. **ACD/MS Fragmenter** predicts fragment ions based on established MS fragmentation rules from the literature. The task of fragmentation prediction can be performed automatically, in seconds, with **ACD/MS Fragmenter**, allowing for the understanding of the fragmentation pathways of a molecule. Key fragmentation pathways and ion structures can be reviewed in a convenient fragmentation tree.
4. **ACD/SpecManager Analytical Data Management System** is a superset of **ACD/ChromManager**, allowing the databasing of mass spectral data as well as chromatographic data in addition to spectral data from spectroscopy techniques such as NMR, IR, UV, Raman, as well as

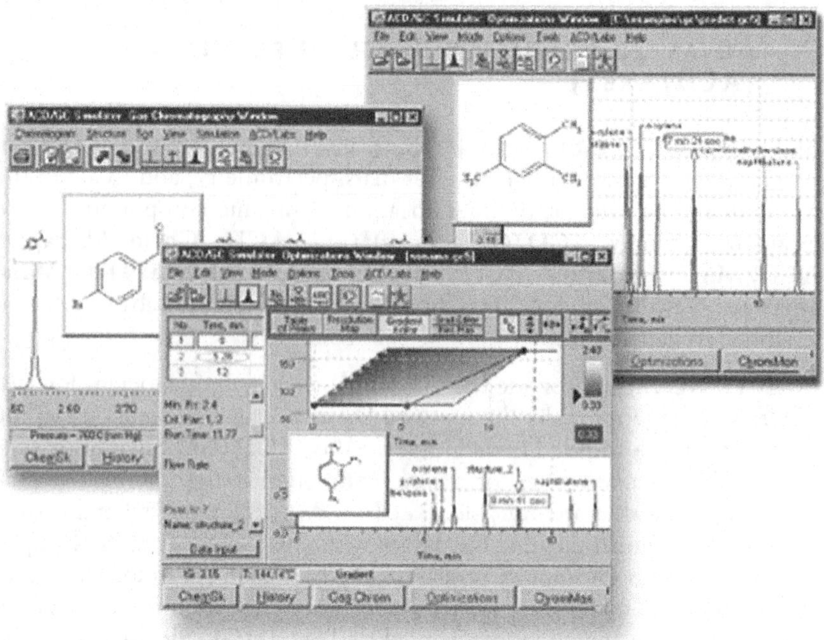

Figure K.15 Screen displays of ACD/GC Simulator.

data from TGA, IMS, XRPD, and so forth. **ACD/SpecManager** is the heart of the offerings from ACD/Labs. This is a true chemistry database product.

5. **ACD/ChemSketch** is a chemical structure drawing program. In addition, it allows the determination of chemical names according to systematic application of a preferred set of International Union of Pure and Applied Chemistry (IUPAC) nomenclature rules for molecules containing no more than 50 atoms, and no more than three rings, with atoms from only H, C, N, P, O, S, F, Cl, Br, I, Li, Na, and K. **ACD/ChemSketch** also interfaces with **ACD/Name**, which generates names from structures not only according to the IUPAC rules but also in agreement with the International Union of Biochemistry and Molecular Biology (IUBMB) and the Chemical Abstracts Service (CAS) rules. In addition, **ACD/Name** will create chemical structures for systematic and semisystematic names of general organic compounds and many natural product derivatives. Another add-on for **ACD/ChemSketch** is **ACD/Name Chemist Version**, which generates systematic names according to IUPAC nomenclature rules and creates chemical structures from systematic, trade and trivial names, and registry

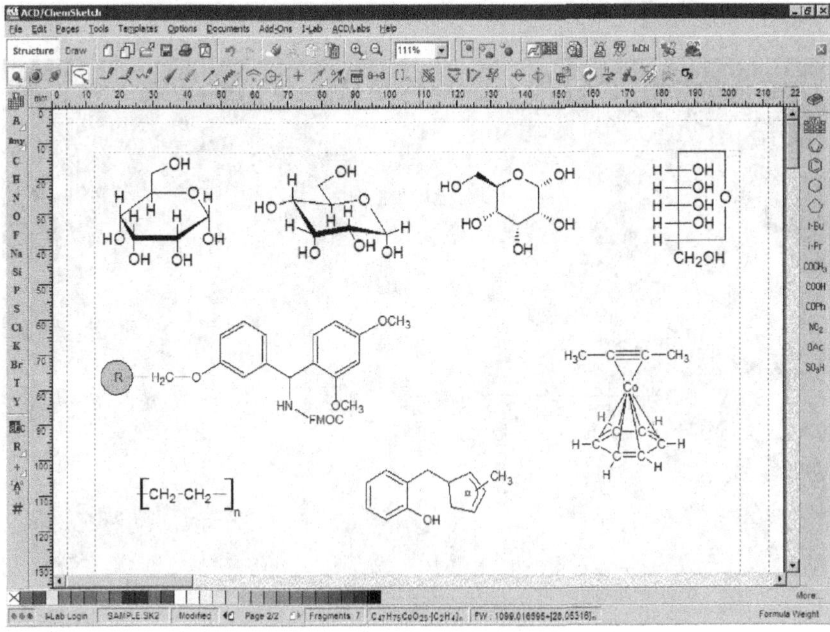

Figure K.16 ACD/ChemSketch structure drawing utility.

numbers inside the **ACD/ChemSketch** interface. There is a free download version of **ACD/ChemSketch** on the ACD Web site; however, this download, like many of the free downloads of software, is restricted to individual and educational use. This software should not be used in company or government labs; only the purchased version of this program should be used in such labs (see Figures K.16 and K.17).

CambridgeSoft Corporation

CambridgeSoft Corp. is a company that is similar to ACD/Labs in that, like ACD/Labs, CambridgeSoft specializes in database products used for chemicals and chemical data. One of CambridgeSoft's products is a chemical structural drawing program—**ChemDraw**. This program can be interfaced with the CambridgeSoft structural and spectral database or used as a standalone program. CambridgeSoft has a free downloadable version of its suite of programs that contain **ChemDraw**; however, this free download has a time limit of 2 weeks.

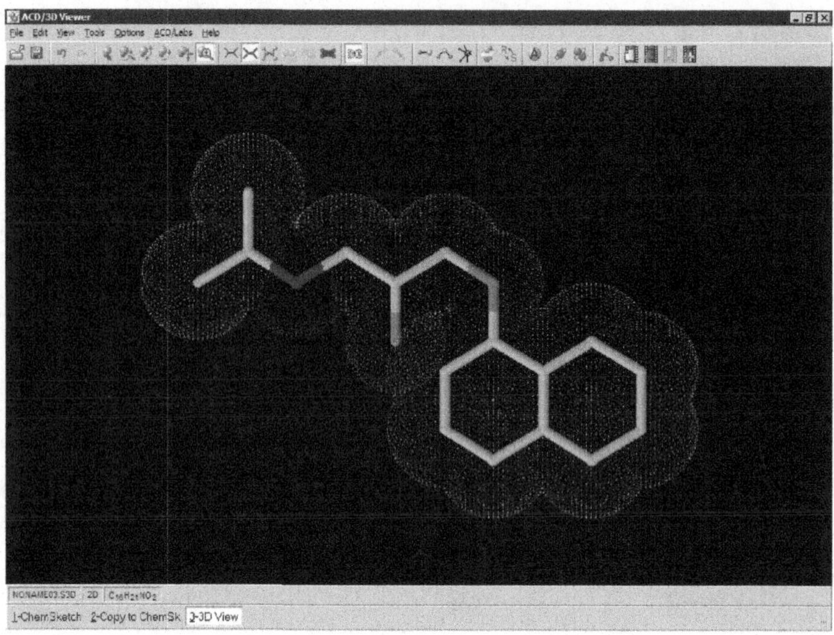

Figure K.17 Another way of presenting a drawn structure using ACD/ChemSketch.

Chemical Structure Drawing Programs

GC/MS users will eventually want to build user-generated mass spectral databases; therefore, they will need a good chemical structure drawing program. In addition to those already referenced above, there are a number of others. **MarvinSketch** from ChemAxon (http://chemaxon.com) Budapest, Hungary; **Symyx Draw**, successor to **ISIS Draw** (http://www.symyx.com), Northern California (Free Download available for academic and/or individual use); and **ChemDoodle** (http://www.chemdoodle.com). The Bio-Soft Net Web site has a number of chemical structure drawing programs listed. **Chem 4-D Draw** from ChemInnovation Software, Inc. (San Diego, CA, USA) (http://www.cheminnovation.com/products/chem4d.asp) is another chemical structural drawing program with a nomenclature feature. This program has a free limited-time downloadable demo. All programs with Free Downloads have Nag-Screens that continually pop up during their use. All have an individual/academic user restriction and should not be used in commercial or government laboratories.

ChemSW

In addition to **MS Tools** (described above), ChemSW (http://www.chemsw.com, Fairfield, CA, USA) has several other programs that are useful to the GC/MS practitioner. They are **ChemSite® Pro**, a molecular modeling program; **GC and GC/MS File Translator Professional**, to convert data files from one instrument manufacturer's format to another; **MassSpec Calculator™ Professional**, for use in the interpretation of EI mass spectra; and **ChromView**, a program used for annotation of chromatograms and spectra for inclusion in reports.

Scientific Instrument Services

Scientific Instrument Services (SIS) (http://www.sisweb.com) offers software for use with mass spectrometry. They are the distributor for many of the ChemSW programs, and they are currently the sole provider of **MASSTransit** by Palisade. MASSTransit is an extremely useful utility for GC/MS users. The program allows the conversion of GC/MS data files from one instrument-company's format to another. It also converts mass spectral libraries from one format to another. This program allows you to use the ChemStation software to look at and to process data files acquired using another instrument manufacturer's GC-MS.

APPENDIX L

GC Installation and Maintenance

The best way to avoid problems with a gas chromatograph is to install the instrument in accordance with the recommendations of the manufacturer and to be meticulous in maintaining the instrument. Each gas chromatograph should have a logbook to assure that maintenance is carried out according to the recommended schedule. This includes changing gas filters and septa on time. A working flame ionization detector (FID) on the gas chromatograph can be useful in isolating a problem and determining whether a specific problem is in the gas chromatograph or the mass spectrometer.

Recommended Steps for Setting Up a Gas Chromatograph

Required tools and supplies

The equipment below is the minimum required in a chromatography laboratory. There are undoubtedly additional tools that may be useful.

A general tool* kit containing:

- Slotted and Phillips screwdrivers of various sizes to fit the screws on the gas chromatograph
- Set of open-end wrenches from 1/32 to 9/16 inch (two each size) plus metric wrenches
- Adjustable 14-inch wrench (a.k.a. crescent wrench) for attaching gas regulators to cylinders
- Allen wrenches (be sure to check to see if parts are metric)
- Pliers and needle-nose pliers
- Small flashlight
- Stopwatch
- Electronic "duster" products that are recommended for cleaning dust in electronics products (spray cans containing Freon™ compounds).

* It is recommended that tools be purchased from a chromatography supply house to assure that they are oil-free.

Gas Chromatography and Mass Spectrometry
DOI: 10.1016/B978-0-12-373628-4.00048-4

The following parts are available in chromatography or capillary start-up kits and may be purchased from the instrument manufacturer or from chromatography supply companies:

- Electronic flow meter
- Magnifier for examining the ends of capillary columns
- Capillary column accessories including small ruler, column cutter, tweezers, septum pullers, graphite/vespel ferrules for capillary columns
- Spare septa and injector inserts
- Leak detector

Carrier, detector, and injector gas supplies (see Section 2.3.3 Carrier Gas Considerations in Chapter 2)

- Tubing cleaned for chromatography applications: copper 1/8 in o.d.
- Tubing cutter
- Swagelock™ fittings to fit the above tubing, plus unions, tee fittings, and caps and plugs
- Filters to remove water, hydrocarbons, and oxygen from the carrier gas (individual or multipurpose filters may be used)
- Gas regulators*
- High-purity (99.999%) helium carrier gas
- Liquid CO_2 or LN_2 for cryogenic cooling of PTV injectors or the column oven

Detector gas supplies

Gases required for GC detectors depend on the individual detector and are summarized below in Table L.1:

Table L.1 Gases for GC detectors

Detector	Makeup gas	Additional gases
TCD	Helium (not necessary for micro-TCD)	None
ECD	Nitrogen or argon/methane	None
FID	Helium or nitrogen	Hydrogen, air
NPD or TSD	Helium or nitrogen	Hydrogen, air
FPD or PFPD	Helium or nitrogen	Hydrogen, air

The supplies for these gases are the same as for the carrier gas with the exception of the hydrogen and air, which require only hydrocarbon filters. Note that gas generators are available for hydrogen, air, and nitrogen.

*Unless the specific history of existing regulators is known, new regulators should be purchased for a new instrument. Never use regulators that have been used for any purpose other than high-purity gas delivery.

Setting Up a Leak-Free System

Connect the regulator to the carrier gas tank; using the copper tubing, connect the filters to the regulator; finally connect the carrier gas to the gas chromatograph. Install the column in the injector but do not install the column in the mass spectrometer or in a detector. Turn on the carrier gas; using a flow meter, verify that there is flow through the septum purge line and the splitter, if applicable. Insert the end of the column into clean isooctane or methanol to verify by the presence of bubbles that there is a vigorous flow through the column. At this point, the system should not be heated.

Pressure-check the system as follows:

Close off the end of the column by inserting it into a septum, close the injector splitter and the septum purge line, and close the regulator on the carrier gas tank. The pressure on the regulator should not drop after 15 minutes. If there is a leak, the pressure will drop. Leaks must be located* and corrected. Verify that all of the fittings are tight; and if the leak persists, pressure-check by isolating various sections; i.e., from the carrier gas tank to the gas chromatograph, then from the gas chromatograph to the injector body, and finally from the injector body to the end of the column.

After all leaks are eliminated, the system should be flushed with carrier gas for approximately 30 minutes and the column baked out as recommended by the column manufacturer. The column may then be connected to a GC detector or to the mass spectrometer. At this time, the mass spectrometer should be turned on to pump it down to the appropriate operating pressure. Once the mass spectrometer is running, the connections in the gas chromatograph can be checked by spraying the fittings with a small quantity of the electronic duster. These products contain Freons™ such as difluoroethane, which are not retained by most stationary phases. If there is a leak in the fittings, the mass spectrometer should respond strongly to these compounds.

Leaks may develop after the system is cooled and heated several times; therefore, the system should be checked periodically for leaks. Periodically, the mass spectrometer data acquisition range may be set to start at m/z 25 to monitor the signal strength of mass spectral peaks at m/z 28 and 32 and ascertain if air leaks have developed.

After the installation is complete, the manufacturer's performance specifications should be verified. Hard copies of mass chromatograms used for these verifications and spectra of the test compound, along with spectra of the compound used for tuning, should be put into the instrument logbook so that the data from a properly functioning instrument can be reviewed if necessary.

* Use isopropanol or methanol as a liquid leak detector and only very sparingly on the fittings outside of the GC. The liquid leak detector known as "Snoop" should not be used on a GC/MS system. An electronic leak detector is highly recommended to check the fittings in the GC.

APPENDIX M

Troubleshooting Common GC Problems

Problems that commonly originate in the gas chromatograph are listed below in Table M.1.

Table M.1 GC troubleshooting

Problem	To solve
No peaks (mass spectrometer (MS) appears to have a normal baseline)	Verify that all of the flows are correct; i.e., the septum purge is not set too high (usually 2–4 mL min^{-1} is recommended), the splitter flow is in a reasonable range, and the column flow is appropriate for the column i.d. (1–2 mL min^{-1} for most capillary columns).
	Is the autosampler needle bent?
	Is the syringe needle clogged (usually with septum particles)?
	Is the injector liner broken?—peaks may be small or nonexistent.
Peaks tail toward the front	The column is overloaded; inject less sample or use a higher split ratio.
Peaks have a rear tail	Verify that nonpolar compounds such as hydrocarbons do not tail. If all compounds tail, there is probably a problem in the column installation or in the injector.
	If only late-eluting compounds tail, there may be a cold spot in the system. Verify that the injector, any transfer lines, and the MS are hot enough to keep the analytes from condensing.
	If only polar compounds tail and injection is in the splitless mode, then inject a more concentrated sample using the split mode (the shorter residence time in the split mode minimizes injector problems). If the polar compounds no longer tail, then the injector is active. Deactivated liners should be used that are not packed with glass wool (the fibers of glass wool can break and expose active sites). Change to a split mode, if possible, using a low split ratio.

(*Continued*)

Gas Chromatography and Mass Spectrometry
DOI: 10.1016/B978-0-12-373628-4.00049-6

Table M.1 (*Continued*)

Problem	To solve
	If the polar compounds tail when injected in the split mode, then the column or the MS is active. Possible solutions: Remove several centimeters from the front of the column; the front of the column may be active after many injections of a dirty sample. Use a more polar column. Derivatize the compounds.
Large-tailing solvent peak in the splitless mode.	The initial column temperature should be 15–20 °C below the boiling point of the solvent. It is best to use a solvent with as high a boiling point as possible, as long as there is separation from the earliest eluting peaks of interest. Verify using the "FlowCalc" software (see Figure 2.8 under Section 2.2.1 in Chapter 2) that the total volume of the injection is not too large for the injector inlet and the pressure in the injector. Use a slow injection speed, as low an injector temperature as possible, and as early a splitter purge time (0.5–0.7 minutes) as possible without losing analyte. If available, try pressure-pulsed injection. If the above measures fail, there is likely to be a problem in the injector causing excessive dead volume. Verify that the glass liner is not broken and that all of the injector parts are tight.
Some thermally labile compounds appear to break down in the GC system.	To minimize breakdown in the injector, avoid hot splitless injection; or, if this is not possible, use a deactivated insert with the smallest i.d. and as low a temperature as possible. If available, use a temperature-programmed injection mode. Use a column that is as short as possible (5 m or less) and a higher carrier gas flow.
Peaks splitting (especially early-eluting peaks).	Peak splitting is caused by the compound not focusing on the front of the column after injection. Two possible causes are as follows: the injection volume is too large and/or the solvent is not compatible with the column phase. Use the "FlowCalc" software mentioned above in this table to determine if the injection volume is too large. Pressure-pulsed injection may eliminate the peak splitting. The solvent polarity should be an approximate match to that of the column. For example, peak splitting usually results if water or methanol is used as the solvent with a nonpolar column.

(*Continued*)

Table M.1 (*Continued*)

Problem	To solve
	If changing the solvent is not feasible, the use of a retention gap should help. This involves installation of up to a meter of deactivated fused silica between the injector and the analytical column. The deactivated fused silica is normally the same i.d. as the column and is attached to the column with a glass connector.
Extraneous or "ghost peaks"	Extra peaks may be due to sample carryover (see next row). Another common cause of extraneous peaks is contamination in the column or injector. A blank run without an injection may remove the contamination. For this reason, a blank run should be made at the beginning of each day, before using the GC-MS for analysis. If the extraneous peak is persistent, it may be due to particles of septa in the injector (peaks at m/z 73) or gases eluting from solid particulates in the system. In this case, it is necessary to replace or clean the injector liner, change the septum and usually remove the first part of the column. If the septum is shredding, the syringe needle may be deformed, exacerbating the problem.
Sample carryover	Sample carryover is a common problem where small peaks from a previous run appear in subsequent runs. It may occur when a syringe is not washed sufficiently between samples. To determine if the syringe is the cause of the carryover, inject a sample with the syringe that has been in use for the analysis; then inject a second sample with a brand new syringe. If it is determined that the syringe is not the problem, the analyst should examine the injector liner and verify that the liner is clean and deactivated. A packed insert may retain some analyte, and it may be necessary to use an unpacked liner. Essentially, it is necessary to examine the system at each point and determine if there are areas with cold spots or with active sites that retain the analyte. It may be necessary to inject blanks between runs to eliminate the problem or to derivatize the compound so that it is less polar.
Poor retention time rsd's	This is not a very common problem. Verify that the septum is intact and all of the injector parts are tight. The gas chromatograph's pneumatics may need to be serviced.

(*Continued*)

Table M.1 (*Continued*)

Problem	To solve
Poor area count rsd's	Rsd's can vary from less than 1% to around 20% and still be acceptable. As expected, manual injection will not give as good reproducibility as automatic injection and there should be better reproducibility with large well-defined peaks than with small barely resolved peaks. Taking these facts into account, if there still appears to be a problem, try the following suggestions: Inject pure solvent to verify that there are no contaminant peaks in the system that coelute with the peaks of interest. Then inject a concentrated sample, followed by a blank, to verify that there is no sample carryover. Finally, extend the run time well beyond the usual run time to verify that there are no late-eluting compounds that are retained in the column. These compounds could be coeluting with peaks of interest in later runs. Are the peaks Gaussian in shape and well resolved from adjoining peaks? If the peaks are tailing or split, see the suggestions for resolving these problems in this table. If resolution from adjoining peaks is poor and cannot be improved, use peak height rather than area for quantitation or adjust the integration parameters in the software to give more consistent results. If using an autosampler, verify that the syringe is in good condition and not clogged. Inject a sample containing a sample with a few large well-resolved peaks to eliminate the autosampler as a cause of the problem. Area counts can vary with injection speed and residence time in the injector before and after injection. Use the "sandwich technique" referred to earlier for manual injection (see Table 2.1 and Figure 2.15 under Section 2.2.1 in Chapter 2). It is very common in GC/MS to use deuterated internal standards. These elute just before the compound of interest, and the relative recovery of the analyte and deuterated internal standard should be very consistent, thus resulting in improved precision.

APPENDIX N

Maintenance, Operating Tips, and Troubleshooting for Mass Spectrometers

Maintenance Logbook

A logbook should be maintained on the operation of each mass spectrometer. In this book, the instrument tuning parameters used to obtain reference mass spectra should be kept. The tuning parameters should include the repeller, lens, and multiplier voltages or settings as well as other parameters such as the gain setting, instrument resolution, source temperature, acquisition rate, accelerating voltage, and so forth. If the instrument is autotuned, print out the autotune report and staple it in the logbook. If a named file is used in running the autotune, record the file's name. It is best to obtain the spectra when the instrument has just met manufacturer specifications and after servicing. The lens settings for optimum performance will change with time and the multiplier voltage may need to be increased to maintain the gain, but this data set makes it possible to follow the performance of the instrument and readily determine when the cause of unsatisfactory GC data is due to the mass spectrometer rather than the method or some other factor. Deteriorating performance is most often an indication that the ion source is dirty.

All of the information regarding chromatographic column changes, septa and injection liner replacements, carrier gas tank replacements, and filter changes should be kept in this logbook. A separate user/sample logbook should also be maintained for each instrument that contains each sample run, the notebook where the data are, who ran the sample, and when the sample was analyzed.

A calendar should be in the front of the maintenance logbook that has the schedule for changing pump oil and filters. In addition to a maintenance logbook, a sample logbook should be maintained that keeps a record of every method and every sample run on the instrument. Sample and maintenance logbooks should contain the names of the people that are performing the operations.

Gas Chromatography and Mass Spectrometry
DOI: 10.1016/B978-0-12-373628-4.00050-2

General Maintenance

As was pointed out in Section 2.2.1 in Chapter 2, split vents should be exhausted to a fume hood or filtered; and, as pointed out in Section 4.3.1 in Chapter 4, mechanical pumps should be exhausted to a fume hood or filtered. If filters, they should be changed on a monthly basis. More often may be required depending on use conditions. These filters should be treated as hazardous waste and should be disposed of properly. Oil-vane rotary pumps should be fitted with a mist eliminator before the exhaust line of filter. The filter in the mist eliminator should also be changed monthly, with the used filters being treated as hazardous waste.

The oil in oil-vane rotary pumps should be changed every 3 months, again depending on the instrument's workload. If the high-vacuum system is a diffusion pump, the diffusion pump oil needs to be replaced every 6 months. Scroll pumps are beginning to make their way into the GC/MS laboratory. If these pumps have been replaced by oil-vane rotary pumps, changing oil and mist eliminators is not an issue; however, these pumps require the same venting or filtering as the oil-vane rotary pumps.

Filters and mist eliminators along with the proper hardware for a gas-tight fit to the pump can be a complex and complicated array of parts. Scientific Instrument Services (Ringoes, NJ, USA) has complete kits of these parts that contain all necessary items for most manufacturers' pumps.

When changing the pump oil, use only the oils recommended by the pump manufacturer. One or two grades up from the recommended oil is not a bad idea. Like the filters described above, the used oil is hazardous waste and should be properly disposed of.

There is a Tygon® tube that connects the high-vacuum system to the mechanical pump. This tube usually has a spiral wire inside of it to keep the tube from collapsing under vacuum. It will be noticed that this tube becomes discolored over time. This is from oil back-streaming and from un-ionized sample passing into the mechanical pump. This material contributes to the chemical background of the systems. When that background begins to interfere with the desired performance of the instrument, replace this tube.

At some point in time, a peak at *m/z* 149 will begin to appear in all the spectra. This peak represents a fragment ion of phthalate plasticizers. A good potential source of unwanted plasticizers in the mass spectrometer is in the tube connecting the high-vacuum system and the mechanical pump.

In addition to filters on pumps and split vents, there will be air filters on various inlets to instrumentation components. These filters need to be checked and cleaned or replaced on a regular basis; at a minimum, every 6 months depending on the laboratory air quality.

Keep the area around the instrument free from clutter. Do not have old sample and solvent bottles sitting around. The work area should provide enough space to be able to make entries into the logbooks and experimental notebooks. Dust tends to accumulate in untouched areas; these areas should be regularly cleaned. Clean dust that may accumulate on power cords or gas lines.

Instrument Shutdown

Many maintenance procedures require the mass spectrometer be shut down and vented to atmospheric pressure; clean the ion source, replace burned-out filaments, replace electron multipliers (EMs), even replace the GC column and the injection port septum or liner in some cases. As was pointed out in Chapters 2 and 4, there are devices that will allow keeping the mass spectrometer under vacuum while doing GC maintenance.

Many of the internal parts of the mass spectrometer are heated in a GC-MS to prevent condensation of high-boiling analytes. When current is passing through the filament, the result is heating of the ion source. Before venting any mass spectrometer, cool it as close to room temperature as possible. This is especially important when using an oil diffusion pump for the high-vacuum component. The oil in this pump is heated to a higher temperature to effect a vacuum. If this oil is not brought to near room temperature before venting the mass spectrometer, then oil vapor can contaminate the *m/z* analyzer resulting in quite a mess. Hot metal surfaces in the mass spectrometer are more likely to attract water molecules, which are hard to get rid of when trying to restore the vacuum.

Although somewhat controversial and considered to be urban legend by some, the authors strongly recommend that the instrument not be vented to room air but to dry nitrogen. Room air can have a high water content, which can coat even room-temperature metal parts. Do not use argon because argon will hide in corners and result in a longer pump downtime when it is time to restore the vacuum. Do not use helium because helium will quickly dissipate, allowing water vapor in. Once the vacuum is broken, work as quickly as possible to do the necessary maintenance and restore the vacuum. Unless necessary, do not expose internal parts to the open air. For instance, if the instrument is shut down to change the mechanical pump and there is no reason to open the analyzer even though it is at atmospheric pressure, do not open it. Anytime the internal parts are exposed to atmosphere, keep them covered so that dust does not have a tendency to fall or accumulate on the parts. A source of dry nitrogen to blow off parts is handy.

Instruments using turbomolecular pumps for the high-vacuum component can attain operating pressures in just a few minutes and can be used if necessary. It will take some time longer for internal parts to reach operating temperature stability. It is recommend that a vented instrument be *pumped down* overnight before attempting to use it. Instruments using oil diffusion pumps should not take any longer to stabilize, but they will not be usable for an hour or more after restarting because of the necessity to heat the diffusion pump oil. Once the instrument is stable, in all likelihood it will be necessary to run the autotune and recalibrate the *m/z* scale.

Simple Troubleshooting

The logbook should also contain a description of problems that are encountered with the GC-MS and the fix that was employed. Over time, the same symptoms are likely to recur. Occasionally, add a reference spectrum of PFTBA to the logbook with the parameters used to obtain the spectrum. This will give an indication of the average operating performance of the instrument. A slow deterioration that is not remedied by cleaning the ion source and not related to the multiplier voltage is an indication that the analyzer needs to be cleaned. For transmission quadrupole instruments, cleaning the prefilter is often sufficient to restore performance. (A prefilter is not available on all quadrupole instruments.) Lenses that are not part of the ion-source assembly should be cleaned at the time the ion source is cleaned.

The most likely cause of problems in any mass spectrometer is associated with the ion source. When a mass spectrometer problem is discovered, a good rule is to clean the ion source and check for shorts or a burned-out filament. For most instruments, it is relatively simple to determine that a filament has burned out because the emission current will be zero. Sometimes, the filament will short to the block. In this case, the emission current will read high when in the CI mode where emission current is read between the filament and the ion-source block. In either case, the source should be cleaned and the filament inspected. If the filament is sagging, it is a good idea to replace it. A good source for replacement ion-source filaments is Scientific Instrument Services, Inc. (SIS), Ringoes, NJ, USA (http://www.sisweb.com). SIS supplies all GC/MS instrument manufacturers with their filaments, and they also offer a filament repair service.

Air leaks are another source of trouble in the mass spectrometer. A simple method of leak detection is to squirt a small volume of acetone or isopropanol on flanges and other areas where leaks could occur. *Caution is advised not to use this procedure near hot surfaces because of the flammability of these compounds.* A second way to test for small leaks is to tune the mass spectrometer to *m/z* 40 and to use argon to test for leaks. The *m/z*-40 peak will increase if argon enters the source. Helium (*m/z* 4) is a better choice, except when helium carrier gas is used in conjunction with the gas chromatograph. A small stream of the gas is aimed at all seals where a leak can occur. If a leak is detected at a seal, it can sometimes be stopped by tightening the seal, but it is better to replace the seal than to overtighten it.

Contaminated Carrier Gas

Specified high-purity helium is supposed to be contaminant free. If condiments are present, filters like the UOP/matsen filter or the heated getter filter from VICI on the helium gas line are supposed to remove them. There is one contaminant in six-nines helium that cannot be filtered out and that is argon. The Ar peak at *m/z* 40 may not be noticeable if the data acquisition is started at (or the data are stored from) *m/z* 40 or higher. The presence of argon can have an adverse effect on chromatographic performance. Argon contamination is usually not a problem in less than six-nines helium; however, each time a new cylinder of helium (regardless of its stated purity) is put on the instrument, an acquisition should be run starting at *m/z* 10 to check for air leaks (peaks at *m/z* 18, H_2O; *m/z* 28, N_2; *m/z* 32, O_2; *m/z* 44, CO_2) and for Ar, *m/z* 40. Air leaks are bad for the GC column stationary phase and the mass spectrometer. Anytime they are found, they should be corrected.

Ion-Source Cleaning

Do not disassemble an *m/z* analyzer such as a transmission quadrupole or a quadrupole ion trap or the ion source of a double-focusing or TOF instrument unless you have watched someone else do this. At a minimum, view a training video before attempting such an operation. If possible, have an instrument company service person do the first ion-source cleaning while you are in close attendance. Experienced laboratory personnel should train junior members of the laboratory and observe them during their first two or three solo ion-source cleanings.

Make sure all necessary tools are available and have been thoroughly cleaned so that there is no oil on them. In addition to the normal tools, a three- or four-prong stone holder (Figure N.1) is useful to facilitate the handling of small parts. Another valuable tool is a magnifying lamp as shown in Figure N.2. Stereomicroscopes can also be very useful in the examination of small parts.

An ample supply of ion-source disposables, such as washers, ceramic and sapphire insulators, small easily lost screws, and so forth, should be at hand and stored in small pull-out plastic drawers that are lined with *Absorbent Lab Paper* described below.

Never touch any surfaces that will be exposed to a vacuum with an ungloved hand. The recommended gloves are *clean* cotton photographer's

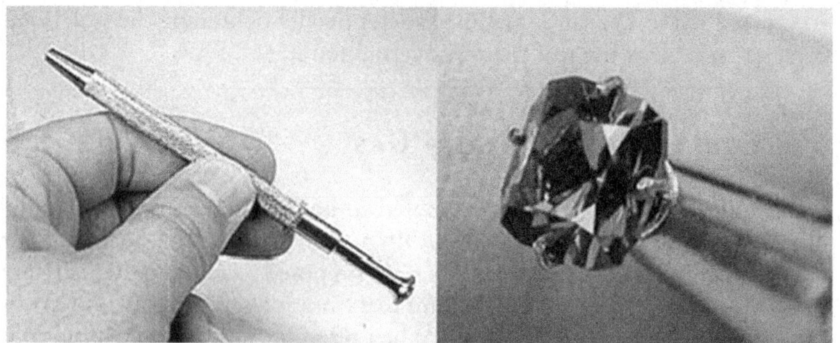

Figure N.1 Stone holder used for handling small parts during ion-source cleaning.

Figure N.2 A fluorescent magnifying desk lamp.

gloves. Disposable nitrile gloves may be substituted. Never let the gloves touch anything that may contaminate them while you are wearing them. Do not use your gloved hand to answer a telephone, pick up a pen and write in a book, scratch yourself, and so forth. In addition to your normal safety glasses, wear a surgical mask to avoid getting moist breath on the parts you are cleaning. On the day that you will break vacuum, do not wear any perfumes or colognes; do not bathe using scented soaps or scented shampoos; do not use any hand or body oils, powders, or other skin treatments; wash your hands with a strong unscented soap like Ivory before beginning.

As can be seen from Figures N.3 and N.4, ion sources can be simple with very few parts or they can be quite complex. Some ion-source designs require electrical isolation of metal components. This is often accomplished with the use of ceramic parts or parts manufactured from sapphires. It is

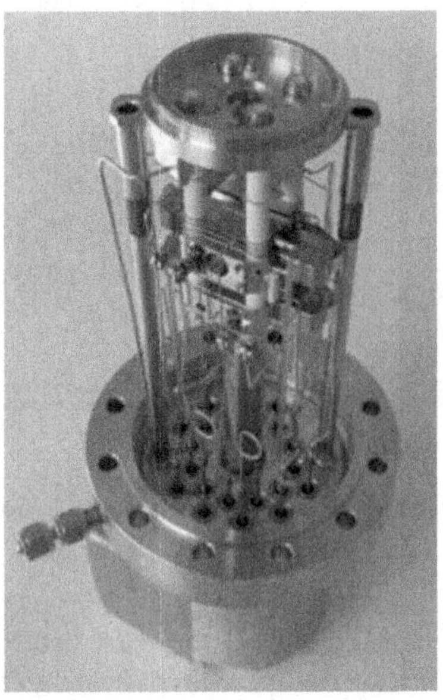

Figure N.3 MAT-212 ion source.

Figure N.4 5975 ion source.

important to have an exploded diagram of an ion source with many small parts before attempting disassembly (Figure N.5). These exploded diagrams are usually part of the instrument's manual set. These manuals are usually provided as PDF files on today's newly purchased instruments; however, older instruments may have only paper manuals. If this is the case, the manual should be scanned to a PDF and the file put on the instrument's computer. All parts should be segregated into small disposable aluminum

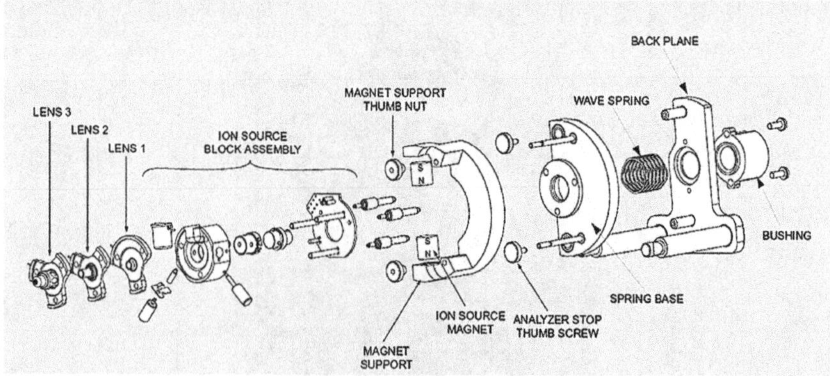

Figure N.5 An exploded assembly view of the ion source from a Polaris Q.

containers. Separate containers should be used for parts before cleaning and after cleaning. The containers used for parts before cleaning should be disposed of; those used for the parts after cleaning can be saved and used as the before-cleaning containers the next time the source is cleaned. The area where the source is to be cleaned should have Absorbent Lab Paper (a hard plastic backing that goes against the counter top with a soft absorbent paper on top, sometimes called *isotope paper*) that has been freshly laid and is taped to hold it steady. A slight slip of hand on an untaped bench-covering can send parts flying, maybe never to be seen again.

A good reference for ion-source cleaning and general mass spectrometer maintenance is *The Mass Spec Handbook of Service*, Volume 2, published by SIS, Ringoes, NJ, USA, edited by John J. Manura and Christopher W. Baker. Although this book dates back to 1993, it contains a great deal of relevant information.

Metal surfaces are often cleaned with a water slurry (paste-like) of alumina. If this is used, the parts should be thoroughly rinsed with deionized water and dried in an oven at least 120 °C. DO NOT use acetone on these parts after rinsing to facilitate drying. Never use organic solvents from a squeeze bottle on any parts that will be under vacuum, as this may deposit a residue of plasticizers on the parts. Ceramic parts can be cleaned in dichromic acid and the sapphires should be fired with a glassblowing torch. Ultrasonic cleaners are often employed with various solvents, aqueous and organic.

SIS has an ion-source cleaning service with a quick turnaround time. If possible, a second ion source is a very handy thing to have so that you can still operate your instrument while cleaning a contaminated source or sending it out to be cleaned. It is important when cleaning metal components that you do not want to scratch them. This is why motorized power tools with metal brushes can be problematic when cleaning the ion source. An interesting article appeared in *Analytical Chemistry* in 1987 (Vol. 59, No. 20, October 15, 1987) by Louis R. Alexander, Vince L. Maggio, Vaughn E. Green, James B. Gill, Elizabeth R. Barnhart, Donald G. Patterson, Jr, and Lance C. Nicolaysen entitled "Vibrating Tumblers as Cleaning Devices for Mass Spectrometer Ion Source Parts" in which a vibrating ammunition case (shell casing) tumbler is utilized with crushed corn cobs and a metal polish.

Electron Multipliers

In addition to EMs supplied with your original instruments, there are several third-party sources of these devices. Before making a switch, be sure to explore all advantages and disadvantages and all options. EMs should only be ordered on an as-needed basis. They will have a tendency to go bad while sitting in storage; therefore, only order one when the EM in your instrument needs to be changed. The EM should last for one to three years depending on the instrument's use. GC/MS instruments are designed to have an optimum *gain*. The gain of an EM is the number of electrons produced from a single ion and is usually about 10^5. As the EM ages, to maintain this 10^5, it is necessary to increase the voltage applied to the EM. As the voltage is increased, the observable electrical noise will increase. Once the electrical noise reaches an unacceptable performance level, it is time to replace the EM. It should not have gone unnoticed that the upper voltage limit was not stated. This is something that differs from instrument manufacturer to instrument manufacturer and from EM type to EM type. The appropriate manufacturer's literature should be consulted for the maximum recommended voltage before replacement. Regardless of the stated value, the instrument performance should be the deciding factor.

One of the most challenging issues with the replacement of the EM is its proper positioning. The EM's performance is highly dependent on its physical position relative to the ion beam exiting the *m/z* analyzer. Often times, the relatively bulky EM is connected to two copper wires which can have considerable flexibility in spite of their sometimes large diameters. In such cases, it is not uncommon to have to break vacuum a second or third time to position the EM so that its previous performance is duplicated. One innovative design originally from K&M Electronics, Inc. (now owned by ITT Power Solutions, West Springfield, MA, USA) is shown in Figure N.6. Once the desired position is achieved, the EM is replaced by an easy plug-in pull-out module. Manufacturers, such as Varian, have designed their own self-aligning EM systems.

Most TOF instruments use multichannel plates for ion detections and some instruments use photomultipliers.

Figure N.6 Removable EM in permanently mounted holder.

CALIBRATION

Calibration is another source of potential problems. The frequency with which an instrument needs to be calibrated for mass accuracy depends on the instrument. For some instruments, it might be necessary to recalibrate every day and check the calibration at least once a day. Other instruments will hold calibration for months. A good suggestion, and a good laboratory practice for all instruments, is to perform a quick calibration check each morning. This can be as simple as injecting a fixed quantity of a volatile solvent into the gas chromatograph or acquiring spectra of PFTBA for 20–30 seconds over a range of *m/z* 10–100 at the normal cal-gas inlet pressure. A brief glance at the resulting mass spectra will confirm that the *m/z* assignment is correct. This procedure also allows some assessment of the total performance of the GC/MS instrument and only takes a few minutes.

In the positive-ion mode, calibration is most frequently performed using perfluorotributylamine (some double-focusing instruments use perfluorokerosine). Calibration of the *m/z* scale is usually a part of the autotune procedure; consult the instrument's manual for the recommended procedure. The calibration reference for *m/z* values will be stored in a calibration file. Do not allow calibration compound to flow into the ion source for too long of a period too often as this can result in ion-source fouling, requiring more frequent ion-source cleaning.

CALIBRATION GAS PEAKS IN THE SAMPLE SPECTRUM

When using PFTBA as the *m/z* scale calibration and as a tuning compound, if peaks at *m/z* 69, 131, 219, and 502 are observed in every mass spectrum in an analysis, check to make sure that the cal-gas valve was not left open or has developed a leak. This problem is not as prevalent on newer instruments that have electronically controlled valves as it was on instruments that had manual valves.

Accurate Mass GC/MS

Identification of unknowns using GC/MS is greatly simplified if accurate mass measurements are made of all the ions in a spectrum so that reasonable elemental compositions of each ion can be assigned. Unfortunately, obtaining a mass measurement that is accurate enough to significantly limit the number of possible elemental compositions requires more than a routine operation. This is usually done with instruments that have high resolving power such as a TOF or a double-focusing mass spectrometer; however, it is possible to generate such data with a transmission quadrupole when using *MassWorks* from *Cerno Biosciences.* Regardless of the technique used, the process is much more time-consuming than a library search of a mass spectrum.

Accurate mass measurement is generally associated with high resolving power. Mass resolution is necessary to eliminate interference of either the ions whose masses will be measured or the internal reference ions by other ions appearing at the same nominal mass. If good chromatography techniques are achieved, interferences from ions of compounds other than the compound being mass measured are essentially eliminated, and the chief cause of interference problems is the overlap of peaks representing ions from the sample and internal reference. For this reason, an internal reference material is chosen in which all the peaks have as large a negative mass defect as possible. If perfluoroalkanes are used, a resolving power (R) of >3,000 ($M/\Delta m$, 10% valley definition) is usually necessary to eliminate peak overlap and poor results.

Because an increase in resolving power causes a decrease in sensitivity in double-focusing instruments, it is best to operate at the lowest resolving power commensurate with good results. Some instrument data systems will allow calibration with an external reference material such as perfluorokerosene and then use of a secondary reference material for the internal mass reference. Tetraiodothiophene, vaporized using the solids probe inlet, is recommended as the secondary reference. The accurate masses are 79.9721, 127.9045, 162.9045, 206.8765, 253.8090, 293.7950, 333.7810, 460.6855, and 587.5900. For a higher mass standard, use hexaiodobenzene. Because the mass defect for these internal reference ions are so large, a resolving power of 2,000 (10% valley) is ample to separate these ions from almost any sample ions encountered in GC/MS.

GC/MS systems capable of high resolving power based on a TOF *m/z* analyzer do not allow for resolving power adjustments; however, the high resolving power at which they do operate does not compromise limits of detection.

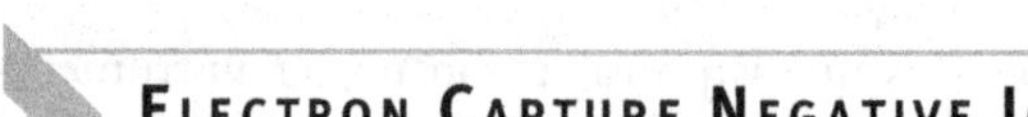

Electron Capture Negative Ionization GC/MS

Electron capture negative ionization (ECNI) can give excellent results for certain types of compounds. Compounds with electronegative substituents and unsaturation can be expected to have a large electron capture cross section and thus work well in the negative-ion detection mode. Frequently, much higher sensitivity is obtained for these compound types in the negative-ion mode than under positive-ion conditions. In addition, the negative-charge molecular ion ($M^{-\bullet}$) is usually very abundant. The method requires that a gas be added to a chemical ionization source at a slightly lower pressure than is used in positive-ion CI. It is best to tune the ion-source pressure and source parameters using a peak from the reference material. This added gas is ionized through an electron ionization process that results in the low-energy ions required for the electron capture ionization. Unfortunately, the gas often employed is methane. This is one of the reasons that ECNI is erroneously called *negative chemical ionization* (*NCI*). Another problem with using any organic reagent gas to generate the thermal energy electrons is that these techniques tend to be dirty, resulting in frequent ion-source cleanings. As was pointed out in the first edition of this book, nitrogen and argon are also acceptable sources of thermal energy electrons and much cleaner. CO_2 is another acceptable reagent gas for the generation of thermal energy electrons.

Calibration in negative-ion mode to at least *m/z* 700 can be achieved using perfluorokerosine. Polyperfluoroisopropylene oxide oligomer (see following structure) distributions have been used to calibrate to *m/z* 5,000 in the negative-ion electron capture mode. To reach *m/z* 5,000,* it was necessary to vaporize the oligomer from a solids probe. This material can be used as an internal reference for accurate mass negative-ion GC/MS. The reference ions have *m/z* values of 31.9898, 68.9952, 84.9901, 116.9963, 162.9818, 184.9837, 250.9754, 282.9817, 328.9671, 350.9691, 416.9607, 448.9670, 494.9524, 516.9544, 614.9524, 682.9397, 780.9377, 848.9251, 946.9230, 1,014.9104, 1,112.9084, 1,180.8957, 1,278.8937, 1,346.8811, 1,444.8790, 1,512.8664, 1,610.8643, 1,678.8517, 1,776.8496, 1,844.8370, and 1,942.8349.

$$F(C(CF_3)(F)-CF_2O)_nC(F)(H)-CF_3$$

*The material used for this work was Freon E-13. The 13 refers to the most abundant oligomer (n = 13) in the above structure.

Table N.1 Common mass spectral peaks representing contaminants in GC/MS

m/z Values of ions	Substance	Possible source
16, 17, 46, 58	CH_4, NH_3, CO_2, *i*-C_4H_{10} and reagent ions thereof	Leaking CI gas valve
18, 28, 32, 44 or 14, 16	H_2O, N_2, O_2, CO_2, or N, O	Residual air and water, air leaks, out-gassing from Vespel ferrules
29, 31, 32	H_3COH (methanol)	Cleaning solvent or sample diluent
31, 51, 69, 100, 119, 131, 169, 181, 214, 219, 264, 376, 414, 426, 464, 502, 576, 614	ions in perfluorotributylamine (PFTBA) a. k. a. FC-43	PFTBA tuning/calibration compound Calibration-compound valve leak
31, 45, 59, 74	$H_5C_2OC_2H_5$ (diethyl ether)	Sample diluent
31, 45, 46	H_3CCH_2OH (ethanol)	Cleaning solvent or sample diluent
39, 41, 54, 67, 82	C_6H_{10} (cyclohexene)	Antioxidation stabilizing compound in dichloromethane
40	Ar	Contaminated He carrier gas
40, 41	H_3CCN (acetonitrile)	Sample diluent
43, 58	*n*-butane	Leak test compound (butane cigarette lighter)
43, 58	$H_3C(CO)CH_3$ (acetone)	Cleaning solvent or sample diluent
44	CO_2	Injector coolant
49, 41, 84, 86, 88	H_2CCl_2 (dichloromethane)	Cleaning solvent or sample diluent
59	$H_3CCH(OH)CH_3$ (isopropanol)	Cleaning solvent or sample diluent
69	Foreline (rough) pump oil	Mechanical pump backstreaming (time to change oil)
69, 81	Base peak and 2nd most intense peak in mass spectrum of squalene	Component in skin moisturizers; results from handling vacuum exposed parts with ungloved hands
73, 147, 207, 281, 295, 355, 429	$((-CH_2)_2SiO-)_n$ dimethyl polysiloxane	Bleed from methyl silicone GC columns and septa
73	$(CH_3)_3Si$ – ions	Residue from trimethylsilyl derivatizing reagents

(*Continued*)

Table N.1 (*Continued*)

m/z Values of ions	Substance	Possible source
74	CS_2 (carbon disulfide)	Sample diluent
77, 94, 115, 141, 168, 170, 262, 354, 446	thermal fragmentation of polyphenyl ethers	Thermal fragmentation of diffusion pump oil
78	C_6H_6 (benzene)	Cleaning solvent or sample diluent
83, 85, 86	$HCCl_3$ (chloroform)	Sample diluent
97, 99 or 97, 99 and 83, 85	$C_2H_3Cl_3$ depending on isomer	1,1,1- or 1,1,2-trichloroethane cleaning solvent
91, 92	$C_6H_5CH_3$ (toluene)	Cleaning solvent or sample diluent
91, and 105, 106	$C_6H_4(CH_3)_2$ (xylene)	Cleaning solvent
151 and 153 and 117 and 119 or 101 and 103	$C_2Cl_3CF_3$	1,1,1- 2,2,2- or 1,1,2- 1,2,2- isomers of trichlorotrifluoroethane; a. k. a. Freon, Freon 113, FC-113, CFC-113 isomer dependent. Cleaning solvent
117, 119, 121	CCl_4 (carbon tetrachloride)	Sample diluent
149, 167, 279	$C_6H_4((CO)OC_8H_{17})_2$	Branched octyl esters of phthalic acid used as plasticizer. Normal octal esters don't show much *m/z* 167
149	$C_6H_4((CO)OC_nH_{2n+1})_2$	Alkyl esters of phthalic acid used as plasticizer from plastic containers
Peaks 14 *m/z* units apart	Non-aromatic hydrocarbons $(-CH_2-)_n$	Hydrocarbon oils from fingerprints, mechanical pump oil, and sample environment or accumulation on the GC column
Peaks 50 *m/z* units apart	Perfluoronated aliphatic ions$(-CF_2-)_n$	Use to calibrate *m/z* scale of magnetic-sector and double-focusing instruments as well as lock mass for exact mass measurements

APPENDIX O

Mixtures for Determining Mass Spectral Resolution

Table O.1 Mixtures for determining resolution

Mixtures for determining resolution	Ratio	Nominal *m/z* value	Resolution (M/ΔM, 10% valley)
Bromobenzene/ dimethylnaphthalene	1:1	*m/z* 156	1,144
Cyclohexane/cyclopentanone	2:1	*m/z* 84	2,300
Styrene/benzonitrile 200 °C inlet temperature required	3:1	*m/z* 103	8,195
Dimethylquinoline/ dimethylnaphthalene	1:2	*m/z* 156	12,400
1-Octene	–	*m/z* 71	15,900
Toluene/xylene	1:10	*m/z* 92	20,922

Gas Chromatography and Mass Spectrometry
DOI: 10.1016/B978-0-12-373628-4.00051-4

APPENDIX P

Cross-Index Chart for GC Stationary Phases

In the first edition of this book, there were nine suppliers of GC columns that had compatible stationary proprietary phases. Several of these companies were acquired and disappeared, changed their names, or went out of business. Chrompack was acquired by and became Varian; HP changed its name to Agilent and acquired J&W, resulting in the disappearance of J&W stationary phases and so forth. Now Varian has been acquired by Agilent; Agilent plans to divest itself of the Varian GC line, which means that the fate of the Chrompack products is unknown. The industry is changing rapidly.

Every vendor provides such a chart in their catalogs, which can be downloaded from their Web site; therefore, such a table, which would in all likelihood be incomplete and out-of-date before this book is printed, is not included; just a reminder to look for such tables when selecting a replacement GC column.

Download the PDF of your vendor's catalog and save the cross-reference pages as a separate file.

Gas Chromatography and Mass Spectrometry
DOI: 10.1016/B978-0-12-373628-4.00052-6

APPENDIX Q

Ions for Determining Unknown Structures

$[M + 1]^+$ **Possible Precursor Compounds**

^{13}C isotope (present at 1.1% for each ^{12}C)
Aliphatic nitriles also usually exhibit an $[M - 1]^+$ peak
Ethers
Sulfides
Aliphatic amines
Alcohols
Esters
Increasing the sample size or decreasing the repeller voltage may increase the relative abundance of the $[M+1]^+$ ion. If the sample pressure is very high, dimers may also be produced.

$M^{+\bullet}$

In general, the relative intensity of the $M^{+\bullet}$ peak decreases in the following order.

1. Aromatics
2. Conjugated olefins
3. Alicyclics
4. Sulfides
5. Unbranched hydrocarbons
6. Ketones
7. Amines
8. Esters
9. Ethers
10. Carboxylic acids
11. Branched hydrocarbons
12. Alcohols

$[M - 1]^+$ **Possible Precursor Compounds and Functionalities**

Dioxolanes
Some amines
Aldehydes*

Gas Chromatography and Mass Spectrometry
DOI: 10.1016/B978-0-12-373628-4.00053-8

Some fluorinated compounds (e.g., $C_6F_5{-}CH_2OCH_2C{\equiv}CH$)
Acetals
Segmented fluoroalcohols (some even lose $^{\bullet}H$ and HF) (e.g., $CF_3CF_2CF_2CH_2CH_2OH$)*.
Aryl–CH_3 groups
N–CH_3 groups
Aliphatic nitriles (also may have $[M+1]^+$ peak)**
Aromatic isocyanates
Aromatic phenols
Certain butenols
Certain fluorinated amines***
e.g., $C_8F_{17}CH_2CH_2CH_2NH_2$
or
$CF_3(CF_2)_2CH_2CH_2CH_2NH_2$

$[M-2]^+$ **Possible Precursor Compounds**
Polynuclear aromatics (e.g., dihydroxyphenanthrene)
Ethylsilanes (dimers to heptamers)

2 **Structural Significance**
H_2

$[M-3]^+$ **Possible Precursor Compounds**
$[M-3]^+$ ions must be fragments from a higher-mass molecular ion (e.g., $[M-CH_3]^+$ and $[M-H_2O]^+$)

$[M-10]^+$

Add 18 to the higher mass and 28 to the next lower mass to determine M. Look for an $[M-44]^+$ peak to see if it is an aldehyde.

$[M-13]^+$

Not from a molecular ion, but could be the loss of a $^{\bullet}CH_3$ (radical) and the loss of a $^{\bullet}C_2H_4$ (radical) or CO from the molecular ion.

$[M-14]^+$ **Possible Precursor Compounds**
Not observed from a molecular ion; however, a mixture of two different compounds may be present where the higher mass ion is MH.
The apparent loss of 14 could also be the loss of 15 from the molecular ion.

14 **Structural Significance**
CH_2 in ketene

---

* Also lose HF

** The $[M-1]^+$ peak is larger than the $[M+1]^+$ peak, especially in *n*-nitriles.

*** May not observe the $M^{+\bullet}$ peak, just the $[M-H]^+$.

$[M-15]^+$ **Possible Precursor Compounds**

Methyl derivatives
tert-Butyl and isopropyl compounds
Trimethylsilyl derivatives
Acetals
Compounds with NC_2H_5 groups
Compounds with aryl–C_2H_5 groups
Saturated hydrocarbons do not lose CH_3 from a straight-chain compound.

15 **Structural Significance**

CH_3 as in CH_3F, $CH_3N{=}NCH_3$, $CH_3OC(O)OCH_3$, $CH_3OC(O)(CH_2)_4C(O)OCH_3$

$CH_3OCH=CH_2$, $CH_3OCF=CF_2$, $H_3C{-}O{-}S(=O)_2{-}O{-}CH_3$

$CH_3O(CO)CH_2(CO)OCH_3$
NH

$[M-16]^+$ **Possible Precursor Compounds**

Aromatic nitro compounds*
N-Oxides
Oximes
Aromatic hydroxylamines
Aromatic amides (loss of NH_2)
Sulfonamides
Sulfoxides
Epoxides
Quinones
Certain diamines (e.g., hexamethylenediamine)

16 **Structural Significance**

O, NH_2, CH_4

$[M-17]^+$ **Possible Precursor Compounds**

Carboxylic acids (loss of OH)
Aromatic compounds with a functional group containing oxygen
Ortho to a group containing hydrogen
Diamino compounds (loss of NH_3)

---

*Nitro compounds show peaks representing $[R-O]^+$ and $[R-NO]^+$ and should have a large *m/z* 30.

Simple aromatic acids
Some amino acid esters
Loss of NH_3 (amines having a four-carbon chain or longer)

17 **Structural Significance**

OH, NH_3

$[M - 18]^+$ **Possible Precursor Compounds**

Primary straight-chain and aromatic alcohols (also look for a large *m/z* 31)*
Alcohol derivatives of saturated cyclic hydrocarbons
Steroid alcohols (e.g., cholesterol)
Straight-chain aldehydes (C_6 and upward)**
Steroid ketones (peak generally <10%)
Carboxylic acids, particularly aromatic acids with a methyl group ortho to the carboxyl group
Aliphatic ethers with one alkyl group containing more than eight carbons
Loss of CD_3 from deuterated TMS derivatives

18 **Structural Significance**

A highly characteristic peak for amines (NH_4)
H_2O

$[M - 19]^+$ **Possible Precursor Compounds**

Fluorocarbons
Alcohols, $[M - (H_2O + H)]^+$

19 **Structural Significance**

Fluorine compounds (*m/z* 19, 31, 50, and 69)
Rearrangement ion characteristic of hydroxyl compounds
Glycols
Acetals
Diols

$[M - 20]^+$ **Possible Precursor Compounds**

Aliphatic alcohols
Alkyl fluorides and fluoroethers***
Fluorosteroids
Fluoroalcohols
Compounds such as $C_8F_{17}CH_2CH_3$ may not show a $M^{+\bullet}$ peak but an $[M - F]^+$ peak as the highest *m/z* value.

20 **Structural Significance**

n-Alkyl fluorides

---

* Primary alcohols: $[M - 18]^+$, $[M - 33]^+$, and $[M - 46]^+$.

** Aldehydes observe $[M - 28]^+$ and $[M - 44]^+$ ions.

*** $M^{+\bullet}$ is usually not observed. In CI, the highest mass observed is $[M + H - HF]^+$.

$[M-21]^+$ **Possible Precursor Compounds**
$[M-(H_2F)]^+$ from segmented fluoroalcohols
e.g.,

$$\left(F_{17}C_8-C(H_3C)=C(OH)H \right)$$

$[M-(H+HF)]^+$ from: $CF_3CH_2CH_2CF_2CF_2CF_3$
Segmented fluoroalcohols are represented by
$CF_3CH_2CH_2OH$, $CF_3CF_2CH_2CH_2OH$,
$CF_3CF_2CF_2CH_2CH_2OH$, etc.

24 **Structural Significance**
Acetylene

25 **Structural Significance**
Acetylene (strong ion observed in spectrum)
Maleic acid
Acrolein
Fluoroacetylene

$[M-26]^+$ **Possible Precursor Compounds**
Loss of C_2H_2 from thiophene
Aromatic compounds
Pyrrole
Bicyclic compounds:
for example,

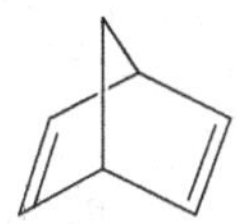

Isocyanides ($R-N\equiv C$)

26 **Structural Significance**
Aliphatic nitriles and dinitriles
HCN
Acrylonitrile
Propanenitrile
Pyrazines
Pyrroles
Acetylene
Maleic acid
Maleic anhydride
Succinic anhydride
Vinyl group (look for *m/z* 40 and 54)

[M – 27]⁺ **Possible Precursor Compounds**

N-containing heterocyclics (see pyridine, etc.)
Simple aromatic amines
Loss of C_2H_3 by double-H rearrangement in ethyl-containing phosphites* and phosphonates*
HCN for unsaturated nitriles

27 **Structural Significance**

Aliphatic nitriles and dinitriles
Alkenes and compounds with unsaturated R groups
CH_2=CH–R
CH_2=CHC(O)R
CH_2=CHC(O)OR, etc.
HCN, CH_2=CH^-

[M – 28]⁺ **Possible Precursor Compounds**

Phenols
Diaryl ethers (especially ethyl ethers)
Quinones
Anthraquinones
Aldehydes (look for $[M-1]^+$, $[M-18]^+$, $[M-28]^+$, and $[M-44]^+$ ions)
Cyclic ketones
Anhydrides
Naphthols
O-containing heterocyclics
$[M-(HCN+H)]^+$ (low-molecular-weight aliphatic nitriles) and isocyanides**
Phenylisocyanate ($[M-CH_2N]^+$)
Phenylenediisocyanate
Chlorophenylisocyanate
N-containing heterocyclic compounds
Ethyl esters
Cycloalkanes
Acid fluorides (e.g., benzoyl fluoride)

28 **Structural Significance**

Lactones
Ethyleneimines
Alkyl amines

Saturated hydrocarbons
Dialkyl aromatic hydrocarbons
Dioxanes
Cyclobutane
Butyrolactone
Butyraldehyde
Succinic acid
1-Tetralone
Carbonyls

$[M-29]^+$ Possible Precursor Compounds

Aromatic aldehydes (should observe $[M-H]^+$ as well as $[M-CHO]^+$)
Simple phenols
Quinones
Ethyl groups or alicyclic compounds
Cyclic ketones and cyclic amino ketones such as pyrilidone, piperidone, and caprilactam
Phenols
Naphthols
Polyhydroxybenzenes
Diaryl ethers
Aliphatic nitriles
Purines $[M-(CH_2{=}NH)]^+$ and imines
Propanals
Indicates an alkyl group (may not be ethyl)
Loss of CHO from compounds such as $CH_3C(O)OCH_2OC(O)CH_3$

29 Structural Significance

Alkanes or compounds with an alkyl groups (*m/z* 43, 57, 71, etc. may also be present)*
Aldehydes
Propionates
Cyclic polyethers
α-Amino acids
Hydroperoxides
C_2H_5, CHO

$[M-30]^+$ Possible Precursor Compounds

Aromatic nitro compounds
Loss of CH_2O from simple aromatic ethers
Lactones**

---

* For branched alkanes such as *tert*-butyl, *m/z* 43 is the dominant ion.

** Produces $[M+H]^+$ using isobutane CI.

Loss of CH_2O from

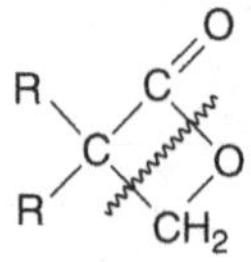

(usually do not see the molecular ion peak)

(see dioxane, dioxolanes, and epichlorohydrin)

Morpholine

30 **Structural Significance**

Cyclic amines unsubstituted on the nitrogen

Primary amines*

Nitro compounds and aliphatic nitrites (should observe $[M-16]^+$ and $[M-30]^+$)

Secondary amides

Formamides

Nitrosamines

Ureas

Caprolactam

NO (aliphatic and aromatic nitro compounds)

N_2H_2, CH_2O, CH_4N, CH_2NH_2, C_2H_6, etc.

Note: In the absence of rearrangement, fragment ions containing an odd number of nitrogen atoms have an even *m/z* value.

$[M-31]^+$ **Possible Precursor Compounds**

Methyl esters—the simultaneous presence of peaks at *m/z* 74 and 87, and $[M-31]^+$ strongly indicates a methyl ester

Methoxy derivatives, including methoximes

Primary aliphatic alcohols and glycols

Loss of $^{\bullet}CH_2OH$ radical

Loss of SiH_3

31 **Structural Significance**

A peak occurs at *m/z* 31 in almost all alcohols and ethyl ethers, as well as in some ketones; *m/z* 31 in the absence of fluorine indicates oxygenated compounds.

Fluorocarbons (with *m/z* 19, 31, 50, and 69)

Ethers and alcohols (*m/z* 31, 45, and 59)**

---

* Amines show a powerful tendency to break beta to the nitrogen atom. This mass is not characteristic only of primary compounds because the ion is formed from other amines by rearrangement.

** If the spectrum is of an alcohol, to identify the molecular weight, it may be necessary to add 18 or 46 to the highest *m/z* value peak observed.

Primary straight-chain alcohols
Primary alcohols bonded at the γ-carbon
Formates
Aliphatic carboxylic acids
Alkyl ethyl ethers
Cyclic polyethers
Phosphorus compounds
Rearrangement peak in dioxanes

$[M-32]^+$ **Possible Precursor Compounds**
o-Methyl benzoates
Loss of methanol
Ortho substituent of methyl esters of aromatic acids
Loss of 31 also should be present.
$[M-(CH_3OH)]^+$ from methyl esters and ethers
$[M-(O+NH_2)]^+$ from sulfonamides
Loss of sulfur from thiophenols and disulfides

32 **Structural Significance**
Hydrazine
Methanol
Fluoroethylene
Deutero compounds
Oxygen
Methyl formate
Carbonyl sulfide

$[M-33]^+$ **Possible Precursor Compounds**
Alcohol derivatives of cyclic hydrocarbons
Thio compounds
Thiocyanates
Short-chain unbranched primary alcohols
$[M-(H_2O+CH_3)]^+$ from hydroxy steroids
Loss of –SH

33 **Structural Significance**
Glycols
Diols
Alcohols
Acetals
CH_2F

$[M-34]^+$ **Possible Precursor Compounds**
Mercaptans (usually aliphatic)

34 **Structural Significance**
CH_3F, H_2S, PH_3 (see triethyl phosphine, etc.)

$[M-35]^+$ **Possible Precursor Compounds**
Chloro compounds

$[M-(H_2O+OH)]^+$ from certain dihydroxy and polyhydroxy compounds

$[M-(CH_3+HF)]^+$ from compounds such as

$(CH_3)_3Si$ O CF_2 O F C CF_3

35 **Structural Significance**

Thioethers

Sulfides

Chloro compounds

$[M-36]^+$ **Possible Precursor Compounds**

n-Alkyl chlorides

(Some alkyl chlorides lose HCl and appear to be butenes; for example, 1,2,3,4-tetrachlorobutane appears to be trichlorobutene)

$[M-(H_2O+H_2O)]^+$ from certain polyhydroxy compounds

36 **Structural Significance**

HCl

$[M-37]^+$ **Possible Precursor Compounds**

$[M-{}^{37}Cl]^+$ from alkyl chlorides

(H_2O+F) from certain fluoroalcohols

37 **Structural Significance**

HC–C≡C–

$[M-38]^+$ **Possible Precursor Compounds**

$[M-2F]^+$ from fluorocarbons

38 **Structural Significance**

C_2N, $-CH_2C{\equiv}C-$

Diallyl sulfide

Malononitrile

Dicyanoacetylene

Furan

$[M-39]^+$ **Possible Precursor Compounds**

$(HF+F)$ (e.g., fluoroalcohols)

HO H

H_3C O CF_2 CF_2 C CF_2 CF_2 O CH_3

39 **Structural Significance**

Furans

Alkenes
Dienes
Cyclic alkenes
Acetylenes
Aromatic compounds, particularly di- and tetrasubstituted compounds
(*m/z* 39, 50, 51, 52, 63, and 65)
C_3H_3, $CH_3C{\equiv}C{-}$, $-CH{=}C{=}CH_2$, $-CH_2C{\equiv}CH$

[M – 40]$^+$ **Possible Precursor Compounds**
Aliphatic dinitriles (loss of CH_2CN)*
Aromatic compounds
Cyclic carbonate compounds
Segmented fluoroalcohols such as (loss of H_2F_2)

$F_{17}C_8$ C=CHOH, H_3C

[M – (HF + HF)]$^+$

40 **Structural Significance**
Butanol
Dinitriles (may have *m/z* 41, 54, and 68)
C_3H_4, CN_2, C_2O, $CH_3C{\equiv}CH$, $CH_2{=}C{=}CH_2$, C_2H_2N, $-CH_2C{\equiv}N$

[M – 41]$^+$ **Possible Precursor Compounds**
Nitriles
Suggests a propyl ester
N-influenced fragmentation, for example,

H_3C, N, CH_3, H_3C, N, CH_3

41 **Structural Significance**
Nitriles and dinitriles (*m/z* 54 also may be present)**
Esters of aliphatic dibasic carboxylic acids
Thioethers

---

* The highest mass may be [M – 28]$^+$ and/or [M – 40]$^+$. (Usually the $M^{+\bullet}$ is not observed.) Large *m/z* 41, 55, and 54 peaks also should be present.

** CH_3CN exists in lower aliphatic nitriles. Higher nitriles have $(CH_2)_nCN$ (e.g., *m/z* 54, 68, etc.).

Isothiocyanates
Primary aliphatic alcohols
Alkenes and compounds with an alkenyl group
Cyclohexyl (*m/z* 41, 55, 67, 81, and 82)
Methacrylates (also look for *m/z* 69)*
C_3H_5, C_2HO, C_2H_3N, CH_3CN, $-CH_2CH{=}CH_2$,
$H_3C{-}CH{=}CH{-}$ and

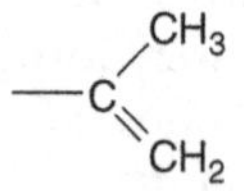

$$-C(CH_3){=}CH_2$$

m/z 69, 41, and 86 suggest segmented fluoromethacrylates**

[M – 42]$^+$ **Possible Precursor Compounds**
Acetates (loss of ketene, see if *m/z* 43 is also present)
N-Acetylated compounds (acetamides)
Enol acetates
Simple purines (loss of cyanamide)
2-Tetralone
Loss of ketene from bicyclic structures (e.g., for camphor)

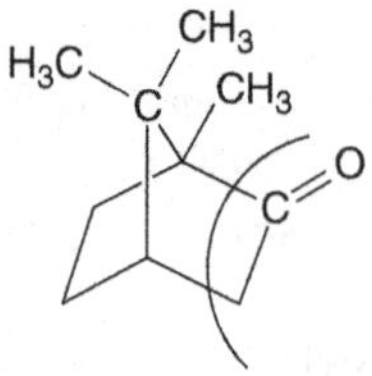

42 **Structural Significance**
Cycloalkanes
Alkenes (<C_6); peaks at *m/z* 42 and 56 suggest alkenes
Ethyleneimines ($H_2C{=}C{=}NH$)
Pteridines
Cyclic ketones (saturated)
THF
Butanediol
Simple purines

---

* May have to add 86 Da to the *m/z* value of the last peak observed to deduce the molecular weight.

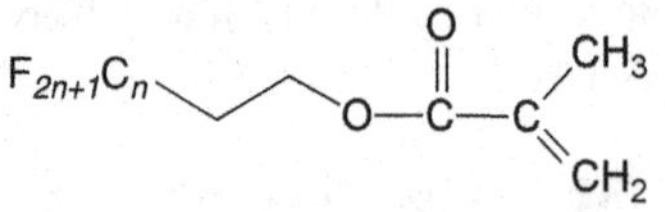

** $F_{2n+1}C_n{-}CH_2CH_2{-}O{-}C({=}O){-}C(CH_3){=}CH_2$

C_3H_6, C_2H_2O, CNO, CH_2N_2, C_2H_4N, $CH_3N{=}CH–$

H_2C, H_2C >N—, $H_3CCH{=}N—$, N—H, H—C—CH_2, $H_2C{=}N$—CH_2

$(CH_3)_2N–$ also should observe *m/z* 44

[M − 43]⁺ **Possible Precursor Compounds**

Uracils: $[M-(C(O)NH)]^+$
Cyclic amides
Propyl derivatives (particularly isopropyl)
Cyclic peptides
Dioxopiperazines
Tertiary amides
Aliphatic nitriles ($>C_6$)
Common loss from cyclohexane rings
Terpenes

43 **Structural Significance**

Alkanes (*m/z* 43, 57, 71, 85, etc. characterize the spectra of alkanes)
Compounds with an alkyl group
Acetates (also observe *m/z* 61)
Alditol acetates (Check for the presence of *m/z* 145, 217, 289, etc.)
Methyl ketones (when looking for larger than methylethyl ketone, look for *m/z* 58 as well).
Cyclic ethers
Vinyl alkyl ethers
N-Alkylacetamides
Isothiocyanates
Aliphatic nitriles
With *m/z* 87 suggests triethyleneglycol diacetate

C_3H_7, $CH_3C{\equiv}O$, $(CH_3)_2CH$, $H_2C{=}CHO$, C_2H_5N, CH_3N_2, C(O)NH, $C_3H_7C(O)OR$, $C_3H_7C(O)NR_2$, $C_3H_7C(O)R$, C_3H_7SR, C_3H_7OR, $C_3H_7OC(O)R$, C_3H_7X, $C_3H_7NO_2$

$[M-44]^+$ **Possible Precursor Compounds**

Aliphatic aldehydes (look for $[M-1]^+$, $[M-18]^+$, $[M-28]^+$, and $[M-44]^+$ ions.)
Compounds with $(CH_3)_2N–$
Aromatic amides
Anhydrides (e.g., phthalic acid anhydride)

44 **Structural Significance**

Aliphatic aldehydes unbranched on the α-carbon (is the base peak for C_4–C_7). (Look for $[M-1]^+$, $[M-18]^+$, $[M-28]^+$, and $[M-44]^+$ ions.)

Primary amines with α-methyl groups

Secondary amines unbranched on the α-carbon

Tertiary amides*

R^1–C(=O)–N(R^2)(R^3)

Vinyl alkyl ethers

Caprolactam derivatives

Cyclic alcohols

Piperazines (C_2H_6N)

CO_2, N_2O, $CF{\equiv}CH$, CH_3CHO, C_2FH, C_3H_8, $OC{-}NH_2$, C_2H_4O, CH_2NHCH_3, $(CH_3)_2N$, CH_2NO, CH_2CHO+H, CH_3CHNH_2, CH_4N_2, $HC{=}NOH$, C_2H_5NH, $-CH_2CH_2NH_2$

$[M-45]^+$ **Possible Precursor Compounds**

Ethoxy derivatives

Ethyl esters

Carboxylic acids (may lose 17, 18, and 45)

Cyanoacetic acid, 8-cyano-1-octanoic acid

45 **Structural Significance**

Aliphatic carboxylic acids

2° alcohols (e.g., *sec*-butyl alcohol, *m/z* 45, 59, and 31)**

Propylene glycol

Di- and triethylene glycols

Isopropyl and *sec*-butyl ethers

---

* Will produce

H_3C–C(=O)–N(R^2)(R^3)

via McLafferty rearrangement if R_1 contains a hydrogen on carbon 4.

** R_2(H)(R_1)C—OH 2° alcohol

Butyrates
Methylalkyl ethers unbranched on the α-carbon
$-C(O)-OH$, $H_2C{=}CF-$, C_2H_5O, CH_3OCH_2, CH_2CH_2OH, C_2H_2F, CH_3NO, C_2H_7N, CH_3SiH_2, $CH_2{=}P^+$, CH_3CHO+H, CHS^*
Oxygenated compounds can be disregarded if peaks are absent at *m/z* 31, 45, 59, etc.

$[M-46]^+$ **Possible Precursor Compounds**
Nitro compounds (NO_2)
Loss of C_2H_5OH from ethyl esters
Aromatic acids with methyl groups ortho to the carboxyl groups
Long-chain unbranched primary alcohols
Loss of C_2H_3F
Straight-chain high-molecular-weight primary alcohols
Loss of water plus C_2H_4
Possibly loss of formic acid (e.g., $C_6H_5CH_2CH_2C(O)OH$)
Cyano acids

46 **Structural Significance**
Nitrates
$CH_2{=}CHF$, CH_3CH_2OH, $HC(O)OH$
CH_3OCH_3, NO_2, C_2FH_3, $CH_3NH{=}NH_2$

$[M-47]^+$ **Possible Precursor Compounds**
Alkylnitro compounds
Acid fluorides
Sulfur compounds (loss of CH_3S)
Loss of CH_3OH+CH_3

47 **Structural Significance**
Thiols (*m/z* 61 and 89 also suggest sulfur-containing compounds)
Acid fluorides
Thioethers (sulfides)
Fluorosilanes such as $FC_6H_4SiH_3$
Acetals (generally containing the ethoxy group)
Fluoroethane
C, O, and F compounds (peak may be small)
Formates (small peak)
CH_3S, $-C(O)-F$, SiF, PO
CH_3CHF, CH_2SH, CH_3PH, CH_2CH_2F

$[M-48]^+$ **Possible Precursor Compounds**
Methyl thioethers
($COH+F$) from pentafluorophenol

---

* Is characteristic of certain sulfur compounds such as thiophenes.

($CHO + F$) from fluoroaldehydes
Aromatic sulfoxides

48 **Structural Significance**
Methyl thioethers
CH_3SH from mercaptans
SO, CH_3SH, –CHCl, C_2H_5F, CH_3S, +H, C_4
Tetraborane

$[M - 49]^+$ **Possible Precursor Compounds**
Chlorinated compounds (loss of CH_2Cl) (e.g., β-chloroisopropylbenzene)

49 **Structural Significance**
Halogenated compounds containing CH_2Cl
Methyl thioethers
CH_2Cl, –C≡C–C≡CH

$[M - 50]^+$ **Possible Precursor Compounds**
Fluorocarbons (loss of CF_2)
Methyl esters of unsaturated acids (loss of $CH_3OH + H_2O$)
Methyl esters of straight-chain hydroxycarboxylic acids except 2-hydroxy acids

50 **Structural Significance**
CF compounds
Compound containing phenyl or pyridyl groups
Chloromethyl derivatives
CF_2

$[M - 51]^+$ **Possible Precursor Compounds**
Loss of CHF_2
HC≡CCN from α,β-unsaturated nitriles

51 **Structural Significance**
Acetylenes
Compounds containing phenyl or pyridyl groups
Compounds containing CHF_2
Small *m/z* 51 peak indicates compounds containing C, H, and F
CHF_2, C_4H_3 (aromatics), SF

52 **Structural Significance**
Butadienes
Acetylenes
Compounds containing phenyl or pyridyl groups
Nitrogen trifluoride
C_2N_2, –CH=CHCN, CH_2=CH–C≡CH, C_2H_4, NF_2
Chromium
Cyanogen, C_2N_2

53 **Structural Significance**
Furans
Cyclobutenes
Dienes

Acetylenes
Pyrazine
Aminophenol
Acrylonitrile
Chloroprene
Hydroquinone
C_3H_3N, C_4H_5, C_2HN_2, C_3HO, NF_2H, etc.
The mass spectra of pyrrole derivatives usually contain prominent ions at *m/z* 53 and 80, and sometimes the *m/z* 67 rearrangement ion.

$[M - 54]^+$ **Possible Precursor Compounds**

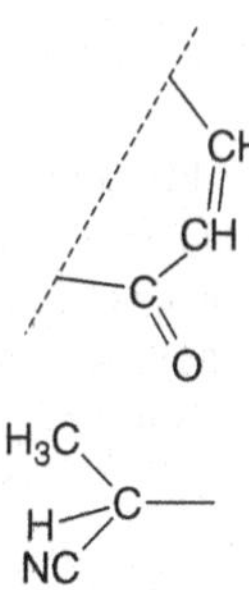

(from branched dinitriles)

54 **Structural Significance**

Unsaturated cyclic hydrocarbons (e.g., vinyl-cyclohexene)
Butadienes
Dinitriles (C_2H_4CN) (dinitriles lose 40 Da and 28 Da and generally give *m/z* 41, 55, and 54 ions.)
Acetylenes
Cyclohexanes
Nitriles (*m/z* 41 also may be present)
Maleic acid
Maleic anhydride
Quinone
$-CH_2CH_2CN$, $C(O)-CN$, C_3H_2O, C_4H_6, $C_2H_2N_2$

$[M - 55]^+$ **Possible Precursor Compounds**

A loss of 55 is possibly the loss of C_4H_7 from esters (double hydrogen rearrangement). The loss suggests a butyl or isobutyl group, especially when *m/z* 56 is also present.
Loss of ($CO + HCN$) from aromatic isocynate
Loss of $H_2C{=}CH-C(O)$

55 **Structural Significance**

Cyclic ketones
Cycloalkanes

Indicates a cyclohexyl ring (*m/z* 55, 83, and 41)
Esters of aliphatic dibasic carboxylic acids
Aliphatic nitriles and dinitriles (see *m/z* 54)
Thioethers
Alkenes
Primary aliphatic alcohols
Olefins have fragment ions at *m/z* 55, 69, 83, etc.
Cyclohexanones
$(CH_3)_2C{=}CH$, $CH_2{=}CH{-}C(O){-}$
Acrylates (*m/z* 55 and 99 suggest glycol diacrylates)

Base Peak	Next Most Intense Peaks				Compound	Highest *m/z* Peak >1%
55	82	54	67	27	1,6-Hexanediol diacrylate (MW = 226 Da)	113
55	196	82	127	126	Pentaerythnitol triacrylate (MW = 298 Da)	225
55	54	27	71	85	1,4-Butylene glycol diacrylate (MW = 198 Da)	126
55	54	71	126	85	1,4-Butylene glycol diacrylate (MW = 198 Da)	126
55	56	70	41	43	*n*-Octyl acrylate (MW = 184 Da)	112
55	27	70	41	56	*n*-Heptyl acrylate (MW = 170 Da)	113
55	56	73	27	41	Butyl acrylate (MW = 128 Da)	99
55	56	73	84	43	*n*-Hexyl acrylate (MW = 156 Da)	99
55	72	59	27	31	2-Ethoxyethyl acrylate (MW = 144 Da)	99, 100, 101
55	82	67	73	54	Cyclohexyl acrylate (MW = 154 Da)	111
55	84	27	69	43	2-Ethylbutyl acrylate (MW = 156 Da)	98
55	45	45	27	29	2-Methoxyethyl acrylate (MW = 130 Da)	87
55	27	85	42	58	Methyl acrylate (MW = 86 Da)	86
55	69	70	43	97	Ethylene glycol diacrylate (MW = 170 Da)	14, 141
55	91	79	107	162	Benzyl acrylate (MW = 162 Da)	162
55	127	56	27	68	2,2-Dimethyl propane diacrylate (MW = 212 Da)	140, 152

$[M-56]^+$ **Possible Precursor Compounds**

Aromatic diisocyanates
Quinones
Ketals of cyclohexanone
Anthraquinones
C_4H_8 from carbonyl compounds
Butyl compounds (e.g., $C_6H_5OC_4H_9$)

56 **Structural Significance**

Butyl esters
Cyclohexylamines
Isocyanates (*n*-alkyl). Peaks associated with M − CO, M − (H + CO), M − (HCN + CO) are generally observed. Also look for a peak at *m/z* 99.
Cycloalkanes
Some *n*-chain alcohols (butanol, hexanol, heptanol, octanol, nonanol, etc.)
Aliphatic nitriles (see *m/z* 54)
$-CH_2-N=C=O$, C_4H_8, C_3H_6N, $(CH_3)_2N=CH_2$, $CH_3N=CHCH_2-$,

H_2C / H_2C (ring) $>NCH_2-$

$[M-57]^+$ **Possible Precursor Compounds**

Isocyanates ($M-CH_3N=C=O$)
Loss of F_3 in perfluorotributylamine
C_4H_9 from TBDMS derivatives

57 **Structural Significance**

Compounds containing alkyl groups, particularly tertiary butyl group
Ethyl ketones
Propionates
Aliphatic nitriles
Isobutylene trimers and tetramers (also observe *m/z* 41 and 97)
Alkanes (*m/z* 29, 43, 57, 71, etc. are also observed)
C_4H_9; $C_2H_5CHCH_3$, $(CH_3)_3C$, etc.
C_3H_5O; $C_2H_5C(O)-$, $CH_3C(O)CH_2-$, etc.
$C_2H_5N=N$

$[M-58]^+$ **Possible Precursor Compounds**

Aliphatic methyl ketones ($CH_3C(O)CH_3$
Simple aromatic nitro compounds (NO + CO)
Straight-chain mercaptans

Thiocyanates and isothiocyanates

58 **Structural Significance**

Methyl alkyl ketones unbranched at the α-carbon (43 + 58)

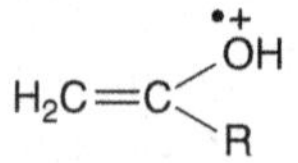

Rearrangement ion of 2-methylaldehydes

n-Propyl-, butyl-, amyl-, and hexylketones

Primary amines with a 2-ethyl group

Tertiary amines with at least one *N*-methyl group

Thiocyanates

Isothiocyanates

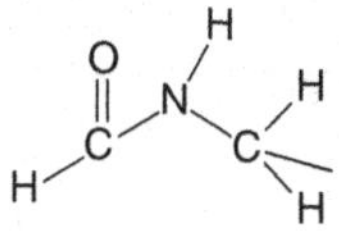

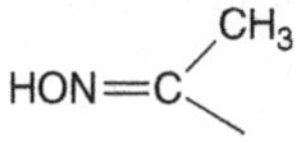

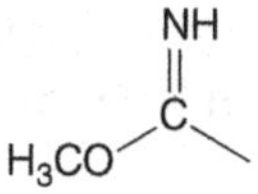

$(CH_3)_2CNH_2$, $(CH_3)_3CH$, C_3H_8N

$(CH_3)_2NCH_2–$ (sometimes an *m/z* 42 peak also is present), $C_2H_5NHCH_2–$

$CH_3C(O)NH–$ (generally observe an *m/z* 60 rearrangement ion)

$CH_3C(O)CH_2 + H$, C_3H_6O, $C_2H_2O_2$, C_2H_4NO, $C_2H_6N_2$, C_3H_8N, $–C(O)–NHCH_3$, CNO_2, C_3H_3F, CH_2N_2O, C_4H_{10}, $HC(O)NHCH_2–$

Peaks at *m/z* 58, 72, and 86 suggest that a C=O group is present (also look for M – CO).

$[M - 59]^+$ **Possible Precursor Compounds**

Loss of $(C(O)OCH_3)$ from methyl esters

Loss of OC_3H_7 from propyl esters

Methyl esters of 2-hydroxycarboxylic acids

Loss of $CH_3C(O)O$ from sugar acetates

59 **Structural Significance**

Tertiary aliphatic alcohols

Methyl esters of carboxylic acids

Esters of *n*-chain carboxylic acids

Silanes

Ethers $(C_2H_5OCH_2–)$

m/z 59, 45, 31, and 103 suggest di- or tri-propylene glycol

Primary straight-chain amides (greater than propionamide)

Amides

OH
$H_2C{=}C{-}NH_2$

(look for *m/z* 59, 44, and 72)

$(CH_3)_2COH$ (tertiary alcohols)

Largest peak in the mass spectrum of propylene glycol ethers

$CH_3OCH_2CH_2$– (diethylene glycol dimethyl ether)

$CH_3CHOHCH_2$–, $C_2H_5OCH_2$–, $(CH_3)_2COH$

$CH_3OC(O)$–, CH_3CHCH_2OH, C_3FH_4

Peaks at *m/z* 31, 45, and 59 reveal oxygen in alcohols, ethers, and ketones.

$[M-60]^+$ **Possible Precursor Compounds**

Acetates

Methyl esters of short-chain dicarboxylic acids

Loss of $-OCH_2CH_2NH_2$

$-C_6H_{10}(CO_2CH_3)_2$

O-Methyl toluates

60 **Structural Significance**

A characteristic rearrangement ion of monobasic carboxylic acids above C_4 (also see *m/z* 73)

Cyano acids (observe $[M-46]^+$ and large *m/z* 41, 54, and 55 peaks)

Esters of nitrous acids

Sugars

Aliphatic nitriles unbranched at the α-carbon

Ethyl valerate has *m/z* 60 and 73 (these ions are normally associated with carboxylic acids)

Cyclic sulfides

$H_2C{-}CH_2$
S

m/z 60 with an *m/z* 58 peak suggests $CH_3C(O)NH+2H$ C_3H_5F C_3H_8O, CH_6N_3, $-CH_2ONO$, CH_4N_2O, CH_2NO_2, C_2H_6NO, C_2H_4S, $-NHCH_2CH_2OH$, $ClC{\equiv}CH$, COS, $C_2H_4O_2$, $C_2H_8N_2$,

$[M-61]^+$ **Possible Precursor Compounds**

Suggests $CH_3N^{\bullet}OCH_3$ and CH_3NHOCH_3

Loss of (Si_2H_5) from

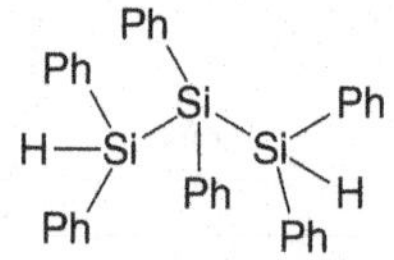

61 **Structural Significance**

m/z 61 is a characteristic rearrangement ion in acetates other than methyl acetate (should observe $[M - 42]^+$ as well as a large *m/z* 43)

Esters of high-molecular-weight alcohols ($CH_3CO_2H_2$)

Base peak in 1,2-difluorobutanes

Acetals

Primary thiols (CH_2CH_2SH)

Thioethers (CH_3SCH_2–)

N-TFA of *n*-butylmethionine

m/z 61 and 89 suggest sulfur-containing compounds (RSR′)

TMS derivatives; for example,

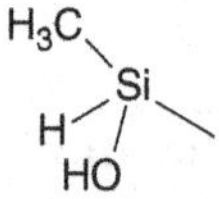

CH_3CH_2CHF–, $CH_3C(O)C + 2H$, CH_3CFCH_3

–$CHOHCH_2OH$, –CH_2CH_2SH, CH_3CHSH

$CH_3NOCH_3 + H$

$[M - 62]^+$ **Possible Precursor Compounds**

Thiols

62 **Structural Significance**

Ethyl thioethers

$F_2C{=}C$–, –$CHCH_2Cl$, –$CF{=}CF$–, $PH_2C_2H_5$, (see triethylphosphine and tributylphosphine)

CH_3CH_2SH, $(CH_3)_2S$

$[M - 63]^+$ **Possible Precursor Compounds**

Methyl esters of dibasic carboxylic acids

63 **Structural Significance**

Aromatic nitro compounds

Acid chlorides (C(O)Cl)

Ethyl thioethers

$CF_2{=}CH$–, –CH_2CH_2Cl, CH_3CHCl, $CH_3S(O)$, SiCl, C_2F_2H, $CH_3P{-}OH$, $CF_2{=}CH$–

$CF_3CH_2CH_2$– structure also provides a large *m/z* 63 peak

$C_2H_5OC(O)$ (e.g., diethyl carbonate)

–CH_2CH_2Cl, –OCH_2CH_2Cl

$[M - 65]^+$ **Possible Precursor Compounds**
Loss of OCH_2Cl
$-SO_2CH_2-$
Certain sulfones (HSO_2)

65 **Structural Significance**
Aromatic compounds (C_5H_5)
Aromatic nitro compounds
Aromatic alcohols
Vinyl furans
CH_3CF_2-, $C_2F_2H_3$

66 **Structural Significance**
Unsaturated nitriles
Dicyanobutones
Methyl pyridines
Ethyl disulfides (HSSH)
Acrylonitrile dimers
CFCl, N_2F_2, $C_2F_2H_4$, C_5H_6

NH CH3

67 **Structural Significance**
Perfluoro acids
Cycloalkyl compounds
Alkadienes (also may have an intense *m/z* 81 ion)
Alkynes
CHClF (in Freon-21), CF_2OH, SOF–, C_5H_7, C_4H_5N,

$H_2C{=}C(CH_3){-}C{=}CH_2$

m/z 67, 81, 95, 109, etc. suggest 1-acetylenes

$[M - 68]^+$ **Possible Precursor Compounds**
Dinitriles

68 **Structural Significance**
Cyclopentanes
Cyclohexenes
Cyclohexanols
Aliphatic nitriles (*m/z* 41 and 54 also may be present), C_4H_6N

	C_5H_8 (cyclopentane derivatives), $-CH_2CH_2CH_2CN$, $NCCH_2C(O)$ CH_2ClF
$[M - 69]^+$	**Possible Precursor Compounds**
	Fluorocarbons (loss of CF_3)
69	**Structural Significance**

Fluorocarbons
Cycloparaffins (e.g., cyclopentyl)
Esters of aliphatic dibasic carboxylic acids
Aliphatic nitriles
Isothiocyanates ($>C_7$)
Methacrylates (*m/z* 41 is also present)*
$C(O)C(CH_3){=}CH_2$, $CH_3CH{=}CHC(O)-$, $(CH_3)_2C{=}CH{-}CH_2-$, CF_3, C_5H_9, PF_2, $-CH{=}CHCH(CH_3)_2$,
A series of peaks at *m/z* 69, 83, 97, 111, etc. (two mass units less than the corresponding paraffins) suggest olefins
m/z 69, 41, and 86 suggest segmented fluoromethacrylates
m/z 69, 77, 65, and 51 are characteristic of segmented fluoroiodides (e.g., $C_8F_{17}CH_2CH_2I$, etc.)

Abundant Ions				Compound	MW
69	41	87	39	Ethylene glycol monomethacrylate	30
69	41	113	112	Ethylene glycol dimethacrylate	198
69	113	41	86	Diethylene glycol dimethacrylate	242
69	41	113	45	Triethylene glycol dimethacrylate	286
69	41	55	43	Trimethylolpropane trimethacrylate	338
69	41	54	55	1,6-Hexanediol dimethacrylate	254

$[M - 70]^+$ **Possible Precursor Compounds**

(CHF_2+F) loss observed in 1-H perfluoro compounds (e.g., $CF_3(CF_2)_7H$, $CF(CF_2)_8H$, $-C_4H_8N$)

70 **Structural Significance**

Pyrrolidines

NH CH3

Aliphatic nitriles
Amyl esters

* To postulate the mass of the molecular ion, add 86 Da to the even-mass ion of a doublet in the high-mass region.

If the mass spectrum has a large peak at *m/z* 42, check 1-pentanol

C_4H_8N, $CH_3N{=}CHCH_2CH{-}$

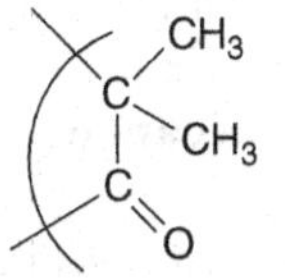

Hexamethyleneimine

Isocyanates ($-CH_2CH_2NCO$)

m/z 70 and 43 suggest a diacetoxybutene

[M − 71]⁺ Possible Precursor Compounds

Loss of $(CH_3)_3CCH_2$

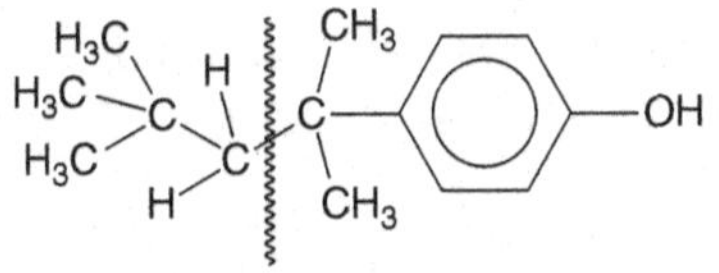

71 **Structural Significance**

THF derivatives

Butyrates

Methyl cyclohexanols

Propyl ketones

Terminal aliphatic epoxides:

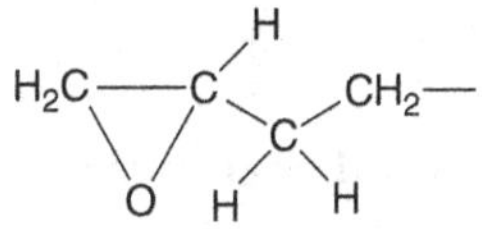

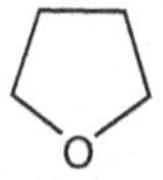

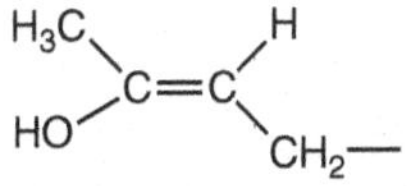

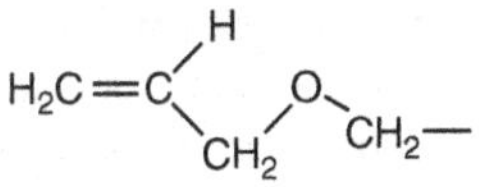

$CH_3CH(OH)CH{=}CH{-}$, $C_3H_7C(O)$

$(CH_3)_2CHC(O)$, $-C(O)OCH{=}CH_2$,

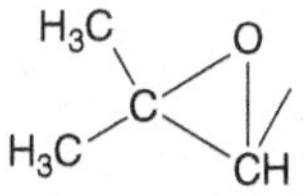

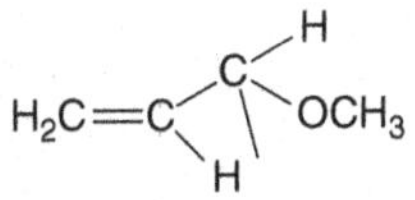

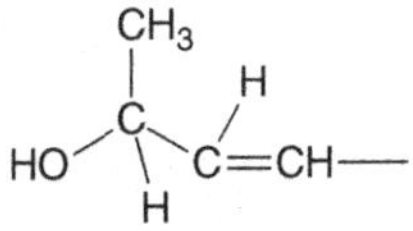

C_5H_{11}, C_4H_9N, NF_3, $(C_2H_5)_2CH$

$[M - 72]^+$ **Possible Precursor Compounds**

Loss of $-NHCH_2CH(CH_3)_2$

Acrylates (loss of acrylic acid)

Loss of C_2O_3 from compounds such as cyclohexane dicarboxylic acid anhydride

72 **Structural Significance**

Amides

Secondary and tertiary amines ($C_3H_{10}N$)

Ethyl ketones

Thiocyanates (CH_2SCN)

Isothiocyanates ($-CH_2-N{=}C{=}S$)

Rearrangement ion of 2-ethylaldehydes

$CH_3C(O)NHCH_2$, $C_2H_5NHCH_2CH_2-$, $CH_3C(O)CHCH_3 + H^*$

$(CH_3)_2C-NHCH_3$, $C_3H_7CHNH_2$, $C_2H_5C(O)CH_2 + H^*$

$(C_2H_5)_2N$, $(CH_3)_2NC(O)-$, CH_3NCH_2-,

$[M - 73]^+$ **Possible Precursor Compounds**

Alditol acetates

Loss of ($C(O)C_2H_5$)

Methyl esters of aliphatic dibasic carboxylic acids ($-CH_2C(O)OCH_3$)

n-Butyl esters (loss of C_4H_9O)**

73 **Structural Significance**

Sugars

Aliphatic acids (*m/z* 73 with a peak at *m/z* 60 suggests acids)

Alcohols (C_3H_7CHOH-)

Ethers ($C_3H_7OCH_2-$)

Ethyl esters ($-C(O)OC_2H_5$)

Methyl esters of dibasic carboxylic acids ($CH_2C(O)OCH_3$)

1,3-Dioxolanes

(*m/z* 73 and 45 suggest dioxolanes)

TBDMS and trimethylsilyl derivatives, $(CH_3)_3Si–$

$H_3C–CH_2–C(H)(H)–OCH_3$ (structure), $H_5C_2–O–\dot{C}(H)–CH_3$, $(CH_3)_2\dot{C}–OCH_3$,

$–CH_2CH_2C(O)OH$

$CH_2OC(O)CH_3$, 1,3-dioxolan-2-yl (ring O–C(H)–O), C_3H_2Cl

[M – 74]$^+$ **Possible Precursor Compounds**

Some butyl esters (C_4H_9OH by rearrangement)

(CH_2CHSCH_3) loss from *N*-TFA, *n*-butyl methionine

74 **Structural Significance**

m/z 74 is a characteristic rearrangement ion of methyl esters of long-chain carboxylic acids in the C_4–C_{26} range. Also look for an *m/z* 87 peak.

Monobasic carboxylic acids with an α-methyl group

Esters of aliphatic dibasic carboxylic acids

Cyclic sulfides

Aliphatic nitriles with an α-methyl group

$–CH_2–N(OCH_3)(CH_3)$, $CH_3NHCH_2CHOH–$, $HOC_2H_4NHCH_2–$,

CH_3CHONO, $–CH_2NHCH_2CH_2OH$, $CH_2{=}C(OH)–OCH_3$,

C_3F_2, $H_2NCHC(O)OH$*, $–N(CH_2OCH_3)(CH_3)$*

[M – 75]$^+$ **Possible Precursor Compounds**

Loss of $–CH_2OCH_2CH_2OH$

75 **Structural Significance**

Dimethyl acetals [$(CH_3O)_2CH–$]

Disubstituted benzene derivatives containing electrophilic substituents C_6H_3

* Also observe *m/z* 74, 73, and 72.

Dichlorobutenes, $CH_2ClCH{=}CH{-}$, $CH_2CH{=}CHCl$
Propionates (generally small)
Sulfides ($C_2H_5SCH_2{-}$)
$(CH_3)_2SiOH$, $-CF{=}CHCH_2OH$, $C_2H_5OC(O)+2H$, $(CH_3)_2CH{-}S{-}$, $CH_3OCH(OH)CH_2{-}$
TMS derivatives of all aliphatic alcohols [$HO^+{=}Si(CH_3)_2$]

[M – 76]⁺ **Possible Precursor Compounds**
Loss of C_2H_4SO from ethyl sulfones
Loss of CF_2CN

76 **Structural Significance**
Benzene derivatives (C_6H_4)
Aliphatic nitrates ($-CH_2ONO_2$)
Propyl thioethers (C_3H_7SH)
$-CF_2CN$, C_6H_4, CS_2, C_3H_5Cl, $N(OCH_3)_2$, $(CH_3)_2NS$*

77 **Structural Significance**
Monosubstituted benzene derivatives containing an electrophilic substituent
$(CH_3)_2SiF$, $-CF_2CH{=}CH_2$
$C_3F_2H_3$ (e.g., $-CHFCF{=}CH_2$, $-CH_2CH{=}CF_2$, $CH_3CF{=}CF{-}$, $-CH{=}CHCHF_2$), $C_6H_5{-}$, $(CH_3)_2P(O)$, $CH_2ClC(O)$,

$C_2H_5Si(H)F$ Cl–C(CH₃)(H₃C)– $H_3CCH_2CHCl{-}$

Alkylbenzene (*m/z* 39, 50, 51, 52, 63, 65, 76, 77, and 91)
$C_nF_{2n+1}CH_2CH_2I$

[M – 78]⁺ **Possible Precursor Compounds**
Loss of benzene from compounds such as $(C_6H_5)_2$ and $CHOCH(C_6H_5)_2$

78 **Structural Significance**
Phenyl tolyl ethers
Compounds containing the pyridyl group

C_2F_2O, $C_3F_2H_4$, [pyridyl structure]

[structure: C–O, C–O ring with C=CF₂]
C_2H_7PO

* *m/z* 76, 42, and 61 suggest $(CH_3)_2NS{-}$

$[M-79]^+$ **Possible Precursor Compounds**
Bromides

79 **Structural Significance**
Cycloalkadienes
Bromides
Pyridines
–CH=CClF, CH_3SS–, –$CF_2C_2H_5$, $H_3CC_5H_4$, Br

N N, O O H_3CS—, H Cl H C—C OH H

$CHF_2C(O)$, $HOP(O)OCH_3$

$[M-80]^+$ **Possible Precursor Compounds**
Alkyl bromides (loss of HBr)

80 **Structural Significance**
Alkyl pyrroles
Substituted cyclohexenes (C_6H_8)
Methyl disulfides (CH_3SS+H)
C_5H_6N, C_2H_2ClF,

NH CH_2—, N ⟶ C_5H_6N (*m/z* 80)

O_2N—⟨benzene⟩—NH_2 ⟶ C_5H_6N (*m/z* 80)

81 **Structural Significance**
Hexadienes (C_6H_9)
Alkyl furans
N-TFA, *n*-butyl histidine
CH_3OCF_2–, CH_3SiF_2, CF_2=CF–, –CH=CH–CH= CH–CH=O,

–CF_2CH_2OH, O CH_2—, , CH_3CFCl–,

C_6H_9, $(HO)_2P+O$ present in phosphate spectra

$[M - 82]^+$ **Possible Precursor Compounds**

Loss of $C_3H_4N_3$,

82 **Structural Significance**

Aliphatic nitriles ($-CH_2CH_2CH_2CH_2CN$) (*m/z* 82, 96, 110, 124, 138, 152, etc. suggest straight-chain nitriles)

Benzoquinones

Piperidine alkaloids

Some fluoroalcohols

m/z 82 and 67 suggest cyclohexyl compounds (see *m/z* 83)

m/z 82 and 182 suggest cocaine (MW = 303 Da)

CCl_2, C_6H_{10}, C_2F_3H, $CHF{=}CF_2$, $(CD_3)_3Si-$, C_5H_8N,

$[M - 83]^+$ **Possible Precursor Compounds**

Dicyclohexylbenzene

83 **Structural Significance**

Aliphatic nitriles

Cyclohexanes (C_6H_{11}) (*m/z* 83, 82, 55, and 41 suggest a cyclohexyl ring)

Thiophene derivatives

Trialkyl phosphites

SO_2F, $CHCl_2$, $C_2H_3F_3$,* CHF_2CHF-, $-CF_2CH_2F$, CF_3CH_2-, C_5H_7O, C_5H_9N-, $(CH_3)_2C{=}CH{-}C(O)$ (e.g., mesityl oxide)**

$CH_3CH{=}CHCH_2C(O)$

$CH_3CH_2CH{=}CHC(O)$

$CH_2{=}CHCH_2CH_2C(O)$

$HP(OH)_3$, phosphonic acid derivatives

* *m/z* 83 and 33 suggest CF_3CH_2-

** *m/z* 55 and 83 suggest this structure.

$[M - 84]^+$ **Possible Precursor Compounds**

Loss of —N(piperidinyl) ($C_5H_{10}N$) observe m/z 84 and 56

84 **Structural Significance**

Aliphatic isocynate $-CH_2CH_2CH_2NCO$

Esters of aliphatic dibasic carboxylic acids (>dimethyl suberate)

Piperidines (m/z 84, 56, etc.) (piperidine, N–H), $C_5H_{10}N$,

Pyrrolidines (pyrrolidine, NH), CH_2Cl_2, $C_5H_{10}N$, (2-methyl-N-methylpyrrolidine, N–CH_3), $CDCl_2$,

C_3H_7–CN–CH_3, C_4H_8Si, (5-methyl-2-pyrrolidinone, O=C, NH), $C_4H_4O_2$

$[M - 85]^+$ **Possible Precursor Compounds**

Glycol diacrylates (M–CH_2=CHC(O)OCH_2–)

Loss of (C_4H_9 + CO) from TBDMS derivatives of amino acids

Caffeine [–C(O)N(CH_3)C(O)–]

85 **Structural Significance**

Butyl ketones (C_4H_9C(O))

N-TFA, *n*-butyl glutamic acid

Piperazines ($C_4H_9N_2$)

Tetrahydropyranyl ethers

Lactones

Methyl-8® derivatives of primary amines

SiF_3, CF_2Cl, CF_3O, POF_2, $C_4H_5O_2$, C_5H_9O, $C_5H_{11}N$, C_6H_{13}, $(CH_3)_2NCH{=}NCH_2$–

$CH_3C(O)C(CH_3)_2$–, (tetrahydrofurfuryl, O, CH_2—), –CH=CH–C(O)OCH_3,

(R-substituted γ-lactone, O, C=O), (tetrahydropyranyl O—R)

(See if a peak at m/z 101 is also present.)

[M − 86]$^+$ **Possible Precursor Compounds**
Loss of $CH_2{=}C(CH_3)C(O)OH$ from methacrylate esters

86 **Structural Significance**
Propyl ketones ($C_3H_7C(O)CH_2 + H$)
α-Methylisothiocyanates
SOF_2, C_4F_2, $(C_2H_5)_2NCH_2$–, C_4H_9–$NHCH_2$, $C_5H_{12}N$
$C_4H_9CHNH_2$, $HC(O)NH$–$C(CH_3)_2$, –$CHCH_3$–$N{=}C{=}S$, –$CH_2NHC_4H_9$
$C_2H_5NHCH_2CH_2CH_2$– (will observe peaks at *m/z* 44, 58, 72 and 86)

[M − 87]$^+$ **Possible Precursor Compounds**
n-Amyl esters
Loss of –$CHCH_3C(O)OCH_3$
Loss of $C(O)OC_3H_7$ from *N*-TFA, isopropyl esters of amino acids

87 **Structural Significance**
Methyl esters of aliphatic acids (*m/z* 74 and 87 indicate a methyl ester)
Glycol diacetates (*m/z* 87 and 43 are characteristic ions)
Methyl dioxolanes
Long-chain methyl esters
Esters of *n*-chain dibasic carboxylic acids
$CH_2CH_2C(O)OCH_3$
$C_4H_7O_2$:

$CH_3C(O)OCH_2CH_2$, $C_2H_5C(O)OCH_2$–$C_5H_{11}O$:
$CH_3CH_2CH(CH_3)OCH_2$–, –$CH(OH)CH_2CH(CH_3)_2$,
–$CH_2OCH_2CH(CH_3)_2$

C_4H_4Cl, C_4H_7S, $[(C_2H_5)_2NCH_2\ H]$, $C(O)OC_3H_7$,

88 **Structural Significance**
Long-chain ethyl esters (look for *m/z* 101)
($C_2H_5OC(O)CH_2 + H$)
$C_3H_6NO_2$, CF_4, $C_4F_2H_2$, –$C(CH_3)_2NCO$
$H + CH(CH_3)C(O)OCH_3$, $C_4H_{10}NO$, C_4H_8S, –$CHNH_2C(O)OCH_3$
Long-chain methyl esters with an α-methyl group

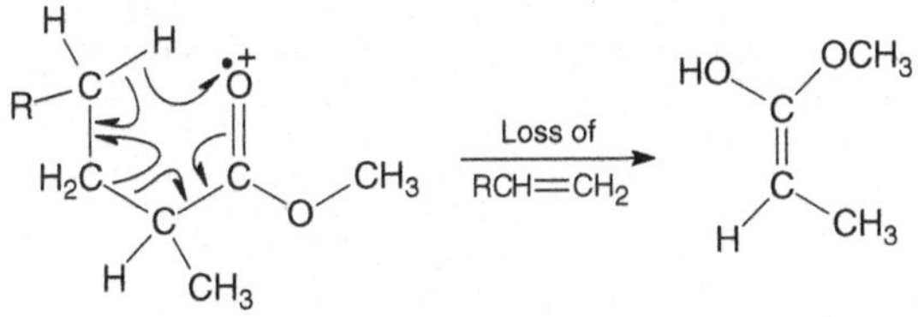

[M – 89]$^+$ **Possible Precursor Compounds**

Loss of $OSi(CH_3)_3$ from TMS derivatives (*m/z* 90 should be more prevalent)

89 **Structural Significance**

Triethylene glycol (also look for *m/z* 45)

O- and N-containing heterocyclic compounds

Characteristic rearrangement ion of butyrates except methyl

Dinitrotoluenes

TMS derivative of primary aliphatic alcohols $[H_2C{=}O^+SiH(CH_3)_2]$*

Sulfur-containing compounds (also expect *m/z* 61)

C_7H_5, C_4H_6Cl, $C_3H_7SCH_2$–, [benzene ring–C], , $C_4F_2H_3$,

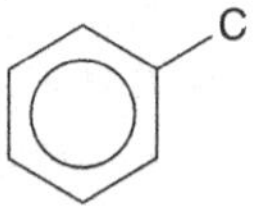

$(H_3CO)_2C(CH_3)$–

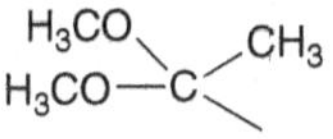

C_4H_9S–, $C_3H_7SCH_2$–, $CH_2{=}CHCF{=}CF$–, –$CH_2OSiH(CH_3)_2$

–$CH(OH)CH_2CH(OH)CH_3$, $(CH_3)_3SiO$–, $[C_3H_7OC(O) + 2H]$, $(CH_3O)_2CCH_3$, $CH_3OSi(CH_3)_2$

–$CF{=}CHCH_2OCH_3$ (should observe *m/z* 45 and 59 as well)

[M – 90]$^+$ **Possible Precursor Compounds**

Loss of $(CH_3)_3SiOH$ from TMS derivatives (may see the loss of 180 Da and/or 270 Da from the molecular ion, depending on the number of OH groups present)

90 **Structural Significance**

O- and N-containing heterocyclics

Aliphatic nitrates with an α-methyl group (CH_3CHONO_2)

---

* Look for homologs in the spectra of 2° and 3°, for example, *m/z* 103 (without a peak at *m/z* 89), *m/z* 117, etc.

Methyl esters of α-hydroxy carboxylic acids
C_6H_4N, C_3FCl, C_6H_5CH, C_7H_6, $(C_2H_5)_2PH$,

R, P=O, CH_3, H_2C, N, H_2C, CH_2

91 **Structural Significance**
Alkyl benzenes (C_7H_7) (*m/z* 104 and/or 117 are also characteristic ions)
Phenols
Aromatic alcohols
Benzyl esters
Alkyl chlorides (C_6–C_{18}); C_4H_8Cl from terminal chloroalkanes
N-TFA, *n*-butyl phenylalanine

$CH_2ClCH_2C(O)$, CH_2—, CH_3

$ClCH_2CHCH_3CH_2$–, $C_4H_5F_2$, $Cl(CH_3)_2CCH_2$–
CF_2=$C(CH_3)CH_2$–, –CF_2–$C(CH_3)CH_2$, *m/z* 43 and 91 suggest 1-chlorodecane

H, H, H_3C, C, C=CF_2, $C_2H_4O_2P$

[M – 92]$^+$ **Possible Precursor Compounds**
Certain esters of dibasic carboxylic acids
Loss of (C_3H_9SiF) for fluorinated sugars

92 **Structural Significance**
β- and γ-monoalkyl pyridines

CH_2—, N

Phosphorus compounds
Salicylates (see *m/z* 138)
Benzyl compounds with a γ-hydrogen

NH—, NH₂ (benzene ring), C=C=O,

$(CH_3)_3SiF$, N_3CF_2–
C_7H_8 double rearrangement ion common to alkylbenzenes
CFClCN
C_3H_8PO

[M – 93]⁺ **Possible Precursor Compounds**
Phenoxy derivatives

–OH

93 **Structural Significance**
Nitrophenols
Fluorocarbons
Salicylates (see *m/z* 138)
Alkyl pyridines

N, CH_2 ∿∿∿ + H, O + H,

OH, C_3H_3FCl: $-CFClCH{=}CH_2$, $CH_3OCF{=}CF-$

C_3F_3: $CF_3-C{\equiv}C-$

$(CH_3O)_2P$, $-C(O)OCH_2Cl$, C_3H_6OCl: $C_2H_5OCHCl-$,

OH, C, ClH_2C, CH_3 C_2H_6SiCl: H_5C_2, Cl, Si, H_3C, Cl, H_3C–Si

94 **Structural Significance**
Alkyl phenyl ethers (except anisole)
Dimethyl pyrroles

C_6H_5OH from Benzopyrans

Alkylpyrazines

, C_2Cl_2, C_3F_3H, $CH_3OP(OH)CH_3$,

CH_3Br, ClCC(O)F from

[M – 95]⁺ **Possible Precursor Compounds**

Methyl esters of aromatic sulfonic acids (SO_2OCH_3)

95 **Structural Significance**

Derivatives of furan carboxylic acids

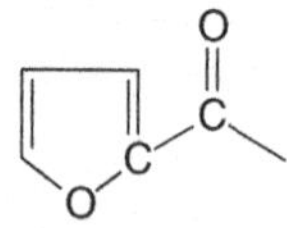

Hydroxymethylcyclohexanes

Methylfurans

Methyl esters of sulfonic acids

Segmented fluoroalcohols: $C_nF_{2n+2}CH_2CH_2OH$ (m/z 31, 95, 69, and 65)

, C_7H_{11}, $C=CH_2$, C_6H_9N, CF_3CN,

$-CF_2CH_2CH_2OH$, $CHF_2CF=CH-$, C_6H_4F, C_6H_7O,

C_3H_5ClF, $C_5H_7N_2$, , $C_2F_2O_2H$,

$CF_3CH=CH-$, C_7H_{11}, $C_3F_3H_2$,

96 **Structural Significance**

Aliphatic nitriles ($-(CH_2)_5CN$) (*m/z* 96, 82, and 110 suggest nitriles)

Piperidines

Esters of dibasic carboxylic acids

Dicycloalkanes

$(CH_3)_2SiF_2$, C_6H_5F, CH_3CH_2CFCl-

C_6H_8O:

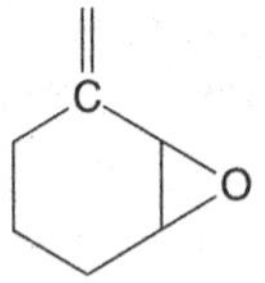

$[M - 97]^+$ **Possible Precursor Compounds**

Loss of ($CF_2{=}CFO$) from fluoroesters, fluoroamides, and fluoronitriles

(e.g., $CF_2{=}CFOCF_2CF(CF_3)OCF_2CF_2C(O)OCH_3$,

$CF_2{=}CFOCF_2CF(CF_3)OCF_2CF_2C(O)NH_2$,

$CF_2{=}CFOCF_2CF(CF_3)OCF_2CF_2CN$)

97 **Structural Significance**

Alkyl thiophenes

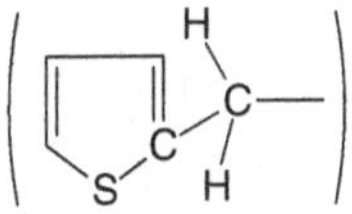

O- and F-containing compounds:

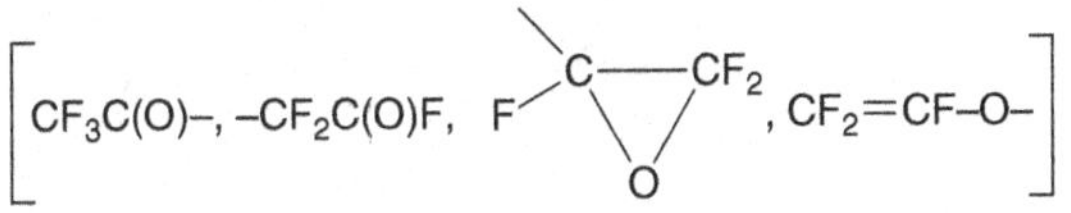

[$CF_3C(O)-$, $-CF_2C(O)F$, F–C(–)–CF_2 (epoxide, O), $CF_2{=}CF{-}O-$]

Aliphatic nitriles ($C_6H_{11}N$) (cyclohexyl C=NH)

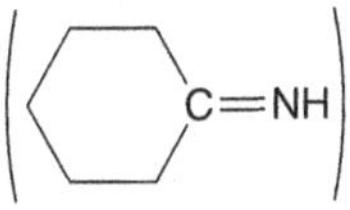

$CF_3{-}N{=}N-$, 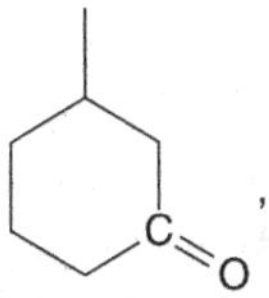 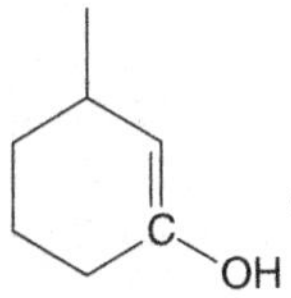 , , $CFCl{=}CF-$,

$CH_2{=}CHCH_2CH(CH_3)C-$, C_7H_{13}, C_2F_2Cl

CH_3CCl_2-, $CH_3CH_2CH_2CH{=}CHC(O)-$

$CH_3P(O)Cl$, C_7H_{13}, $CH_3P(OH)_3$, $CH_2ClCHCl-$

[M − 98]$^+$ **Possible Precursor Compounds**

Loss of $-(C_7H_{14})$

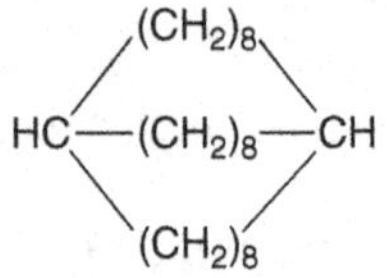

HC—$(CH_2)_8$—CH (bridged by two further $(CH_2)_8$ chains)

98 **Structural Significance**

Piperidine alkaloids (*N*-alkyl)

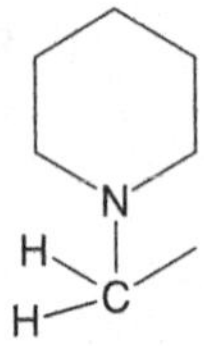

Dicarboxylic esters
Alkyl thiophenes
Dicarboxylic acids (e.g., palmitic and fumaric acids)
Bis(hexamethylene)triamine

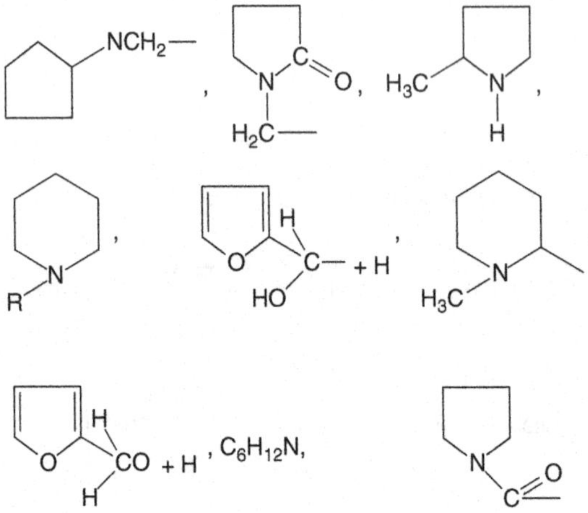

NCH_2—, H_2C—, H_3C, N, H, R, N, H, C— + H, HO, H_3C, N, H, CO + H, H, $C_6H_{12}N$, N, C=O

[M − 99]$^+$ **Possible Precursor Compounds**

Loss of $(CD_3)_3SiOH$ from deuterated TMS derivatives

99 **Structural Significance**

Maleates (>methyl)*

---

*Look for a fragment ion with *m/z* 29 and an ion with *m/z* 99 + 28 = 127 for diethyl-; a fragment ion with *m/z* 41 and 99 + 40 = 139 for dipropyl-; and so forth.

N-TFA, *n*-butyl aspartic acid
Acrylates (*m/z* 99 and 55 suggest glycol acrylates)
Isocyanates (C_5H_9NO) (also look for *m/z* 56)
Ethylene ketals of cyclic compounds (e.g., steroids)
Amyl ketones ($CH_3(CH_2)_4C(O)$)
Pentafluorobenzene
γ-Lactones, $CH_2{=}CH{-}C(O)OCH_2CH_2$
$CF_3OCH_2{-}$
CF_3CHOH, C_5F_2H, $CH_3C(O)C_2H_4C(O)$, C_5H_9NO,

, $C_6H_{11}O{-}$,

$C_6H_{13}N$, , ,

$CH_2{=}CHC(O)NHCH_2NH{-}$,

(Methyl-8® derivatives of amides) → $-C(O)N{=}CH{-}N(CH_3)$

[M – 100]⁺ **Possible Precursor Compounds**
Loss of C_2F_4 from fluorine compounds

100 **Structural Significance**
Perfluoroalkenes and perfluorocycloalkanes (C_2F_4)
$(C_2H_5)_2NC(O)$, $-C_5H_{11}CHNH_2$, $C_4H_9C(O)CH_2 + H$,

$+ H$, C_5H_5Cl, $(CH_3)_2SiN_3$, $(CH_3)_2CHN(CH_3)C(O)$

m/z 86, 100, 114, 128, 142, 156, etc. suggest diamines. Look for a large *m/z* 30 peak.

[M – 101]⁺ **Possible Precursor Compounds**

Loss of $-C(CH_3)_2C(O)OCH_3$

N-TFA, *n*-butyl amino acids

Loss of $(-C(O)OC_4H_9)$

101 **Structural Significance**

Ethyl esters $(-CH_2CH_2C(O)OC_2H_5)$

Butyl esters $(-C(O)OC_4H_9)$

Succinates (e.g., dicyclohexyl succinate)

Hexafluoropropylene oxide (HFPO) dimers and trimers

Fluorochloro compounds having $CFCl_2$

Malonic acids $C(O)CH_2C(O)OCH_3$

Cyclic sulfides

$C_4H_5O_3$, CF_3CFH-, $CFCl_2$, CHF_2CF_2-, $C_6H_{13}O$, $C_5H_9O_2$, $CH_3(CH_2)_4CHOH-$, $CH_3SC(O)CH{=}CH-$, $-CH_2CH_2OC(O)C_2H_5$,

$CH_3C(O)OCHC_2H_5$, (H₃C)(H₇C₃)(HOH₂C)C—, PSF_2, CF_3S, PCl_2,

(H₃C)(H₃C)(H₃COC(=O))C—, tetrahydropyran-2-yl-O—

[M – 102]⁺ **Possible Precursor Compounds**

Loss of acetic anhydride from sugar acetates

102 **Structural Significance**

Quinolines

Long-chain propyl esters $(C_3H_7OC(O)CH_2+H)$

$C_6H_4(CN)-$, $C_6H_5C{\equiv}CH$ (e.g., phenylmaleic anhydride)

$CH_3OC(O)C(CH_3)_2+H$
(e.g., $CH_3OC(O)C(CH_3)_2C(O)OCH_3$)

$(CH_3)_3SiNHCH_2-$, $-CHNH_2-C(O)OC_2H_5$

[M – 103]⁺ **Possible Precursor Compounds**

Loss of $[CH_2OSi(CH_3)_3]$ from the molecular ion of TMS Ketohexoses and 1,2,6-hexanetriol

103 **Structural Significance**

Alkyl indoles (characteristic rearrangement ion)

Cinnamates $(C_6H_5CH{=}CH-)$

Valerates (>methyl)

Double rearrangement of protonated carboxylic acids

Trimethylsilyl derivatives [$CH_2{=}O^+Si(CH_3)_3$] of primary aliphatic alcohols*
[1,1-Diethoxyalkanes ($(C_2H_5O)_2CH-$]
$(CH_3)_2CHSCH_2CH_2-$, C_8H_7, $C_4H_9OC(O)+2H$,

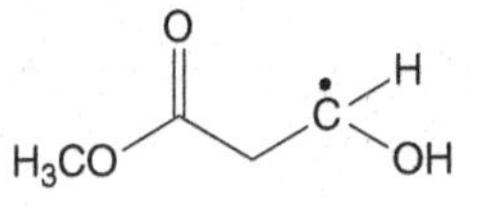

**, ,

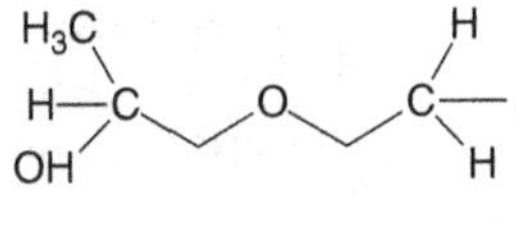

$CH_3CH_2C(OCH_3)_2-$

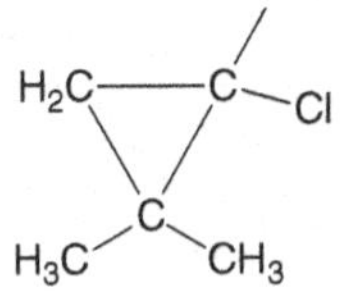

104 **Structural Significance**
Tetralins (tetrahydronaphthalene)
m/z 104 and 158 (see phenylcyclohexene)
CF_3Cl, N_2F_4, SiF_4, C_8H_8, $C_2H_5CHONO_2$,

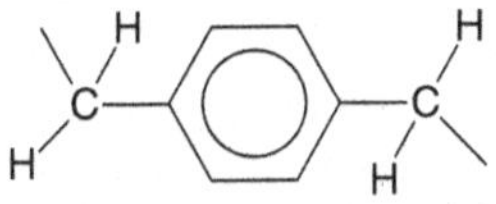

, $CH_3SCH_2CH_2CHNH_2-$,

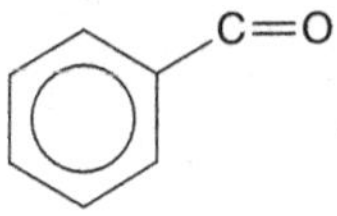

, or

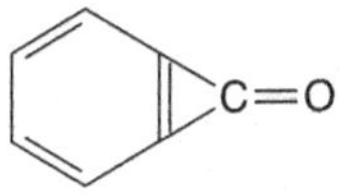

m/z 104 and 91 or 117 are characteristic of some alkybenzenes

$[M-105]^+$ **Possible Precursor Compounds**
Esters of aliphatic dibasic carboxylic acids
Benzoin ethers (e.g., benzoin isopropyl, benzoin isobutyl)

Loss of from

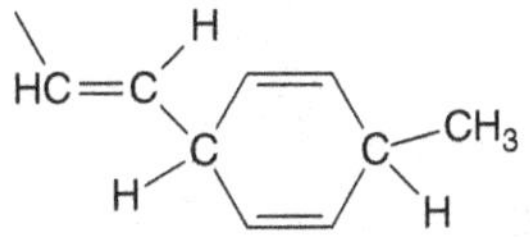

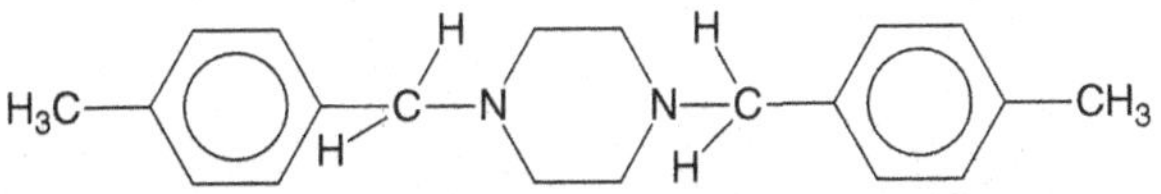

---

* Look for homologs in the spectra of 2° and 3°, for example, *m/z* 117 (without a peak at *m/z* 103), *m/z* 131, etc.

** From dimethyl malate.

Loss of $[CH_3+(CH_3)_3SiOH]$ from TMS derivatives of aldohexoses, etc.

105 **Structural Significance**

Benzoates

Aromatic alcohols

Azobenzenes

$(C_2H_5)_2SiF$, $C_6H_5C{=}O$, $C_6H_5N{=}N$, C_4F_3, $(CH_3O)_2SiCH_3$
$C_6H_5CHCH_3$, $C_6H_5CH_2CH_2$, C_7H_7N,

[structures: H_3C–C_6H_4–CH_2–, Cl(H)C=C(–)–C(=O)–OH]

$[M-106]^+$ **Possible Precursor Compounds**

Esters of long-chain dibasic carboxylic acids

106 **Structural Significance**

Substituted *N*-alkylanilines $(C_6H_5NHCH_2–)$

Alkyl pyridines: [2-pyridyl–CH_2–CH_2–], C_7H_8N, C_4F_3H,

[3-methyl-5-pyridyl–CH_2–], H_2N–C_6H_4–CH_2–,

[C_6H_5–CH(NH_2)–], [C_6H_5–NH–CH_2–],

$(CH_3)_2NP(O)CH_3$, [2-pyridyl–C(=O)–]

$[M-107]^+$ **Possible Precursor Compounds**

Loss of $(CH_3C_6H_4O–)$ as in $(CH_3C_6H_4O)_3P$

$(CH_3+C_3H_9SiF)$ from TMS derivative of fluorinated sugars, for example,

$$F_3C-CH(OH)-C(CF_3)(OH)-C(CF_3)(OH)-CH(OH)-CF_3$$

107 **Structural Significance**

Alkyl phenols ($HOC_6H_4CH_2$–), pyrethrin II (MW = 372 Da)
See benzoin isopropyl ether (MW = 254 Da)

$C_6H_4CH_2O$–, $C_6H_5CH(OH)$–, $-CF_2C_4H_9$,

$-CH_2CH_2OCH_2CH_2Cl$, $C_6H_5SiH_2$
$CH_3(CH_2)_3CH{=}CH{-}C{\equiv}C$– or $CH_3(CH_2)_3{-}C{\equiv}C{-}CH{=}CH$–
$C_6H_5OCH_2$–, $CH_3C_6H_4O$–, $-C_6H_4OCH_3$

$[M-108]^+$ **Possible Precursor Compounds**

Loss of H_2SiPhH from $Ph-SiH_2-SiH(Ph)-SiH_2-Ph$

$CH_3NH-C_6H_4-OH$

108 **Structural Significance**

Tolyl ethers ($CH_3C_6H_4O+H$)
Alkyl pyrroles
Benzothiazoles (C_6H_4S)
$H_2NC_6H_4O$–, C_6H_5SH, $C_6H_5CH_2O$– + H,

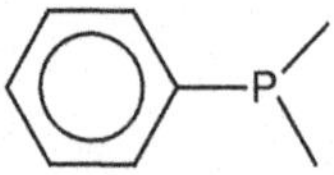

*

$C_6H_5CH_2OH$,

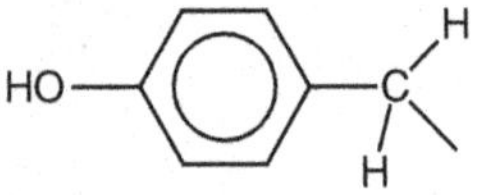

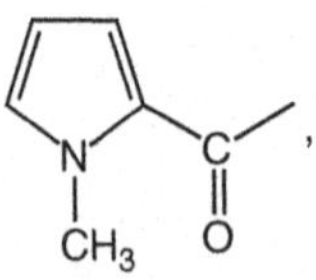

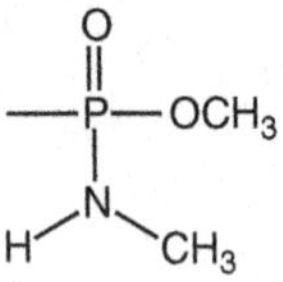

109 **Structural Significance**

Purines ($C_5H_7N_3$)

Certain perfluoroketones (C_3F_3O)

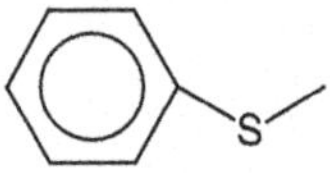

C_3F_2Cl, HOC_6H_4O–, $CH_3OCF{=}CFO$–,

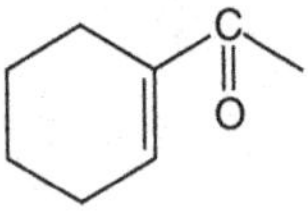

, $CF_2{=}CFC(O)$, $C_4H_4F_3$, C_8H_{13}, C_3F_2Cl,

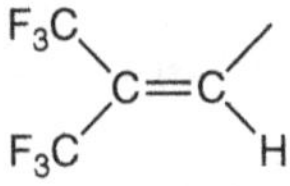

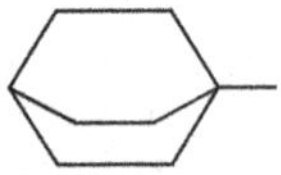

, $CH_3CCl{=}CCl$–,

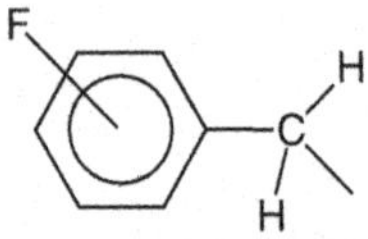

+ H,

–$CCl_2CH{=}CH_2$, 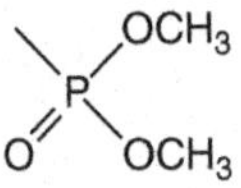 $HOP(O)OC_2H_5$

110 **Structural Significance**

Esters of aliphatic dibasic carboxylic acids

Aliphatic cyanides** $(CH_2)_6CN$, $C_7H_{12}N$, C_6F_2, C_2Cl_2O, $C_7H_{10}O$, $CH_3OPOH(O)CH_3$, $C_4H_5F_3$,

---

* See triphenylphosphine, MW = 262 Da.

** *m/z* 110, 96, and 82 suggest nitriles.

$CH_2{=}CCF_3{-}CH_3$,

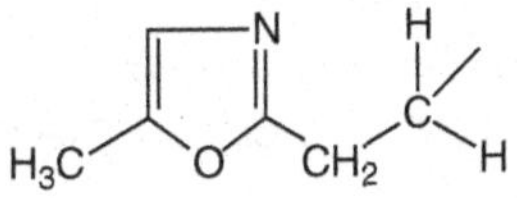

[M – 111]$^+$ **Possible Precursor Compounds**

Suggests an octyl group (C_8H_{16})

111 **Structural Significance**

Methyl alkyl thiophenes

Adipates, $C_6H_7O_2$

Esters of aliphatic dibasic carboxylic acids

Long-chain methyl esters of carboxylic acids

$CH_3CH_2CCl_2{-}$, $-C_6H_5Cl$, 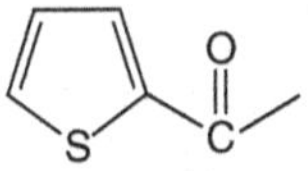 C_8H_{15},

$C_7H_{11}O$, CFBrH, $CF_3CH(CH_3)CH_2{-}$, $C_3H_5Cl_2$

$C_2H_5OP(O)C_2H_5$, $CHCl_2C(O)$, $-CF_2CH_2C(O)F$

$CF_3C(O)CH_2{-}$, $CF_3C{=}N{-}NH_2$,

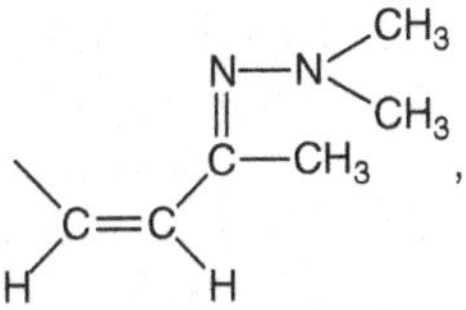

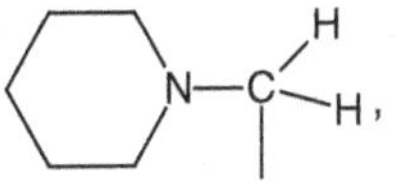

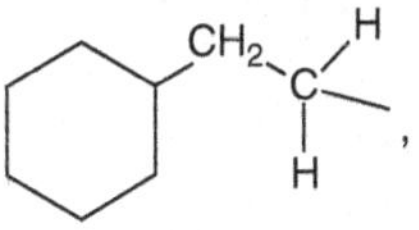

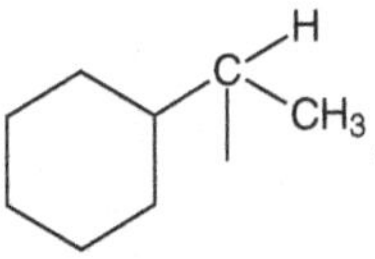

$-CF_2C(Cl){=}CH_2$, $C_2H_5P(OH)_3$

[M – 112]$^+$ **Possible Precursor Compounds**

Loss of

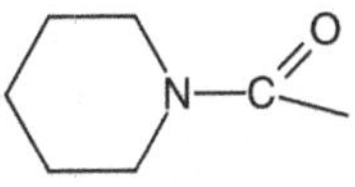

R′–H (suggests an octyl group)

Loss of $(CH_2)_5NCO$ from isocyanates

112 **Structural Significance**

N-Alkylcyclohexylamines

Dibutylhexamethylenediamine
2,7-Dioxo-1,8-diazacyclotetradecane
$C_7H_{14}N$, C_3F_4, $C_6H_{12}Si$, C_6H_5Cl, $C_6H_8O_2$, $(CH_3)_2P(O)Cl$

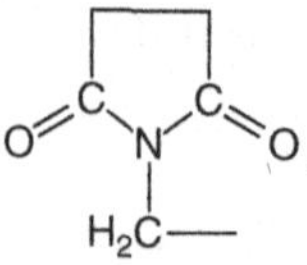

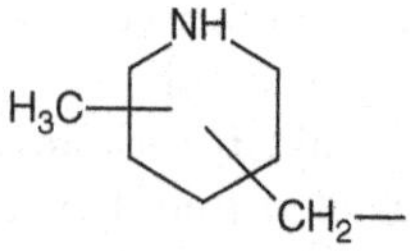

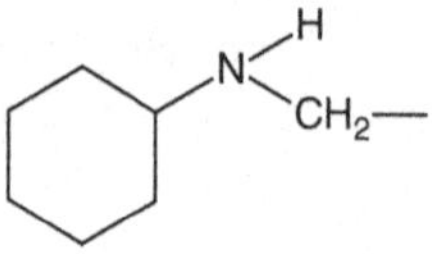

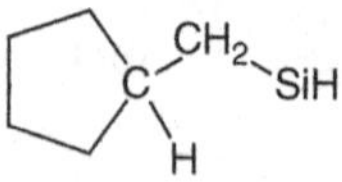

113 **Structural Significance**
Methyl esters of Δ^2-fatty acids
$-CH_2CH_2CH{=}CHC(O)OCH_3$
Methacrylates ($CH_2{=}C(CH_3)-C(O)OCH_2CH_2-$)
Certain diketones
N-Methylsuccinimide
Cyclic ethers ($C_7H_{13}O$)
$C_4H_9C(O)CH_2CH_2-$, $(CH_3)_2CHC(O)CH_2C-$,
$CH_3CH_2C(O)CH_2CH_2C(O)$,

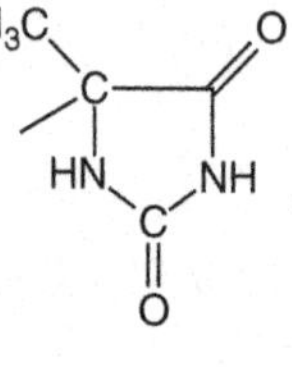

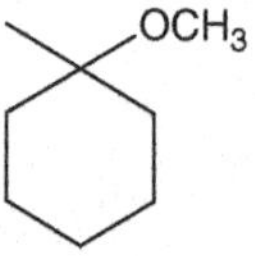

$CH_3C(O)CHCH_2C(O)CH_3$,
$-CH_2C(CH_3){=}CH-C(O)OCH_3$, $CF_3CH{=}CF-$,
$-CF_2CH{=}CF_2$, C_2FCl_2, $CH_2ClCHClCH_2-$,
C_2F_2ClO, C_3F_4H, $C_6H_3F_2$, $C_6H_{13}C(O)$,
C_8H_{17}, $C_7H_{15}N$, $C_6H_{13}N_2$, $C_6H_{11}NO$, $C_6H_9O_2$,
$CF_3C(O)NH_2$,

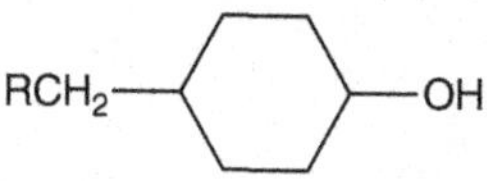

RCH₂–(cyclohexane)–OH, $C_7H_{13}O$, CH_3SiCl_2

$CF_2{=}CFS{-}$, $CH_2{=}CCH_3C(O)NHCH_2NH{-}$

114 **Structural Significance**

Steroid alkaloids:

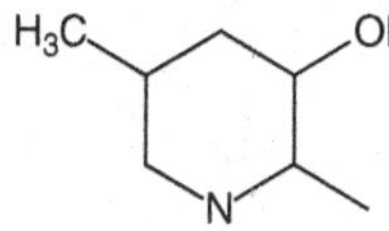

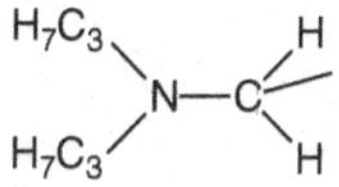

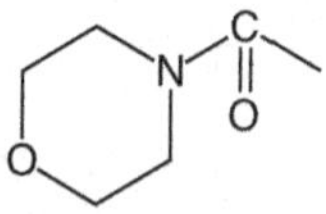

$C_7H_{16}N$, $C_6H_4F_2$, $CHCl{-}CClF$, $CF_3C(O){-}OH$, $-CF_2N{=}CF_2$, $-CH_2CH_2C({=}NOH)C_3H_7$, $-CF_2{-}CF{=}NF$, C_2F_4N

115 **Structural Significance**

Sugar acetates

Glutarates

Alkyl indoles (other than methyl)

C_9H_7 ion from naphthalenes (may produce a double-charge ion at *m/z* 57.5)

Creatinine-TRITMS (MW = 329 Da)

Quinolines (C_9H_{11})

$-C(O)CH_2{-}C(O)OC_2H_5$, $C_2H_2FCl_2$, $CHCl_2CHF{-}$, $C_7H_{15}O$, $CFCl_2CH_2{-}$, $(C_2H_5)_3Si{-}$, $(CH_3)_2CH{-}C(O)OCH_2CH_2{-}$,

$(CH_3)_3CSi(CH_3)_2{-}$, $CH_3CHCH{=}C(OH)OC_2H_5$,

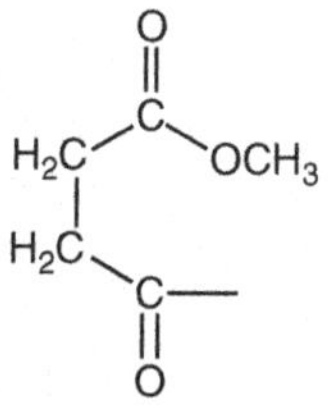

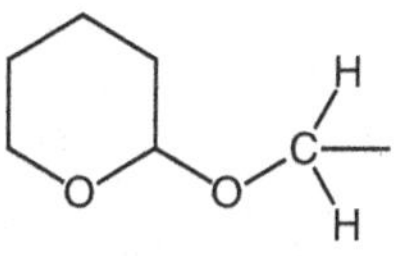

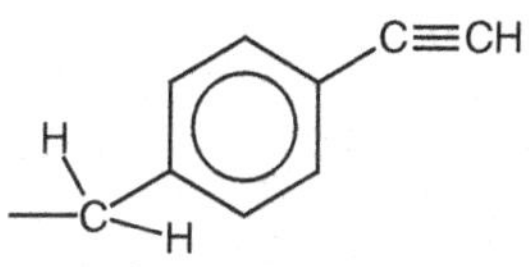

[M – 116]$^+$ **Possible Precursor Compounds**

Loss of $CH_2{=}CHC(O)O(CH_2)_2OH$ from some acrylates

116 **Structural Significance**

Amino alcohols and ethers ($C_6H_{14}NO$)

N-TFA, *n*-butyl glutamic acid

$-CH_2C_6H_4CN$, C_2F_3Cl, $-CF_2CFCl$,

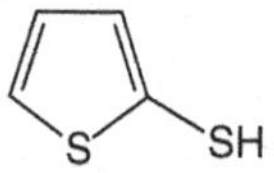

$C_3H_7C(O){-}OC_2H_5$, C_5H_5OCl,

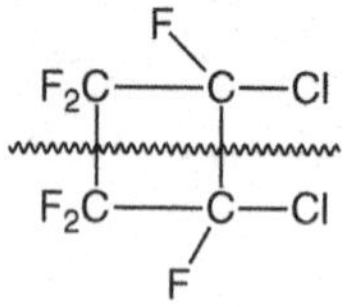

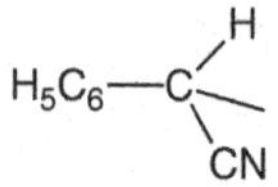

[M – 117]$^+$ **Possible Precursor Compounds**

Loss of CCl_3 (e.g., DDT)

Loss of $-COOTMS$

Loss of CF_3CHFO from $CF_3CHFOCF_2CF(CF_3)OCF_2CF_2CN$

117 **Structural Significance**

Alkyl indans

Styrenes (C_9H_9), caproates (>methyl)

CCl_3, C_2F_4OH, C_5F_3, $C_6H_5CH_2CH{=}CH-$,
$CH_3C_6H_4CH{=}CH-$, $-CF_2OCHF_2$, $-CF_2CHClF$,
$POCl_2$, $C_6H_5CH{=}CH{-}CH_2-$, C_2HF_3Cl,
$CF_2ClCHF-$, $CH_3OCHCH_2C(O)OCH_3$, C_2HF_4O,
$-C(O)OSi(CH_3)_3$, $-CH_2CH_2OSi(CH_3)_3$

Intense peaks at *m/z* 117 and 104 (characteristic of alkyl benzenes)

If a peak is at *m/z* 117 in the spectrum of a TMS derivative, it probably indicates that a carboxyl group is present.

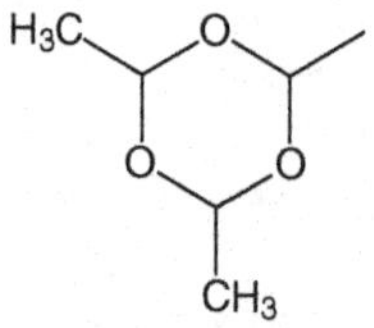

118 **Structural Significance**

$C_4H_9OCH_2CH_2OH$, $(CH_3)_2NSi(CH_3)_2O–$

$N{=}C$ CH_3,

H_3C $C{=}C$ H H from H_3C C—CH_2 118 C—S=O F_3C CF_3 O

119 **Structural Significance**

Fluorocarbons ($CF_3CF_2–$)

Toluates ($CH_3C_6H_4C{\equiv}O$)

Alkyl benzenes: for example,

CH_3 C CH_3

Pyrrolizidine alkaloids

N N N NH

C H O , H_3C C H O ,

—CH_2

C CH_3 O

120 **Structural Significance**

Salicylates

Pyrrolizidine alkaloids

Flavones

Isoflavones

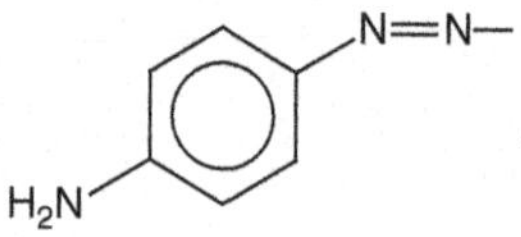

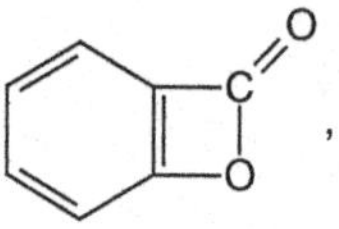

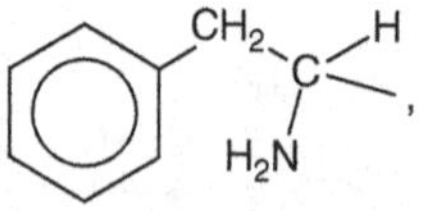

$C_5H_{13}PO$*,

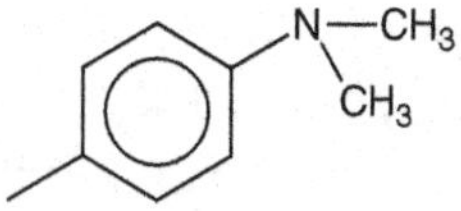

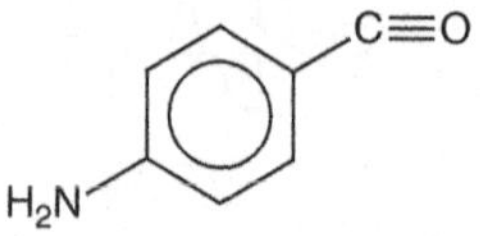

$C_6H_5CH_2CHNH_2$

121 **Structural Significance**

Salicylates (also see *m/z* 138)

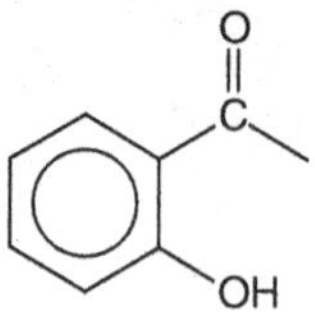

Alkyl phenols

Terpenes

$CH_3OC_6H_4CH_2$–, C_8H_9O, $(CH_3O)_3Si$–, $(C_2H_5)_2SiCl$, $(CH_3)_2CBr$

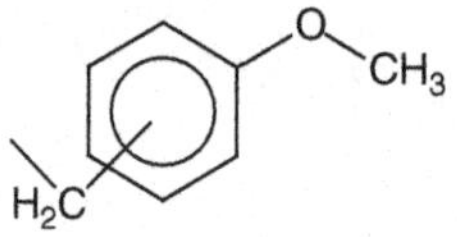

C_3H_7S–$CHSH$, $C_6H_5CHOCH_3$

$(CH_3O)_2Si(CH_3)O$–,

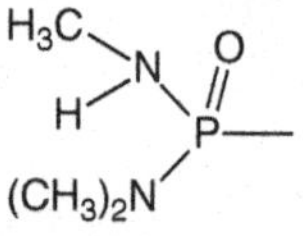

$HOC_6H_4CHCH_3$

* From tributylphosphine oxide.

122 **Structural Significance**

$C_6H_5CO_2+H$, Δ^3-ketosteroids

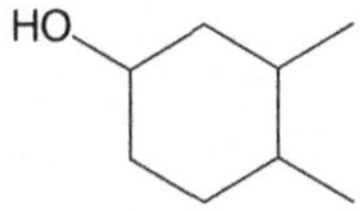

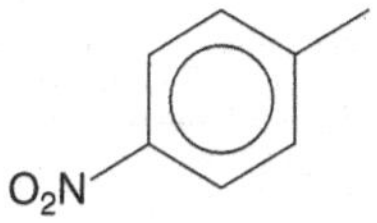

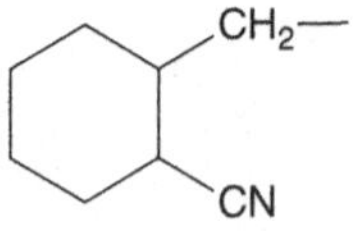

123 **Structural Significance**

Pyrethrin I

Benzoates (by double hydrogen rearrangement. The ester must be ethyl or higher.)

Trichlorobutenes

$CF_3CH{=}CHC(O)-$, $(HO)_2C_6H_3CH_2-$, $C_6H_5CH_2S-$

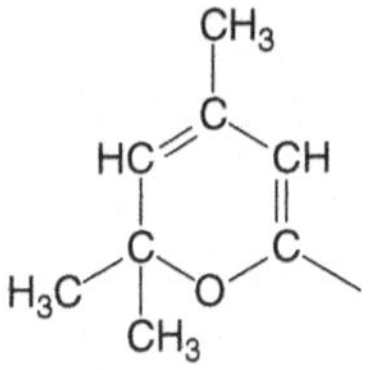

$CH_3C_6H_4S-$, $FC_6H_4C(O)-$, $C_8H_{11}O$,

$C_2H_3SO_4$, $-(CH_2)_2SO_3CH_3$,

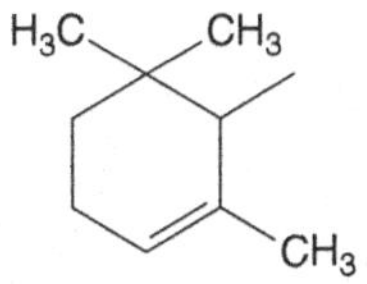

$C_6H_5SCH_2-$, $CH_2FC_6H_4CH_2$, $-OS(O)_2OCH_2CH-$

124 **Structural Significance**

Suggests an alkaloid in certain cases

Aliphatic cyanides (also may see *m/z* 96, 110, 138, 152, etc.)

Testosterone (also may see *m/z* 124, 288, 246)

C_4F_4, $(CH_3O)_2P(O)CH_3$, $C_8H_{12}O$,

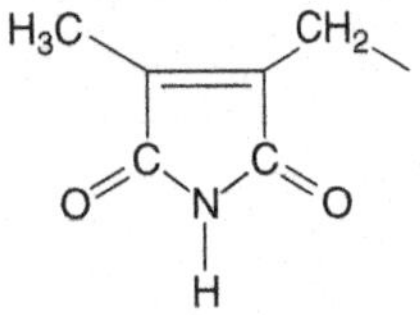

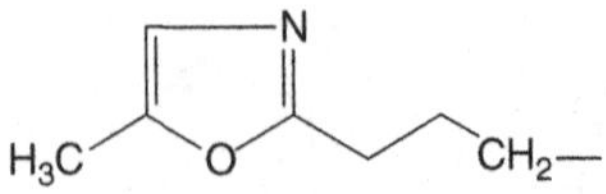

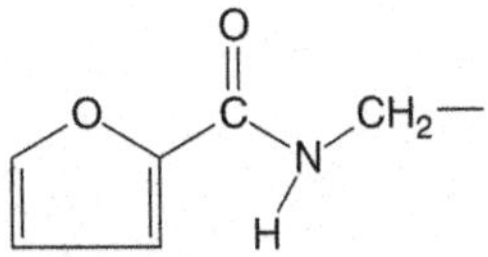

125 **Structural Significance**

C_6H_5SO (e.g., the base peak in diphenylsulfone) (should also observe *m/z* 141)

C_4F_4H, $ClC_6H_4CH_2$–, C_6H_5CHCl–, $(CH_3O)_2P{=}S$, $C_7H_9O_2$,

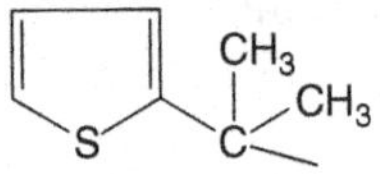

126 **Structural Significance**

$CF_3C(O)NHCH_2$–,

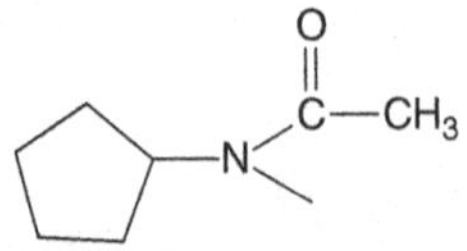

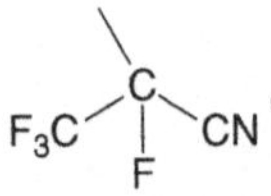

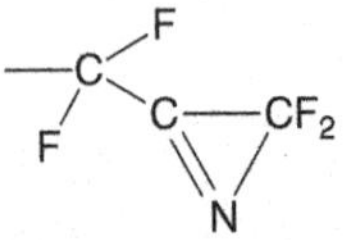

–CF_2CF_2CN,

$C_4F_4H_2$,

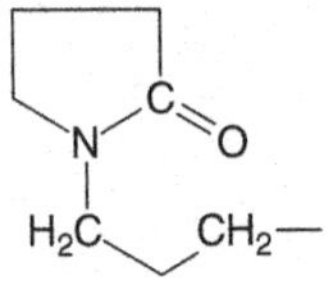

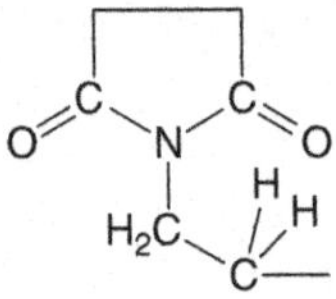

–$(CH_2)_6NCO$,

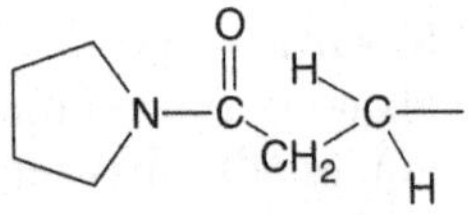

$[M - 127]^+$ **Possible Precursor Compounds**

Iodides

127 **Structural Significance**

Iodo compounds

Naphthyl compounds:

I, CF_3SCN, $C_4F_4H_3$, $CH_2{=}CHCF_2CF_2{-}$,

$C_6H_5{-}C(CN){=}C{=}$ $C_8H_{15}O$, C_9H_{19}, $C_6H_5CF_2{-}$

$C(O)OCH_2CF_3$, $C_7H_{15}N_2$, R–C(H)(H)–C–H … C–OH,

$(H_3CO)_2P(OH)_2$

[M – 128]+ **Possible Precursor Compounds**

Loss of HI from iodides

128 **Structural Significance**

Quinolines

Naphthalene

FC—CF2 / FC—O

$-CH_2CH_2C(O)N(C_2H_5)_2$, C_3F_4O, $-CF_2CF_2C(O)-$

C_3F_3Cl, HI, $C_2N_2F_4$, $C_3H_3Cl_2F$, ClC_6H_4O+H,

$-CH_2CH(CH_3)CH_2N(CH_3)C_3H_7$,

$-CH_2CH(CH_3)CH_2N(CH_3)C(O)CH_3$

[M – 129]+ **Possible Precursor Compounds**

$(CH_3)_3Si\overset{+}{O}{=}CH{-}CH{=}CH_2$

(from the A ring in TMS derivatives of steroids)

129 **Structural Significance**

Adipates:

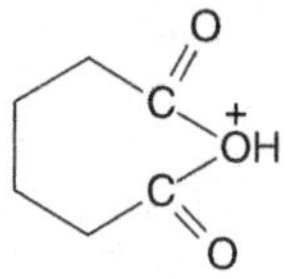

$(CH_3)_3Si{-}O{-}CH{-}CH{=}CH_2$

C_2FCl_2O, C_6F_3, C_2Cl_3, $C_3H_4FCl_2$, $C_6H_5CH{=}CHCN$, $-CF_2Br$, $C_{10}H_9$, $CF_3CH{=}CCl{-}$, $H_2N(CH_2)_6NHCH_2{-}$ $CH_3OC(O)C(CH_3)_2C(O)$, $-CH_2C(O){-}CH_2CH_2C(O){-}OCH_3$

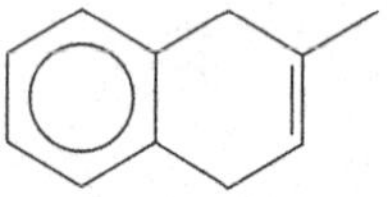

130 **Structural Significance**

Indoles

C_6H_4FCl, $(CH_3)_2CH{-}CH{=}C(OH)OC_2H_5$,

$C_6H_{10}O_3$, C_9H_8N, $-CH_2{-}N$

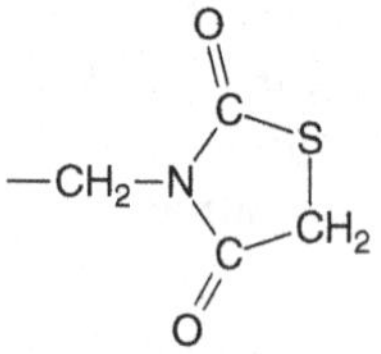

$-CF_2{-}CF{=}NCl$, 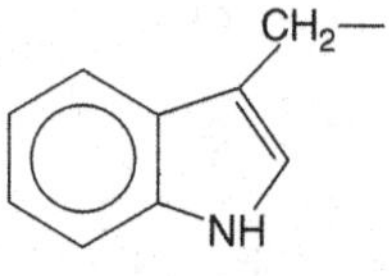 ,

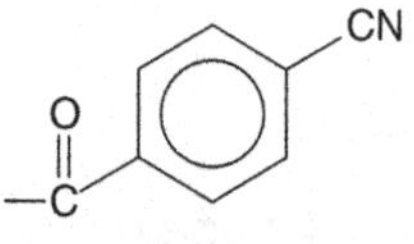

$(CH_3)_2NCH{=}COC_2H_5$

131 **Structural Significance**

Cinnamates ($C_6H_5CH{=}CHC{\equiv}O$)

Fluorocarbons (C_3F_5)

$-CF_2{-}CF{=}CF_2$, $CF_3CF{=}CF{-}$, $CF_3C{=}CF_2$, $CHF_2CF_2OCH_2{-}$, $C_6H_2F_3$, $CH_3OCF_2CF_2{-}$

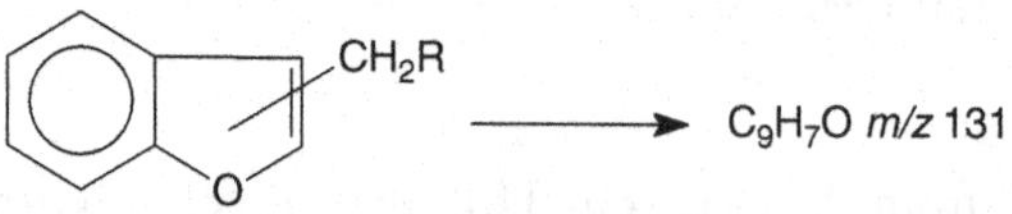

$\longrightarrow$ C_9H_7O m/z 131

132 **Structural Significance**

Benzimidazoles

$C_6H_3F_3$, $F_2C{=}CCl_2$

$(CH_3)_3SiOCH_2NCH_3$, $C_9H_{10}N$,

$CH_3P(O)Cl_2$

$[M - 133]^+$ **Possible Precursor Compounds**

133 **Structural Significance**

Acetanilides

$CH_3SCH{-}C(O)OC_2H_5$, $CH_3C_6H_4{-}C(CH_3)_2$

, $CF_3CH_2CF_2-$,

$CF_3CF_2CH_2-$, CHF_2CHFCF_2-, $HC{\equiv}C-$,

$-SiCl_3$

134 **Structural Significance**

$CH_3C_5H_4Mn-$

$CH_3C(CF_2)CH_2NHC(O)$, $-CF_2NHCF_3$

$C_6H_5C(O)NHCH_2$

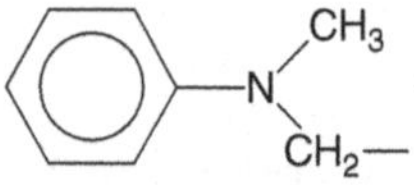

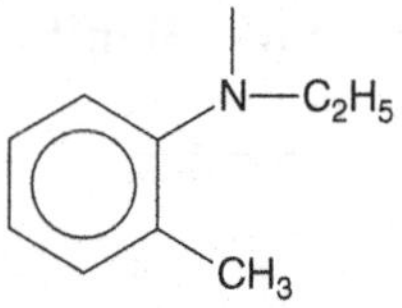

135 **Structural Significance**

Adamantanes ($C_{10}H_{15}$), C_6–C_{18} *n*-alkylbromides $(CH_2)_4Br$ (with an isotope peak at *m/z* 137)

$C_6H_5C(OH)C_2H_5$, $C_6H_5C(O)OCH_2$–

PTH-amino acids,

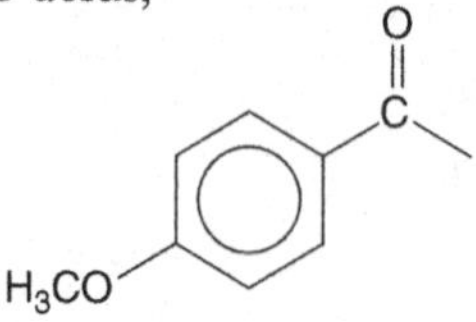

H₃CO

–CF_2CF_2Cl, CF_3CFCl–, CF_3CF_2O–,

CF_3OCF_2–, C_8H_7S, $C_6H_5Si(CH_3)_2$, $C_6H_5Si(H)_2CH_2CH_2$–,

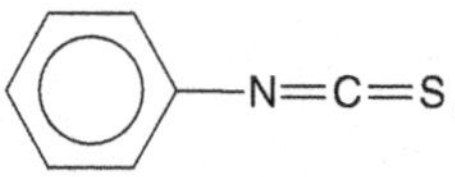

–N=C=S, H_2C ... CH_2—

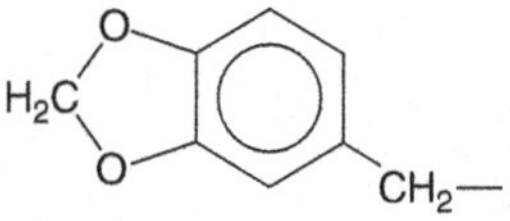

$[(CH_3)_2N]_2P(O)$

136 **Structural Significance**

$O_2N(C_4H_6)CH_2$–, C_5F_4, $CH_2N(CH_2CH_2CN)_2$ (also should observe peak at *m/z* 54)

$[M - 137]^+$ **Possible Precursor Compounds**

$C_9H_{13}O$

137 **Structural Significance**

Decalins

C_5F_4H, –$(CH_2)_3SO_3CH_3$, $(C_2H_5O)_2P(O)$

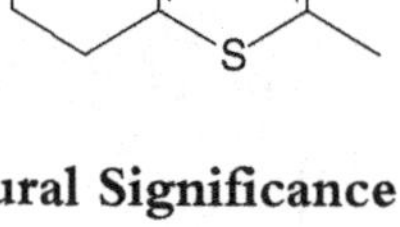

138 **Structural Significance**

Salicylate and $[M + H]^+$ *m/z* 139

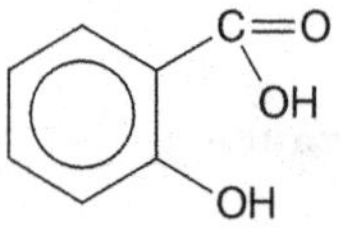

Nitriles $[(CH_2)_8CN]$ (also may see ions at *m/z* 96, 110, 124, etc.)

Base peak in dicyclohexylamine and nitroanilines

C_5H_5FeOH, C_2F_6, $(CH_2{=}CHCH_2)_2Fe$ (*m/z* 110, 124, 138, and 152 suggest nitriles; may observe $[M - 1]^+$ instead of $M^{+\bullet}$)

139 **Structural Significance**

Salicylates (protonated salicylic acid) also may have an intense peak at *m/z* 138, *N*-TFA, *n*-butyl serine

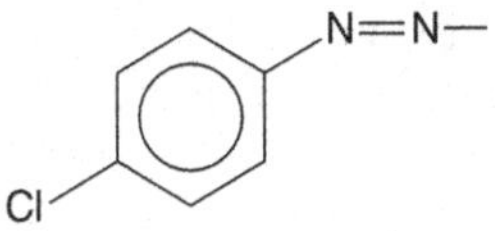

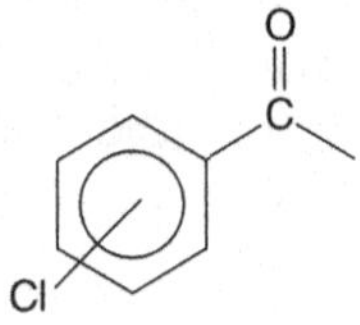

$C_5F_4H_3$, $CF_3CH(CH_3)CH_2C(O)$, $CF_2CF_2C(CH_3){=}C{-}$

140 **Structural Significance**

Methyl esters of dibasic carboxylic acids > dimethyl suberate (usually have a series of peaks at *m/z* 84, 98, 112, 126, 140, etc.)

N-TFA butyl ester of alanine

C_4F_4Cl, C_3Cl_3,

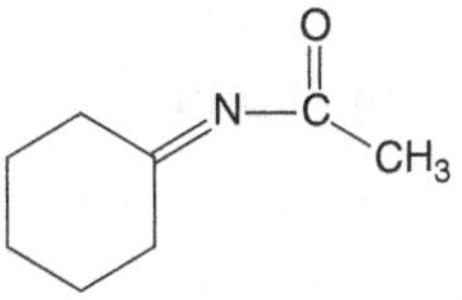

141 **Structural Significance**

Naphthalenes [$C_{11}H_9$ (alkyl naphthalenes should also have *m/z* 91)]

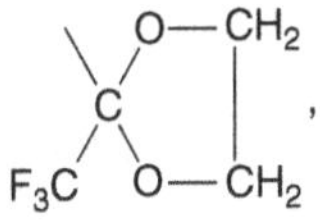

, $C_6H_5SS{-}$,

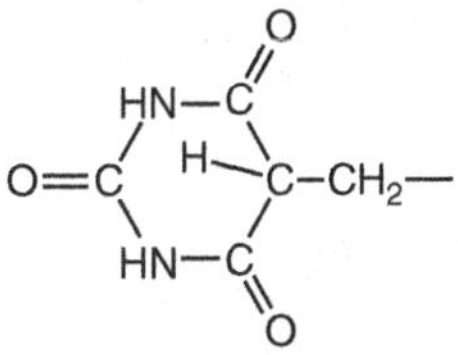

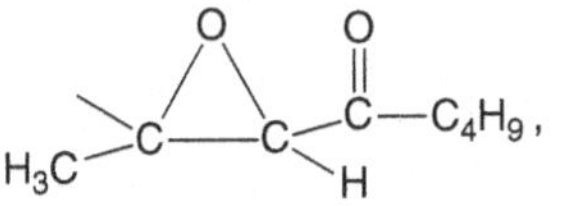

$-C_4H_9$, C_7F_3, $C_4H_7Cl_2O$,

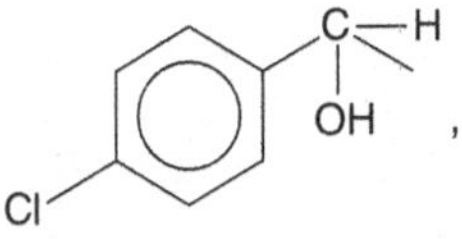

, $C_8H_{17}N_2$,

$C_6H_5SO_2$ (also should observe *m/z* 125)

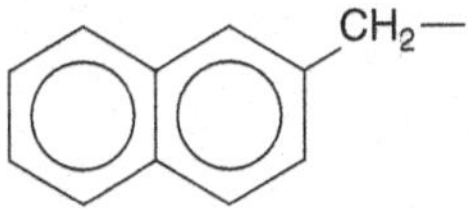

$-CF_2CF{=}CHC(O)F$, $-CF_2CF{=}CHCH_2SH$

142 **Structural Significance**
$(C_4H_9)_2NCH_2–$
$C_{10}H_8N$ (from quinolines)
$–CF_2CF_2N_3$

143 **Structural Significance**
Indole and indoline alkaloids
$–C(O)(CH_2)_4CO_2CH_3$ from dimethyl adipate
$(CH_3OC(O)C=CHC(O)CH_3)$
C_4F_5, $C_8H_{15}O_2$, $–C(Cl)=C(Cl)CH_2Cl$, $–CH=C(Cl)CHCl_2$,

$–C(Cl)=CHCHCl_2$, $(C_4H_9)_2SiH$,

O=C(–O–C_2H_5)–C(=CH_2)–CH_2–O–CH_2–

144 **Structural Significance**
Indoles

$(H_3C)_2C(NH–)–C(=O)–OC_3H_7$

$H_3C–CH(–)$– (indol-3-yl, NH)

[M – 145]$^+$ **Possible Precursor Compounds**
The loss of CF_3CF_2CN has been seen from a compound whose probable structure is $(CF_3)_2C=CFC(CF_3)=CFN(CH_3)_2$

145 **Structural Significance**
Tetralins

$CF_3OCH_2CF_2$

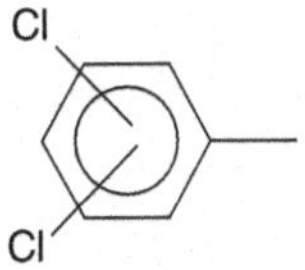

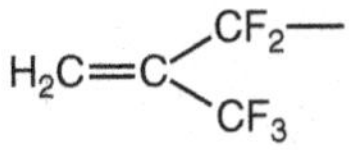

Cl, Cl, , $H_2C{=}C(CF_2-)(CF_3)$, CFClBr, $CF_3C_6H_4-$,

$CF_3CH{=}CHCF_2-$, $C_4F_5H_2$, , $CCl_3CH_2CH_2-$,

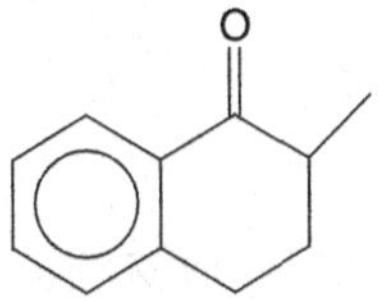

$-CCl_2CH_2CH_2Cl$, , ,

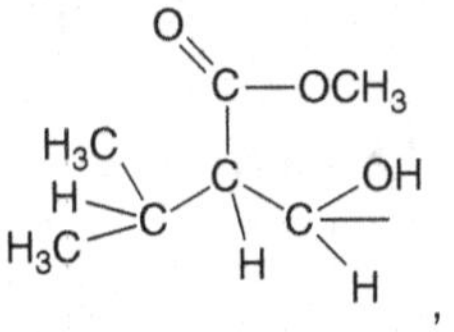

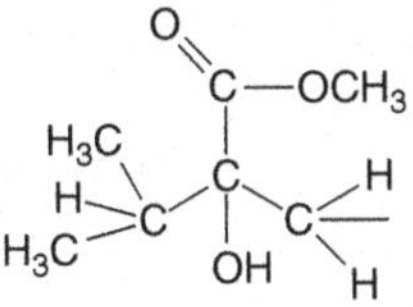

146 **Structural Significance**

$CF_3CF_2C{=}NH$

CFClBrH, $C_6H_5CF_3$, isomers of

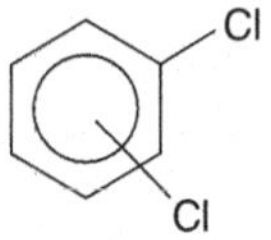

Cl, Cl

147 **Structural Significance**

$CF_3C(O)CF_2-$, C_3F_5O, $CF_3CCl{=}CF-$, C_3F_4Cl

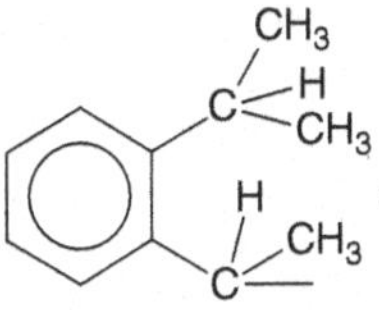

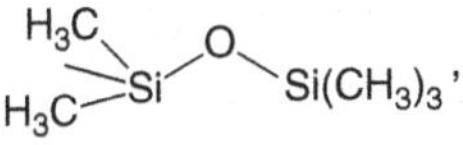

, $(H_3C)_2Si{-}O{-}Si(CH_3)_3$, $C_5H_{11}OSiO_2$

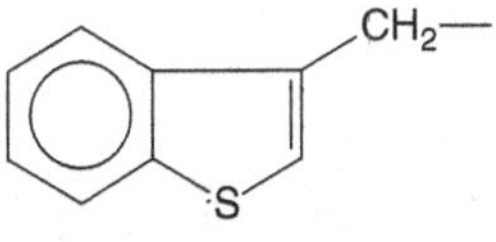

$C_{10}H_{11}O$

148 **Structural Significance**

Aminopurines

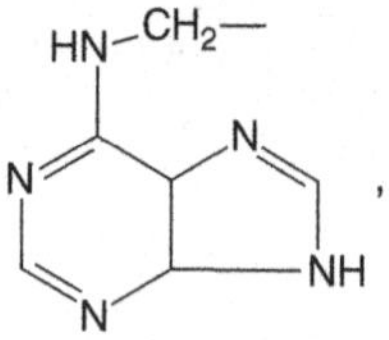

C_6F_4, C_2FCl_3, $C_6H_3F_2Cl$,

$C_6H_5CH_2CHN(CH_3)_2$,

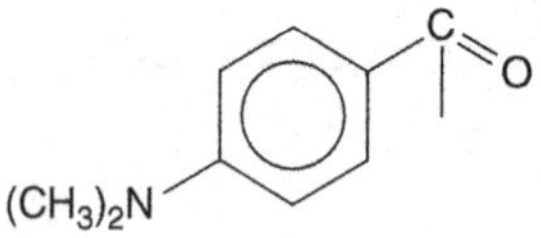

149 **Structural Significance**

Esters of phthalic acid*:

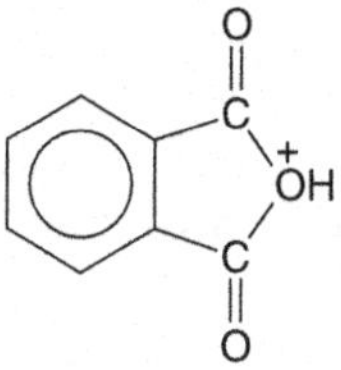

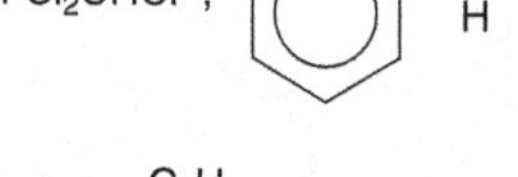

$-C_6H_4-CO_2C_2H_5$

$CF_3CH_2OCF_2-$, $C_2H_5C_6H_4C(O)-$, C_6HF_4, $-CF_2SO_3F$,

FCl_2CHCl-, , $-CF_2CH_2CF_2Cl$,

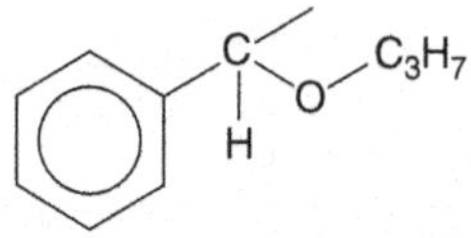

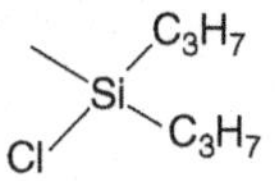

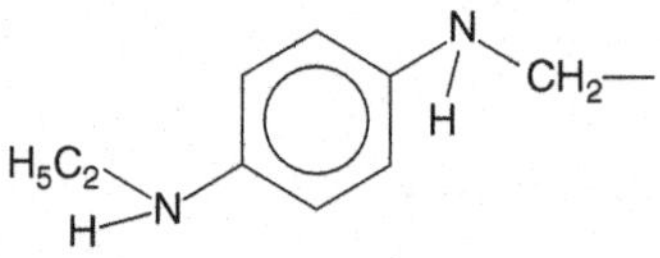

,

$C_9H_{13}N_2$

150 **Structural Significance**

Steroid alkaloids

$C_3F_6H_2F_4$, $CF_3CF{=}CF_2$, C_3F_6

$[M - 151]^+$ **Possible Precursor Compounds**

Loss of $CF_3CFClO-$

---

* *m/z* 149, 167, and 279 suggest dioctyl phthalate or di(2-ethylhexyl) phthalate.

151 **Structural Significance**

$C_9H_{15}C(O)$, $C_5H_{11}CH{=}CH{-}CH{=}CH{-}C(O)$

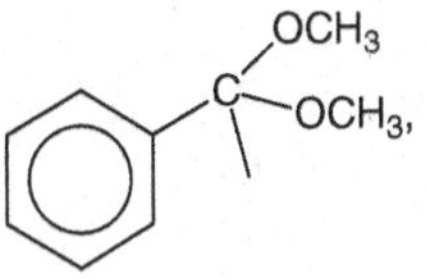

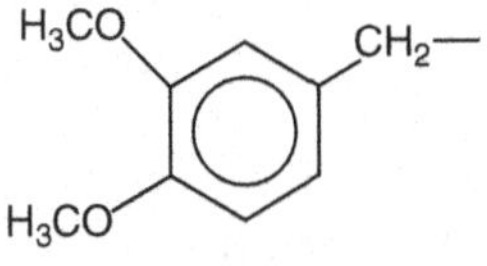

OCH_3, OCH_3, H_3CO, CH_2—, H_3CO

$(CF_3)_2CH-$, CF_3CCl_2-, $CFCl_2CF_2-$, $CF_2ClCFCl-$, $C_2F_3Cl_2$, $C_6H_3F_4$

152 **Structural Significance**

$(CF_3)_2N-$, C_2F_5NF

$[M - 153]^+$ **Possible Precursor Compounds**

N-TFA, *n*-butylthreonine

153 **Structural Significance**

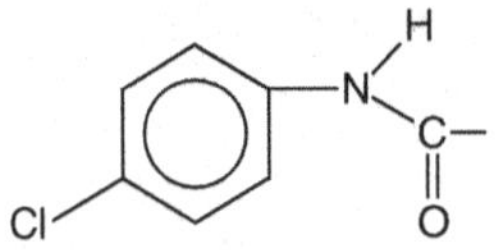

H, N, C—, O, Cl (formation of isocyanate gives *m/z* 153)

154 **Structural Significance**

O, C, H, O—, F_3C, N, C, CH_3, CH_3

S—, O_2N

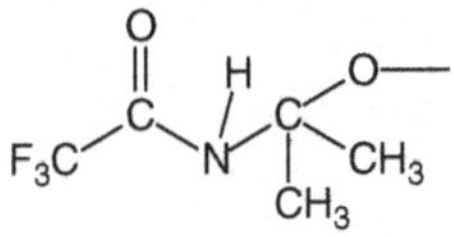

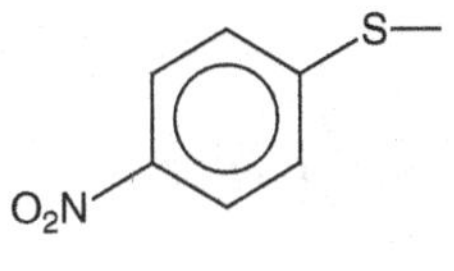

$[M - 155]^+$ **Possible Precursor Compounds**

Loss of $C_8H_{17}+C_3H_6$ from steroids (alkyl chain at C17)

Loss of

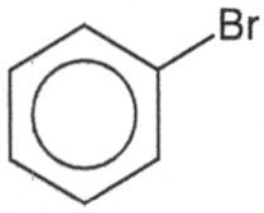

Br

155 **Structural Significance**

$CF_3C(O)OC(CH_3)_2-$, C_5F_5, $CH_3C_6H_4-SO_2-$,

CH_2CH_2I, $C_4H_9SiCl_2-$

156 **Structural Significance**

$(C_4H_9)_2NC(O)$,

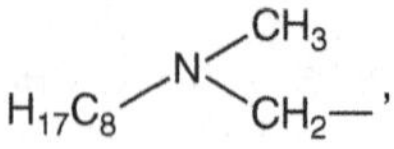

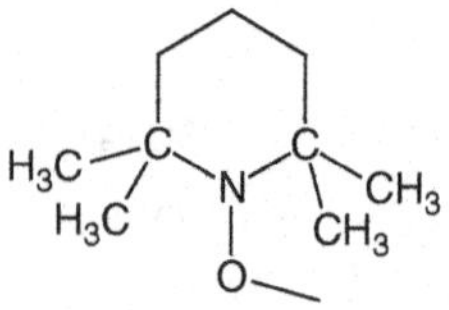

C_5F_5H, $C_9H_{18}NO$

157 **Structural Significance**

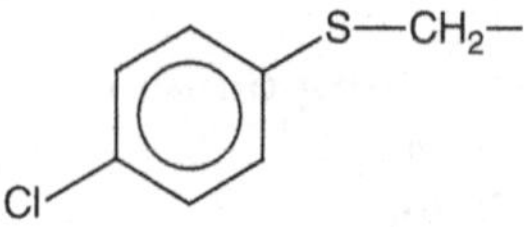

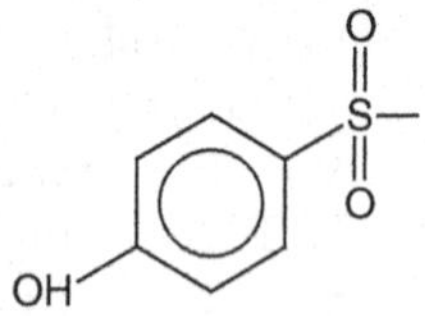

$(CH_3C(O)O)_2CHCH{=}CH–$, $C_5F_5H_2$, $ClC_6H_4NO_2$

158 **Structural Significance**

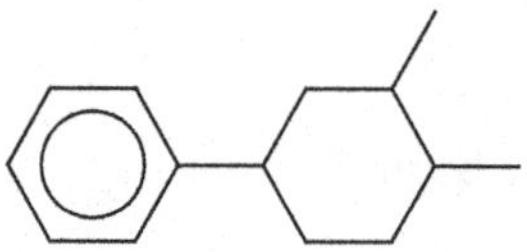

$C_{12}H_{14}$

$CH_2–C(O)CH_2CH_2CH_2CH_2C(O)OCH_3$

Trichlorobutenes, $C_4H_5Cl_3$

$[M - 159]^+$ **Possible Precursor Compounds**

The loss of $–C(O)OSi(CH_3)_2C(CH_3)_3$ is characteristic of TBDMS derivatives of amino acids

159 **Structural Significance**

$C_6H_5CCl_2–$,

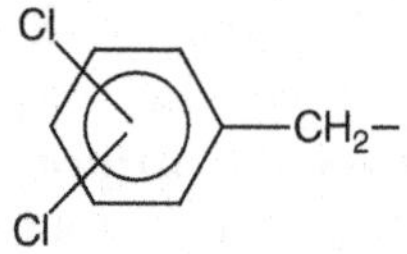

$C_6H_5SCF_2–$, $(C_4H_9O)_2CH–$, 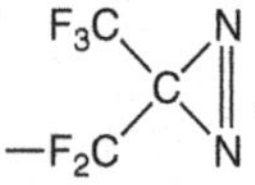 $–CF_2CF_2C(O)OCH_3$,

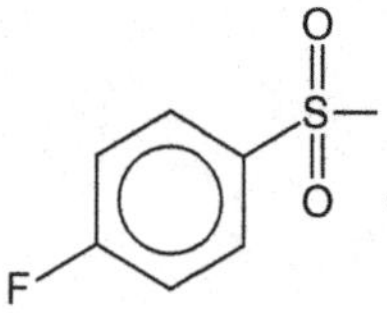

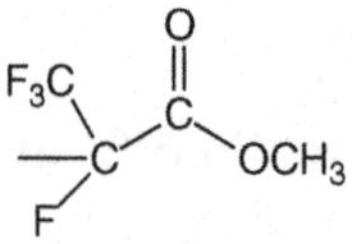

160 **Structural Significance**

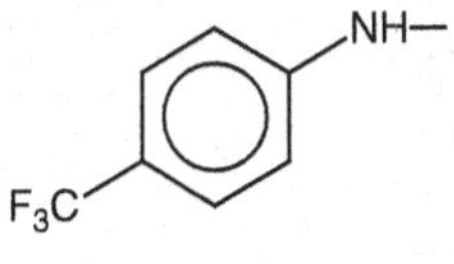

$[(CH_3)_2N]_3Si–$,

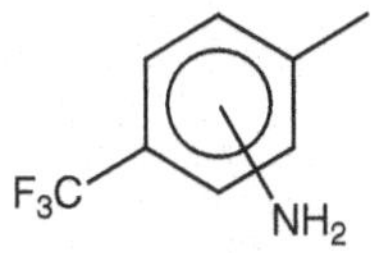

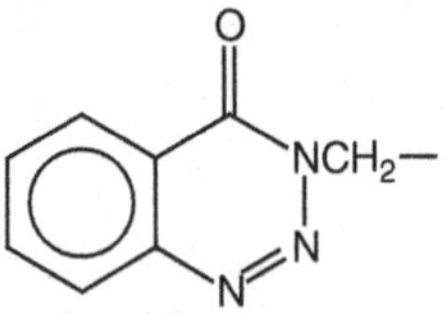

161 **Structural Significance**

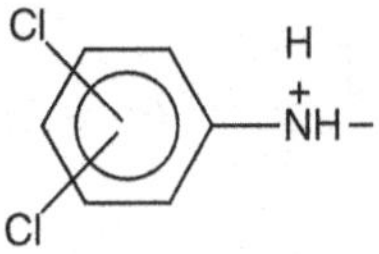

$CCl_2Br–$, $CHF_2CF_2OCH_2OCH_2–$

162 **Structural Significance**

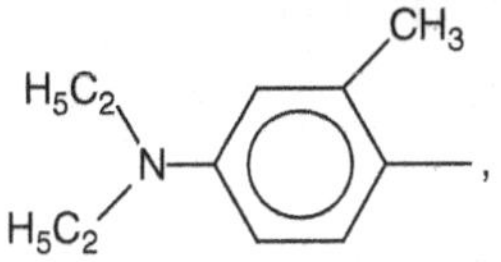

C_4F_6,

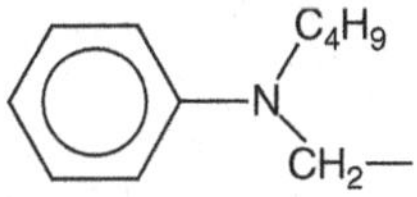

163 **Structural Significance**

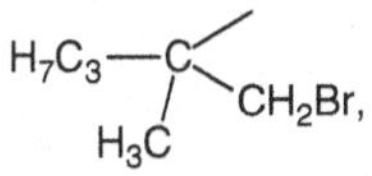

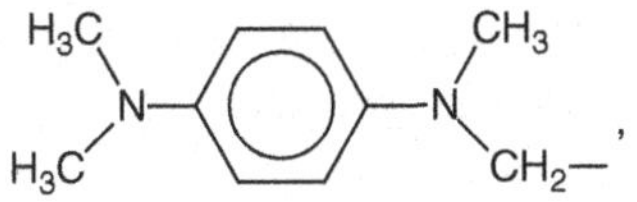

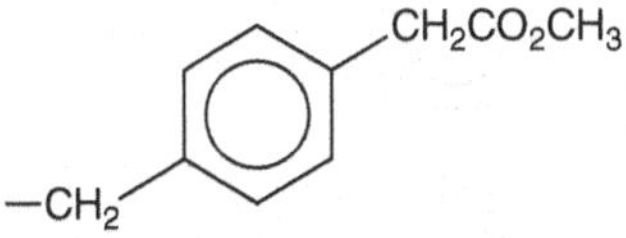

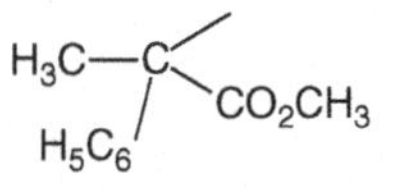

$CF_3CF_2CH_2OCH_2–$, $CHF_2CHFCF_2OCH_2–$
$CF_3CH_2CF_2OCH_2–$, $C_3F_3Cl_2$, C_4F_6H, C_3F_4ClO,

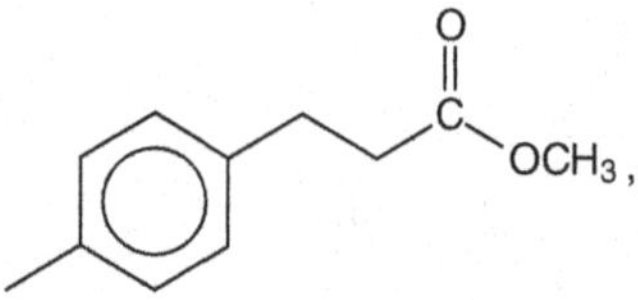

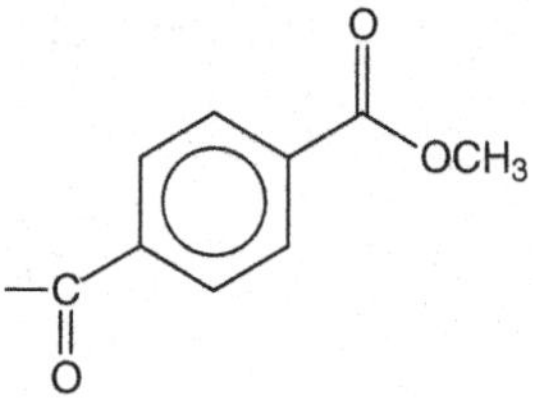

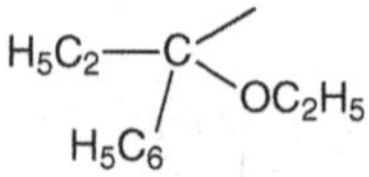

$H_5C_2-C(C_6H_5)(-)OC_2H_5$

164 **Structural Significance**

CCl_3CCl-, $-CF_2N{=}CFCF_3$, $C_4F_6H_2$, $C_6H_3FCl_2$, $C_7H_{10}Cl_2$

165 **Structural Significance**

Dinitrotoluenes

$C_{13}H_9$, 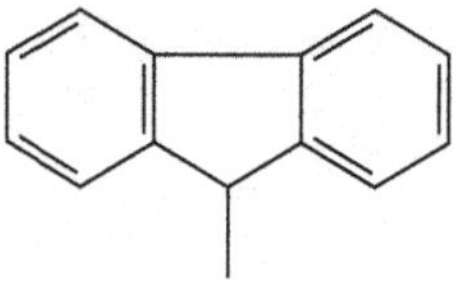 , CCl_3CHCl, $-Sn(CH_3)_3$

(Sn isotope peaks at *m/z* 161, 163, and 165)

166 **Structural Significance**

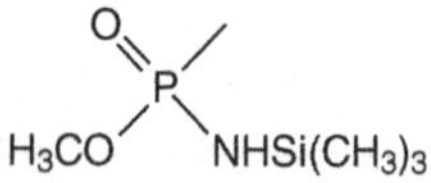

$H_3CO-P(=O)(-)NHSi(CH_3)_3$, $C_6H_2F_3Cl$,

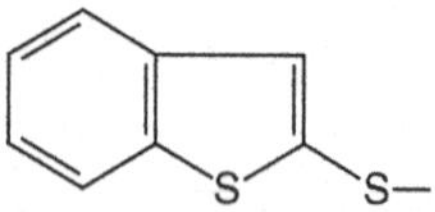

167 **Structural Significance**

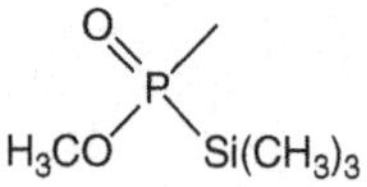

$H_3CO-P(=O)(-)Si(CH_3)_3$, $CF_3CFHOCF_2-$, CCl_3CF_2-, C_6F_5-,

$CFCl_2CFCl-$, CF_2ClCCl_2-, $(C_6H_5)_2CH$, $C_2F_2Cl_3$

Certain phthalates:

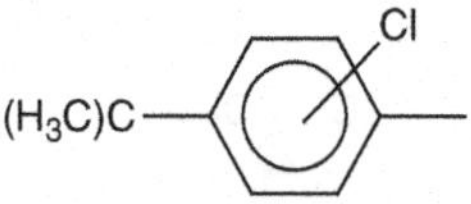

$(H_3C)C-C_6H_3Cl-$

168 **Structural Significance**

Dichlorophenyl phenyl ethers

$CF_3C(O)NH(CH_2)_4-$, $C_6H_9F_3ON$, C_6F_5H

[M − 169]⁺ **Possible Precursor Compounds**

Loss of C_3F_7 from perfluoro compounds

169 **Structural Significance**

$H_3C-C(=O)-CH_2$-furan-$CH(OH)-$, $CF_3CF_2CF_2-$, $C_6F_5H_2$

$C_{10}H_{21}C(O)$

170 **Structural Significance**

$(CH_3)_2N$-naphthyl-, C2F4Cl2,

Cl-(O_2N)-phenyl-CH_2-, $(C_5H_{11})_2NCH_2-$

[M − 171]⁺ **Possible Precursor Compounds**

Loss of

[F-phenyl-$S(=O)_2-$ + F]

171 **Structural Significance**

HO-naphthyl-C(=O)−

172 **Structural Significance**

2-pyridyl-$CH_2CH_2-P(OH)(OH)-$ + H

173 **Structural Significance**

$C_8H_{17}Si(CH_3)_2(OH)$ [structure], $Cl_2C_6H_3N{=}N{-}$ [structure], $CF_3C_6H_4C(O)$,

$C_8H_5F_3O$, $C_2H_5OC(O)CH(CH_3)CH_2C(O)OC_2H_5$ [structure], $CF_3C_6H_4N{=}N{-}$

174 **Structural Significance**

$-CH_2N(SiC_3H_9)_2$* loss is observed when

$HO(H)C(-)CH_2N(TMS)_2$ [structure]

or $-(CH_2)_3N(TMS)_2$ is present

175 **Structural Significance**

C_5F_6H, $C_6H_5C(CH_3){=}CHSi(CH_3)_3$ [structure],

$C_6H_5C(CH_3){=}CH{-}Si(CH_3)_2{-}$ [structure],

$ClC_6H_4SO_2{-}$ [structure], $H_3COC(O)CH_2C(OH)(CH_2CH_3)C(O)OCH_3$ [structure]

176 **Structural Significance**

$F_3C{-}CF(CN){-}CF_3$ [structure], $C_5F_6H_2$

---

* Apparently two TMS groups can add to the amino group, especially for $-(CH_2)_nN$ when $n > 2$.

177 **Structural Significance**

SnC_4H_9 (*m/z* 173, 175, 177)

$C_6H_5NHC(O)CH_2C(O)CH_3$, $C_5H_2F_6$, F_3C O—CHF C O—CHF ,

($C_4H_2F_5O_2$), O C OC$_2$H$_5$ O C , $(C_4H_9)_2SiCl$

178 **Structural Significance**

C_4F_5Cl, $CF_2{=}CFCF{=}CFCl$

179 **Structural Significance**

$C_6H_5CHOSi(CH_3)_3$, $C_3F_3Cl_2O$, $(CF_3)_2CHC(O)$–, C_4HF_6O

$(CF_3)_2C{=}C{\bullet}{-}OH$

$C_{10}H_{11}O_3$

180 **Structural Significance**

$C_6H_3Cl_3$

$C_6H_5CH_2OC(O)CO_2H$

181 **Structural Significance**

C_4F_7

182 **Structural Significance**

$(C_6H_5)_2Si$– (see tetraphenylsilane)

$[M-183]^+$ **Possible Precursor Compounds**

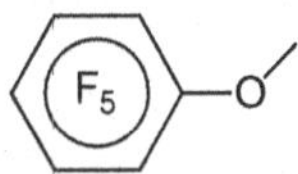

183 **Structural Significance**

C_2FCl_4, $-CF_2CF_2SO_2F$, $-CFCl_2CCl_2-$, $(C_6H_5)_2COH$, $(C_6H_5)_2SiH$

184 **Structural Significance**

O C H$_3$C N C$_8$H$_{17}$ CH$_2$—

185 **Structural Significance**

$C_8H_{11}NO_2S$

Tributyl citrate (MW = 360 Da)

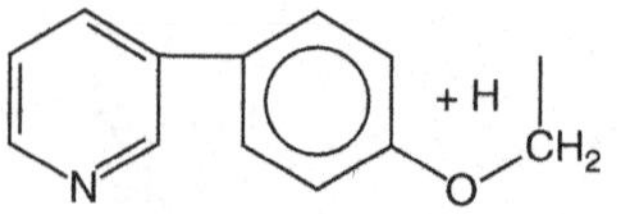

, $(C_2H_5O)_2PS_2$,

$(CH_3)_2SnCl$, (*m/z* 181, 183, and 185; isotopes for Sn)

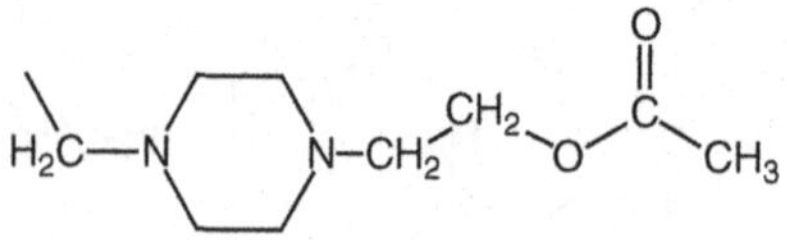

$(C_6H_5)_2CF$–,

$C_9H_{19}N_3O$, $C_{11}H_{21}O_2$, $C_{10}H_{17}O_3$, $C_4H_9OC(O)(CH_2)_4C(O)$–
(*m/z* 185 and 129 suggest dibutyl adipate)

186 **Structural Significance**

C_6F_6

187 **Structural Significance**

$C_{13}H_{15}O$, C_6F_6H

188 **Structural Significance**

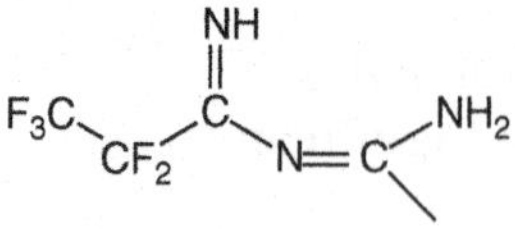

189 **Structural Significance**

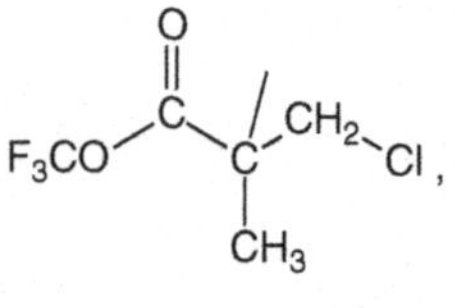

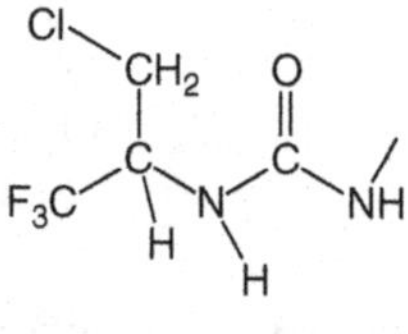

191 **Structural Significance**

m/z 191, 204, and 217 suggest a TMS hexose ($CH[OSi(CH_3)_3]_2$)

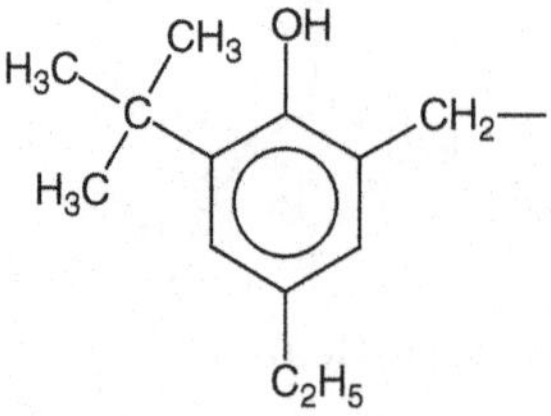

193 **Structural Significance**

$-SnSi(CH_3)_3$ (*m/z* 193, 191, 189)

C_5F_7, $(C_2H_5O)_3SiOCH_2-$

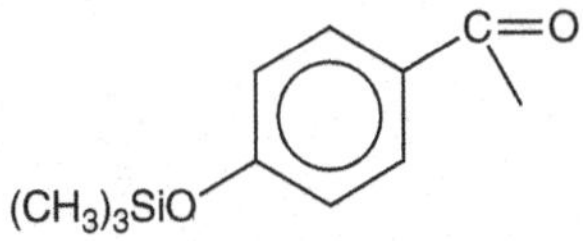

195 **Structural Significance**

$C_3F_2Cl_3O$, $C_5F_7H_2$,

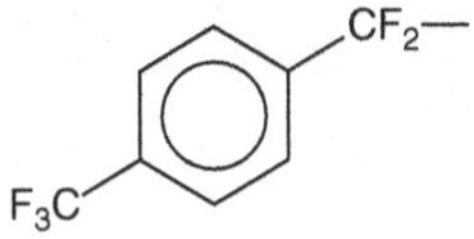

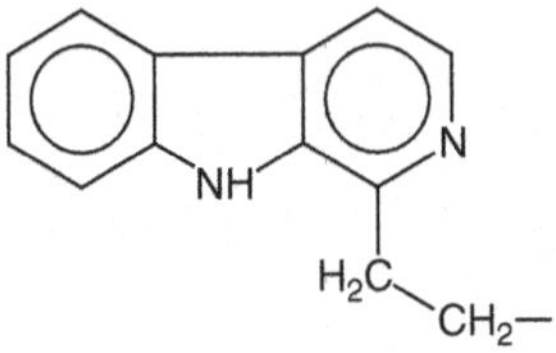

196 **Structural Significance**

$C_5H_3F_7$,

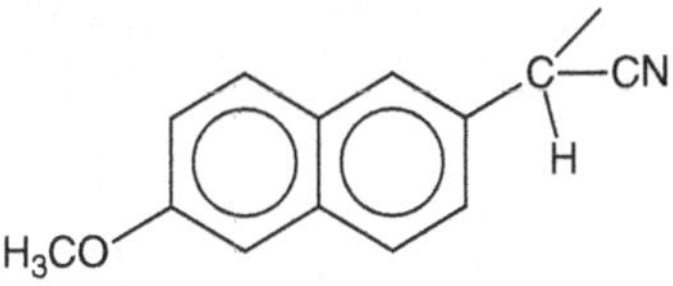

197 **Structural Significance**

C_4F_7O, 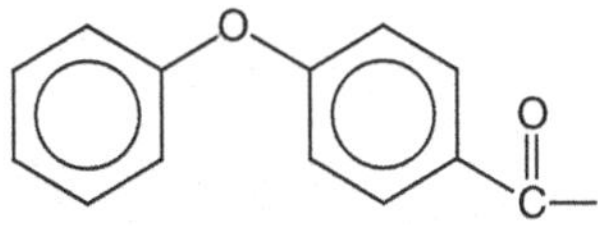 , $C_{13}H_9O_2$

$[M - 198]^+$ **Possible Precursor Compounds**

Loss of two CD_3SiOH groups from deuterated TMS derivative (e.g., cholic acid)

198 **Structural Significance**

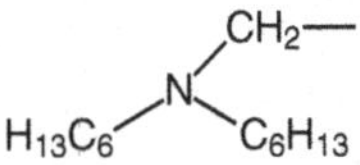

199 **Structural Significance**

$(C_4H_9)_3Si–$

$CF_3CF_2CF_2OCH_2–$

200 **Structural Significance**

C_4F_8

Index

Printed in the United States
By Bookmasters